STUDY GUIDE AND SOLUTION MANUAL

PAULA YURKANIS BRUICE
University of California
Santa Barbara

ESSENTIAL
ORGANIC CHEMISTRY

Third Edition

PEARSON

Editor in Chief: Jeanne Zalesky

Marketing Manager: Will Moore

Team Lead, Project Management Biology, Chemistry,
Environmental Science, and Geo Science: David Zielonka

Project Manager: Sarah Shefveland

Compositor: PreMediaGlobal

Cover Designer: Seventeenth Street Studios

Cover Image Credit: Andrew Johnson/E+/Getty Images

Credits and acknowledgments borrowed from other sources and reproduced, with permission, in this textbook appear on the appropriate page within the text.

1 2 3 4 5 6 7 8 9 10—**EB**—18 17 16 15

www.pearsonhighered.com   ISBN-10: 0-13-386725-0 ISBN-13: 978-0-13-386725-1

CONTENTS

to my students

I am very grateful to Ron Bishop of SUNY Oneonta, who reworked all the problems to make this book as error free as possible. I am solely responsible for any errors that may remain. If you find any, please email me so they can be fixed in a future printing.

Try to work as many problems as possible, so you can truly enjoy the wonderful world of organic chemistry.

Paula Yurkanis Bruice
pybruice@chem.ucsb.edu

. The mass spectrum of an ether is shown here. Determine the molecular formula of the ether that would produce this spectrum and then draw possible structures for it.

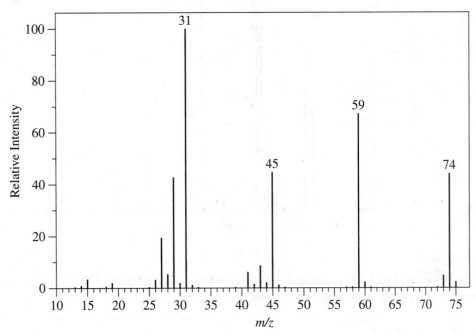

. The mass spectra of pentane and isopentane are shown here. Determine which spectrum belongs to which compound.

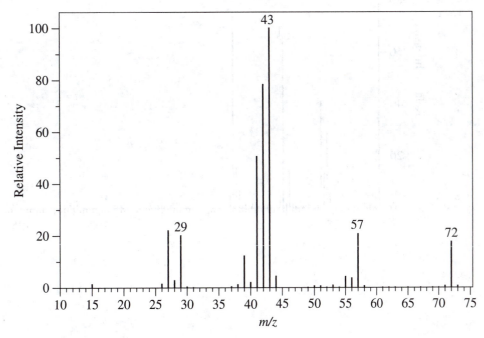

1

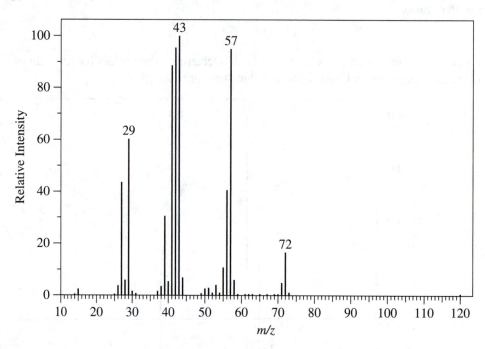

3.   An unknown acid underwent a reaction with 1-butanol. The product of the reaction gave the mass spectrum shown here. What is the product of the reaction, and what acid was used?

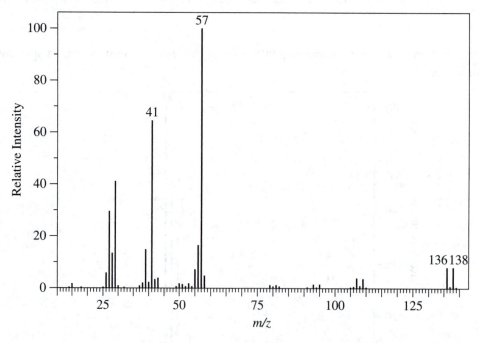

Identify the compound with molecular formula $C_9H_{10}O_3$ that gives the following IR and ^1H NMR spectra.

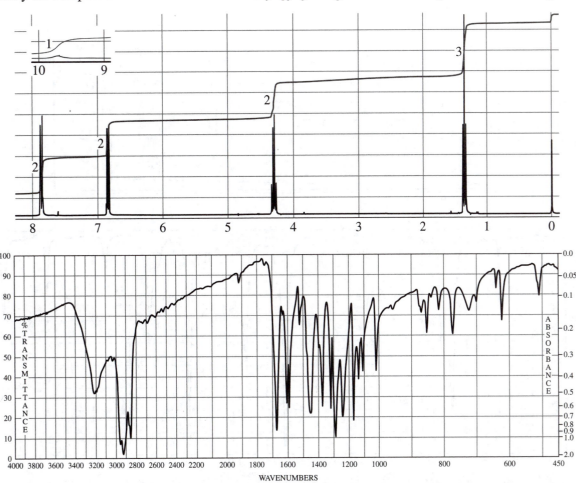

5. Identify the compound with molecular formula $C_5H_{11}Br$ that gives the following ^1H NMR spectrum.

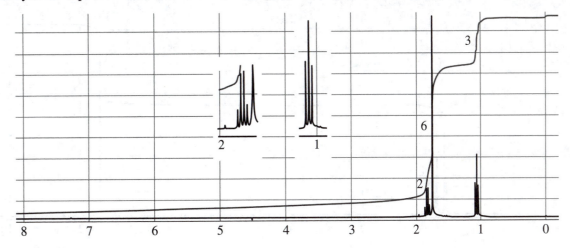

6.   Identify the compound with molecular formula $C_6H_{12}O$ that gives the following IR and ^1H NMR spectra.

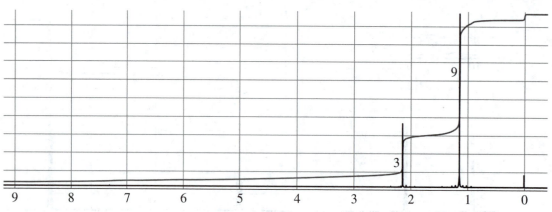

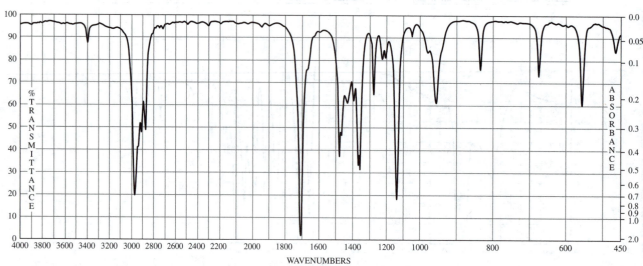

7.   Identify the compound with molecular formula C_7H_7Cl that gives the ^1H NMR spectrum shown here.

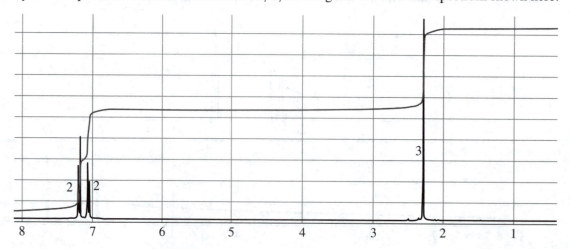

. Identify the compound with molecular formula $C_6H_{14}O$ that gives the following IR and ^1H NMR spectra.

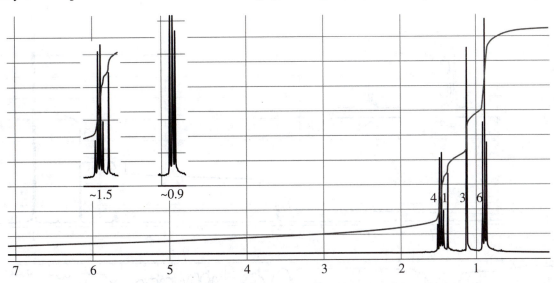

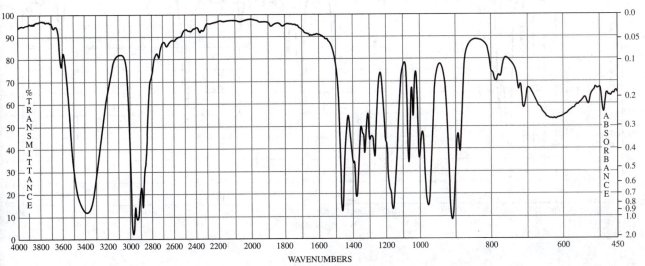

9. Identify the compound with molecular formula C_4H_9Br that gives the following ^1H NMR spectrum.

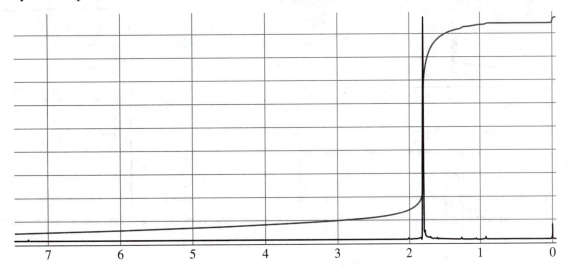

10.   Identify the compound with molecular formula C_6H_{12} that gives the following IR and ^1H NMR spectra.

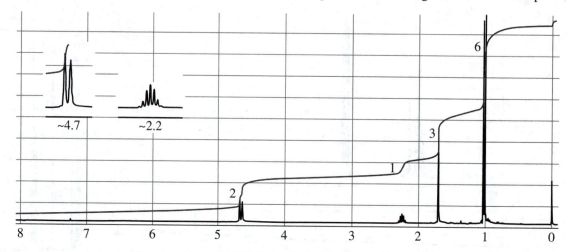

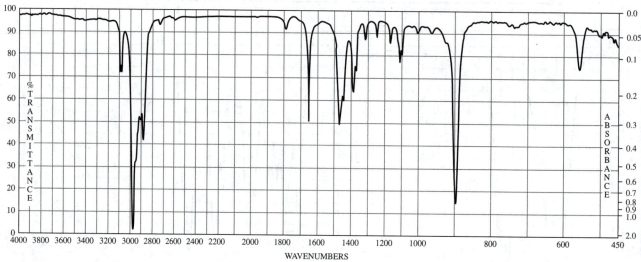

WAVENUMBERS

11.   Identify the compound with molecular formula $C_8H_{10}O$ that gives the following ^1H NMR spectrum.

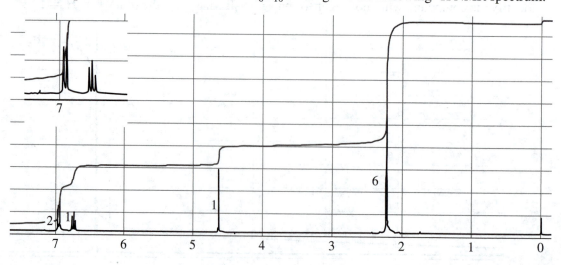

12. Identify the compound with molecular formula C_4H_7ClO that gives the following IR and ^1H NMR spectra.

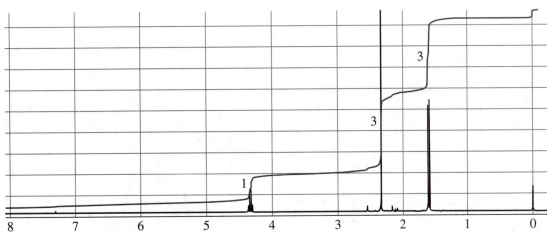

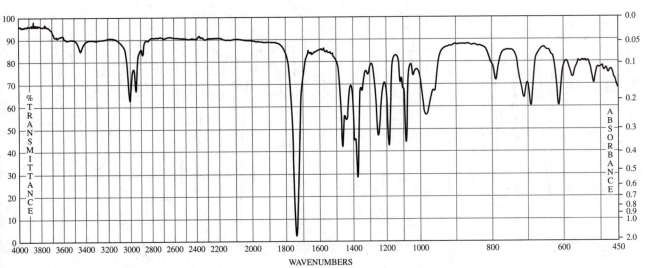

13. Identify the compound with molecular formula $C_8H_8Br_2$ that gives the following ^1H NMR spectrum.

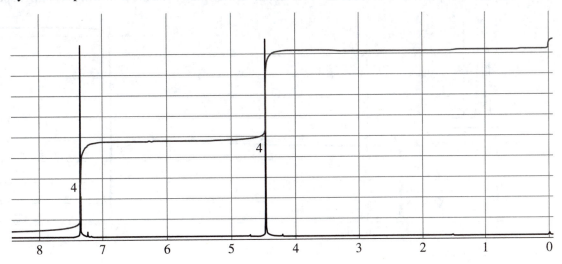

14. Identify the compound with molecular formula C_4H_6O that gives the following IR and ^1H NMR spectra.

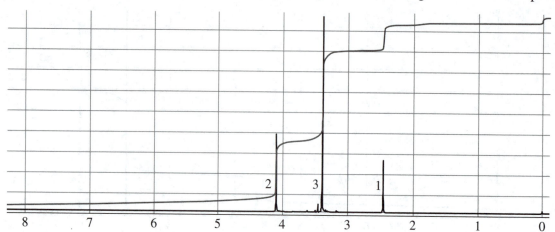

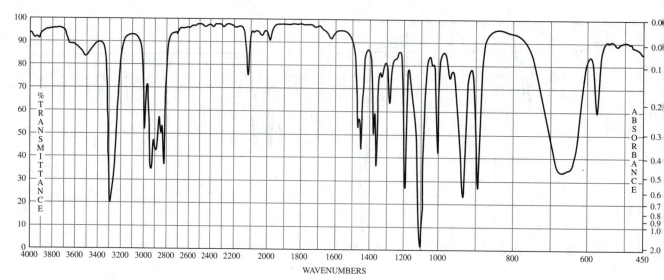

WAVENUMBERS

15. Identify the compound with molecular formula C_7H_8BrN that gives the following ^1H NMR spectrum.

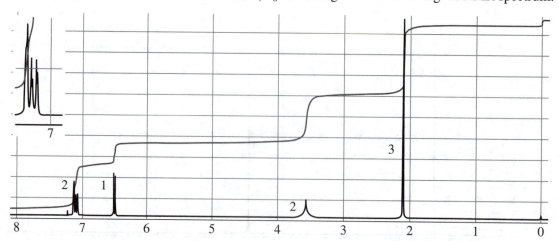

16. The ^{1}H NMR spectra of 1-chloro-3-iodopropane and 1-bromo-3-chloropropane are shown here. Which compound gives which compound spectrum?

a.

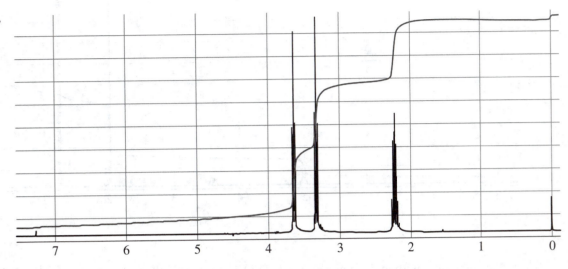

b.

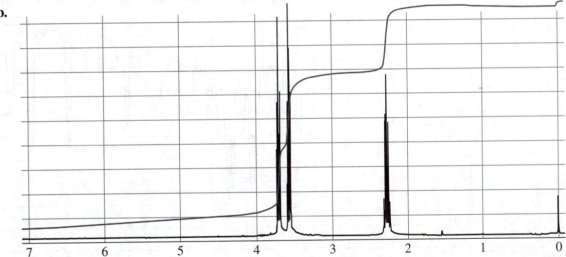

17. Identify the compound with molecular formula C_4H_8O that gives the following ^{1}H NMR spectrum.

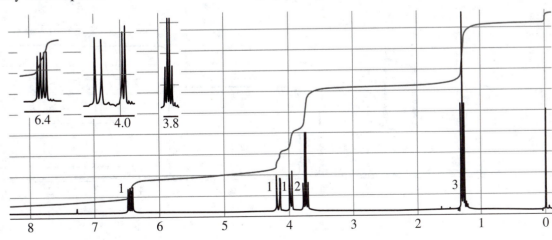

18.   Identify the compound with molecular formula $C_5H_{10}O_2$ that gives the following ^1H NMR spectrum.

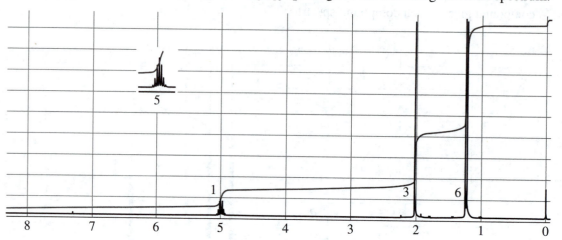

19.   Identify the compound with molecular formula $C_8H_7O_2Br$ that gives the following IR and ^1H NMR spectra.

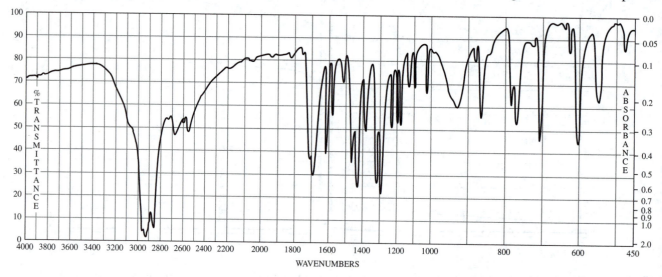

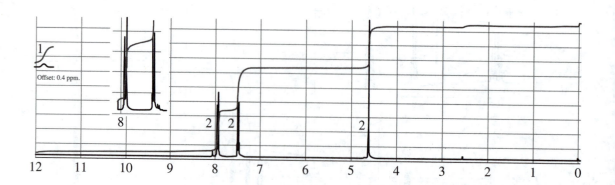

20. Identify the compound with molecular formula C_7H_6O that gives the following ^1H NMR spectrum.

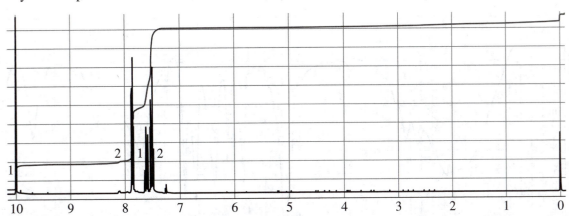

21. The two ^1H NMR spectra shown here are given by constitutional isomers with molecular formula C_3H_7Br. Identify each isomer.

a.

b.

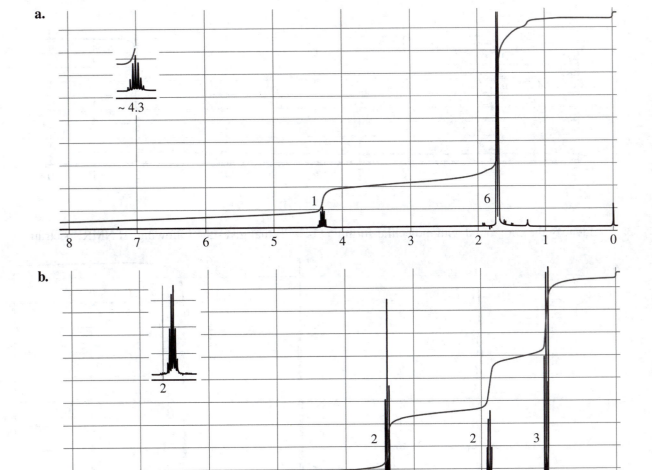

22.   Identify the compound with molecular formula $C_4H_7BrO_2$ that gives the following IR and ^1H NMR spectra.

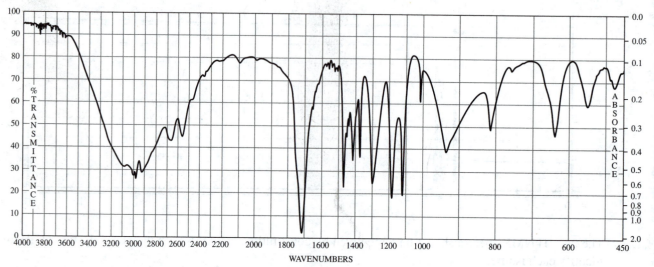

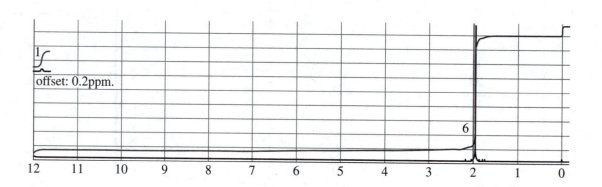

23.   Identify the compound with molecular formula $C_3H_6Cl_2$ that gives the following ^1H NMR spectrum.

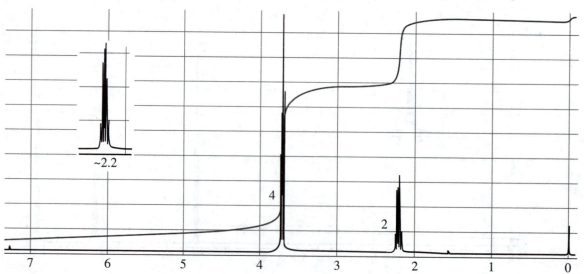

24.   Identify the compound with molecular formula C_3H_6O that gives the ^1H NMR spectrum shown here.

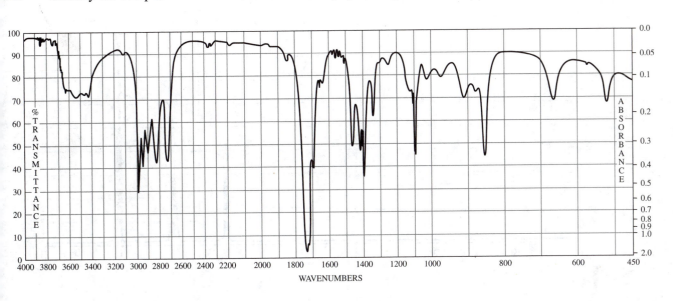

25.   Identify the compound with molecular formula $C_4H_8O_2$ that gives the following ^1H and ^{13}C NMR spectra (bottom and top of the figure, respectively).

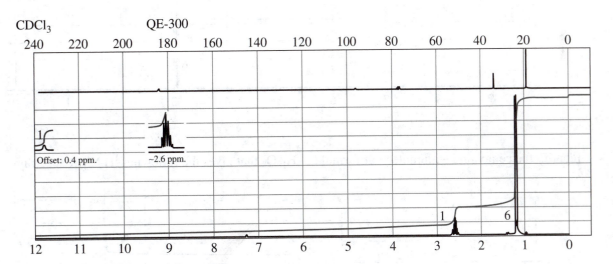

26.   Identify the compound with molecular formula $C_4H_8O_2$ that gives the following ^1H NMR spectrum.

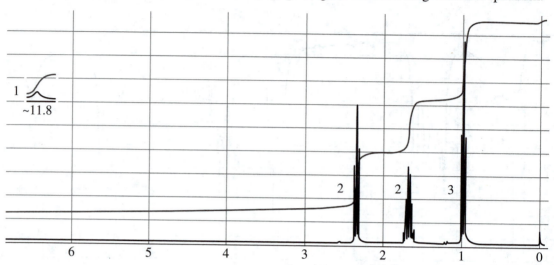

27.   Identify the compound with molecular formula $C_9H_{11}NO$ that gives the following ^1H NMR spectrum.

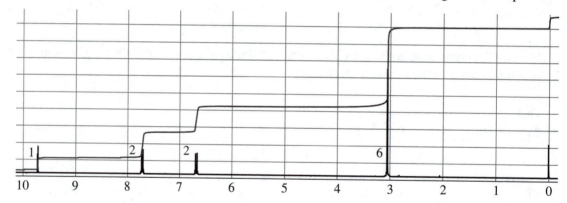

28.   Identify the compound with molecular formula $C_9H_{10}O_2$ that gives the following ^1H NMR spectrum.

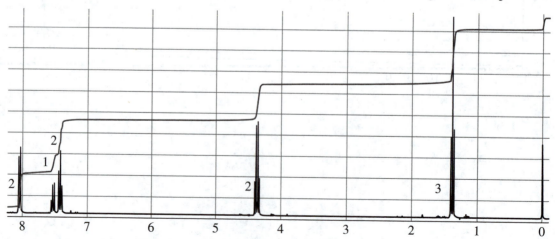

29. The ^{13}C NMR and ^1H NMR spectra of 1, 2-, 1, 3-, and 1, 4-ethylmethylbenzene are shown here. Determine which spectrum belongs to which compound.

a.

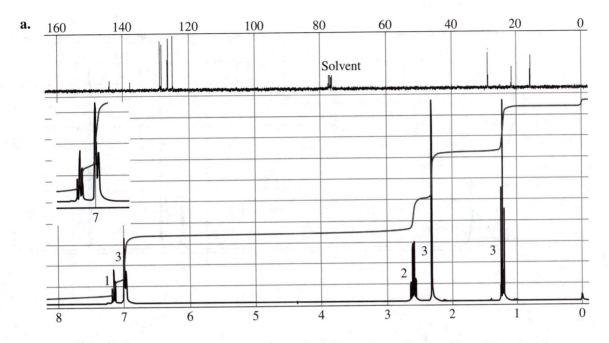

b.

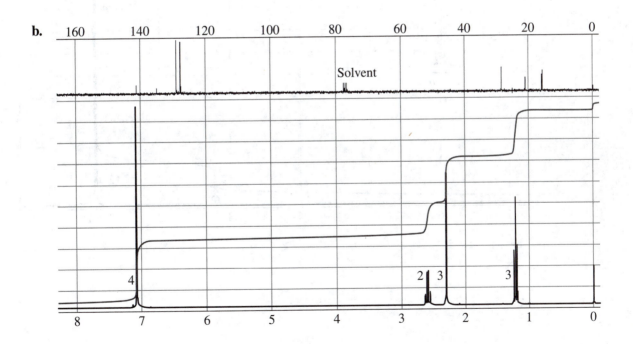

c.

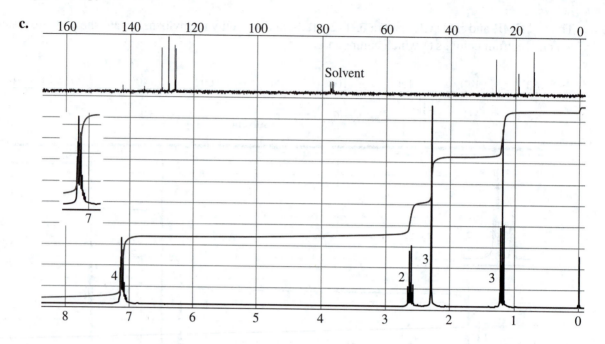

30. Identify the compound with molecular formula $C_7H_{14}O$ that gives the following ^1H NMR spectrum.

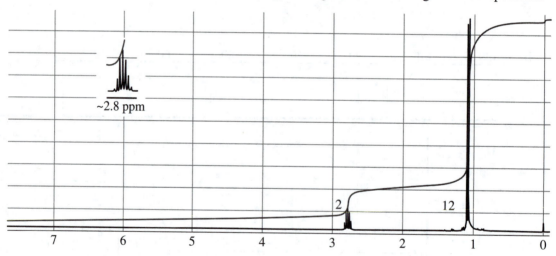

31. Identify the compound with molecular formula C_5H_{10} that gives the following ^1H NMR spectrum.

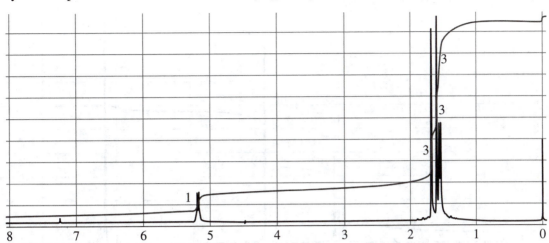

32. Identify the compound with molecular formula C_3H_4O that gives the following IR and ^1H NMR spectra.

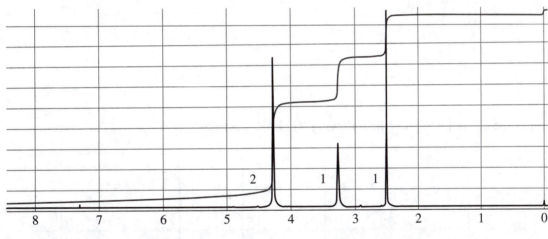

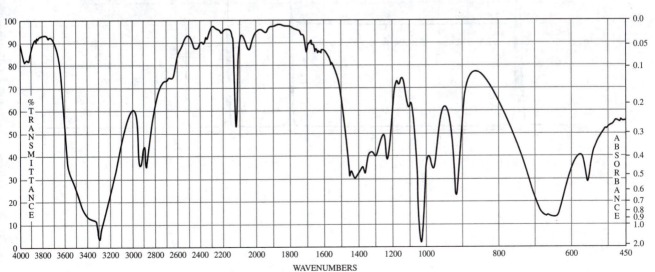

33.   Identify the compound with molecular formula C_4H_7Cl that gives the following ¹H NMR spectrum.

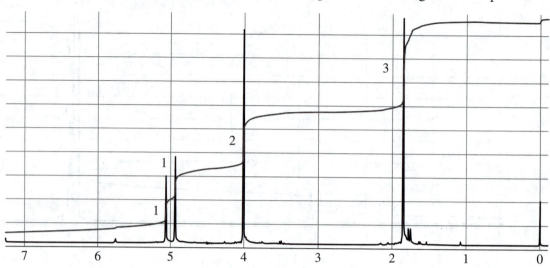

34.   Identify the alcohol that gives the following ¹H NMR spectrum.

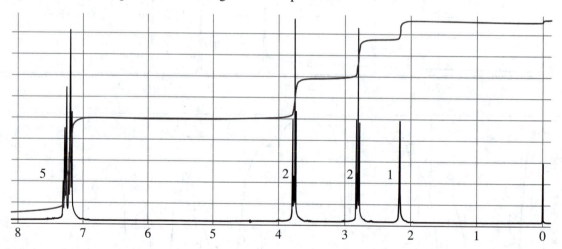

35.   Identify the alcohol that gives the following ^1H NMR spectrum.

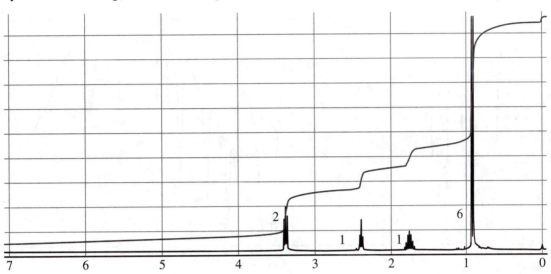

36.   Identify the compound with molecular formula $C_5H_{12}O_2$ that gives the following ^1H NMR spectrum.

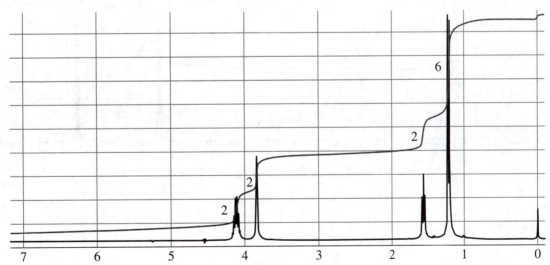

37.   Identify the compound with molecular formula C_3H_7NO that gives the following IR and ^1H NMR spectra.

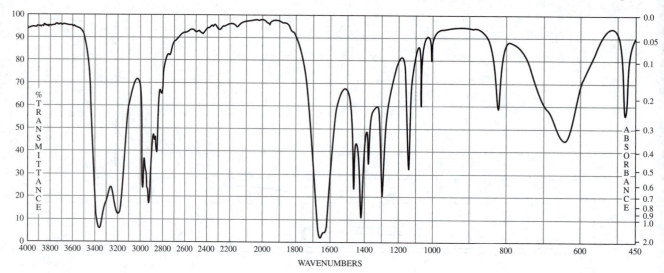

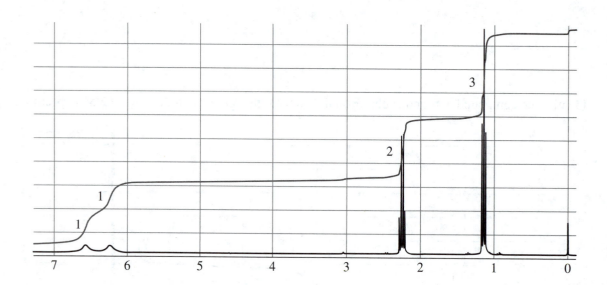

38. The ^{1}H NMR spectra shown here are given by constitutional isomers of propylamine (C_3H_9N). Identify the isomer that gives each spectrum.

a.

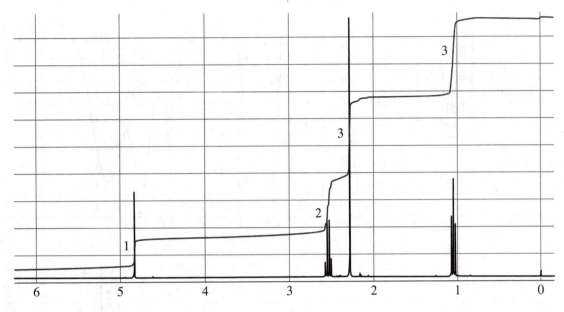

b.

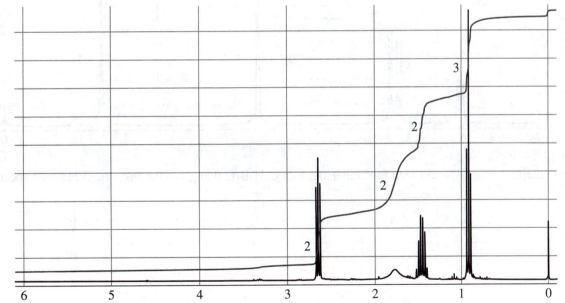

39.   Identify the compound with molecular formula C_3H_7N that gives the following IR and ^{1}H NMR spectra.

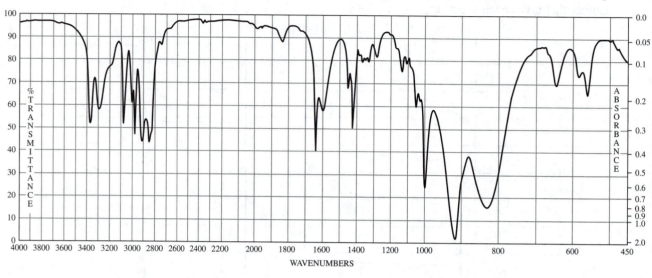

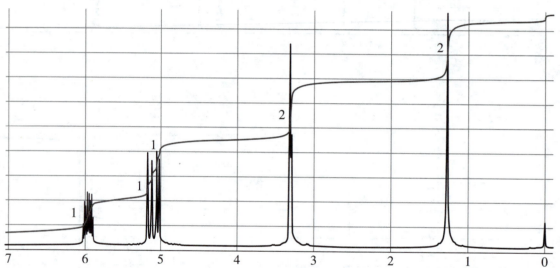

40.   Identify the compound with molecular formula C_4H_9BrO that gives the following ^{1}H NMR spectrum.

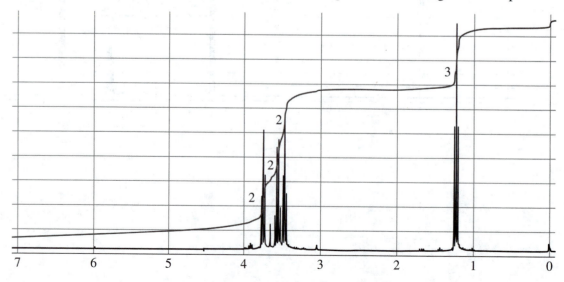

41.   Identify the compound with molecular formula $C_7H_8O_2$ that gives the following IR and NMR spectra.

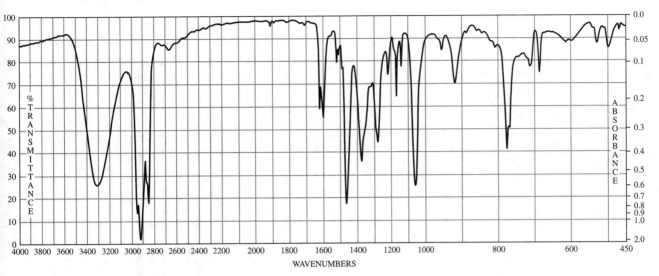

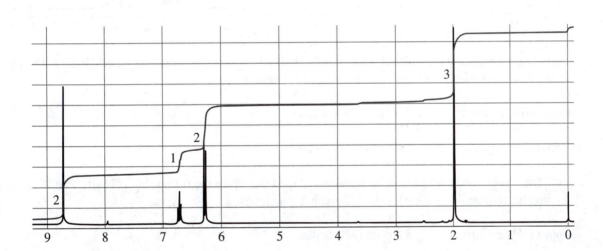

42.   Identify the alcohol that gives the following ^{1}H NMR spectrum.

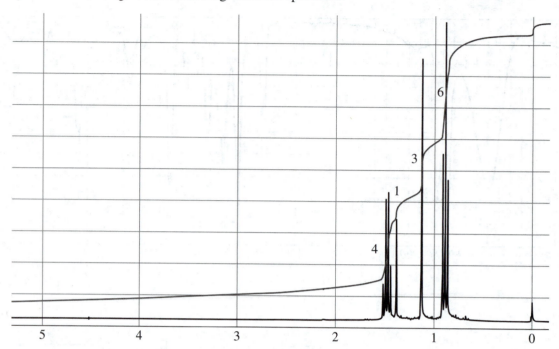

43.   Identify the compound with molecular formula $C_6H_{12}O_2$ that gives the following ^{1}H NMR data. The number of hydrogens responsible for each signal is given in parentheses.

1.1 ppm (6H) doublet   2.2 ppm (2H) quartet

1.7 ppm (3H) triplet   3.8 ppm (1H) septet

44.   Identify the compound with molecular formula $C_9H_{10}O$ that gives the following ^{1}H NMR data. The number of hydrogens responsible for each signal is given in parentheses.

1.4 ppm (2H) multiplet   3.8 ppm (2H) triplet

2.5 ppm (2H) triplet   6.9–7.8 ppm (4H) multiplet

45.   Identify the compound with molecular formula $C_{11}H_{22}O$ that gives the following ^{1}H NMR data.

1.1 ppm (18H) singlet   2.2 ppm (4H) singlet

46.   Identify the compound with molecular formula C_4H_8O that gives the following ^{1}H NMR data.

1.6 ppm (4H) multiplet   3.8 ppm (4H) triplet

47.   Identify the compound with molecular formula $C_{15}H_{14}O_3$ that gives the following ^{1}H NMR data.

3.8 ppm (6H) singlet   7.7 ppm (4H) doublet

7.3 ppm (4H) doublet

Answers to Spectroscopy Problems

1. First, we must first identify the molecular ion. The molecular ion, the peak that represents the intact starting compound, has an $m/z = 74$. Now we can use the rule of 13 to determine the molecular formula.

$$\frac{74}{13} = 5 \text{ carbons with } 9 \text{ left over}$$

From the rule of 13, we end up with a molecular formula of C_5H_{14}. Because the compound is an ether, we know that it has one oxygen, so we must add one O and subtract one C and four Hs from the molecular formula. The resulting molecular formula is shown below:

$$C_4H_{10}O$$

There are three ethers that have this molecular formula: methyl propyl ether, diethyl ether, and isopropyl methyl ether.

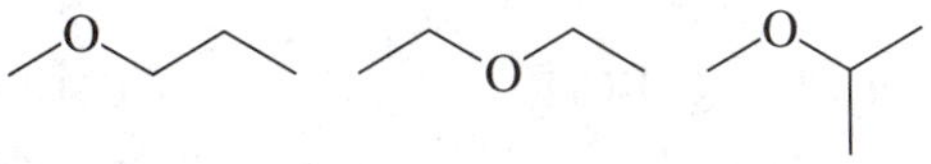

2. First, we need to determine what the most abundant cationic fragments would be for each compound.

The possible fragments for **pentane** are as follows:

1 $\left[CH_3-CH_2CH_2CH_2CH_3\right]^{\cdot+}$ $\longrightarrow$ $\cdot CH_3$ + $^+CH_2CH_2CH_2CH_3$
   $m/z = 57$

2 $\left[CH_3CH_2-CH_2CH_2CH_3\right]^{\cdot+}$ $\longrightarrow$ $\cdot CH_2CH_3$ + $^+CH_2CH_2CH_3$
   $m/z = 43$

3 $\left[CH_3CH_2-CH_2CH_2CH_3\right]^{\cdot+}$ $\longrightarrow$ $\cdot CH_2CH_2CH_3$ + $^+CH_2CH_3$
   $m/z = 29$

4 $\left[CH_3-CH_2CH_2CH_2CH_3\right]^{\cdot+}$ $\longrightarrow$ $\cdot CH_2CH_2CH_2CH_3$ + $^+CH_3$
   $m/z = 15$

The most abundant fragments are going to be the result of bond cleavages that produce the most stable cations and radicals. Fragments from patterns 2 and 3 should be the most common, because both the cations and the radicals formed are primary.

Fragmentation pattern 2 is expected to give the base peak (the most stable fragment). The cation formed in pattern 2 ($m/z = 43$) will be more stable than the cation formed in pattern 3 ($m/z = 29$), because the longer fragment will result in greater inductive stabilization (the longer chain will do a better job of stabilizing the carbocation).

Fragments from fragmentation patterns 1 and 4 are expected to be less common. They each form a primary species, like fragmentation patterns 2 and 3, but the second species they form is a methyl fragment (either a radical or a carbocation), which is less stable than the second species formed in fragmentation patterns 2 and 3.

Only four fragmentation patterns are shown for **isopentane.** Fragmentations that would result in only a primary fragment and a methyl fragment have been excluded, because these would be less abundant than those shown here.

$$
1 \quad \left[\begin{array}{c} CH_3 \\ | \\ CH_3CHCH_2CH_3 \end{array} \right]^{\cdot+} \longrightarrow \cdot CH_3 \; + \; CH_3\overset{+}{C}HCH_2CH_3
$$
$$
m/z = 57
$$

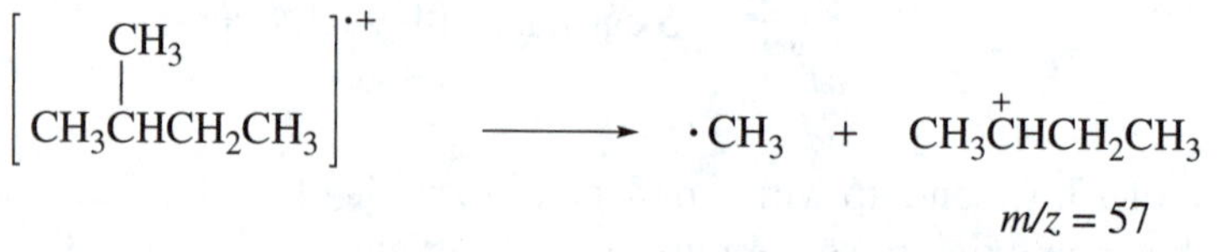

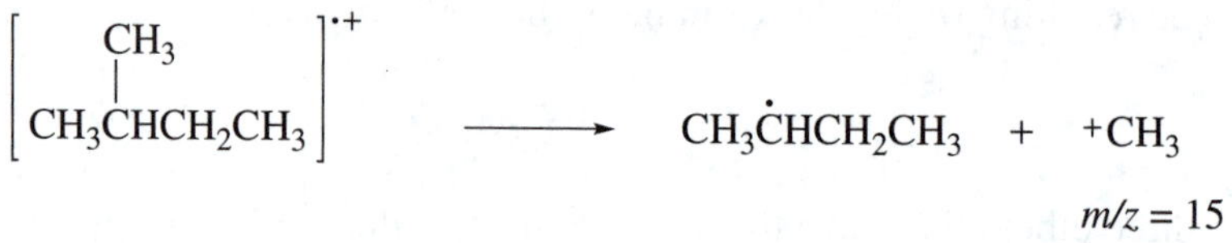

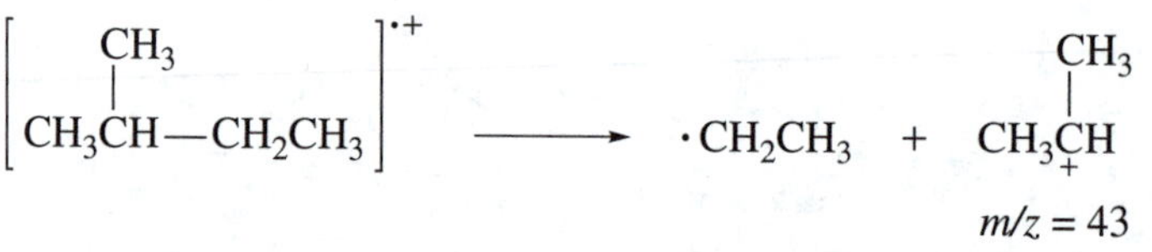

$$
2 \quad \left[\begin{array}{c} CH_3 \\ | \\ CH_3CHCH_2CH_3 \end{array} \right]^{\cdot+} \longrightarrow CH_3\overset{\cdot}{C}HCH_2CH_3 \; + \; {}^{+}CH_3
$$
$$
m/z = 15
$$

$$
3 \quad \left[\begin{array}{c} CH_3 \\ | \\ CH_3CH-CH_2CH_3 \end{array} \right]^{\cdot+} \longrightarrow \cdot CH_2CH_3 \; + \; \begin{array}{c} CH_3 \\ | \\ CH_3\overset{+}{C}H \end{array}
$$
$$
m/z = 43
$$

$$
4 \quad \left[\begin{array}{c} CH_3 \\ | \\ CH_3CH-CH_2CH_3 \end{array} \right]^{\cdot+} \longrightarrow \begin{array}{c} CH_3 \\ | \\ CH_3\overset{\cdot}{C}H \end{array} \; + \; {}^{+}CH_2CH_3
$$
$$
m/z = 29
$$

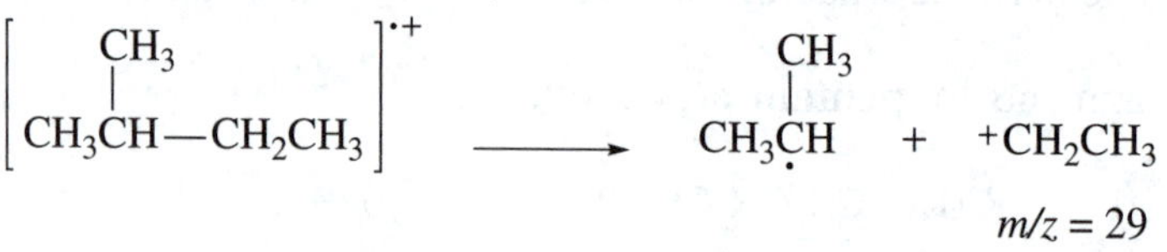

The four major fragments occur at the same m/z values (57, 43, 29, and 15) as those found for pentane, but their relative intensities are different.

Fragmentation patterns 1 and 3 are expected to produce the most abundant fragments because they both form a secondary cation, and the stability of the cation is more important than the stability of the radical in determining the most abundant fragments. Therefore, we expect a base peak with $m/z = 43$ (because the secondary cation is accompanied by a primary radical) and a less intense peak with $m/z = 57$ (because the secondary cation is accompanied by a methyl radical).

Both spectra show a base peak at $m/z = 43$. The major difference in the two spectra is the intensity of the peak with $m/z = 57$. The spectrum of isopentane should show a more intense peak because it is due to a secondary cation, whereas the peak with $m/z = 57$ in the spectrum of pentane is due to a primary cation.

Thus, **pentane** gives the first mass spectrum and **isopentane** gives the second.

The mass spectrum has two peaks of identical height with m/z values $= 136$ and 138, indicating the presence of bromine in the product. (Bromine has two isotopes of equal abundance with weights of 79 and 81 amu.)

Now we need to think about the type of reaction that occurred.

Under acidic conditions, the starting material (butyl alcohol) will be protonated.

The protonated alcohol now has a leaving group that can be replaced by a nucleophile. Because we know that bromine is present in the product, we can assume that bromide ion is the incoming nucleophile.

We now know that the product of the reaction is **1-bromobutane.** The acid, which must be the source of the nucleophile, is **HBr.**

The two signals near 7 and 8 ppm are due to the hydrogens of a benzene ring. Because these signals integrate to 4 protons, the benzene ring must be disubstituted. Since both signals are doublets, the protons that give each signal must each be coupled to one proton ($N + 1 = 1 + 1 = 2$). Thus, the substituents must be at the 1- and 4-positions. The proton or set of protons that gives the signal at the lowest frequency is labeled *a,* the next lowest *b,* and so on.

By subtracting the six Cs and four Hs of the benzene ring from the molecular formula, we know that the two substituents contain three Cs, six Hs, and three Os ($C_9H_{10}O_3 - C_6H_4 = C_3H_6O_3$).

A triplet (1.4 ppm) that integrates to three protons and a quartet (4.2 ppm) that integrates to 2 protons are characteristic of an ethyl group. Because the signal for the CH_2 group of the ethyl substituent appears at a relatively high frequency, we know that it is attached to an electronegative atom (in this case, an O).

The presence of the CH_3CH_2O group consumes more of the remaining molecular formula ($C_3H_6O_3 - C_2H_5O = CHO_2$). There is one remaining NMR signal, a singlet (9.8 ppm) that integrates to 1 proton.

To help with the identification, we turn to the IR spectrum. The broad absorption near $3200\ cm^{-1}$ indicates the O—H stretch of an alcohol (or a phenol); the proton of this OH group would give the broad NMR signal at 9.8 ppm. The strong absorption at $1680\ cm^{-1}$ indicates the presence of a C=O group.

Now that all the fragments of the compound have been identified, we can put them together. The com pound is the one shown here.

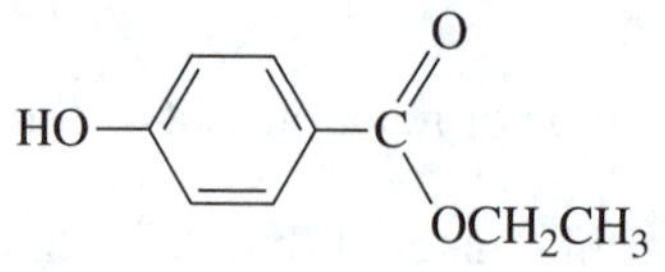

5.   The signals at 1.1 and 1.8 ppm have been magnified and are shown as insets on the spectrum (the **2** and represent the ppm scale) so that you can better see the splitting. The triplet (1.1 ppm) that integrates to protons and the quartet (1.8 ppm) that integrates to 2 protons are characteristic of an ethyl group. (The peal to the right of the quartet is actually the beginning of the adjacent signal that integrates to 6 protons.)

The singlet (1.7 ppm) that integrates to 6 protons indicates that there are two methyl groups in the sam environment. Because the signal is a singlet, the carbon to which they are attached cannot be bonded to an hydrogens. The only atom not accounted for in the molecular formula is Br.

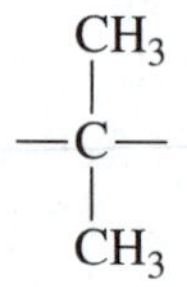

$$
\begin{array}{c}
CH_3 \\
| \\
-C- \\
| \\
CH_3
\end{array}
$$

Therefore, the ethyl group and the bromine must be the two substituents that are attached to the carbon Thus, the compound is **2-bromo-2-methylbutane.**

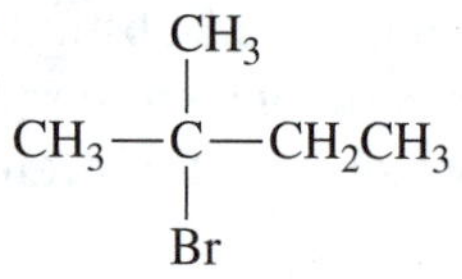

$$
\begin{array}{c}
CH_3 \\
| \\
CH_3-C-CH_2CH_3 \\
| \\
Br
\end{array}
$$

6.   A major clue comes from the IR spectrum. The strong absorption at ~1710 cm^{-1} indicates the presence of a carbonyl (C=O) group. Because the compound has only one oxygen, we know that it must be a aldehyde or a ketone. The absence of absorptions at 2820 and 2720 cm^{-1} tells us that the compound is no an aldehyde.

The absorptions at 2880 and 2970 cm^{-1} are due to C—H stretches of hydrogens attached to sp^3 carbons.

The ^{1}H NMR spectrum has two unsplit signals. One integrates to 9 protons, and the other to 3 protons. A signal that integrates to nine protons suggests a *tert*-butyl group, and a signal that integrates to 3 proton suggests a methyl group. The fact that they are both singlets indicates that they are on either side of the carbonyl group. Thus, the compound is the one shown here.

$$
\begin{array}{c}
CH_3\ O \\
|\ \ \ || \\
CH_3-C-C-CH_3 \\
| \\
CH_3
\end{array}
$$

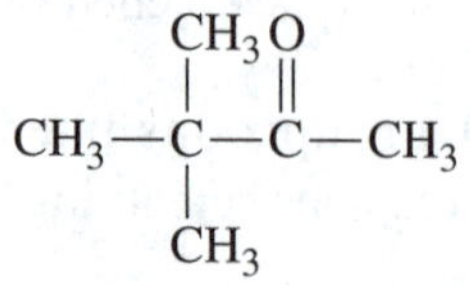

That the methyl group shows a signal at ~2.1 ppm reinforces this conclusion, because that is where methyl group attached to a carbonyl group is expected to occur.

The singlet (2.3 ppm) that integrates to three protons is due to the methyl group.

The signals in the 7–8 ppm region that integrate to 4 protons are due to the protons of a disubstituted benzene ring. Because both signals are doublets, we know that each proton is coupled to one adjacent proton. Thus, the compound has a 1,4-substituted benzene ring.

Therefore, the compound is the one shown here.

$$Cl-\langle \bigcirc \rangle-CH_3$$

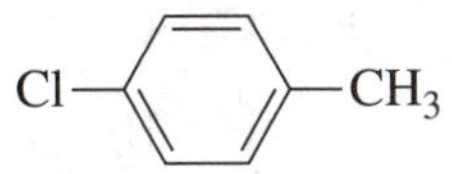

8. The strong and broad absorption in the IR spectrum at 3400 cm^{-1} indicates a hydrogen-bonded O—H group. The absorption bands between 2800 and 3000 cm^{-1} indicate hydrogens bonded to sp^3 carbons.

There is only one signal in the ^{1}H NMR spectrum that integrates to 1 proton, so it must be due to the hydrogen of the OH group. The singlet that integrates to 3 protons can be attributed to a methyl group that is attached to a carbon that is not attached to any hydrogens.

Since the other two signals show splitting, they represent coupled protons (that is, protons on adjacent carbons). The quartet and triplet combination indicates an ethyl group. Since the quartet and triplet integrate to 6 and 4 protons, respectively, the compound must have two ethyl groups.

The identified fragments of the molecule are as follows:

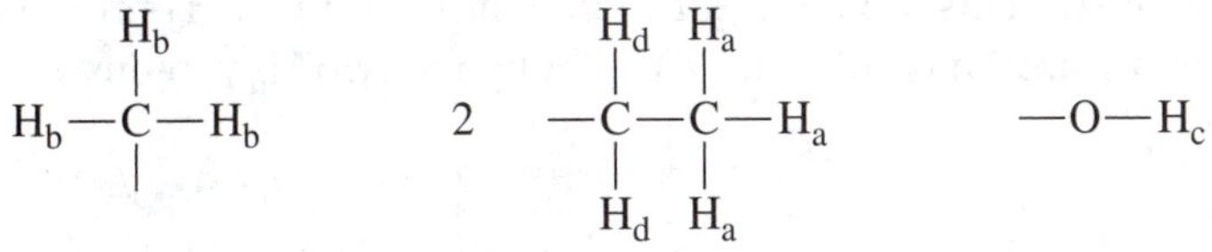

$$H_b-\overset{\overset{\displaystyle H_b}{|}}{\underset{\displaystyle |}{C}}-H_b \qquad 2\;\; -\overset{\overset{\displaystyle H_d}{|}}{\underset{\displaystyle H_d}{C}}\overset{\overset{\displaystyle H_a}{|}}{\underset{\displaystyle H_a}{C}}-H_a \qquad -O-H_c$$

When these fragments are subtracted from the molecular formula, only one carbon remains. Therefore, this carbon must connect the four identified fragments. The compound is the one shown here.

OH

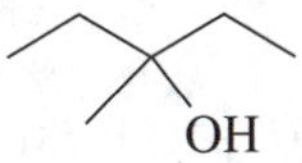

9. The ^{1}H NMR spectrum contains only one signal, so only one type of hydrogen is present in the molecule. Because the compound has 4 carbons and 9 identical hydrogens, the compound must be ***tert*-butyl bromide.**

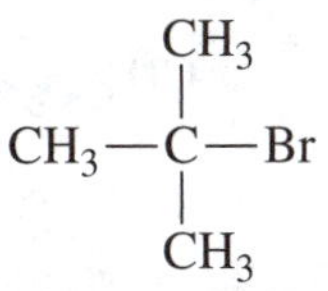

$$CH_3-\overset{\overset{\displaystyle CH_3}{|}}{\underset{\displaystyle CH_3}{C}}-Br$$

10. The molecular formula indicates that the compound is a hydrocarbon with one degree of unsaturation. The IR spectrum can tell us whether the degree of unsaturation is due to a cyclic system or a double bond. The absorption of moderate intensity near 1660 cm^{-1} indicates a C=C stretch. The absorption at ~3100 cm^{-1}, due to C—H stretches of hydrogens attached to sp^2 carbons, reinforces the presence of the double bond.

The two relatively high-frequency singlets (4.7 ppm) are given by vinylic protons. Because the signal integrates to 2 protons, we know that the compound has two vinylic protons. Because the signals are not split, the vinylic protons must not be on adjacent carbons. Thus, they must be on the same carbon.

The singlet (1.8 ppm) that integrates to 3 protons must be a methyl group. Because it is a singlet, the methyl group must be bonded to a carbon that is not attached to any protons.

The doublet (1.1 ppm) that integrates to 6 protons and the septet (2.2 ppm) that integrates to 1 proton are characteristic of an isopropyl group.

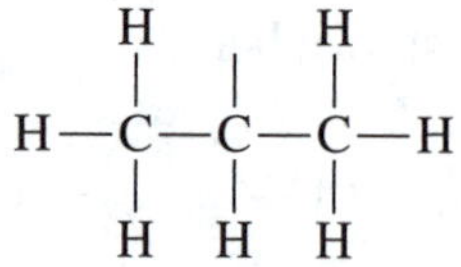

isopropyl group

Now that we know that the compound has a methyl group, an isopropyl group, and two vinylic hydrogens attached to the same carbon, the compound must be the one shown here.

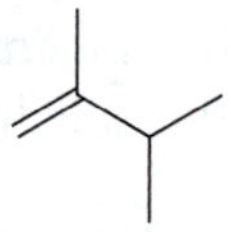

11. The signals in the ^1H NMR spectrum between 6.7 and 6.9 ppm indicate the presence of a benzene ring. Because the signals integrate to 3 protons, it must be a trisubstituted benzene ring.

The triplet (6.7 ppm) that integrates to 1 proton and the doublet (6.9 ppm) that integrates to 2 protons tell us that the three substituents are adjacent to one another. (The H_d protons are split into a doublet by the H_c proton, and the H_c proton is split into a triplet by the two H_d protons.)

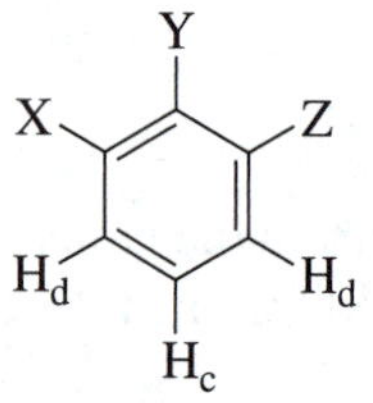

Subtracting the trisubstituted benzene (C_6H_3) from the molecular formula leaves C_2H_7O unaccounted for. The singlet (2.2 ppm) that integrates to 6 protons indicates that there are two methyl groups in identical environments. Now only OH is left from the molecular formula. The singlet at 4.6 ppm is due to the proton of the OH group. The compound is the one shown here.

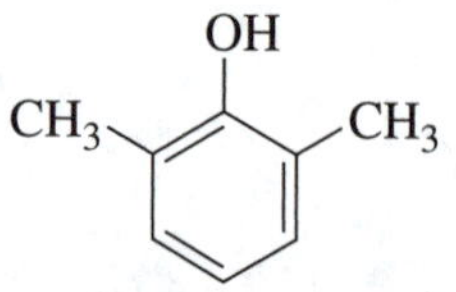

12. A major clue to the compound's structure comes from the IR spectrum. The strong absorption at ~ 1740 cm^{-1} indicates the presence of a carbonyl (C=O) group. Since the compound has only one oxygen, the compound must be an aldehyde or a ketone. The absence of absorptions at 2820 and 2720 cm^{-1} tells us that the compound is not an aldehyde.

The NMR spectrum shows a singlet (2.3 ppm) that integrates to 3 protons, indicating that it is due to a methyl group. The chemical shift of the signal (hydrogens attached to carbons adjacent to carbonyl carbons typically have shifts between 2.1 and 2.3 ppm) and the fact that the signal is a singlet suggest that the methyl group is directly attached to the carbonyl group.

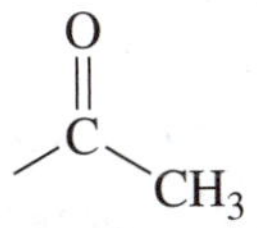

The two remaining signals are split, indicating that the protons that give these signals are attached to adjacent carbons. Because the signal at 4.3 ppm is a quartet, we know that the proton that gives this signal is bonded to a carbon that is attached to a methyl group. The other signal (1.6 ppm) is a doublet, so the proton that gives this signal is bonded to a carbon that is attached to one hydrogen.

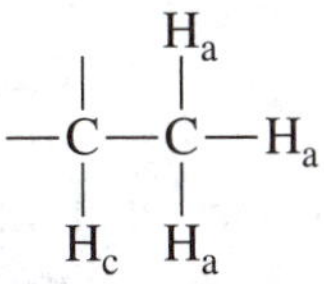

When these two fragments are subtracted from the molecular formula, only a Cl remains.

The only possible arrangement has the alkyl group directly bonded to the other side of the carbonyl group and the chlorine on the last available bond. The relatively high-frequency chemical shift of the quartet (4.3 ppm) reinforces this assignment, because it must be attached to an electronegative atom. Thus, the compound is the one shown here.

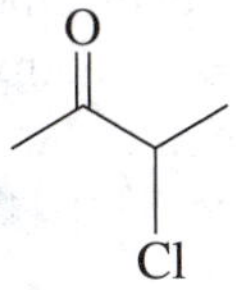

3.   Given the simplicity of the ^{1}H NMR spectrum, the product must be highly symmetrical.

The singlet (7.4 ppm) that integrates to 4 protons is due to benzene-ring protons. Because there are four aromatic protons, we know that the benzene ring is disubstituted. Because the signal is a singlet, we know that the four protons are chemically equivalent. Therefore, the two substituents must be the same and they must be on the 1- and 4-positions of the benzene ring.

Subtracting the disubstituted benzene ring from the molecular formula, only $C_2H_4Br_2$ remains. Thus, each substituent must contain 1 carbon, 2 hydrogens, and 1 bromine. The compound that gives the spectrum, therefore, is the one shown here.

14.   The molecular formula indicates that the compound has two degrees of unsaturation. The weak absorption at ~2120 cm^{-1} is due to a carbon–carbon triple bond, which accounts for the two degrees of unsaturation. The intense and sharp absorption at 3300 cm^{-1} is due to the C—H stretch of a hydrogen attached to an sp carbon. The intensity and shape of this absorption distinguishes it from an alcohol (intense and broad) and an amine (weaker and broad). Thus, we know that the compound is a terminal alkyne.

The absorptions between 2800 and 3000 cm^{-1} are due to the C—H stretch of hydrogens attached to sp^3 carbons.

All three signals in the ^{1}H NMR spectrum are singlets, indicating that none of the protons that give these signals have any neighboring protons. The singlet (2.4 ppm) that integrates to 1 proton is the proton of the terminal alkyne.

$$-C\equiv C-H$$

The two remaining signals (3.4 and 4.1 ppm) that integrate to 3 protons and 2 protons, respectively, can be attributed to a methyl group and a methylene group. When the alkyne fragment and the methyl and methylene groups are subtracted from the molecular formula, only an oxygen remains.

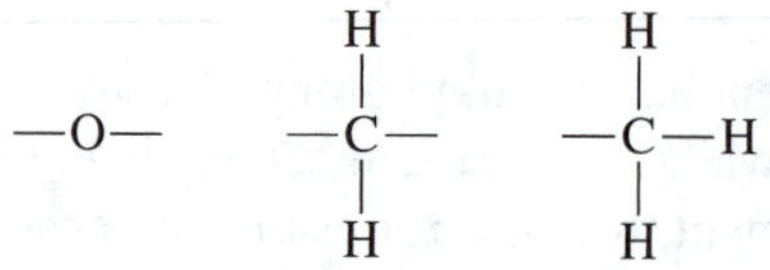

The arrangement of these groups can be determined by the splitting and the chemical shift of the signals. Since each signal is a singlet, the methyl and methylene groups cannot be adjacent or they would split each other's signal. Since the terminal alkyne and the methyl group must be on the ends of the molecule, the only possible arrangement is shown below. Thus, the compound is the one shown here.

$$CH_3OCH_2C\equiv CH$$

Notice that both the methyl and methylene groups show strong deshielding because of their direct attachment to the oxygen. The methylene hydrogens are also deshielded by the neighboring alkyne.

15.   The signals in the ^{1}H NMR spectrum between 6.5 and 7.2 ppm indicate the presence of a benzene ring. Since the signals integrate to 3 protons, it must be a trisubstituted benzene ring.

The singlet (2.1 ppm) that integrates to three protons must be a methyl group; 2.1 ppm is characteristic of protons bonded to a benzylic carbon.

When the trisubstituted benzene ring (C$_6$H$_3$) and the methyl group (CH$_3$) are subtracted from the molecular formula, NH$_2$Br is all that remains. Thus, the three substituents must be a methyl group, bromine, and an amino group (NH$_2$). The amino group gives the broad singlet (3.6 ppm) that integrates to 2 protons. Hydrogens attached to nitrogens and oxygens typically give broad signals.

The substitution pattern for the trisubstituted benzene can be determined from the splitting patterns. Because the signal (6.5 ppm) that integrates to 1 proton is a doublet, we know that the proton that gives this signal has only one neighboring proton. Looking at the magnification of the signal at 7.1 ppm, we see that it is actually two separate signals. One is a singlet, and therefore, it is attached to a carbon that is separated by substituents from the carbons that are attached to protons. The other signal is a doublet that integrates to 1 proton; because it gives a doublet, we know that it is next to the proton that gives the doublet at 6.5 ppm.

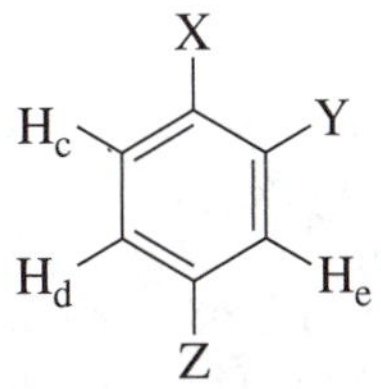

To determine the relative positions of the substituents, the chemical shifts must be analyzed. Bromine is the most electronegative substituent and, therefore, must be adjacent to the two protons that give signals at 7.1 ppm. Thus, Z is Br. The amino group donates its lone-pair electrons into the ring, so it shields benzene-ring protons. Thus, the signal at 6.5 ppm is from a proton in close proximity to the amino group. Therefore, X must be the amino group.

The compound that gives the spectrum is shown here.

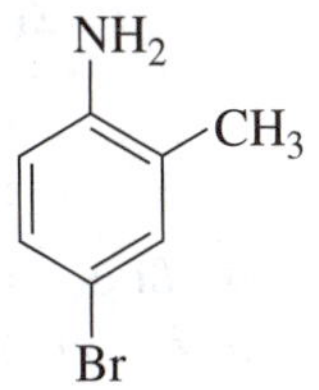

6.   The two compounds that produce the spectra have the following structures.

$$ClCH_2CH_2CH_2Br \qquad ClCH_2CH_2CH_2I$$
1-bromo-3-chloropropane   1-chloro-3-iodopropane

The number of signals (three) and the splitting patterns are identical for each compound. The only difference is variations in the chemical shift due to the different electronegativities of bromine and iodine.

Because chlorine is more electronegative than bromine or iodine, the protons bonded to the carbon that is attached to chlorine will have the most deshielded signal (that is, the signal that occurs at the highest frequency). This is the triplet that occurs at 3.7 ppm in both spectra.

The spectra differ in the signal that occurs at 3.4 ppm in the top spectrum and the one that appears at 3.6 ppm in the second spectrum. Because bromine is more electronegative than iodine, the protons bonded to the carbon that is attached to bromine will occur at a higher frequency than the protons bonded to the carbon that is attached to iodine.

Thus, **1-chloro-3-iodopropane** gives the top spectrum, and **1-bromo-3-chloropropane** gives the bottom spectrum.

7.   The molecular formula shows that the compound has one degree of unsaturation, indicating a cyclic compound, an alkene, or a carbonyl group.

A cyclic system containing an oxygen (a cyclic ether) would have the most deshielded signal at ~3.5 ppm, which would be due to the hydrogens attached to the carbon adjacent to the oxygen. Therefore, a cyclic ether would not give a signal at 6.4 ppm, so it can be ruled out.

Protons attached to a carbon adjacent to a carbonyl group show a signal at ~2.1 ppm. Since there is n
signal in that region, a carbonyl group can also be ruled out.

Vinylic protons would account for the signals in the 3.9–4.2 ppm range that integrate to 2 protons, so w
can conclude that the compound is an alkene. Because a highly deshielding oxygen is also present, th
high-frequency signal (6.4 ppm) is not unexpected.

The triplet (1.3 ppm) that integrates to 3 protons and the quartet (3.8 ppm) that integrates to 2 proton
indicate the presence of an ethyl group. The fact that the quartet is deshielded suggests that the ethyl
methylene group is attached to the oxygen.

The highly deshielded doublet of doublets (6.4 ppm) that integrates to 1 proton suggests that the proto
that gives this signal is attached to an sp^2 carbon that is attached to the oxygen. The fact that the signal
a doublet of doublets indicates that it is split by each of two nonidentical protons on the adjacent carbor
Thus, the compound is the one shown here.

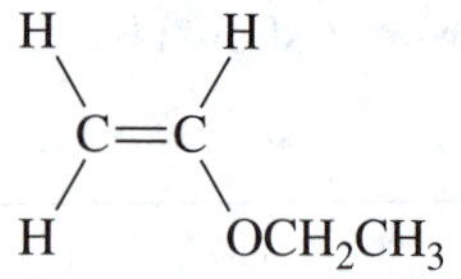

The identification is confirmed by the two doublets (~4.0 and 4.2 ppm) that each integrate to 1 proton
When those signals are magnified, we can see that each is really a doublet of doublets.

18.   The doublet (1.2 ppm) that integrates to 6 protons and the septet (5.0 ppm) that integrates to 1 proton ar
characteristic of an isopropyl group. (The two methyl groups are split by a single proton, and the singl
proton is split by six protons.)

The remaining signal (a singlet at a tiny bit more than 2.0 ppm) that integrates to 3 protons indicates a
unsplit methyl group.

When the isopropyl and methyl groups are subtracted from the molecular formula, one carbon and tw
oxygens are left over. Thus, the compound has the following fragments:

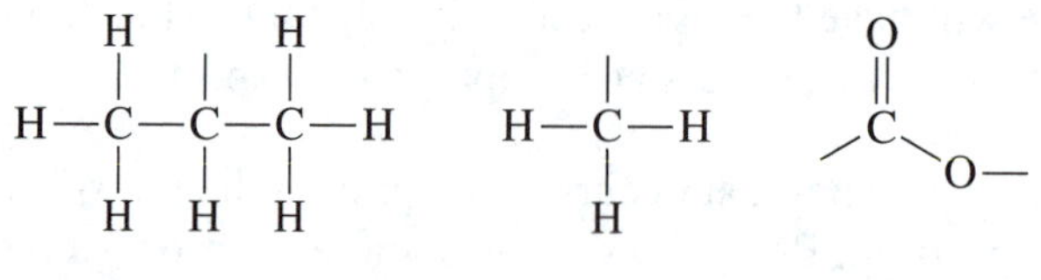

There are two ways these fragments can be pieced together. Because the most deshielded signal in th
spectrum (the one at 5.0 ppm) is the proton bonded to the central carbon of the isopropyl group, that carbo
must be directly attached to the oxygen. Thus, the compound is **isopropyl acetate.**

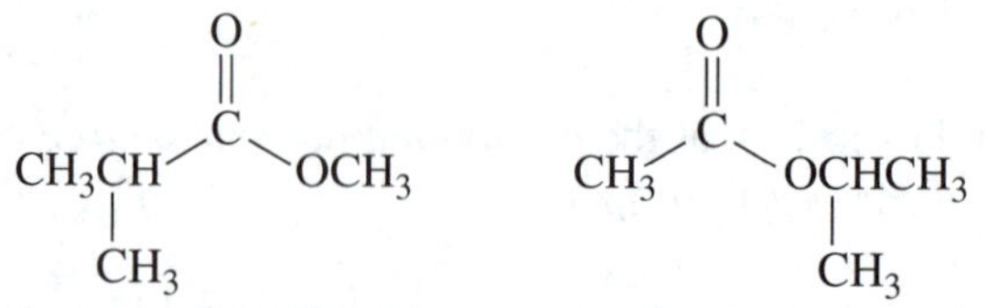

methyl 2-methylpropanoate   isopropyl acetate

19. The IR spectrum shows an absorption at $\sim 1700 \text{ cm}^{-1}$ for a C=O stretch and a very broad absorption ($2300-3300 \text{ cm}^{-1}$) for an O—H stretch, indicating that the compound is a carboxylic acid. Intermolecular hydrogen bonding explains the broad nature of this peak and also the broader-than-expected carbonyl peak. The proton of the carboxylic acid gives a singlet at 12.4 ppm in the NMR spectrum.

The two doublets (7.5 and 7.9 ppm) that each integrate to 2 protons indicate a 1,4-disubstituted benzene ring.

Subtracting the disubstituted benzene ring and the COOH group from the molecular formula leaves CH_2Br, so we know that the second substituent is a bromomethyl group; it gives the singlet at ~ 4.7 ppm.

Therefore, the compound that gives the spectrum is the one shown here.

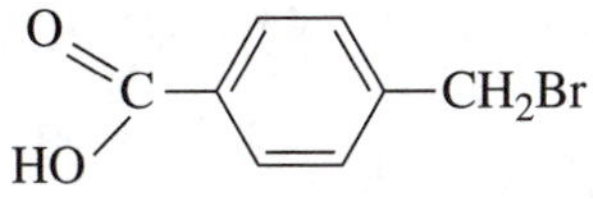

20. The signals with chemical shifts in the range of 7–8 ppm are due to benzene-ring protons. Because the three signals integrate to a total of 5 protons, we know that the benzene ring is monosubstituted.

The singlet at 10 ppm indicates the hydrogen of an aldehyde or a carboxylic acid. Since there is only one oxygen in the molecular formula, we know the compound is an aldehyde. Thus, the compound has an aldehyde group attached to a benzene ring.

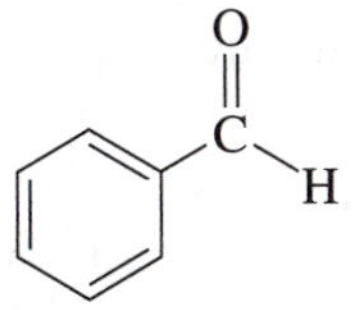

21. In the first spectrum, the doublet (~ 1.7 ppm) that integrates to 6 protons and the septet (~ 4.2 ppm) that integrates to 1 proton indicate an isopropyl group. When the isopropyl group is subtracted from the molecular formula, only a Br remains. Thus, the compound is **2-bromopropane.**

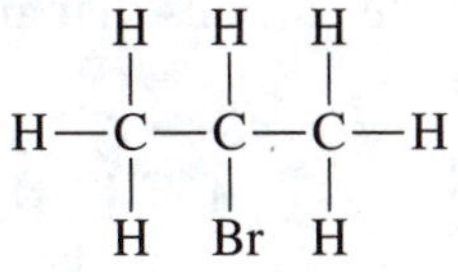

In the second spectrum, the triplet (~ 1.0 ppm) that integrates to 3 protons is a methyl group that is attached to a methylene group. The triplet (~ 3.4 ppm) that integrates to 2 protons is a methylene group that is also attached to a methylene group; the highly deshielded nature of the signal indicates that the carbon is attached to an electronegative group. Thus, the compound is **1-bromopropane.**

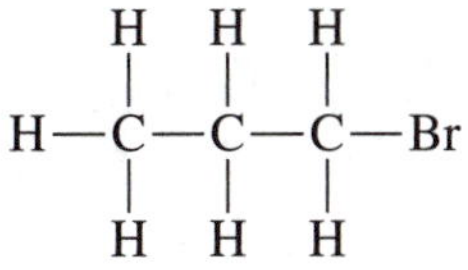

The structure is confirmed by the multiplet (~ 1.8) that integrates to 2 protons; the signal is split by both the adjacent methyl and methylene groups.

Notice that the pattern of a triplet that integrates to 3 protons, a multiplet that integrates to 2 protons, and a triplet that integrates to 2 protons is characteristic of a propyl group.

22.   The strong and sharp absorption in the IR spectrum at ~1720 cm^{-1} indicates the presence of a carbonyl group. The broad absorption centered at 3000 cm^{-1} tells us that the carbonyl-containing compound is a carboxylic acid. The broad singlet (12.0 ppm) in the NMR spectrum (shown as offset by 0.2 ppm from where it is placed on the spectrum) confirms the presence of a carboxylic acid group.

The only other signal in the NMR spectrum is a singlet (2.0 ppm) that integrates to six protons, indicating two methyl groups in the same environment. Because the signal is a singlet, the methyl groups must be attached to a carbon that is not attached to a proton. Because we know that the compound has only four carbons and contains a bromine, the compound must be the one shown here.

$$\begin{array}{c} \text{CH}_3 \\ \\ \text{CH}_3 \end{array} \!\! \text{C} \!\!\! \begin{array}{c} \text{O} \\ \| \\ \text{C} \\ \\ \text{Br} \end{array} \!\!\! \text{OH}$$

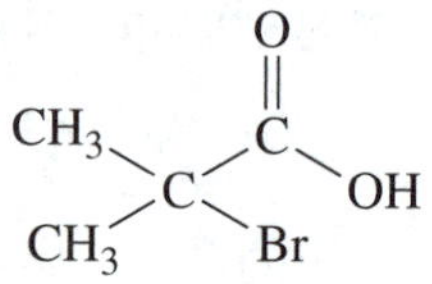

23.   The quintet (~2.2 ppm) that integrates to 2 protons indicates that the protons that give this signal have four identical neighboring protons. A carbon cannot be bonded to four protons and still be able to bond to anything else, so the two protons that give the quintet must be bonded to a carbon that is attached to two methylene groups in the same environment.

The triplet (~3.8 ppm) that integrates to 4 protons must be the signal for the four protons of the two methylene groups. The two methylene groups must be on either side of a carbon that is bonded to two protons (that is, the protons that give the quintet).

$$\begin{array}{ccccc} \text{H} & & \text{H} & & \text{H} \\ | & & | & & | \\ -\text{C} & - & \text{C} & - & \text{C}- \\ | & & | & & | \\ \text{H} & & \text{H} & & \text{H} \end{array}$$

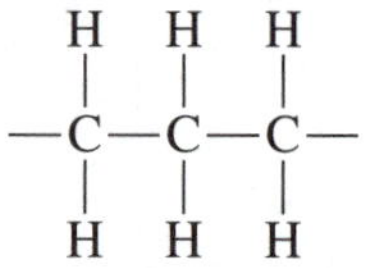

There are two bonds left unaccounted for, so this is where the two chlorines shown in the molecular formula go. Therefore, the compound is **1,3-dichloropropane.** The highly deshielded nature of the signal at 3.8 ppm for the protons bonded to the carbons that are attached to chlorines is further evidence that the chlorines are attached to these carbons.

$$\text{ClCH}_2\text{CH}_2\text{CH}_2\text{Cl}$$

24.   The IR spectrum shows a strong and sharp absorption at ~1720 cm^{-1}, indicating a carbonyl (C=O) group. The two absorptions at 2720 and 2820 cm^{-1} are characteristic of an aldehyde; they are due to the C—H stretch of the bond joining the carbonyl carbon and the hydrogen. Because the compound that gives the spectrum has three carbons, it must be the one shown here.

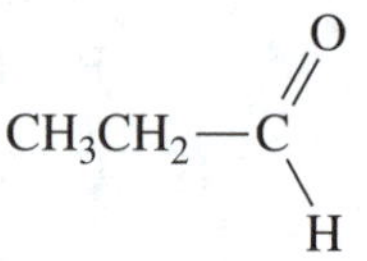

$$\text{CH}_3\text{CH}_2 \!-\! \text{C} \!\!\! \begin{array}{c} \text{O} \\ \text{\textbackslash\textbackslash} \\ \\ \text{H} \end{array}$$

5.   The short signal at ~185 ppm in the ^{13}C NMR spectrum suggests the presence of the carbonyl group of a carboxylic acid.

The broad singlet (12.2 ppm) in the ^{1}H NMR spectrum that integrates to 1 proton confirms that the compound contains a carboxylic acid group.

The doublet (1.2 ppm) that integrates to 6 protons and the septet (2.6 ppm) that integrates to 1 proton are characteristic of an isopropyl group.

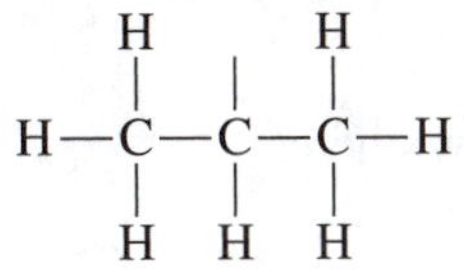

Therefore, the compound is the one shown here.

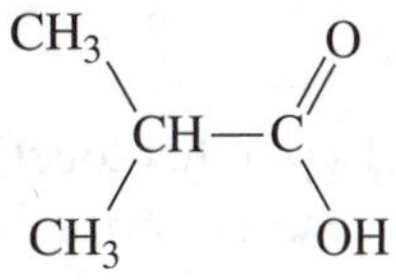

6.   The breath of the singlet (11.8 ppm) that integrates to 1 proton indicates a hydrogen that is attached to an oxygen. The chemical shift of the signal indicates that it is due to the OH group of a carboxylic acid.

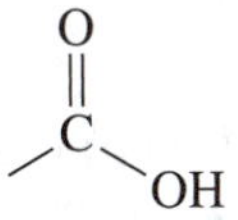

The triplet (~0.9 ppm) that integrates to 3 protons is a methyl group that is attached to a methylene group. The triplet (~2.3 ppm) that integrates to 2 protons indicates a methylene group that is also attached to a methylene group; the chemical shift of this signal indicates that the protons that give this signal are closest to the electron-withdrawing carboxylic acid group. The multiplet at 1.7 ppm that integrates to 2 protons is given by the two protons that split the other two signals into triplets.

We can conclude that the compound responsible for the spectrum is the one shown here.

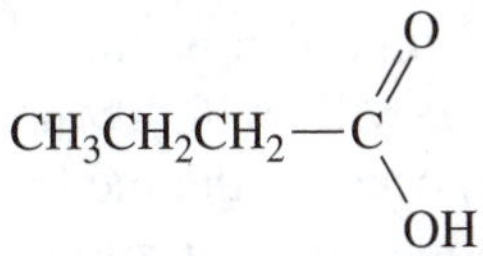

$$CH_3CH_2CH_2-C\overset{O}{\underset{OH}{\parallel}}$$

27.   The singlet (~9.7 ppm) that integrates to 1 proton and the molecular formula that contains one oxygen suggest that an aldehyde is present.

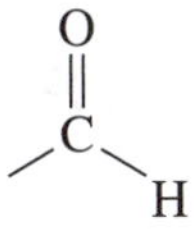

The signals at 7.7 and 6.7 ppm are due to benzene-ring protons. The fact that they are both doublets that integrate to 2 protons tells us that substituents are on the 1- and 4-positions of the benzene ring.

If the aldehyde group and the disubstituted ring are subtracted from the molecular formula, we find tha
the second substituent contains 2 carbons, 6 hydrogens, and 1 nitrogen. The remaining NMR signa
(~3.0 ppm) is a singlet that integrates to 6 hydrogens. These must be due to two methyl groups in the same
environment. The nitrogen must be between the two methyl groups or else they would split each other'
signals. The nitrogen causes the signal for the methyl groups to appear at a higher frequency than wher
methyl groups normally appear.

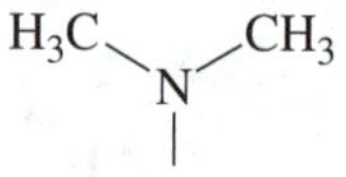

$$H_3C \diagdown \underset{N}{} \diagup CH_3$$

Thus, the compound is the one shown here.

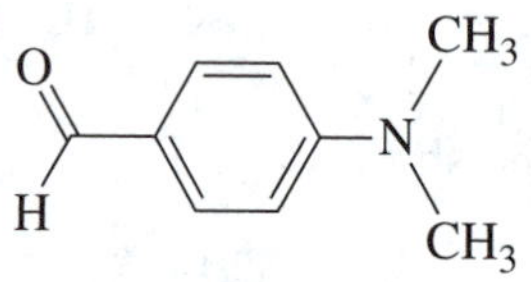

28. The three signals between 7.4 and 8.1 ppm that together integrate to 5 protons indicate a monosubstitute
benzene ring. Subtracting the monosubstituted ring (C_6H_5) from the molecular formula leaves $C_3H_5O_2$ t
be accounted for.

The two oxygens in the molecular formula tell us that the compound is an ester, because a broad single
between 10 and 12 ppm that would indicate a carboxylic acid is not present. The remainder of the molecul
contains two carbons and five hydrogens.

The two remaining signals, a triplet (1.4 ppm) that integrates to 3 protons and a quartet (4.4 ppm) that in
tegrates to 2 protons, are characteristic of an ethyl group. The three known segments can now be joined i
one of two ways:

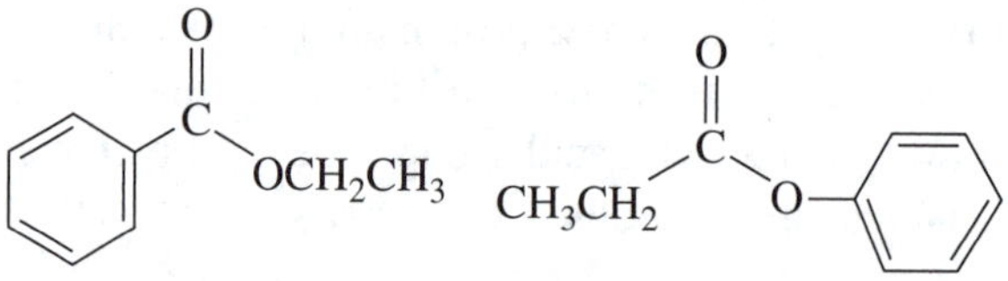

ethyl benzoate   phenyl propanoate

The choice between the two compounds can be made by looking at the chemical shift of the methylene
protons. In the ethyl ester, the signal will be highly deshielded by the adjacent oxygen. In the phenyl ester, the signal will be at ~2.1 ppm, because the methylene protons would be next to the carbonyl group
Because the chemical shift of the methylene protons is 4.4 ppm, we know that the compound is **ethy
benzoate.**

29. Because all three spectra are given by ethylmethylbenzenes, the low-frequency signals in both the ^{1}H NMR and ^{13}C NMR spectra can be ignored because they belong to the methyl and ethyl substituents. The key to determining which spectrum belongs to which ethylmethylbenzene can be found in the aromatic region of the ^{1}H NMR and ^{13}C NMR spectra.

4-ethylmethylbenzene   3-ethylmethylbenzene   2-ethylmethylbenzene

The aromatic region of the ^{13}C NMR spectrum of 4-ethylmethylbenzene will show four signals because it has four different ring carbons.

The aromatic region of the ^{13}C NMR spectrum of 3-ethylmethylbenzene will show six signals because it has six different ring carbons.

The aromatic region of the ^{13}C NMR spectrum of 2-ethylmethylbenzene will also show six signals because it has six different ring carbons.

We now know that spectrum **(b)** is the spectrum of 4-ethylmethylbenzene, because its ^{13}C NMR spectrum has four signals and the other two compounds will show six signals.

To distinguish between 2-ethylmethylbenzene and 3-ethylmethylbenzene, we need to look at the splitting patterns in the aromatic regions of the ^{1}H NMR spectra. Analysis of the aromatic region for spectrum **(c)** is difficult because the signals are superimposed. Analysis of the aromatic region for spectrum **(a)** provides the needed information. A triplet (7.2 ppm) that integrates to 1 proton is clearly present. This means that spectrum (a) is 3-ethylmethylbenzene, because 2-ethylmethylbenzene would not show a triplet. Therefore spectrum (c) is 2-ethylmethylbenzene.

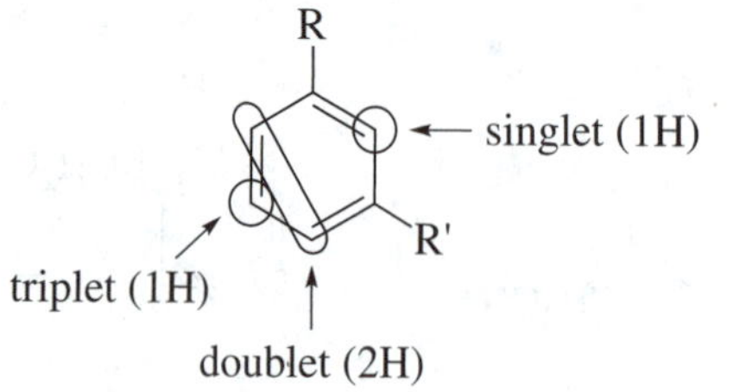

splitting pattern for 3-methylethyl benzene 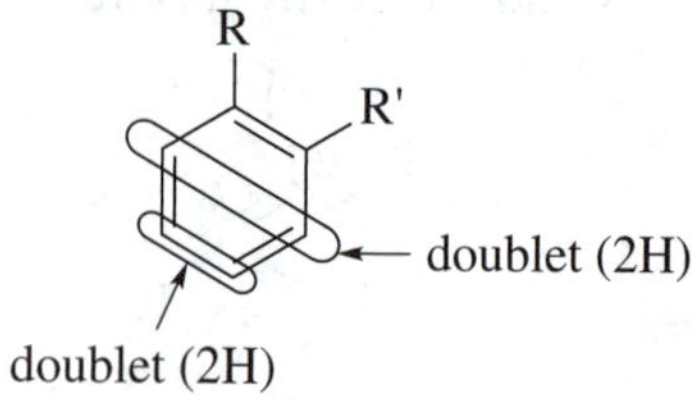   splitting pattern for 2-methylethyl benzene

The final assignments are as follows:

(a) 3-ethylmethylbenzene   (b) 4-ethylmethylbenzene   (c) 2-ethylmethylbenzene

30.   The simplicity of the NMR spectrum of a compound with 7 carbons and 14 hydrogens indicates that the compound must be symmetrical. From the molecular formula, we see that it has one degree of unsaturation. The absence of signals near 5 ppm rules out an alkene. Because the compound has an oxygen, the degree of unsaturation may be due to a carbonyl group.

The doublet (1.1 ppm) that integrates to 12 protons and the septet (2.8 ppm) that integrates to 2 protons suggest the presence of two isopropyl groups.

If two isopropyl groups are subtracted from the molecular formula, we find that the remainder of the molecule is composed of one carbon and one oxygen. Thus, the compound is the one shown here.

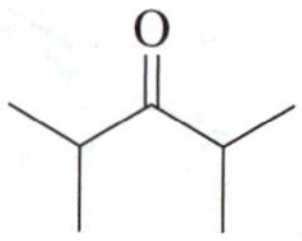

31.   The molecular formula tells us that the compound has one degree of unsaturation. The multiplet (5.2 ppm) that integrates to 1 proton is due to a vinylic proton (that is, it is attached to an sp^2 carbon). Thus, the degree of unsaturation is due to a carbon–carbon double bond. Because there is only one vinylic proton, we can assume that the alkene is trisubstituted.

Three additional signals are present that each integrate to 3 protons, suggesting that all three signals are due to methyl groups. The alkene, therefore, is the one shown here.

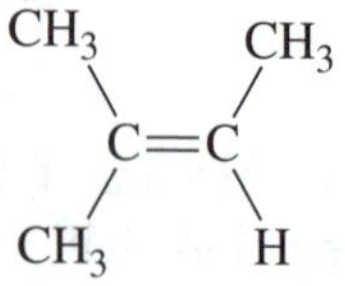

Notice that two of the three signals given by the methyl groups are singlets and one is a doublet. The methyl group that gives the doublet is bonded to the carbon that is attached to the vinylic proton. The other two methyl groups are bonded to the other sp^2 carbon.

32. A medium-intensity absorption at ~2120 cm^{-1} indicates the presence of a carbon–carbon triple bond. The sharp absorption at 3300 cm^{-1} is due to the C—H stretch of a hydrogen bonded to an *sp* carbon. Thus, the compound is a terminal alkyne.

$$-C\equiv C-H$$

The intense and broad peak centered at 3300 cm^{-1} is evidence of an O—H group.

The NMR spectrum can be used to determine the connectivity between the groups. The signal (2.5 ppm) that integrates to 1 proton is due to the proton of the terminal alkyne.

The singlet (3.2 ppm) that integrates to 1 proton must be due to the proton of the OH group.

The singlet (4.2 ppm) that integrates to 2 protons must be due to a methylene group that connects the triply bonded carbon to the OH group. The compound is **2-propyn-1-ol.**

$$H-C\equiv C-CH_2-OH$$

This arrangement explains the absence of any splitting and the highly deshielded nature of the signal for the methylene group.

33. The two singlets (4.9 and 5.1 ppm) that each integrate to 1 proton are vinylic protons. Therefore, we know that the compound is an alkene.

The singlet (2.8 ppm) that integrates to 3 protons is a methyl group. The deshielding results from its being attached to an *sp*^{2} carbon.

If we subtract the two vinylic protons, the two *sp*^{2} carbons of the alkene, and the methyl group from the molecular formula, we are left with CH_2Cl. Thus, a chloromethyl group is the fourth substituent of the alkene and gives the singlet (4.9 ppm) that integrates to 2 protons. Its deshielding is due to the proximity to the electronegative chlorine.

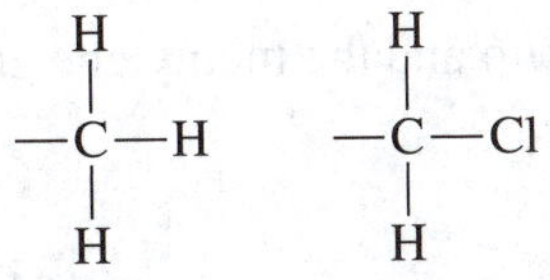

$$\begin{array}{cc} H & H \\ | & | \\ -C-H & -C-Cl \\ | & | \\ H & H \end{array}$$

Now we need to determine the substitution pattern of the alkene. The absence of splitting indicates that the two vinylic protons must be attached to the same carbon.

Thus, the compound is the one shown here.

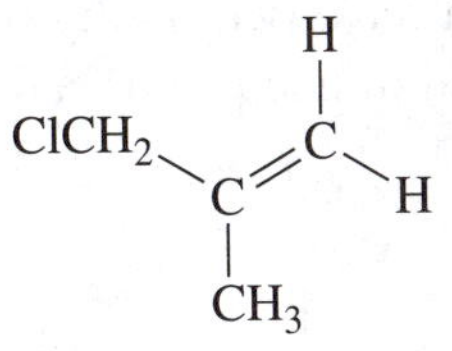

34.   The only signal that integrates to 1 proton is the singlet at 2.2 ppm. This must be due to the OH group of the alcohol.

The signals centered around 7.3 ppm are those given by benzene-ring protons. Because they integrate to 5 protons, the benzene ring must be monosubstituted.

The two triplets (2.8 and 3.8 ppm) that each integrate to 2 protons suggest two adjacent methylene groups. Both signals are fairly deshielded, indicating an electronegative atom nearby.

The fragments identified at this point are a monosubstituted benzene ring, an OH group, and two adjacent methylene groups.

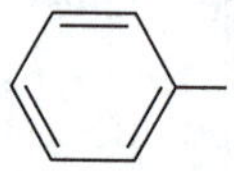

There are no other signals in the NMR spectrum, so the compound must be the one shown here.

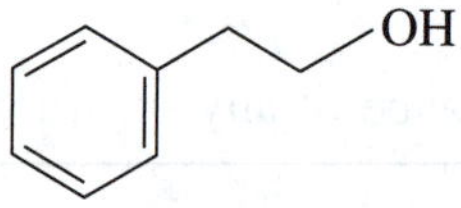

35.   The doublet (0.9 ppm) that integrates to 6 protons and the multiplet (1.8 ppm) that integrates to 1 proton suggest the presence of an isopropyl group.

Because we are told that the compound is an alcohol, the other signal that integrates to 1 proton (the triplet at 2.4 ppm) must be due to the OH proton. The fact that the signal is a triplet indicates that the OH group is probably attached to a methylene group.

The signal for the methylene group must be the remaining signal at 3.4 ppm, because it integrates to 2 protons. The relatively high-frequency chemical shift confirms that the methylene group is attached to the oxygen.

Putting together the isopropyl group and the methylene group that is attached to an OH group identifies the compound as the one shown here.

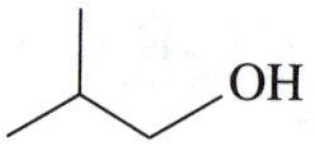

36.   The protons that are responsible for the doublet (1.2 ppm) that integrates to 6 protons must be adjacent to a carbon that is attached to only one proton. Because the spectrum does not have a signal that integrates to one proton, the compound must have two methyl groups in the same environment, and each must be adjacent to a carbon that is attached to one proton, and those two single protons must also be in identical environments.

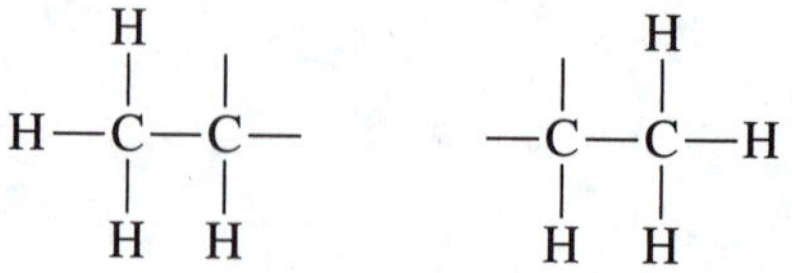

Because the compound must be symmetrical, the two oxygens in the compound must be due to two OH groups in identical environments. The hydrogens of the OH groups give a singlet (3.8 ppm) that integrates to 2 protons.

$$
\begin{array}{cc}
\text{H} & \text{OH} \\
| & | \\
\text{H}-\text{C}-\text{C}- \\
| & | \\
\text{H} & \text{H}
\end{array}
\qquad
\begin{array}{cc}
\text{OH} & \text{H} \\
| & | \\
-\text{C}-\text{C}-\text{H} \\
| & | \\
\text{H} & \text{H}
\end{array}
$$

The protons that give the triplet (2.6 ppm) must be bonded to a carbon that is adjacent to a total of two protons. Because the triplet integrates to 2 protons, it must be due to a methylene group that connects the two pieces.

$$
\begin{array}{ccccc}
\text{H} & \text{OH} & \text{H} & \text{OH} & \text{H} \\
| & | & | & | & | \\
\text{H}-\text{C}-&\text{C}-&\text{C}-&\text{C}-&\text{C}-\text{H} \\
| & | & | & | & | \\
\text{H} & \text{H} & \text{H} & \text{H} & \text{H}
\end{array}
$$

This structure is confirmed by the relatively high-frequency multiplet (4.2 ppm) that is given by the protons attached to the carbons that are attached to the OH groups. The signal for these protons is split by both the adjacent methyl group and the adjacent methylene group.

37. The absorption in the IR spectrum at ~1650 cm^{-1} could be due to either a carbonyl group or an alkene. Its strength and breadth tells us that it is probably due to a carbonyl (C=O) group. The strong and broad absorption at ~3300 cm^{-1} that contains two broad peaks suggests two N—H bonds; thus, an NH$_2$ group is present. When these two groups are subtracted from the molecular formula, all that is left is C$_2$H$_5$.

The triplet (~1.1 ppm) that integrates to 3 protons and the quartet (2.2 ppm) that integrates to 2 protons indicate the presence of an ethyl group; this accounts for the C$_2$H$_5$ fragment. Thus, all the fragments of the compound have been identified: C=O, NH$_2$, and CH$_3$CH$_2$. The compound, therefore, is **propanamide.**

The presence of an amide explains the lower-than-normal frequency of the C=O stretch in the IR spectrum. The breadth of the N—H stretches confirms that these are amide N—H stretches and not amine N—H stretches. The broad singlets (6.2 and 6.6 ppm) in the NMR spectrum are given by the protons attached to the nitrogen. The protons resonate at different frequencies, because the C—N bond has partial double bond character, which causes the protons to be in different environments.

38. The singlet (2.3 ppm) in the first spectrum that integrates to 3 hydrogens must be due to an isolated methyl group.

The triplet (1.1 ppm) that integrates to 3 protons and the quartet (2.5 ppm) that integrates to 2 protons are characteristic of an ethyl group.

The singlet (4.8 ppm) that integrates to one proton must be due to a single hydrogen attached to nitrogen.

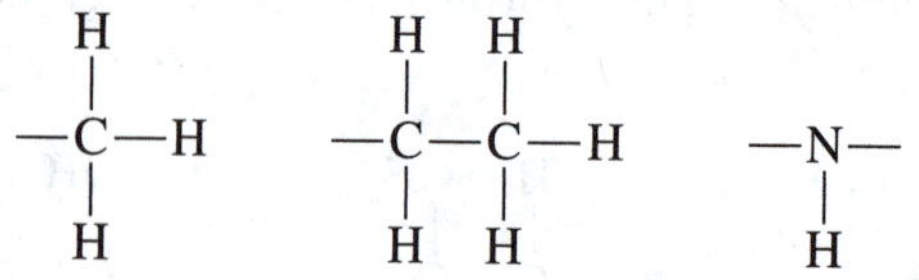

Now that the three fragments have been identified, we know that the compound is **ethylmethylamine.**

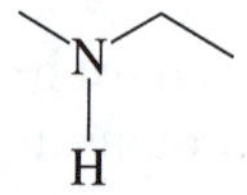

The second spectrum shows that a broad singlet (2.8 ppm) must be due to hydrogens that are attached to nitrogens. Because the signal integrates to 2 protons, we know that the compound is a primary amine.

The triplet (0.8 ppm) that integrates to 3 protons is due to a methyl group that is adjacent to a methylene group. The triplet (2.7 ppm) that integrates to 2 protons must also be adjacent to a methylene group. The multiplet (1.5 ppm) that integrates to 2 protons is the methylene group that splits both the methyl and methylene groups. The two triplets and multiplet are characteristic of a propyl group.

Therefore, the compound is the one shown here.

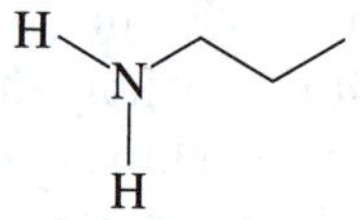

39.   The relatively weak absorption in the IR spectrum at $\sim 1650 \text{ cm}^{-1}$ tells us it is probably due to a carbon–carbon double bond. This is reinforced by the presence of absorptions at $\sim 3080 \text{ cm}^{-1}$, indicating $C-H$ bond stretches of hydrogens attached to sp^2 carbons.

The shape of the two absorptions at $\sim 3300 \text{ cm}^{-1}$ suggests the presence of an NH_2 group of a primary amine. (Compare these to the shape of the $N-H$ stretches of an NH_2 group of an amide in Problem 36.)

The three signals in the NMR spectrum between 5.0 and 6.0 ppm that integrate as a group to 3 protons indicate that there are three vinylic protons. Therefore, we know that the alkene is monosubstituted.

The two remaining signals in the NMR spectrum are a doublet (3.3 ppm) and a singlet (1.3 ppm) that each integrate to 2 protons. Because splitting is not typically seen with protons attached to nitrogens, we can identify the singlet at 1.3 ppm as due to the two amine protons. The doublet must be due to a methylene group that is attached to an sp^2 carbon and split by a vinylic proton that is attached to the same carbon. The compound, therefore, is the one shown here.

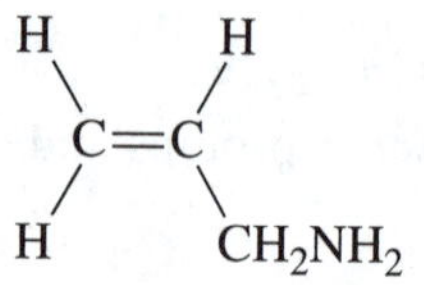

Now we can understand why the signal at 5.9 ppm is a multiplet. This vinylic proton is split by the methylene group and the two vinylic protons attached to the adjacent carbon. The signals for the these two vinylic protons are both doublets because each is split by the single proton attached to the adjacent sp^2 carbon.

40. The molecular formula tells us that the compound does not have any degrees of unsaturation. Therefore, the oxygen must be the oxygen of either an ether or an alcohol. Because there are no signals that integrate to one proton, we can conclude that the compound is an ether.

The triplet (1.2 ppm) that integrates to 3 protons and the quartet (3.5 ppm) that integrates to 2 protons suggest the presence of an ethyl group. The high-frequency chemical shift of the ethyl's methylene group and the fact that it shows splitting only by the three protons of the methyl group indicate that the ethyl group is next to the oxygen.

The two remaining signals are both triplets (3.4 and 3.5 ppm), and each integrates to 2 protons. Thus, the signals are due to two adjacent methylene groups. Because both signals occur at high frequencies, both must be attached to electron-withdrawing atoms.

Because the molecular formula tells us that the compound contains a bromine, we can conclude that the compound is the one shown here.

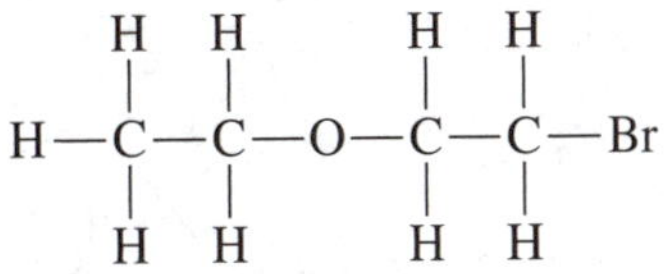

$$\begin{array}{ccccccccc} & H & H & & & H & H & & \\ & | & | & & & | & | & & \\ H - & C - & C - & O - & C - & C - & Br \\ & | & | & & & | & | & & \\ & H & H & & & H & H & & \end{array}$$

41. The IR spectrum shows a strong and broad absorption at $\sim 3300 \text{ cm}^{-1}$, indicating that the compound is an alcohol or a phenol.

The signals in the NMR spectrum between 6 and 7 ppm indicate the presence of a benzene ring. Because these signals integrate to a total of 3 protons, the benzene ring must be trisubstituted.

Because the signal at 6.3 ppm is a doublet, it must be adjacent to one proton; and since the signal at 6.7 ppm is a triplet, it must be adjacent to two protons. Thus, the three benzene-ring protons must be adjacent to one another.

The singlet at 8.7 ppm is the only signal in the spectrum that can be attributed to the proton of the OH group. Because the signal integrates to 2 protons, the compound must have two OH groups in the same environment.

The singlet (2.0 ppm) that integrates to 3 protons indicates that the compound has a methyl group that is not adjacent to a carbon that is attached to any hydrogens.

Therefore, we know that the three substituents that are attached to adjacent carbons on the benzene ring are two OH groups and a methyl group. Because the OH groups are in the same environment, the compound must be the one shown here.

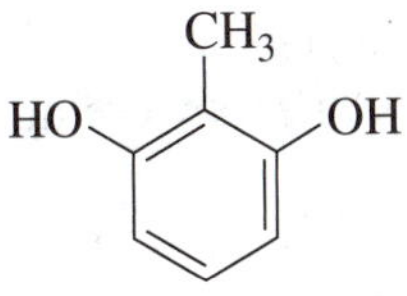

42. We are told that the compound is an alcohol. Because the singlet (1.4 ppm) is the only signal that integrate to 1 proton, it must be the signal given by the OH group.

We know that a triplet that integrates to 3 protons and a quartet that integrates to 2 protons are characteris tic of an ethyl group. In this case, the triplet (0.8 ppm) integrates to 6 protons and the quartet integrates to 4 protons. Therefore, the compound must have two ethyl groups in identical environments.

The only other signal in the spectrum is the singlet (1.2 ppm) that integrates to 3 protons. This signal mus be due to a methyl group that is bonded to a carbon that is not attached to any hydrogens.

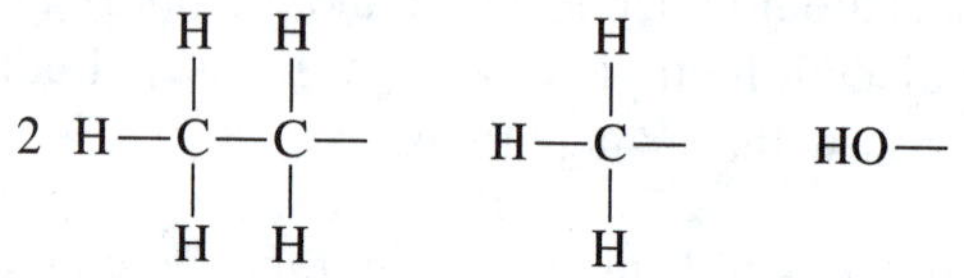

Because the NMR spectrum does not show any additional signals, the compound must be the one shown here.

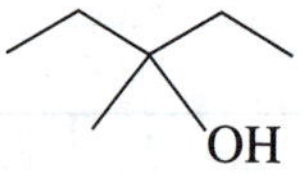

43. The doublet (1.1 ppm) that integrates to 6 protons and the septet (3.8 ppm) that integrates to 1 proton sug- gest an isopropyl group. The triplet (1.7 ppm) that integrates to 3 protons and the quartet (2.2 ppm) that integrates to 2 protons suggest an ethyl group. When these two groups are subtracted from the molecu- lar formula, all that remains is CO_2. The observed splitting patterns tell us that the isopropyl and ethyl groups are isolated from one another. We can conclude then that the compound is an ester. There are two possibilities:

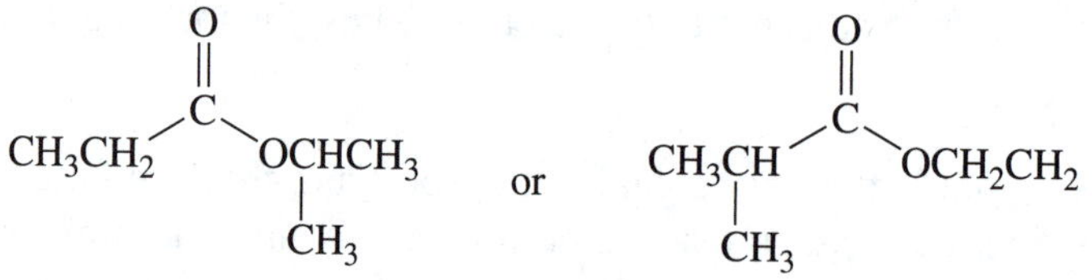

isopropyl propanoate   ethyl 2-methylpropanoate

Because the highest-frequency signal (the septet) is given by the CH of the isopropyl group, we know that the CH is attached to an oxygen. Therefore, we know that the compound that gives the NMR data is **iso- propyl propanoate.**

44. The multiplet (6.9–7.8 ppm) that integrates to 4 protons indicates a disubstituted benzene ring. Because the signal is a multiplet, we know that the substituents are on either the 1- and 2-positions or the 1- and 3-po- sitions. If the substituents were on the 1- and 4-positions, either one singlet (if the two substituents were identical) or two doublets (if the substituents were not identical) would be observed.

Three signals (1.4, 2.5, 3.8 ppm) each integrate to 2 protons. The fact that the signals are two triplets and a multiplet suggests that the compound has three adjacent methylene groups. (The methylene groups on the ends would be triplets, and the one in the middle would be a multiplet.)

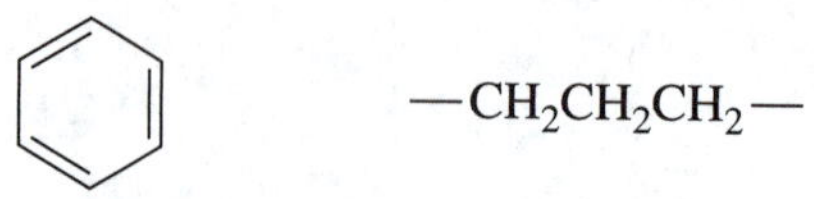

$-CH_2CH_2CH_2-$

From the molecular formula, we know that the compound has an oxygen. The triplet at 3.8 ppm indicates that that particular methylene group is next to the oxygen. The two fragments that we have identified account for the entire molecular formula. Thus, the compound must have the structure shown here.

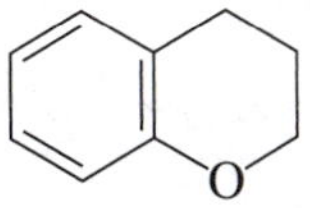

45. The molecular formula indicates that the compound has one degree of unsaturation. The NMR spectrum does not show any signals in the area expected for vinylic protons, so the compound must be either a ketone or a cyclic ether. A cyclic ether would be expected to have protons on adjacent carbons, so the signals would show splitting. Because the two signals in the spectrum are both singlets, the compound must be a ketone.

The fact that the compound has 11 carbons and 22 hydrogens but gives only two singlets in the NMR spectrum indicates that the compound must be symmetrical.

We know that a *tert*-butyl group gives a singlet that integrates to 9 protons. The symmetry of the molecule leads us to conclude that the singlet that integrates to 18 protons is due to two *tert*-butyl groups. We can then assume that the singlet that integrates to 4 protons is due to two nonadjacent methylene groups.

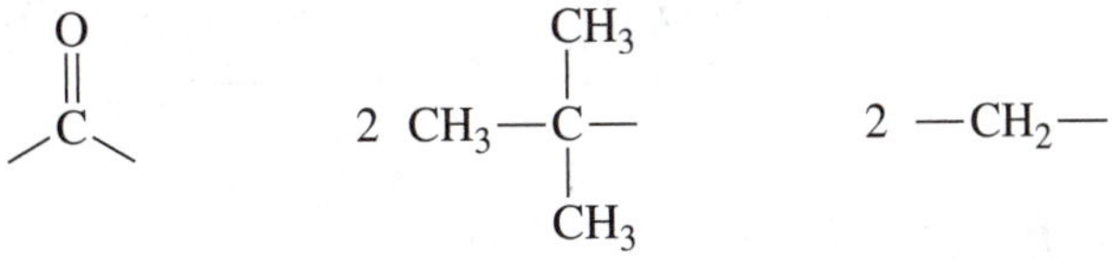

These fragments account for all atoms in the molecular formula. Therefore, the compound must be the one shown here.

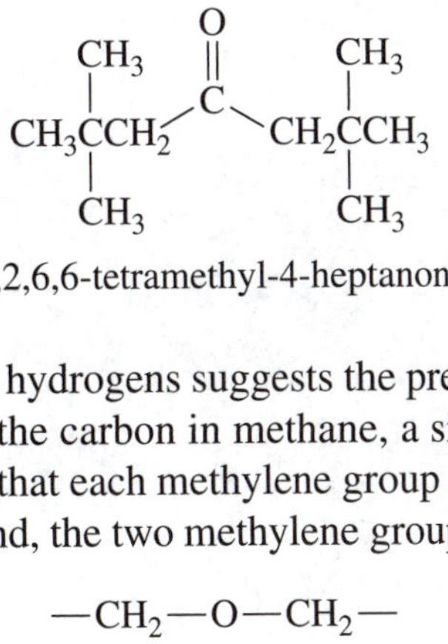

2,2,6,6-tetramethyl-4-heptanone

46. The signal (3.8 ppm) that integrates to 4 hydrogens suggests the presence of two methylene groups in identical environments because, other than the carbon in methane, a single carbon cannot be attached to four hydrogens. The chemical shift suggests that each methylene group must be attached to an oxygen. Because there is only one oxygen in the compound, the two methylene groups must be attached to the same oxygen.

$$-CH_2-O-CH_2-$$

The signal (1.6 ppm) that integrates to hydrogens also suggests the presence of two methylene groups in identical environments. The four methylene groups and the oxygen account for all the atoms in the molecular formula. Thus, the compound must be the cyclic ether shown here.

The fact that the signal at the higher frequency is a triplet and the other signal is a multiplet confirms this structure.

47. The doublet (7.3 ppm) that integrates to 4 protons and the doublet (7.7 ppm) that also integrates to 4 protons are given by benzene-ring protons. Because a benzene ring does not have 8 protons, there must be two benzene rings in the compound. The doublets indicate that the benzene rings have substituents at the 1- and 4-positions and, because each doublet integrates to 4 protons, the two substituents on each of the benzene rings must be the same.

The singlet (3.8 ppm) that integrates to 6 protons suggests the compound has two methyl groups in an identical environment. The chemical shift of the singlet indicates that each is attached to an electronegative atom. The molecular formula indicates that the electronegative atom is an oxygen.

$$2 \quad -OCH_3$$

When the two disubstituted benzene rings and the two CH_3O groups are subtracted from the molecular formula, all that remains is CO. Thus, a carbonyl group must connect the two benzene rings.

CHAPTER 1
Remembering General Chemistry: Electronic Structure and Bonding

Important Terms

atomic number	a number that tells us how many protons (or electrons) a neutral atom has.
atomic orbital	an orbital associated with an atom; the three-dimensional area around a nucleus where electrons are most likely to be found.
atomic weight	the average mass of the atoms in the naturally occurring element.
bond dissociation energy	the amount of energy required to break a bond in a way that allows each of the atoms to retain one of the bonding electrons, or the amount of energy released when a bond is formed.
bond length	the internuclear distance between two atoms at minimum energy (maximum stability).
carbanion	a species containing a negatively charged carbon.
carbocation	a species containing a positively charged carbon.
condensed structure	a structure that does not show some (or all) of the covalent bonds.
core electrons	electrons in filled shells.
covalent bond	a bond created as a result of sharing electrons.
dipole	a separation of positive and negative charges.
dipole moment (μ)	a measure of the separation of charge in a bond or in a molecule.
double bond	a bond composed of a sigma bond and a pi bond.
electronegative	describes an element that readily acquires an electron.
electronegativity	the tendency of an atom to pull electrons toward itself.
electrostatic attraction	an attractive force between opposite charges.
electrostatic potential map (potential map)	a map that allows you to see how electrons are distributed in a molecule.
equilibrium constant	the ratio of products to reactants at equilibrium.
formal charge	the number of valence electrons (the number of nonbonding electrons + half the number of bonding electrons).

free radical (radical)	a species with an unpaired electron.
hybrid orbital	an orbital formed by hybridizing (mixing) atomic orbitals.
hydride ion	a negatively charged hydrogen (a hydrogen atom with an extra electron).
hydrogen ion (proton)	a positively charged hydrogen (a hydrogen atom without its electron).
ionic bond	a bond formed as a result of the attraction of opposite charges.
ionic compound	a compound composed of a positive ion and a negative ion.
ionization energy	the energy required to remove an electron from an atom.
isotopes	atoms with the same number of protons but a different number of neutrons.
Kekulé structure	a model that represents the bonds between atoms as lines.
Lewis structure	a model that represents the bonds between atoms as lines or dots and the lone-pair electrons as dots.
lone-pair electrons (nonbonding electrons)	valence electrons not used in bonding.
mass number	the number of protons plus the number of neutrons in an atom.
nonbonding electrons	valence electrons not used in bonding.
nonpolar covalent bond	a bond formed between two atoms that share the bonding electrons equally.
octet rule	a rule that states that an atom will give up, accept, or share electrons in order to achieve a filled outer shell (or an outer shell that contains eight electrons) and no electrons of higher energy. Because a filled second shell contains eight electrons, this is known as the octet rule.
orbital	the volume of space around the nucleus where an electron is most likely to be found.
orbital hybridization	mixing of atomic orbitals.
organic compound	a compound that contains carbon.
pi (π) bond	a bond formed as a result of side-to-side overlap of p orbitals.
polar covalent bond	a bond formed between two atoms that do not share the bonding electrons equally.
potential map (electrostatic potential map)	a map that allows you to see how electrons are distributed in a molecule.

proton (hydrogen ion)	a positively charged hydrogen ion.
radical (free radical)	a species with an unpaired electron.
sigma (σ) bond	a bond with a symmetrical distribution of electrons about the internuclear axis.
single bond	a single pair of electrons shared between two atoms.
tetrahedral bond angle	the bond angle (109.5°) formed by an sp^3 hybridized central atom.
tetrahedral carbon	an sp^3 hybridized carbon; a carbon that forms covalent bonds using four sp^3 hybrid orbitals.
trigonal planar carbon	an sp^2 hybridized carbon.
triple bond	composed of a sigma bond and two pi bonds.
valence electron	an electron in the outermost shell.

52   Chapter 1

Solutions to Problems

1.   The atomic number = the number of protons.
The mass number = the number of protons + the number of neutrons.
All isotopes have the same atomic number; in the case of oxygen it is 8. Therefore:

> The isotope of oxygen with a mass number of 16 has 8 protons and 8 neutrons.
> The isotope of oxygen with a mass number of 17 has 8 protons and 9 neutrons.
> The isotope of oxygen with a mass number of 18 has 8 protons and 10 neutrons.

2.   All four atoms have two core electrons (in their filled shell); the valence electrons are in the outer shell.
Notice that because the four atoms in the question are in the same row of the periodic table, they have the
same number of core electrons.
a. 3   **b.** 5   **c.** 6   **d.** 7

3.   They each have seven valence electrons.

4.   The atomic numbers can be found in the periodic table at the back of the book. Notice that elements in the
same column of the periodic table have the same number of valence electrons, and their valence electrons
are in similar orbitals.

a.   carbon (atomic number = 6; 2 core, 4 valence): $1s^2 2s^2 2p^2$
silicon (atomic number = 14; 10 core, 4 valence): $1s^2 2s^2 2p^6 3s^2 3p^2$

b.   oxygen (atomic number = 8; 2 core, 6 valence): $1s^2 2s^2 2p^4$
sulfur (atomic number = 16; 10 core, 6 valence): $1s^2 2s^2 2p^6 3s^2 3p^4$

c.   nitrogen (atomic number = 7; 2 core, 5 valence): $1s^2 2s^2 2p^3$
phosphorus (atomic number = 15; 10 core, 5 valence): $1s^2 2s^2 2p^6 3s^2 3p^3$

d.   magnesium (atomic number = 12; 10 core, 2 valence): $1s^2 2s^2 2p^6 3s^2$
calcium   (atomic number = 20; 18 core, 2 valence): $1s^2 2s^2 2p^6 3s^2 3p^6 4s^2$

5.   **a.** Potassium is in the first column of the periodic table; therefore, like lithium and sodium, which are also
in the first column, potassium has one valence electron.
b. It occupies a $4s$ orbital.

6.   The polarity of a bond can be determined by the difference in the electronegativities of the atoms sharing
the bonding electrons. The greater the difference in electronegativity, the more polar the bond.
a. $Cl—CH_3$ **b.** $H—OH$ **c.** $H—F$ **d.** $Cl—CH_3$

7.   **a.** KCl has the most polar bond, because its two bonded atoms have the greatest differences in electronegativity. The electronegativity differences in the four listed compounds are as follows:

$$KCl \quad 3.0 - 0.8 = 2.2$$
$$LiBr \quad 2.8 - 1.0 = 1.8$$
$$NaI \quad 2.5 - 0.9 = 1.6$$
$$Cl_2 \quad 3.0 - 3.0 = 0$$

b. Cl_2 has the least polar bond, because the two chlorine atoms share the bonding electrons equally.

8. Solved in the text.

9. To answer this question, compare the electronegativities of the two atoms sharing the bonding electrons using Table 3.

a. $\overset{\delta-\ \ \delta+}{HO-H}$ **c.** $\overset{\delta+\ \ \ \delta-}{H_3C-NH_2}$ **e.** $\overset{\delta-\ \ \delta+}{HO-Br}$ **g.** $\overset{\delta+\ \delta-}{I-Cl}$

b. $\overset{\delta-\ \ \delta+}{F-Br}$ **d.** $\overset{\delta+\ \ \delta-}{H_3C-Cl}$ **f.** $\overset{\delta-\ \ \delta+}{H_3C-Li}$ **h.** $\overset{\delta+\ \ \delta-}{H_2N-OH}$

(Notice that if atoms are in the same row of the periodic table, the atom farthest to the right is the most electronegative atom; if atoms are in the same column, the one closest to the top of the column is the most electronegative atom.)

10.   **a.** LiH and HF are polar (they have a red end and a blue end).
   b. A potential map marks the edges of the molecule's electron cloud. The electron cloud is largest around the H in LiH, because that H has more electrons around it than do the Hs in the other molecules.

11. By answering this question, you will see that a formal charge is a book-keeping device. It does *not neces-sarily* tell you which atom has the greatest electron density or is the most electron deficient.

   a. oxygen   **c.** oxygen
   b. oxygen (it is more red)   **d.** hydrogen (it is the deepest blue)

Notice that in hydroxide ion, the atom with the formal negative charge **is** the atom with greater electron density. In the hydronium ion, however, the atom with the formal positive charge **is not** the most electron deficient atom.

12. formal charge = number of valence electrons
           − (number of lone-pair electrons + half the number of bonding electrons)

In all four structures, every H is singly bonded and thus has a formal charge = $1 - (0 + 1) = 0$.
Similarly, all CH_3 carbon atoms have four bonds and a formal charge = $4 - (0 + 4) = 0$.
The formal charges on the remaining atoms follow:

a. 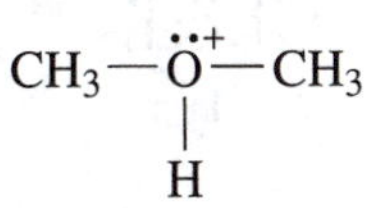 $CH_3-\overset{\ \ +}{\underset{|}{\overset{..}{O}}}-CH_3$
   $\quad\ \ |$
   $\quad\ \ H$
formal charge on O
$6 - (2 + 3) = +1$

b. 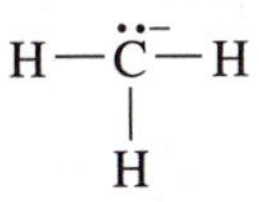 $H-\overset{..-}{\underset{|}{C}}-H$
   $\quad\ \ |$
   $\quad\ \ H$
formal charge on C
$4 - (2 + 3) = -1$

c. 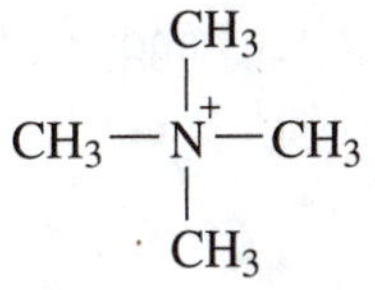 $\overset{\displaystyle CH_3}{\underset{\displaystyle CH_3}{\overset{|}{\underset{|}{CH_3-\overset{+}{N}-CH_3}}}}$
formal charge on N:
$5 - (0 + 4) = +1$

d. 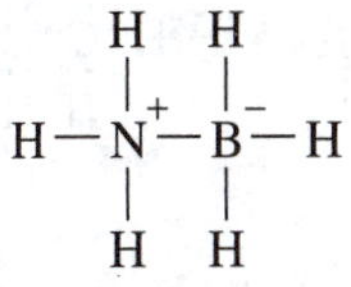 $\overset{\displaystyle H\ \ \ H}{\underset{\displaystyle H\ \ \ H}{\overset{|\ \ \ |}{\underset{|\ \ \ |}{H-\overset{+}{N}-\overset{-}{B}-H}}}$
formal charge on
$N: 5 - (0 + 4) = +1$
$B: 3 - (0 + 4) = -1$

13.   The bond between two atoms can be shown by a pair of dots or by a line, so there are two ways each of the answers can be written.

a.

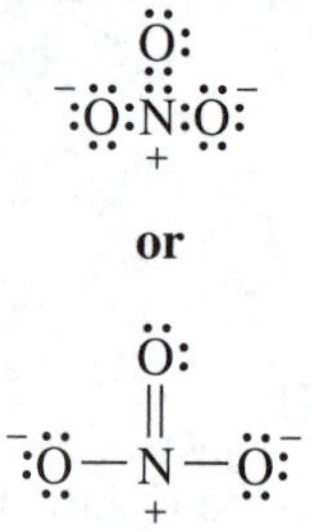

:Ö:
:Ö:N:Ö:⁻
  ⁺

or

    Ö:
    ‖
:Ö—N—Ö:⁻
    ⁺

c.

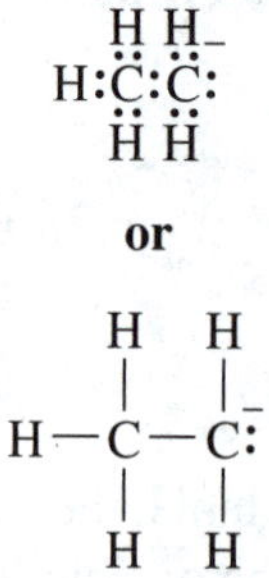

H H⁻
H:C:C:
H H

or

    H   H
    | |
H—C—C:⁻
    | |
    H   H

e.

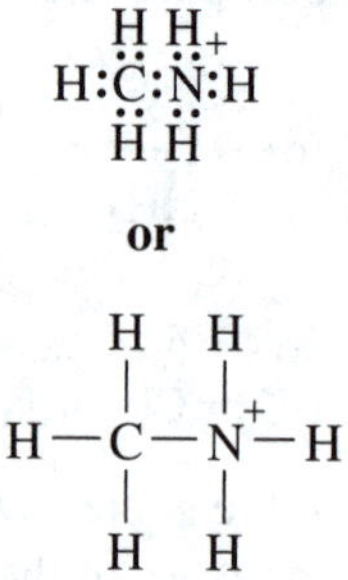

H H₊
H:C:N:H
H H

or

    H   H
    | |
H—C—N⁺—H
    | |
    H   H

b. 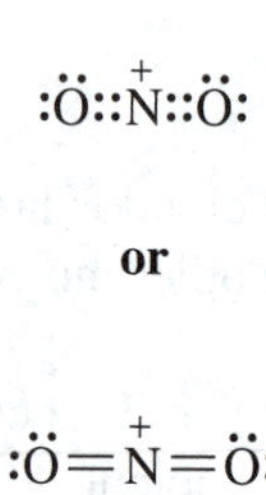   :Ö::N::Ö:

or

:Ö=N⁺=Ö:

d.

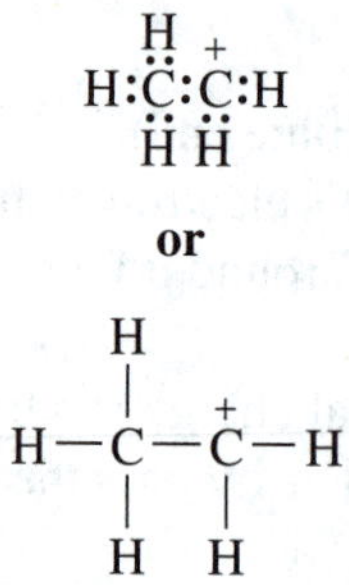

H ₊
H:C:C:H
H H

or

      H
      |
H—C—C⁺—H
      |
   H   H

14.   **a.**

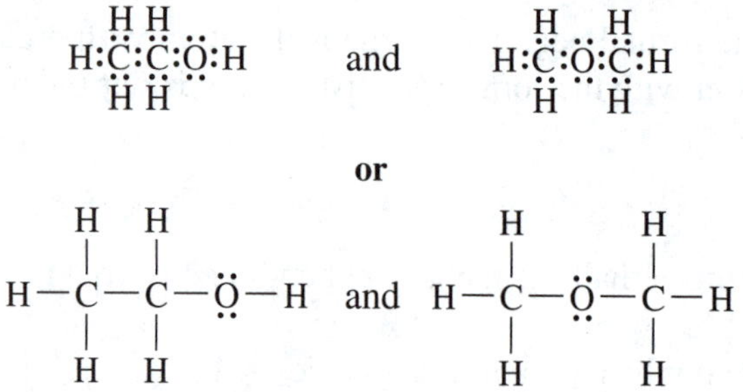

H H
H:C:C:Ö:H
H H

and

H   H
H:C:Ö:C:H
H   H

or

    H   H
    | |
H—C—C—Ö—H
    | |
    H   H

and

    H   H
    | |
H—C—Ö—C—H
    | |
    H   H

b.

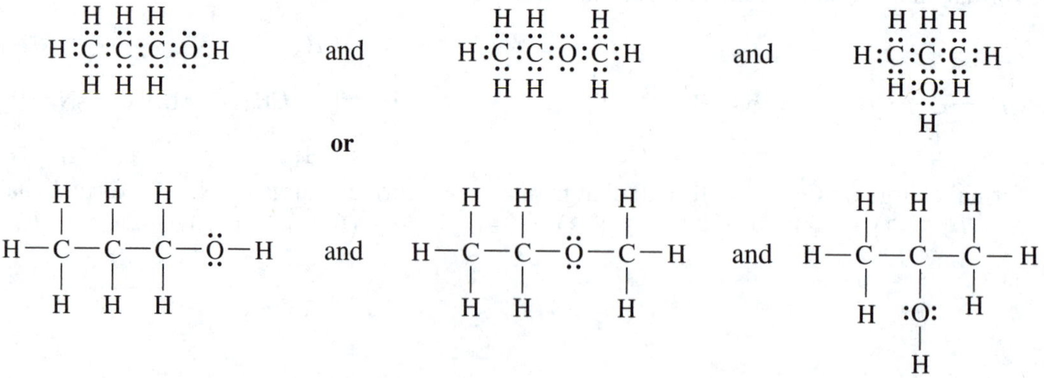

H H H
H:C:C:C:Ö:H
H H H

and

H H   H
H:C:C:Ö:C:H
H H   H

and

H H H
H:C:C:C:H
H:O: H
  H

or

    H   H   H
    | | |
H—C—C—C—Ö—H
    | | |
    H   H   H

and

    H   H   H
    | | |
H—C—C—Ö—C—H
    | | |
    H   H   H

and

    H   H   H
    | | |
H—C—C—C—H
    | | |
    H :O: H
        |
        H

15. Because the compounds are neutral, a halogen will have three lone pairs, an oxygen will have two, a nitrogen will have one, and carbon or a hydrogen will have no lone pairs.

a. $CH_3CH_2\ddot{N}H_2$ c. $CH_3CH_2\ddot{\underset{\cdot\cdot}{O}}H$ e. $CH_3CH_2\ddot{\underset{\cdot\cdot}{C}}l:$

b. $CH_3\ddot{N}HCH_3$ d. $CH_3\ddot{\underset{\cdot\cdot}{O}}CH_3$ f. $H\ddot{\underset{\cdot\cdot}{O}}\ddot{N}H_2$

16. a. $CH_3CH_2CH_2Cl$ b. $CH_3\overset{\overset{\displaystyle O}{\|}}{C}OCH_2CH_3$ c. $CH_3CH_2\overset{\overset{\displaystyle O}{\|}}{\underset{\underset{\displaystyle CH_3}{|}}{C}}NCH_2CH_3$ d. $CH_3CH_2C\equiv N$

17. a. the (green) chlorine atom   c. the (blue) nitrogen atoms
  b. the (red) oxygen atoms   d. the (black) carbon atoms and (gray) hydrogen atoms

18. a.

$$H-\overset{\overset{\displaystyle H}{|}}{\underset{\underset{\displaystyle H}{|}}{C}}-\overset{..}{N}-\overset{\overset{\displaystyle H}{|}}{\underset{\underset{\displaystyle H}{|}}{C}}-\overset{\overset{\displaystyle H}{|}}{\underset{\underset{\displaystyle H}{|}}{C}}-\overset{\overset{\displaystyle H}{|}}{\underset{\underset{\displaystyle H}{|}}{C}}-H$$

c.

$$H-\overset{\overset{\displaystyle H}{|}}{\underset{\underset{\displaystyle H}{|}}{C}} \quad \overset{\overset{\displaystyle H-\overset{\overset{\displaystyle H}{|}}{\underset{\underset{\displaystyle H}{|}}{C}}-H}{}}{\underset{\underset{\displaystyle H-\overset{\overset{\displaystyle H}{|}}{\underset{\underset{\displaystyle H}{|}}{C}}-H}{}}{C}}-\ddot{\underset{\cdot\cdot}{B}}r:$$

b.

$$H-\overset{\overset{\displaystyle H}{|}}{\underset{\underset{\displaystyle H}{|}}{C}}\quad\overset{\overset{\displaystyle H}{|}}{\underset{\underset{\displaystyle \overset{\overset{\displaystyle H}{|}}{\underset{\underset{\displaystyle H}{|}}{C}}}{}}{C}}-\ddot{\underset{\cdot\cdot}{C}}l:$$

d.

$$H-\overset{\overset{\displaystyle H}{|}}{\underset{\underset{\displaystyle H}{|}}{C}}\quad\overset{\overset{\displaystyle H-\overset{\overset{\displaystyle H}{|}}{\underset{\underset{\displaystyle H}{|}}{C}}-H}{}}{\underset{\underset{\displaystyle H-\overset{\overset{\displaystyle H}{|}}{\underset{\underset{\displaystyle H}{|}}{C}}-H}{}}{C}}\quad\overset{\overset{\displaystyle H}{|}}{\underset{\underset{\displaystyle H}{|}}{C}}-\overset{\overset{\displaystyle H}{|}}{\underset{\underset{\displaystyle H}{|}}{C}}-\overset{\overset{\displaystyle H}{|}}{\underset{\underset{\displaystyle H}{|}}{C}}\quad\overset{\overset{\displaystyle :O:}{\|}}{C}-H$$

19. The carbon-carbon bonds form as a result of sp^3—sp^3 overlap.
  The carbon-hydrogen bonds form as a result of sp^3—s overlap.

20. Solved in the text.

21. a. One s orbital and **three** p orbitals form **four** sp^3 orbitals.
  b. One s orbital and **two** p orbitals form **three** sp^2 orbitals.
  c. One s orbital and **one** p orbital form **two** sp orbitals.

22. a(1). Solved in the text.
  b(1). Solved in the text.

  a(2). The carbon forms four bonds, and each chlorine forms one bond.

$$:\ddot{\underset{\cdot\cdot}{C}}l-\overset{\overset{\displaystyle :\ddot{C}l:}{|}}{\underset{\underset{\displaystyle :\ddot{C}l:}{|}}{C}}-\ddot{\underset{\cdot\cdot}{C}}l:$$

b(2). The carbon uses sp^3 orbitals to form the bonds with the chlorine atoms, so the bond angles are all 109.5°.

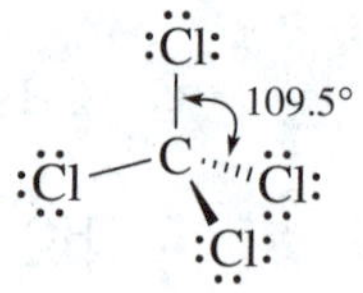

a(3). The first attempt at drawing a Lewis structure shown below shows that carbon does not have a complete octet and does not form the needed number of bonds.

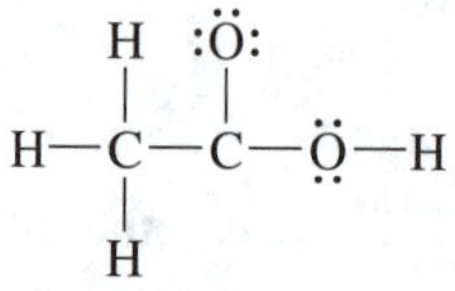

Using one of oxygen's lone pairs to put a double bond between the carbon and oxygen solves both problems.

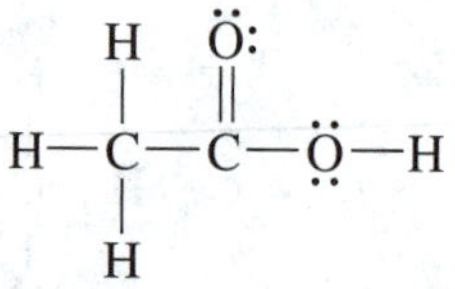

b(3). The sp^3 hybridized CH_3 carbon has 109.5° bond angles, and the sp^2 hybridized $CH=O$ carbon has 120° bond angles.

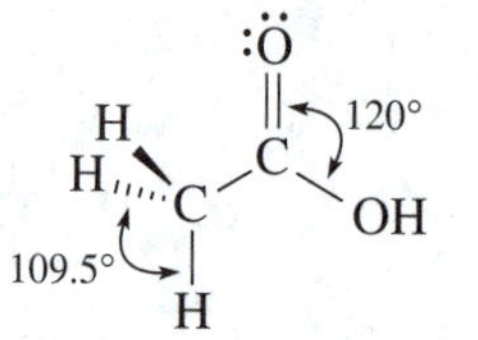

a(4). In order to fill their octets and form the required number of bonds, carbon and nitrogen must form a triple bond.

$$H—C\equiv N:$$

b(4). Because the carbon is sp hybridized, the bond angle is 180°.

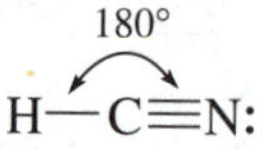

23. **a.** 120°   **b.** 120°
c. Because the carbon is sp^3 hybridized and it has a lone pair, you can predict that the bond angle is similar to that in NH_3 (107.3°).

24. The nitrogen atom has the greatest electron density.
The hydrogens are the bluest atoms. Therefore, they have the least electron density. In other words, they have the most positive (least negative) electrostatic potential.

25. Water is the most polar—it has a deep red area and the most intense blue area.
Methane is the least polar—it is all nearly the same color (green), with no red or blue areas.

26. Solved in the text.

27. Electrons in atomic orbitals farther from the nucleus form **longer** bonds; they also form **weaker** bonds due to less electron density in the region of orbital overlap. Therefore:

    a. **relative lengths** of the bonds in the halogens are: $Br_2 > Cl_2$
        relative strengths of the bonds are: $Cl_2 > Br_2$

    b. **relative lengths**: $CH_3 - Br > CH_3 - Cl > CH_3 - F$
        relative strengths: $CH_3 - F > CH_3 - Cl > CH_3 - Br$

28.   **a. longer:**   **1.** $C - I$ **2.** $C - Cl$ **3.** $H - Cl$
     b. stronger: **1.** $C - Cl$ **2.** $C - C$ **3.** $H - F$

29. Solved in the text.

30. We know that the σ bond is stronger than the π bond, because the σ bond in ethane has a bond dissociation energy of 90.2 kcal/mol, whereas the bond dissociation energy of the double bond ($\sigma + \pi$) in ethene is 174.5 kcal/mol, which is less than twice as strong.

Because the σ bond is stronger, we know that it has more effective orbital–orbital overlap.

31.

32.   **a.**

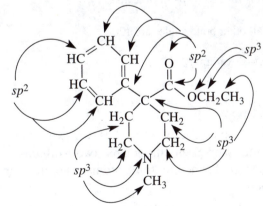

  b.

33. The bond angle depends on the central atom.

    a. sp^3 nitrogen with no lone pair: 109.5°   **c.** sp^3 carbon with no lone pair: 109.5°
    b. sp^3 nitrogen with a lone pair: 107.3°   **d.** sp^3 carbon with no lone pair: 109.5°

34.   **a.   CCl$_4$**

The carbon in CCl$_4$ is bonded to four atoms, so it uses four sp^3 orbitals.

Each carbon-chlorine bond is formed by the overlap of an sp^3 orbital of carbon with an sp^3 orbital o[f] chlorine. Because the four sp^3 orbitals of carbon orient themselves to get as far away from each othe[r] as possible, the bond angles are all **109.5°**.

bond angles $=$ 109.5°

b.   The single-bonded carbon and the single-bonded oxygen use sp^3 orbitals.
Each hydrogen uses an *s* orbital.

the H—C—H and H—C—O bond angles are ~109.5°
the C—O—H bond angle is ~104.5°

c.   HCOOH

The double-bonded carbon and the double-bonded oxygen in HCOOH each uses sp^2 orbitals; thus, the bonds around the double-bonded carbon are all 120°. The single-bonded oxygen forms bonds using sp^3 orbitals, whereas each hydrogen uses an *s* orbital. The predicted C—O—H bond angle is based on an sp^3 oxygen that has a reduced bond angle due to the two lone pairs. The bond angle is predicted to be similar to that in H$_2$O (104.5°).

the σ bond is formed by sp^2—sp^2 overlap
the π bond is formed by *p*—*p* overlap
104.5°; all other bond angles are 120°

d.   N$_2$

The triple bond consists of one σ bond and two π bonds. Each nitrogen has two *sp* orbitals: one is used to form the σ bond, and the other contains the lone pair. Each nitrogen has two *p* orbitals that are used to form the two π bonds. A bond angle is the angle formed by three atoms. Therefore, there are no bond angles in this two-atom-containing compound.

:N≡N: the σ bond is formed by *sp*—*sp* overlap
each π bond is formed by *p*—*p* overlap

5. The electrostatic potential map of ammonia is not symmetrical in the distribution of the charge—the nitrogen is more electron rich and, therefore, more red than the three hydrogens. Therefore, its shape, which indicates charge distribution, is not symmetrical.

The electrostatic potential map of the ammonium ion is symmetrical in the distribution of the charge, so its shape is symmetrical. Its symmetry results from the fact that nitrogen forms a bond with each of the four hydrogens and the four bonds point to the corners of a regular tetrahedron. The nitrogen in the ammonium ion has significantly lower electron density than the nitrogen in ammonia because the lone pair has formed a bond to hydrogen.

6. **a** and **d** have a dipole moment of zero, because they are symmetrical molecules.

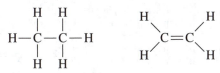

7. **a.** H:Ö:C:Ö:H or H—Ö—C—Ö—H

b. :Ö:C:Ö: or :Ö—C—Ö:

c. H:C:H or H—C—H

d. Ö::C::Ö or Ö=C=Ö

8. **a.** CH_3CH_3 **b.** CH_3F

9. If an atom is sp^3 hybridized, the bond angle will depend on the number of lone pairs it has: none = 109.5°; one = 107.3°; two = 104.5°.

   a. sp^3, 107.3°   **c.** sp^3, 109.5°   **e.** both C and N are sp, 180°   **g.** sp^3, 107.3°

   b. sp^3, 107.3°   **d.** sp^2, 120°   **f.** all Cs are sp^3, 109.5°

10. formal charge = the number of valence electrons – (the number of lone-pair electrons + half the number of bonding electrons)

   a. formal charge = 6 – (6 + 1) = 6 – 7 = – 1

      H:Ö:⁻

   b. formal charge = 6 – (5 + 1) = 6 – 6 = 0

      H:Ö·

   c. formal charge = 5 – (4 + 2) = 5 – 6 = – 1

      H—N⁻—H

   d. formal charge = 4 – (2 + 2) = 4 – 4 = 0

      H—C—H

11. **a.** $CH_3CH_2CH_3$ **b.** $CH_3CH = CH_2$ **c.** $CH_3C\equiv CCH_3$ or $CH_3CH_2C\equiv CH$

42. The hybridization of the central atom determines the bond angle. If the hybridization is sp^3, the number of lone pairs on the central atom determines the bond angle.

   a. 109.5°   **b.** 104.5°*   **c.** 107.3°   **d.** 107.3°

*104.5° is the correct prediction based on the bond angle in water.
However, the bond angle is actually somewhat larger (108.2°), because the bond opens up to minimize the interaction between hydrogen and the electron cloud of the relatively bulky CH_3 group.

43.

                            $1s$ $2s$ $2p_x$ $2p_y$ $2p_z$ $3s$ $3p_x$ $3p_y$ $3p_z$ $4s$

   a. Ca   ↑↓ ↑↓ ↑↓ ↑↓ ↑↓ ↑↓ ↑↓ ↑↓ ↑↓ ↑↓ $1s^2\ 2s^2\ 2p^6\ 3s^2\ 3p^6\ 4s^2$

   b. Ca^{2+} ↑↓ ↑↓ ↑↓ ↑↓ ↑↓ ↑↓ ↑↓ ↑↓ ↑↓ $1s^2\ 2s^2\ 2p^6\ 3s^2\ 3p^6$

   c. Ar   ↑↓ ↑↓ ↑↓ ↑↓ ↑↓ ↑↓ ↑↓ ↑↓ ↑↓ $1s^2\ 2s^2\ 2p^6\ 3s^2\ 3p^6$

   d. Mg^{2+} ↑↓ ↑↓ ↑↓ ↑↓ ↑↓ $1s^2\ 2s^2\ 2p^6$

44. **a.** H:C̈:N̈:H or H—C—N̈—H   **c.** H:N̈:N̈:H or H—N̈—N̈—H

   b. H—Ö—N̈=Ö

   d. H:N̈:Ö:⁻ or H—N̈—Ö:⁻

45. "**d**" is the only one written correctly

   a. $CH_3CH_2CH_3$ **c.** $(CH_3)_3CCH_3$ **e.** $CH_3CH_2CH_3$

   b. CH_4 **d.** $(CH_3)_2CHCH_2CH_3$ **f.** $CH_3CH_2CH_2CH_3$

46. The greater the electronegativity between the two bonded atoms, the more polar the bond.

   a. C—F   >   C—O   >   C—N
   b. C—Cl   >   C—Br   >   C—I
   c. H—O   >   H—N   >   H—C
   d. C—N   >   C—H   >   C—C

47. **a.** $CH_3CH{=}CH_2$ (the $CH{=}$ carbon marked sp^2)   **c.** CH_3CH_2OH (O marked sp^3)   **e.** $CH_3OCH_2CH_3$ (O marked sp^3)

   b. CH_3CCH_3 with O=, O marked sp^2 **d.** $CH_3C{\equiv}N$ (N marked sp)

48.

a.
```
    H O
    | ||
H—C—C—H
    |
    H
```

d.
```
        H
        |
    H   O   H
    |   |   |
H—C———C———C—H
    |   |   |
    H H—C—H H
        |
        H
```

b.
```
    H       H
    |       |
H—C—O—C—H
    |       |
    H       H
```

e.
```
    H H H
    | | |
H—C—C—C—C≡N
    | | |
    H O H
      |
      H
```

c.
```
    H O
    | ||
H—C—C—O—H
    |
    H
```

f.
```
            H               H
            |               |
    H   H H—C—H H H H—C—H H
    |   |   |     |   |     |
H—C———C———C———C———C———C—H
    |   |         |   |
    H H—C—H H   H H—C—H H
        |           |
        H           H
```

49.   Notice that none of the atoms in "**c**" has a formal charge.

a.
```
    H H
    | |
H—C—C:⁻
    | |
    H H
```

b.
```
    H H
    | |
H—C—C⁺
    | |
    H H
```

c.
```
    H H
    | |
H—C—C·
    | |
    H H
```

d.
```
    H :Ö:⁻ H
    |   |   |
H—C—C—C—H
    |   |   |
    H   H   H
```

50.   **a.**   120°   **b.**   180°

51.   **a.** $\overset{\rightarrow+}{H_3C—Br}$ **c.** $\overset{\leftarrow+}{HO—NH_2}$ **e.** $\overset{\rightarrow+}{H_3C—OH}$

b. $\overset{\leftarrow+}{H_3C—Li}$ **d.** $\overset{\rightarrow+}{I—Br}$ **f.** $\overset{\leftarrow+}{(CH_3)_2N—H}$

52.   The open arrow in the structures points to the shorter of the two indicated bonds in each compound.

1. $\overset{sp^3\ sp^2\ sp^2\quad \Downarrow}{CH_3CH=CHC\equiv CH}$
        $\underset{sp\ sp}{}$

3. $\overset{\quad\quad\quad\quad\quad\quad \Downarrow}{CH_3NHCH_2CH_2N=CHCH_3}$
        $\underset{sp^3\ sp^3\ sp^3\ \ sp^3\ sp^2\ \ sp^2\ sp^3}{}$

2. $\overset{\Longrightarrow}{CH_3}\overset{O\ sp^2}{\underset{\quad sp^3}{CCH_2OH}}$
        $\underset{sp^3\ {\Large\diagup}\ sp^3}{\quad sp^2}$

4. $Br—CH_2CH_2CH_2\overset{\Downarrow}{—}Cl$
        $\underset{sp^3\ sp^3\ sp^3}{}$

For **1**, **2**, and **3**:   A triple bond is shorter than a double bond, which is shorter than a single bond.

For **4**:   Cl forms a bond using a $3sp^3$ orbital, and Br forms a bond using a $4sp^3$ orbital. Therefore, the C—Cl bond is shorter.

53.

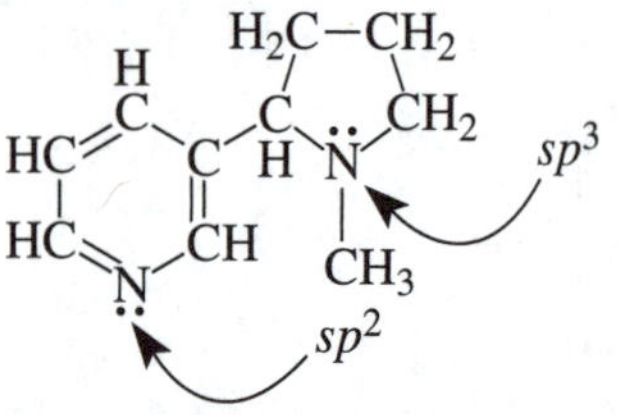

54.   formal charge = number of valence electrons

                                − (number of lone-pair electrons + half the number of bonding electrons)

In all four compounds, H has a single bond and is neutral, and each C has four bonds and is neutral. Thus, the indicated formal charge is for O or N.

a.

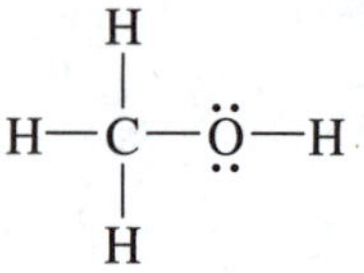

formal charge
$6 - (4 + 2) = 0$

b.

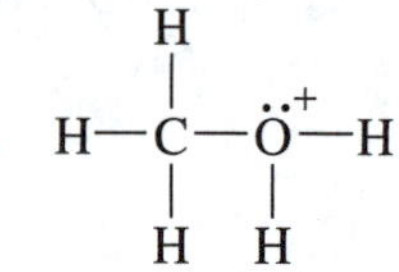

formal charge
$6 - (2 + 3) = +1$

c.

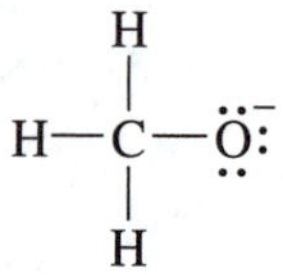

formal charge
$6 - (6 + 1) = -1$

d.

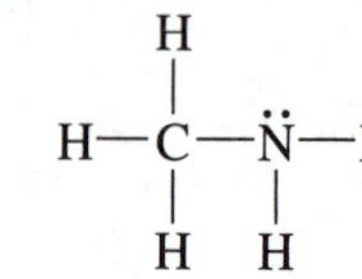

formal charge
$5 - (2 + 3) = 0$

55.

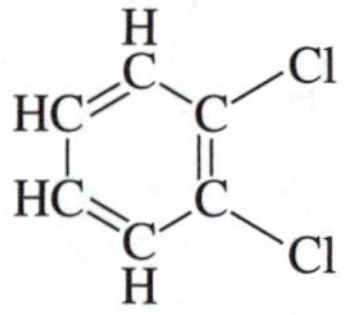

      1   2 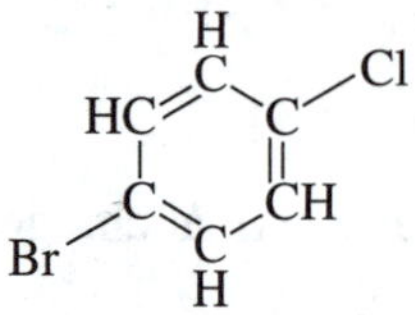   3 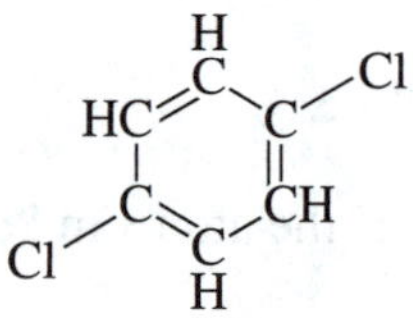   4

highest dipole   lowest dipole
moment   moment

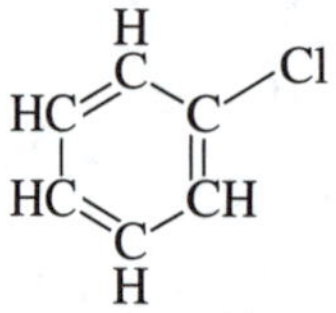

56.

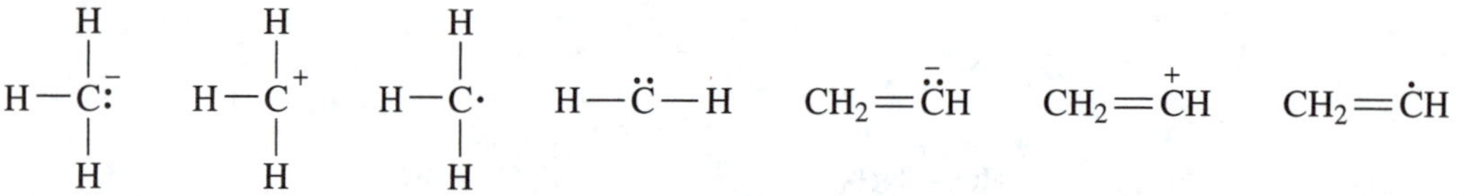

57.   **a.** If the central atom is sp^3 hybridized and it does not have a lone pair, the molecule will have a tetrahedral bond angle (109.5°). Therefore, only $^+NH_4$ has tetrahedral bond angles. The following species are close to being perfectly tetrahedral: H_2O, H_3O^+, NH_3, $^-CH_3$. However, they all have bond angles slightly smaller than 109.5°.

    b. $^+CH_3$ and BF_3

58.   In an alkene, six atoms are in the same plane: the two sp^2 carbons and the two atoms that are bonded to each of the two sp^2 carbons. The other atoms in the molecule will not be in the same plane with these six atoms.

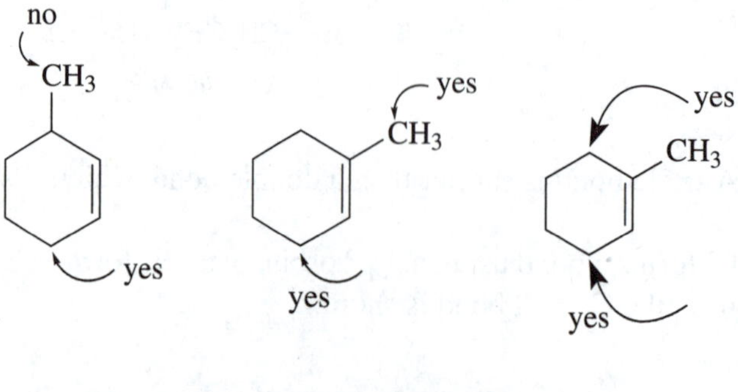

If you put stars next to the six atoms that lie in a plane in each molecule, you will be able to see more clearly whether the indicated atoms lie in the same plane.

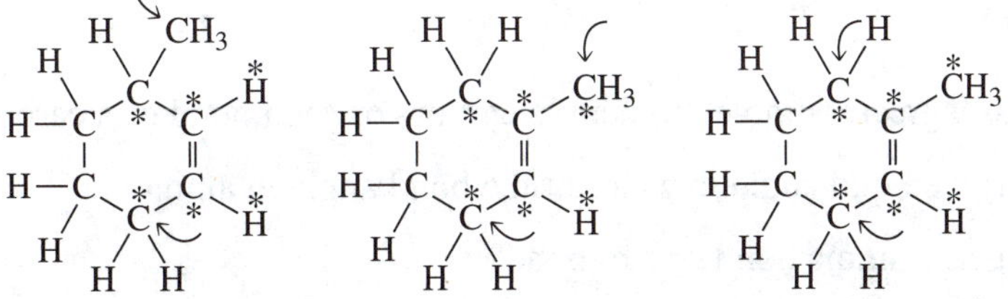

59. The bond between oxygen and sodium is ionic. All the other bonds are covalent. Thus, sodium methoxide has four covalent bonds.

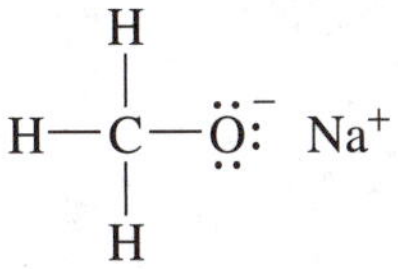

60. **a.** An H—H bond is formed by s—s overlap. A C—C bond is formed by sp^3—sp^3 overlap. Because an s orbital is closer to the nucleus than is an sp^3 orbital, an H—H bond is shorter than a C—C bond.

   b. A C—H bond is formed by sp^3—s overlap. Therefore, a C—H bond will be longer than an H—H bond (O.74Å) but shorter than a C—C bond (1.54 Å).

61. $CHCl_3$ has the larger dipole moment, because the three chlorines are withdrawing electrons in the same general direction, whereas in CH_2Cl_2 two chlorines are withdrawing electrons in the same general direction.

62. CH_3CH_2Cl has the longer C—Cl bond, because it is formed by the overlap of an sp^3 orbital of Cl with an sp^3 orbital of C, whereas the C—Cl bond in $CH_2{=}CHCl$ is formed by the overlap of an sp^3 orbital of Cl with an sp^2 orbital of C. (The more the s character, the shorter and stronger the bond.)

63. Because the triple-bonded carbons are sp hybridized, the bond angle indicated by the arrow needs to be close to 180° if the compound is to be a stable compound; it is impossible to have a 180° bond angle in a six-membered ring. (Try to make a molecular model.)

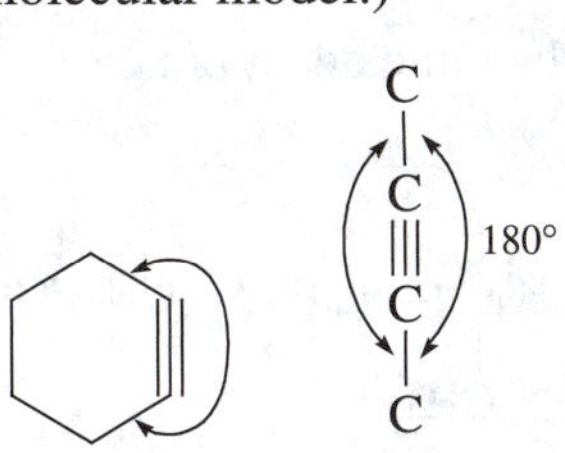

64. The structure on the right has a dipole moment of 2.95 D, because the two Cls are withdrawing electrons in the same general direction, so their electron-withdrawing effect is additive.

The two possible structures are shown here.

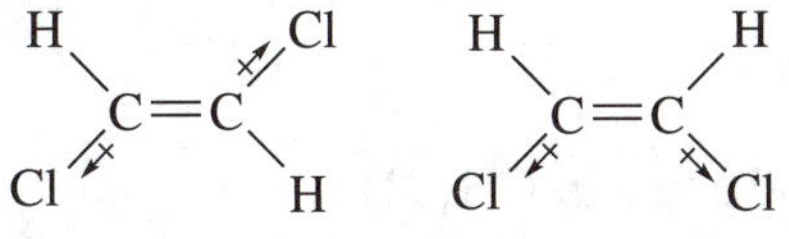

The structure on the left has a dipole moment of 0 D, because the two Cls are withdrawing electrons in opposite directions, so the electron-withdrawing effect of one cancels the electron-withdrawing effect of the other.

Chapter 1 Practice Test

1.   Answer the following:

   a.   Which bond has a greater dipole moment: a carbon–oxygen bond or a carbon–fluorine bond?

   b.   Which is larger: the bond angle in water or the bond angle in ammonia?

   c.   Which is stronger: a sigma bond or a pi bond?

2.   What is the hybridization of the carbon atom in each of the following compounds?

$$^{+}CH_3 \qquad ^{-}CH_3 \qquad \cdot CH_3$$

3.   Draw the Lewis structure for H_2CO_3.

4.   Circle the compounds below that have a dipole moment $= 0$.

$$CH_2Cl_2 \qquad CH_3CH_3 \qquad CH_3Cl \qquad H_2C{=}O \qquad CCl_4$$

5.   Which compound has greater bond angles: H_3O^+ or $^{+}NH_4$?

6.   Draw the structure for each of the following:

   a.   methyl cation

   b.   a hydride ion

   c.   a bromine radical

7.   Draw the structure of a compound that contains five carbons, two of which are sp^2 hybridized and three of which are sp^3 hybridized.

8.   What is the hybridization of each of the indicated atoms?

$$CH_3CH_2C{\equiv}N \qquad CH_3C{=}NCH_3 \qquad CH_3\overset{\overset{\displaystyle O}{\|}}{C}CH_3 \qquad O{=}C{=}O$$
$$\qquad\qquad\qquad\qquad \underset{CH_3}{|}$$

9.   a.   What orbitals do carbon's electrons occupy before promotion?

   b.   What orbitals do carbon's electrons occupy after promotion and before hybridization?

   c.   What orbitals do carbon's electrons occupy after hybridization?

10.   Answer the following:

   a.   What is the H—C—O bond angle in CH_3OH?

   b.   What is the C—O—H bond angle in CH_3OH?

11.   For each of the following compounds, indicate the hybridization of the atom to which the arrow is pointing.

$$\overset{\displaystyle O}{\overset{\displaystyle \|}{HCOH}} \quad HC\equiv N \quad CH_3OCH_3 \quad CH_3CH=CH_2$$

12.   Indicate whether each of the following statements is true or false.

a.   A triple bond is shorter than a double bond.   T   F

b.   The oxygen–hydrogen bonds in water are formed by the overlap
       of an sp^2 orbital of oxygen with an s orbital of hydrogen.   T   F

c.   A double bond is stronger than a single bond.   T   F

d.   A tetrahedral carbon has bond angles of 107.5°.   T   F

Answers to Chapter 1 Practice Test

1. **a.** a carbon–fluorine bond   **b.** the bond angle in ammonia   **c.** a sigma bond

2. $^{+}CH_3$ $^{-}CH_3$ $^{\cdot}CH_3$ 3. $H:\ddot{O}:\overset{\displaystyle :\ddot{O}:}{C}:\ddot{O}:H$
    sp^2 sp^3 sp^2

4. CH_2Cl_2 (CH_3CH_3) CH_3Cl $H_2C{=}O$ (CCl_4)   5. $^{+}NH_4$

6. **a.** $^{+}CH_3$ **b.** $H:^{-}$ **c.** $:\ddot{Br}^{\cdot}$

7. $CH_3CH_2CH_2CH{=}CH_2$ **or** $CH_3CH_2CH{=}CHCH_3$ **or** $CH_3CHCH{=}CH_2$
    $\qquad\qquad\qquad\qquad\qquad\qquad\qquad\qquad\qquad\qquad\qquad\quad |$
    $\qquad\qquad\qquad\qquad\qquad\qquad\qquad\qquad\qquad\qquad\quad CH_3$

8. $CH_3CH_2C{\equiv}N$ $\overset{sp\ \ sp}{}$ $CH_3C{=}NCH_3$ $\overset{sp^2\ sp^2}{}$ $CH_3\overset{O}{\underset{\ }{C}}CH_3$ sp^2, sp^2 $O{=}C{=}O$ $\overset{sp\ \ sp^2}{}$
    $\qquad\qquad\qquad\qquad\qquad\qquad\quad |$
    $\qquad\qquad\qquad\qquad\qquad\quad CH_3$

9. **a.** $1s^2\,2s^2\,2p_x2p_y$ **b.** $1s^2\,2s\,2p_x\,2p_y\,2p_z$ **c.** $1s^2\,2sp^34$

10. **a.** $109.5°$ **b.** $104.5°$

11. $H\overset{O}{\underset{\ }{C}}OH$ sp^2 $HC{\equiv}N$ sp CH_3OCH_3 sp^3 $CH_3CH{=}CH_2$ sp^3

12. **a.** A triple bond is shorter than a double bond.   T

    b. The oxygen–hydrogen bonds in water are formed by the
    overlap of an sp^2 orbital of oxygen with an s orbital of hydrogen.   F

    c. A double bond is stronger than a single bond.   T

    d. A tetrahedral carbon has bond angles of $107.5°$.   F

CHAPTER 2
Acids and Bases: Central to Understanding Organic Chemistry

Important Terms

acid (Brønsted acid)	a species that loses a proton.
acid–base reaction	a reaction in which an acid donates a proton to a base.
acid dissociation constant	a measure of the degree to which an acid dissociates.
acidity	a measure of how easily a compound gives up a proton.
base (Brønsted base)	a species that gains a proton.
basicity	a measure of the tendency of a compound to share its electrons with a proton.
Brønsted acid	a species that loses a proton.
Brønsted base	a species that gains a proton.
buffer solution	solution of a weak acid and its conjugate base.
conjugate acid	the species formed when a base accepts a proton.
conjugate base	the species formed when an acid loses a proton.
delocalized electrons	electrons that are shared by three or more atoms (that is, they do not belong to a single atom, nor are they shared in a bond between two atoms).
equilibrium constant	the ratio of products to reactants at equilibrium.
inductive electron withdrawal	the pull of electrons through sigma bonds by an atom or by a group of atoms.
pH	the pH scale used to describe the acidity of a solution ($pH = -\log [H^+]$).
pK_a	a measure of the tendency of a compound to lose a proton ($pK_a = -\log K_a$, where K_a is the acid dissociation constant).
proton	a positively charged hydrogen ion.
proton transfer reaction	a reaction in which a proton is transferred from an acid to a base.
resonance	having delocalized electrons.
resonance contributors	structures with localized electrons that together approximate the true structure of a compound with delocalized electrons.
resonance hybrid	the actual structure of a compound with delocalized electrons.

67

Solutions to Problems

1. CO_2 and CCl_4 are not acids, because neither has a proton that it can lose.

2.   **a.** $HCl + NH_3 \rightleftharpoons Cl^- + {}^+NH_4$

   b. $H_2O + {}^-NH_2 \rightleftharpoons HO^- + NH_3$

3.   The conjugate acid is obtained by adding an H^+ to the species.

   a. (1) ${}^+NH_4$ **(2)** HCl **(3)** H_2O **(4)** H_3O^+

   The conjugate base is obtained by removing an H^+ from the species.

   b. (1) ${}^-NH_2$ **(2)** Br^- **(3)** NO_3^- **(4)** HO^-

4.   **a.** The lower the pK_a, the stronger the acid, so the compound with $pK_a = 5.2$ is the stronger acid.
   b. The greater the dissociation constant, the stronger the acid, so the compound with a dissociation constant $= 3.4 \times 10^{-3}$ is the stronger acid.

5.   Because butyric acid has a greater pK_a value than vitamin C, butyric acid is a weaker acid than vitamin C.

6.   **a.** $HO^- + H^+ \rightleftharpoons H_2O$

   b. $HCO_3^- + H^+ \rightleftharpoons H_2CO_3 \longrightarrow H_2O + CO_2$

   c. $CO_3^{2-} + 2H^+ \rightleftharpoons H_2CO_3 \longrightarrow H_2O + CO_2$

7.   If the pH is <7, the body fluid is acidic; if the pH is >7, the body fluid is basic.
   a. basic   **b.** acidic   **c.** basic

8.   Remember that a proton will be picked up by an atom that has one or more lone pairs. Notice that two oxygens have lone pairs in part **e.**

   a. $CH_3CH_2\overset{+}{O}H_2$ **b.** CH_3CH_2OH **c.** $CH_3\overset{\displaystyle O}{\underset{}{\overset{\|}{C}}}OH$ **d.** $CH_3CH_2\overset{+}{N}H_3$ **e.** $CH_3CH_2\overset{\displaystyle {}^+OH}{\underset{}{\overset{\|}{C}}}$

9.   <u>If the lone pairs are not shown:</u>

   a. CH_3OH as an acid: $CH_3OH + NH_3 \rightleftharpoons CH_3O^- + \overset{+}{N}H_4$

   CH_3OH as a base: $CH_3OH + HCl \rightleftharpoons CH_3\overset{+}{\underset{\underset{H}{|}}{O}}H + Cl^-$

   b. NH_3 as an acid: $NH_3 + CH_3O^- \rightleftharpoons {}^-NH_2 + CH_3OH$

   NH_3 as a base: $NH_3 + HBr \rightleftharpoons \overset{+}{N}H_4 + Br^-$

If the lone pairs are shown:

a. CH$_3$OH as an acid: CH$_3$ÖH + N̈H$_3$ ⇌ CH$_3$Ö:$^-$ + $^+$N̈H$_4$

CH$_3$OH as a base: CH$_3$ÖH + HC̈l: ⇌ CH$_3$ÖH$^+$ + :C̈l:$^-$
                                                                                      |
                                                                                      H

b. NH$_3$ as an acid: N̈H$_3$ + CH$_3$Ö:$^-$ ⇌ :N̈H$_2$ + CH$_3$ÖH

NH$_3$ as a base: N̈H$_3$ + HB̈r: ⇌ $^+$N̈H$_4$ + :B̈r:$^-$

0.   **a.** ~40   **b.** ~15   **c.** ~5   **d.** ~10

1.   **a.** CH$_3$COO$^-$ is the stronger base.
Because CH$_3$COOH is the weaker acid, it has the stronger conjugate base.

b. $^-$NH$_2$ is the stronger base.
Because NH$_3$ is the weaker acid, it has the stronger conjugate base.

c. H$_2$O is the stronger base.
Because H$_3$O$^+$ is the weaker acid, it has the stronger conjugate base.

2.   The conjugate acids of the given bases have the following relative strengths:

CH$_3$ÖH$_2^+$ > CH$_3$–C(=O)–OH > CH$_3$N̈H$_3^+$ > CH$_3$OH > CH$_3$NH$_2$

The bases, therefore, have the following relative strengths, since the weakest acid has the strongest conjugate base.

CH$_3$N̈H$^-$ > CH$_3$O$^-$ > CH$_3$NH$_2$ > CH$_3$–C(=O)–O$^-$ > CH$_3$OH

3.   Methanol is the acid because it has a pK_a value of ~15, whereas methylamine is a much weaker acid with a pK_a value of about 40.

14. Notice that in each case, the equilibrium favors reaction of the stronger acid to form the weaker acid.

a. CH_3OH + HO^- ⇌ CH_3O^- + H_2O

$pK_a = 15.5$ $pK_a = 15.7$

CH_3OH + H_3O^+ ⇌ $CH_3\overset{+}{O}H_2$ + H_2O

$pK_a = -1.7$ $pK_a = -2.5$

$$CH_3-\overset{\overset{\displaystyle O}{\|}}{C}-OH + HO^- \rightleftharpoons CH_3-\overset{\overset{\displaystyle O}{\|}}{C}-O^- + H_2O$$

$pK_a = 4.8$ $pK_a = 15.7$

$$CH_3-\overset{\overset{\displaystyle O}{\|}}{C}-OH + H_3O^+ \rightleftharpoons CH_3-\overset{\overset{\displaystyle \overset{+}{O}H}{\|}}{C}-OH + H_2O$$

$pK_a = -1.7$ $pK_a = -6.1$

CH_3NH_2 + HO^- ⇌ $CH_3\bar{N}H$ + H_2O

$pK_a = 40$ $pK_a = 15.7$

CH_3NH_2 + H_3O^+ ⇌ $CH_3\overset{+}{N}H_3$ + H_2O

$pK_a = -1.7$ $pK_a = 10.7$

b. HCl + H_2O ⇌ H_3O^+ + Cl^-

$pK_a = -7$ $pK_a = -1.7$

NH_3 + H_2O ⇌ $^+NH_4$ + HO^-

$pK_a = 15.7$ $pK_a = 9.4$

15. Because a strong acid is more likely to lose a proton than is a weak acid, the equilibrium will favor loss of a proton from the strong acid and formation of the weak acid.

a. $HC\equiv CH$ + HO^- ⇌ $HC\equiv C^-$ + H_2O

$pK_a = 25$ $pK_a = 15.7$

b. $HC\equiv CH$ + $^-NH_2$ ⇌ $HC\equiv C^-$ + NH_3

$pK_a = 25$ $pK_a = 36$

c. $^-NH_2$ would be a better base, because when it removes a proton, the equilibrium favors the products. When HO^- removes a proton, the equilibrium favors the reactants.

16. Each of the following bases will remove a proton from acetic acid in a reaction that favors products, because each of these bases will form an acid that is a weaker acid than acetic acid.

$$HO^- \quad CH_3NH_2 \quad HC\equiv C^-$$

The other three choices will form an acid that is a stronger acid than acetic acid.

17. Recall that the weakest acid has the strongest conjugate base.

$$^-CH_3 > {}^-NH_2 > HO^- > F^-$$

18.   Again, recall that the weakest acid has the strongest conjugate base.

$$CH_3\bar{C}H_2 \; > \; H_2C=\bar{C}H \; > \; HC\equiv C^-$$

19.   The species on the right is the stronger acid, because its hydrogen is attached to an sp^2 oxygen, which is more electronegative than the sp^3 oxygen to which the hydrogen in the protonated alcohol is attached.

20.   **a. A** $HC\equiv CH \; + \; CH_3\bar{C}H_2 \; \rightleftharpoons \; HC\equiv C^- \; + \; CH_3CH_3$

         B $H_2C=CH_2 \; + \; HC\equiv C^- \; \rightleftharpoons \; H_2C=\bar{C}H \; + \; HC\equiv CH$

         C $CH_3CH_3 \; + \; H_2C=\bar{C}H \; \rightleftharpoons \; CH_3\bar{C}H_2 \; + \; H_2C=CH_2$

    b.   Only A, because only A has a reactant that is a stronger acid than the acid that is formed in the product.

21.   The smaller the ion, the stronger it is as a base.

$$F^- \; > \; Cl^- \; > \; Br^- \; > \; I^-$$

22.   **a.** oxygen   **b.** H_2S **c.** CH_3SH

The size of an atom is more important than its electronegativity in determining stability. Therefore, even though oxygen is more electronegative than sulfur, H_2S is a stronger acid than H_2O, and CH_3SH is a stronger acid than CH_3OH.

Because the sulfur atom is larger, the electrons in its conjugate base are spread out over a greater volume, which stabilizes it. The more stable the base, the stronger its conjugate acid.

23.   The stronger acid will have its proton attached to the more electronegative atom (if the atoms are about the same size) or to the larger atom (if the atoms are not the same size).

    a. HBr   **b.** $CH_3CH_2CH_2\overset{+}{O}H_2$ **c.** $CH_3CH_2CH_2OH$ **d.** $CH_3CH_2CH_3SH$

24.   Remember that the stronger the acid, the weaker (or more stable) its conjugate base,

    a. Because HI is the strongest acid, I^- is the most stable (weakest) base.

    b. Because HF is the weakest acid, F^- is the least stable base.

    c. Because HF is the weakest acid, F^- is the strongest base.

25.   Compare the acid strengths of the conjugate acids, recalling that a weaker acid has a stronger conjugate base.

    a. HO^-, because H_2O is a weaker acid than H_3O^+.

    b. NH_3, because $^+NH_4$ is a weaker acid than H_3O^+.

    c. CH_3O^-, because CH_3OH is a weaker acid than CH_3COOH.

    d. CH_3O^-, because CH_3OH is a weaker acid than CH_3SH.

26.   **a.** $CH_3OCH_2CH_2OH$ because its conjugate base has its negative charge stabilized by electron withdrawal by the CH_3O group.

    b. $CH_3CH_2CH_2\overset{+}{O}H_2$ because oxygen is more electronegative than nitrogen.

    c. $CH_3CH_2OCH_2CH_2OH$ because the electron-withdrawing oxygen is closer to the OH group.

    d. $CH_3CH_2\overset{\overset{\textstyle O}{\|}}{C}OH$ because the electron-withdrawing C=O is closer to the OH group.

27. The weaker acid has a stronger conjugate base.

$$\text{a.} \quad \underset{\underset{\text{Br}}{|}}{\text{CH}_3\text{CHCO}^-}\!\!\overset{\overset{\text{O}}{\|}}{} \qquad \text{b.} \quad \underset{\underset{\text{Cl}}{|}}{\text{CH}_3\text{CHCH}_2\text{CO}^-}\!\!\overset{\overset{\text{O}}{\|}}{} \qquad \text{c.} \quad \text{CH}_3\text{CH}_2\overset{\overset{\text{O}}{\|}}{\text{C}}\text{O}^- \qquad \text{d.} \quad \text{CH}_3\overset{\overset{\text{O}}{\|}}{\text{C}}\text{CH}_2\text{CH}_2\text{O}^-$$

28. Solved in the text.

29.

    a. Because the atom (P) to which each of the OH groups is attached is also attached to two electronegative oxygens.

    b. The middle OH group is the weakest of the remaining acidic groups. It is an alcohol ($pK_a \sim 15$), and the atom (C) to which it is attached is not attached to any strongly electronegative atoms. (The protonated amino group has a pK_a value of ~ 10.)

30. When a sulfonic acid loses a proton, the electrons left behind are shared by three oxygens. In contrast, when a carboxylic acid loses a proton, the electrons left behind are shared by only two oxygens. The sulfonate ion, therefore, is more stable than the carboxylate ion.

a sulfonate ion

a carboxylate ion

The more stable the base, the stronger is its conjugate acid. Thus, the sulfonic acid is a stronger acid than the carboxylic acid.

31. If the solution is more acidic than the compound's pK_a value, the compound will be in its acidic form (with its proton).

If the solution is more basic than the compound's pK_a value, the compound will be in its basic form (without its proton).

    a. CH_3COO^- **c.** H_2O **e.** $\overset{+}{\text{N}}\text{H}_4$ **g.** NO_2^-

    b. $\text{CH}_3\text{CH}_2\overset{+}{\text{N}}\text{H}_3$ **d.** Br^- **f.** $\text{HC}\equiv\text{N}$ **h.** NO_3^-

32. **a.** **1.** neutral
         2. neutral
         3. charged
         4. charged
         5. charged
         6. charged

b. **1.** charged
       2. charged
       3. charged
       4. charged
       5. neutral
       6. neutral

c. **1.** neutral
       2. neutral
       3. neutral
       4. neutral
       5. neutral
       6. neutral

33. **a.** The $^+NH_3$ group withdraws electrons, which increases the acidity of the COOH group.

b. Both the COOH group and the $^+NH_3$ group will be in their acidic forms, because the solution is more acidic than both of their pK_a values.

$$CH_3CH(^+NH_3)\!-\!C(\!=\!O)\!-\!OH$$

c. The COOH group will be in its basic form, because the solution is more basic than its pK_a value. The $^+NH_3$ group will be in its acidic form, because the solution is more acidic than its pK_a value.

$$CH_3CH(^+NH_3)\!-\!C(\!=\!O)\!-\!O^-$$

d. Both the COOH group and the $^+NH_3$ group will be in their basic forms, because the solution is more basic than both of their pK_a values.

$$CH_3CH(NH_2)\!-\!C(\!=\!O)\!-\!O^-$$

e. No, alanine can never be without a charge. To be without a charge would require a group with a pK_a value of 9.69 to lose a proton before a group with a pK_a value of 2.34. This clearly cannot happen. A weak acid cannot have a stronger tendency to lose a proton than a stronger acid has.

34. **a.** Solved in the text.

b. For the carboxylic acid to dissolve in water, it must be charged (in its basic form), so the pH will have to be greater than 6.8. For the amine to dissolve in ether, it will have to be neutral (in its basic form), so the pH will have to be greater than 12.7 to have essentially all of it in the neutral form. Therefore, the pH of the water layer must be greater than 12.7.

c. To dissolve in ether, the carboxylic acid will have to be neutral, so the pH will have to be less than 2.8 to have essentially all the carboxylic acid in the acidic (neutral) form. To dissolve in water, the amine will have to be charged, so the pH will have to be less than 8.7 to have essentially all the amine in the acidic form. Therefore, the pH of the water layer must be less than 2.8.

35. **a.** The basic form of the buffer removes a proton from the solution.

$$CH_3COO^- + H^+ \rightleftharpoons CH_3COOH$$

b. The acidic form of the buffer loses a proton in order to remove a hydroxide from the solution.

$$CH_3COOH + HO^- \rightleftharpoons CH_3COO^- + H_2O$$

36.   Solved in the text.

37.   All four are Brønsted acids (proton-donating acids). Therefore, they react with HO^- by giving a proton to it.

    a. $CH_3OH \ + \ HO^- \ \rightleftharpoons \ CH_3O^- \ + \ H_2O$

    b. $\overset{+}{N}H_4 \ + \ HO^- \ \rightleftharpoons \ NH_3 \ + \ H_2O$

    c. $CH_3\overset{+}{N}H_3 \ + \ HO^- \ \rightleftharpoons \ CH_3NH_2 \ + \ H_2O$

    d. $CH_3COOH \ + \ HO^- \ \rightleftharpoons \ CH_3COO^- \ + \ H_2O$

38.   **a.** $CCl_3CH_2OH \ > \ CHCl_2CH_2OH \ > \ CH_2ClCH_2OH$

    b. The greater the number of electron-withdrawing chlorine atoms equidistant from the OH group, the stronger the acid. (Notice that the larger the K_a, the stronger the acid.)

39.   The stronger base has the weaker conjugate acid.

    a. HO^- **b.** $CH_3\overset{-}{N}H$ **c.** CH_3O^- **d.** Cl^- **e.** CH_3COO^- **f.** $CH_3CHBrCOO^-$

40.   **a.**

    b.

41.   **a.** $CH_3CH_2\underset{\underset{\textstyle Cl}{|}}{C}HCOOH \ > \ CH_3\underset{\underset{\textstyle Cl}{|}}{C}HCH_2COOH \ > \ ClCH_2CH_2CH_2COOH \ > \ CH_3CH_2CH_2COOH$

    b. An electron-withdrawing substituent makes the carboxylic acid more acidic, because it stabilizes its conjugate base by decreasing the electron density around the oxygen atom. (Remember that the larger the K_a, the stronger the acid.)

    c. The closer the electron-withdrawing chloro substituent is to the acidic proton, the more it can decrease the electron density around the oxygen atom, so the more it stabilizes the conjugate base and increases the acidity of its conjugate acid.

42.   **a.** $HOCH_2CH_2\overset{+}{N}H_3$ **b.** $^-OCH_2CH_2NH_2$

43.   O is more electronegative than N, which is more electronegative than C. Therefore, the alcohol is more acidic than the amine, which is more acidic than the alkane.
S is larger than O, so CH_3CH_2SH is more acidic than CH_3CH_2OH.

    $CH_3CH_2SH \ > \ CH_3CH_2OH \ > \ CH_3CH_2NH_2 \ > \ CH_3CH_2CH_3$

44. If the solution is more acidic than the pK_a of the compound, the compound will be in its acidic form (with the proton).

If the solution is more basic than the pK_a of the compound, the compound will be in its basic form (without the proton).

a. at pH = 3 CH_3COOH **b.** at pH = 3 $CH_3CH_2\overset{+}{N}H_3$ **c.** at pH = 3 CF_3CH_2OH

at pH = 6 CH_3COO^- at pH = 6 $CH_3CH_2\overset{+}{N}H_3$ at pH = 6 CF_3CH_2OH

at pH = 10 CH_3COO^- at pH = 10 $CH_3CH_2\overset{+}{N}H_3$ at pH = 10 CF_3CH_2OH

at pH = 14 CH_3COO^- at pH = 14 $CH_3CH_2NH_2$ at pH = 14 $CF_3CH_2O^-$

45. In all four reactions, the products are favored at equilibrium. (Recall that the equilibrium favors formation of the weaker acid.)

a. $CH_3COOH + CH_3O^- \rightleftharpoons CH_3COO^- + CH_3OH$

b. $CH_3CH_2OH + {}^-NH_2 \rightleftharpoons CH_3CH_2O^- + NH_3$

c. $CH_3COOH + CH_3NH_2 \rightleftharpoons CH_3COO^- + CH_3\overset{+}{N}H_3$

d. $CH_3CH_2OH + HCl \rightleftharpoons CH_3CH_2\overset{+}{O}H_2 + Cl^-$

46. **a.** $HC{\equiv}CCH_2OH > CH_2{=}CHCH_2OH > CH_3CH_2CH_2OH$

b. These three compounds differ only in the group that is attached to CH_2OH. The more electronegative the group attached to CH_2OH, the stronger the acid, because inductive electron withdrawal stabilizes the conjugate base; the more stable the base, the stronger its conjugate acid. An sp carbon is more electronegative than an sp^2 carbon, which is more electronegative than an sp^3 carbon.

47. In each compound, the nitrogen atom is the most apt to be protonated, because it is the stronger base.

a. $CH_3{-}\underset{\underset{\displaystyle OH}{|}}{CH}{-}CH_2\overset{+}{N}H_3$ **b.** $CH_3{-}\underset{\underset{\displaystyle {}^+NH_3}{|}}{\overset{\overset{\displaystyle CH_3}{|}}{C}}{-}OH$ **c.** $CH_3{-}\underset{\underset{\displaystyle {}^+NH_3}{|}}{\overset{\overset{\displaystyle CH_3}{|}}{C}}{-}CH_2OH$

48. The hydrogen bonded to the oxygen is the most acidic hydrogen in Tenormin.

49. **A and C** because, in each case, the acid is stronger than the acid (H_2O) that will be formed as a product.

50. If the reaction is producing protons, the basic form of the buffer will pick up the protons. At the pH at which the reaction is carried out (pH = 10.5), a protonated methylamine/methylamine buffer with a $pK_a = 10.7$ will have a larger percentage of the buffer in the needed basic form than will a protonated ethylamine/ethylamine buffer with a pK_a of 11.0.

51. **a.** $CH_2{=}CHCOOH$ because an sp^2 carbon is more electronegative than an sp^3 carbon

b.   because an oxygen can withdraw electrons inductively

c.   HC≡CCOOH   because an *sp* carbon is more electronegative than an *sp²* carbon

d.   because an *sp²* nitrogen is more electronegative than an *sp³* nitrogen

52.   The first pK_a is lower than the pK_a of acetic acid because the middle COOH group of citric acid has oxygen-containing groups that acetic acid does not have. These groups withdraw electrons inductively and thereby stabilize the conjugate base.

53.   Because the pH of the blood (~ 7.4) is greater than the pK_a value of the buffer (6.1), more buffer species in blood are in the basic form than in the acidic form. Therefore, the buffer is better at neutralizing excess acid.

54.   Charged compounds will dissolve in water, and uncharged compounds will dissolve in ether. The acidic forms of carboxylic acids and alcohols are neutral, and the basic forms are charged. The acidic forms of amines are charged, and the basic forms are neutral.

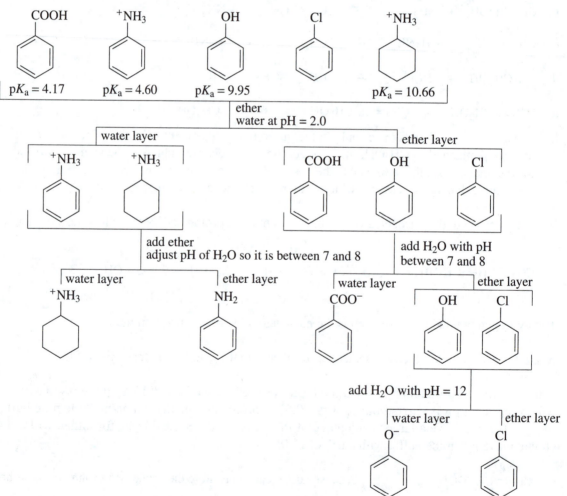

Chapter 2 Practice Test

1. Which compound is the stronger acid?

a. $CH_3CH_2CH_2OH$ or $CH_3CH_2CH_2NH_2$

b. HBr   or   HI

c.

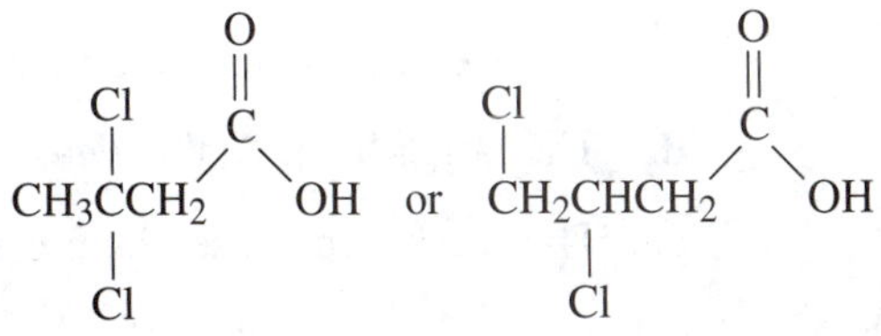

$$CH_3\overset{Cl}{\underset{Cl}{C}}CH_2\overset{O}{\overset{\|}{C}}OH \quad or \quad \overset{Cl}{CH_2}CHCH_2\overset{O}{\overset{\|}{C}}OH$$

d. CH_4 or NH_3

2. Which compound is the stronger base?

a. $CH_3CH_2NH_2$ or CH_3CH_2OH **b.** F^- or I^-

3. Draw a circle around the strongest base and draw a square around the weakest base.

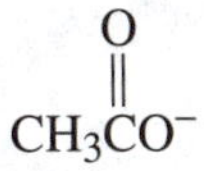

$$CH_3\overset{O}{\overset{\|}{C}}O^- \quad CH_3O^- \quad CH_3OH \quad CH_3\overset{-}{N}H \quad CH_3NH_2$$

4. The following compounds are drawn in their acidic forms, and their pK_a values are given. Draw the form in which each compound would predominantly exist at pH = 8.

CH_3COOH CH_3CH_2OH $CH_3\overset{H}{\overset{+}{O}}H$ $CH_3CH_2\overset{+}{N}H_3$

$pK_a = 4.8$ $pK_a = 15.9$ $pK_a = -2.5$ $pK_a = 11.2$

5. **a.** Write the acid–base reaction that occurs when methylamine is added to water.

b. Does the above reaction favor reactants or products?

6. **a.** What is the conjugate base of NH_3?

b. What is its conjugate acid?

7. **a.** What products would be formed from the following reaction?

$$CH_3OH \quad + \quad \overset{+}{N}H_4 \quad \rightleftharpoons$$

b. Does the reaction favor reactants or products?

8. A compound has a $K_a = 6.3 \times 10^{-9}$. What is its approximate pK_a (that is, between which two integers)?

9. Label the compounds in order of decreasing acidity. (Label the most acidic compound #1.)

CH_3CH_2OH $CH_3CH_2NH_2$ CH_3CH_2SH $CH_3CH_2CH_3$

10. You are planning to carry out a reaction at pH = 4 that releases protons. Would it be better to use a 1.0 M formic acid/sodium formate buffer or a 1.0 M acetic acid/sodium acetate buffer?
(The pK_a of formic acid is 3.75; the pK_a of acetic acid is 4.76.)

11.   Indicate whether each of the following statements is true or false.

   a. HO^- is a stronger base than $^-NH_2$.   T   F

   b. CH_3CH_3 is more acidic than $H_2C=CH_2$.   T   F

   c.   The weaker the acid, the more stable the conjugate base.   T   F

   d.   The larger the pK_a, the weaker the acid.   T   F

   e.   The weaker the base, the more stable it is.   T   F

Answers to Chapter 2 Practice Test

1.
   a. $CH_3CH_2CH_2OH$
   b. HI
   c.

$$CH_3CCH_2 \overset{\overset{\displaystyle Cl}{|}}{\underset{\underset{\displaystyle Cl}{|}}{}} \overset{\overset{\displaystyle O}{\|}}{C} OH$$

   d. NH_3

2.
   a. $CH_3CH_2NH_2$
   b. F^-

3.
   $CH_3\overset{\overset{\displaystyle O}{\|}}{C}O^-$ CH_3CO^- $\boxed{CH_3OH}$ $\bigcirc\kern-3.5em CH_3\bar{N}H$ CH_3NH_2

4.
   CH_3COO^- CH_3CH_2OH CH_3OH $CH_3CH_2\overset{+}{N}H_3$

5.
   a. $CH_3NH_2 + H_2O \rightleftharpoons CH_3\overset{+}{N}H_3 + HO^-$ **b.** reactants

6.
   a. $^-NH_2$
   b. $\overset{+}{N}H_4$

7.
   a. $CH_3OH + {}^+NH_4 \rightleftharpoons CH_3\overset{+}{\underset{\underset{\displaystyle H}{|}}{O}}H + NH_3$ **b.** reactants

8.
   between 8 and 9

9.
   CH_3CH_2OH $CH_3CH_2NH_2$ CH_3CH_2SH $CH_3CH_2CH_3$
     2   **3**   **1**   **4**

10.
   formic acid/sodium formate

11.
   a. HO^- is a stronger base than $^-NH_2$.   F

   b. CH_3CH_3 is more acidic than $H_2C{=}CH_2$.   F

   c. The weaker the acid, the more stable the conjugate base.   F

   d. The larger the pK_a, the weaker the acid.   T

   e. The weaker the base, the more stable it is.   T

TUTORIAL

pH, pK_a, and Buffers

First, we will see how the pH of solutions of acids and bases can be calculated. We will look at three different kind of solutions.

1. A solution made by dissolving a strong acid or a strong base in water.

2. A solution made by dissolving a weak acid or a weak base in water.

3. A solution made by dissolving a weak acid and its conjugate base in water. Such a solution is known as a **buffer solution.**

Before we start, we need to review a few terms.

An acid is a compound that loses a proton, and a base is a compound that gains a proton.

The degree to which an acid (HA) dissociates is described by its acid dissociation constant (K_a).

$$HA \rightleftharpoons H^+ + A^-$$

$$K_a = \frac{[H^+][A^-]}{[HA]}$$

The strength of an acid can be indicated by its acid dissociation constant (K_a) or by its pK_a value.

$$pK_a = -\log K_a$$

The stronger the acid, the **larger** its acid dissociation constant and the **smaller** its pK_a value.

For example, an acid with an acid dissociation constant of 1×10^{-2} (p$K_a = 2$) is a stronger acid than one with an acid dissociation constant of 1×10^{-4} (p$K_a = 4$).

While pK_a is used to describe the strength of an acid, pH is used to describe the acidity of a solution. In other words, pH describes the concentration of hydrogen ions in a solution.

$$pH = -\log [H^+]$$

The smaller the pH, the more acidic the solution:
acidic solutions have pH values <7; a neutral solution has a pH $= 7$; and basic solutions have pH values >7.

A solution with a pH $= 2$ is more acidic than a solution with a pH $= 4$.
A solution with a pH $= 12$ is more basic than a solution with a pH $= 8$.

Determining the pH of a Solution

To determine the pH of a solution, the concentration of hydrogen ion $[H^+]$ in the solution must be determined.

Strong Acids

A strong acid is one that dissociates completely in solution. Strong acids have pK_a values <1.

Because a strong acid dissociates completely, the concentration of hydrogen ions is the same as the concentration of the acid: a 1.0 M HCl solution contains 1.0 M $[H^+]$; a 1.5 M HCl solution contains 1.5 M $[H^+]$. Therefore, to determine the pH of a strong acid, the $[H^+]$ value does not have to be calculated; it is the same as the molarity of the strong acid.

Solution	$[H^+]$	pH
1.0 M HCl	1.0 M	0
1.0×10^{-2} M HCl	1.0×10^{-2} M	2.0
6.4×10^{-4} M HCl	6.4×10^{-4} M	3.2

Strong Bases

Strong bases are compounds such as NaOH or KOH that dissociate completely in water.

Because they dissociate completely, the $[HO^-]$ is the same as the molarity of the strong base.

pOH describes the basicity of a solution. The smaller the pOH, the more basic the solution, just like the smaller the pH, the more acidic the solution.

$$pOH = -\log [HO^-]$$

$[HO^-]$ and $[H^+]$ are related by the ionization constant for water (K_w).

$$K_w = [H^+][HO^-] = 10^{-14}$$
$$pH + pOH = 14$$

Solution	$[HO^-]$	pOH	pH
1.0 M NaOH	1.0 M	0	$14.0 - 0 = 14.0$
1.0×10^{-4} M NaOH	1.0×10^{-4} M	4.0	$14.0 - 4.0 = 10.0$
7.8×10^{-2} M NaOH	7.8×10^{-2} M	1.1	$14.0 - 1.1 = 12.9$

Weak Acids

A weak acid does not dissociate completely in solution. Therefore, $[H^+]$ must be calculated before the pH can be determined.

Acetic acid (CH_3COOH) is an example of a weak acid. It has an acid dissociation constant of 1.74×10^{-5} (p$K_a = 4.76$). The pH of a 1.00 M solution of acetic acid can be calculated as follows:

$$CH_3COOH \rightleftharpoons H^+ + CH_3COO^-$$

$$K_a = \frac{[H^+][CH_3COO^-]}{[CH_3COOH]}$$

Each molecule of acetic acid that dissociates forms one proton and one acetate ion. Therefore, the concentration of protons in solution equals the concentration of acetate ions. Each has a concentration that can be represented by x. The concentration of acetic acid, therefore, is the concentration we started with minus x.

$$1.74 \times 10^{-5} = \frac{(x)(x)}{1.00 - x}$$

The denominator $(1.00 - x)$ can be simplified to 1.00 because 1.00 is much greater than x. (When we actually calculate the value of x, we see that it is 0.004, and $1.00 - 0.004 = 1.00$.)

$$1.74 \times 10^{-5} = \frac{x^2}{1.00}$$

$$x = 4.17 \times 10^{-3}$$

$$pH = -\log 4.17 \times 10^{-3}$$

$$pH = 2.38$$

Formic acid (HCOOH) has a pK_a value of 3.75. The pH of a 1.50 M solution of formic acid can be calculated as follows:

$$HCOOH \rightleftharpoons H^+ + HCOO^-$$

$$K_a = \frac{[H^+][HCOO^-]}{[HCOOH]}$$

A compound with a p$K_a = 3.75$ has an acid dissociation constant of 1.78×10^{-4}.

$$1.78 \times 10^{-4} = \frac{(x)(x)}{1.50 - x} = \frac{x^2}{1.50}$$

$$x^2 = 1.50(1.78 \times 10^{-4})$$

$$x^2 = 2.67 \times 10^{-4}$$

$$x = 1.63 \times 10^{-2}$$

$$pH = -\log(1.63 \times 10^{-2})$$

$$pH = 1.79$$

Weak Bases

When a weak base is dissolved in water, it accepts a proton from water, creating hydroxide ion.

Determining the concentration of hydroxide allows the pOH to be determined, and this in turn allows the pH to be determined.

The pH of a 1.20 M solution of sodium acetate can be calculated as follows:

$$CH_3COO^- + H_2O \rightleftharpoons CH_3COOH + HO^-$$

$$\frac{K_w}{K_a} = \frac{[HO^-][CH_3COOH]}{[CH_3COO^-]}$$

$$\frac{1.00 \times 10^{-14}}{1.74 \times 10^{-5}} = \frac{(x)(x)}{1.20 - x}$$

$$5.75 \times 10^{-10} = \frac{x^2}{1.20}$$

$$x^2 = 6.86 \times 10^{-10}$$

$$x = 2.62 \times 10^{-5} = [HO^-]$$

$$pOH = -\log 2.62 \times 10^{-5}$$

$$pOH = 4.58$$

$$pH = 14.00 - 4.58$$

$$pH = 9.42$$

Notice that by setting up the equation equal to K_w/K_a, we can avoid the introduction of a new term (K_b), because $K_w/K_a = K_b$.

Buffer Solutions

A buffer solution is a solution that maintains nearly constant pH in spite of the addition of small amounts of H^+ or HO^-. That is because a buffer solution contains both a weak acid and its conjugate base. The weak acid can donate a proton to any HO^- added to the solution, and the conjugate base can accept any H^+ that is added to the solution, so the addition of small amounts of HO^- or H^+ does not significantly change the pH of the solution.

A buffer can maintain nearly constant pH in a range of one pH unit on either side of the pK_a of the conjugate acid. For example, an acetic acid/sodium acetate mixture can be used as a buffer in the pH range 3.76–5.76 because acetic acid has a p$K_a = 4.76$; methylammonium ion/methylamine can be used as a buffer in the pH range 9.7–11.7 because the methylammonium ion has a p$K_a = 10.7$.

The pH of a buffer solution can be determined from the following equation. This equation comes directly from the expression defining the acid dissociation constant.

$$pK_a = pH + \log\frac{[HA]}{[A^-]}$$

The pH of an acetic acid/sodium acetate buffer solution (pK_a of acetic acid $= 4.76$) that is 1.00 M in acetic acid and 0.50 M in sodium acetate is calculated as follows:

$$p K_a = pH + \log \frac{[HA]}{[A^-]}$$

$$4.76 = pH + \log \frac{1.00}{0.50}$$

$$4.76 = pH + \log 2$$

$$4.76 = pH + 0.30$$

$$pH = 4.46$$

Remember that compounds exist primarily in their acidic forms in solutions that are more acidic than their pK_a values and primarily in their basic forms in solutions that are more basic than their pK_a values. Therefore, it could have been predicted that the above solution will have a pH less than the pK_a of acetic acid, because there is more conjugate acid than conjugate base in the solution.

There are three ways a buffer solution can be prepared:

1. Weak Acid and Weak Base

A buffer solution can be prepared by mixing a solution of a weak acid with a solution of its conjugate base.

The pH of a formic acid/sodium formate buffer (pK_a of formic acid $= 3.75$) solution prepared by mixing 25 mL of 0.10 M formic acid and 15 mL of 0.20 M sodium formate is calculated as follows:

The equation below shows that the number of millimoles (mmol) of each of the buffer components can be determined by multiplying the number of milliliters (mL) by the molarity (M).

$$M = \text{molarity} = \frac{\text{moles}}{\text{liters}} = \frac{\text{millimoles}}{\text{milliliters}}$$

Therefore,

$$25 \text{ mL} \times 0.10 \text{ M} = 2.5 \text{ mmol formic acid}$$

$$15 \text{ mL} \times 0.20 \text{ M} = 3.0 \text{ mmol sodium formate}$$

Notice that in the following equation, we use mmol for $[HA]$ and $[A^-]$ rather than molarity (mmol/mL), because both the acid and the conjugate base are in the same solution, so they have the same volume. Therefore, volumes will cancel in the equation.

$$p K_a = pH + \log \frac{[HA]}{[A^-]}$$

$$3.75 = pH + \log \frac{2.5}{3.0}$$

$$3.75 = pH + \log 0.83$$

$$3.75 = pH - 0.08$$

$$pH = 3.83$$

It could have been predicted that the above solution would have a pH greater than the pK_a of formic acid, because there is more conjugate base than conjugate acid in the solution.

Weak Acid and Strong Base

A buffer solution can be prepared by mixing a solution of a weak acid with a strong base such as NaOH. The NaOH reacts completely with the weak acid, thereby creating the conjugate base needed for the buffer solution. For example, if 20 mmol of a weak acid and 5 mmol of a strong base are added to a solution, the 5 mmol of strong base will react with 5 mmol of weak acid, creating 5 mmol of weak base and leaving behind 15 mmol of weak acid.

The pH of a solution prepared by mixing 10 mL of a 2.0 M solution of a weak acid with a pK_a of 5.86 with 5.0 mL of a 1.0 M solution of sodium hydroxide can be calculated as follows:

When the 20 mmol of HA and the 5.0 mmol of HO⁻ are mixed, the 5.0 mmol of strong base will react with 5.0 mmol of HA, with the result that 5.0 mmol of A⁻ will be formed and 15 mmol (20 mmol − 5.0 mmol) of HA will be left unreacted.

$$10 \text{ mL} \times 2.0 \text{ M} = 20 \text{ mmol HA} \longrightarrow 15 \text{ mmol HA}$$

$$5.0 \text{ mL} \times 1.0 \text{ M} = 5.0 \text{ mmol HO}^- \longrightarrow 5.0 \text{ mmol A}^-$$

$$pK_a = pH + \log \frac{[HA]}{[A^-]}$$

$$5.86 = pH + \log \frac{15}{5}$$

$$5.86 = pH + \log 3$$

$$5.86 = pH + 0.48$$

$$pH = 5.38$$

Weak Base and Strong Acid

A buffer solution can be prepared by mixing a solution of a weak base with a strong acid such as HCl. The strong acid will react completely with the weak base, thereby forming the conjugate acid needed for the buffer solution.

The pH of an ethylammonium ion/ethylamine buffer (pK_a of $CH_3CH_2\overset{+}{N}H_3$ = 11.0) prepared by mixing 30 mL of 0.20 M ethylamine with 40 mL of 0.10 M HCl can be calculated as follows:

$$30 \text{ mL} \times 0.20 \text{ M} = 6.0 \text{ mmol RNH}_2 \longrightarrow 2.0 \text{ mmol RNH}_2$$

$$40 \text{ mL} \times 0.10 \text{ M} = 4.0 \text{ mmol H}^+ \longrightarrow 4.0 \text{ mmol R}\overset{+}{N}H_3$$

Notice that 4.0 mmol H⁺ reacts with 4.0 mmol RNH₂, forming 4.0 mmol R$\overset{+}{N}$H₃ and leaving behind 2.0 mmol RNH₂.

$$pK_a = pH + \log \frac{[HA]}{[A^-]}$$

$$11.0 = pH + \log \frac{4.0}{2.0}$$

$$11.0 = pH + \log 2.0$$

$$11.0 = pH + 0.30$$

$$pH = 10.7$$

Fraction Present in the Acidic or the Basic Form

A common question asked is what fraction of a buffer will be in a particular form at a given pH; either what fraction will be in the acidic form or what fraction will be in the basic form. This is an easy question to answer if you remember the following formulas that are derived at the end of this section:

$$\text{fraction present in the acidic form } = \frac{[H^+]}{K_a + [H^+]}$$

$$\text{fraction present in the basic form } = \frac{K_a}{K_a + [H^+]}$$

What fraction of an acetic acid/sodium acetate buffer (pK_a of acetic acid = 4.76; K_a = 1.74 × 10^{-5}) is present in the acidic form at pH = 5.20; $[H^+]$ = 6.31 × 10^{-6}?

$$\frac{[H^+]}{K_a + [H^+]} = \frac{6.31 \times 10^{-6}}{(1.74 \times 10^{-5}) + (6.31 \times 10^{-6})}$$

$$= \frac{6.31 \times 10^{-6}}{(17.4 \times 10^{-6}) + (6.31 \times 10^{-6})}$$

$$= \frac{6.31 \times 10^{-6}}{23.7 \times 10^{-6}} = \frac{6.31}{23.7}$$

$$= 0.26$$

What fraction of a formic acid/sodium formate buffer (pK_a of formic acid = 3.75; K_a = 1.78 × 10^{-4}) is present in the basic form at pH = 3.90; $[H^+]$ = 1.26 × 10^{-4}?

$$\frac{K_a}{K_a + [H^+]} = \frac{1.78 \times 10^{-4}}{(1.78 \times 10^{-4}) + (1.26 \times 10^{-4})}$$

$$= \frac{1.78 \times 10^{-4}}{3.04 \times 10^{-4}} = \frac{1.78}{3.04}$$

$$= 0.586$$

$$= 0.59$$

The formulas describing the fraction present in the acidic or basic form are obtained from the definition of the acid dissociation constant.

$$K_a = \frac{[H^+][A^-]}{[HA]}$$

To derive the equation for the fraction present in the acidic form, we need to define $[A^-]$ in terms of $[HA]$, so that we will have only one unknown in the equation.

$$[A^-] = \frac{K_a[HA]}{[H^+]}$$

$$\text{fraction present in the acidic form} = \frac{[HA]}{[HA] + [A^-]} = \frac{[HA]}{[HA] + \dfrac{K_a[HA]}{[H^+]}} = \frac{1}{1 + \dfrac{K_a}{[H^+]}}$$

$$= \frac{[H^+]}{K_a + [H^+]}$$

To derive the equation for the fraction present in the basic form, we need to define $[HA]$ in terms of $[A^-]$, so that we can get rid of the $[HA]$ term.

$$K_a = \frac{[H^+][A^-]}{[HA]}$$

$$[HA] = \frac{[H^+][A^-]}{K_a}$$

$$\text{fraction present in the basic form} = \frac{[A^-]}{[HA] + [A^-]} = \frac{[A^-]}{[A^-] + \dfrac{[H^+][A^-]}{K_a}}$$

$$= \frac{1}{1 + \dfrac{[H^+]}{K_a}}$$

$$= \frac{K_a}{K_a + [H^+]}$$

Preparing Buffer Solutions

The type of calculations just shown can be used to determine how to make a buffer solution.

For example, how can 100 mL of a 1.00 M buffer solution with a pH = 4.24 be prepared if you have available to you 1.50 M solutions of acetic acid (pK_a = 4.76; K_a = 1.74 $\times$ 10^{-5}), sodium acetate, HCl, and NaOH?

First, we need to determine what fraction of the buffer will be present in each form at pH = 4.24; $[H^+]$ = 5.75 $\times$ 10^{-5}. We will start by calculating the fraction of the buffer present in the acidic form.

$$\frac{[H^+]}{K_a + [H^+]} = \frac{5.75 \times 10^{-5}}{(1.74 \times 10^{-5}) + (5.75 \times 10^{-5})}$$

$$= \frac{5.75 \times 10^{-5}}{7.49 \times 10^{-5}}$$

$$= 0.77$$

If a 1.00 M buffer solution is desired, the buffer must be 0.77 M in acetic acid and 0.23 M in sodium acetate.

There are three ways to make such a buffer solution:

1. **By mixing the appropriate amounts of acetic acid and sodium acetate in water, and adding water to obtain a final volume of 100 mL.**

    The amount of acetic acid needed is $[CH_3COOH] = 0.77\ M$

    $$M = \frac{mmol}{mL} = \frac{x\ mmol}{100\ mL} = 0.77\ M$$

    $$x = 77\ mmol$$

    Therefore, we need to have 77 mmol of acetic acid in the final solution.

    To obtain 77 mmol of acetic acid from a 1.50 M solution of acetic acid,

    $$\frac{77\ mmol}{y\ mL} = 1.50\ M$$

    $$y = 51.3\ mL$$

    Notice that the formula $M = mmol/mL$ was used twice. The first time, it was used to determine the number of mmol of acetic acid that is needed in the final solution. The second time, it was used to determine how that number of mmol can be obtained from an acetic acid solution of a known concentration.

    The amount of sodium acetate needed is $[CH_3COO^-] = 0.23\ M$

    $$\frac{x\ mmol}{100\ mL} = 0.23$$

    $$x = 23\ mmol$$

    To obtain 23 mmol of sodium acetate from a 1.50 M solution of sodium acetate,

    $$\frac{23\ mmol}{y\ mL} = 1.50\ M$$

    $$y = 15.3\ mL$$

    The desired buffer solution can be prepared using   51.3 mL 1.50 M acetic acid

    15.3 mL 1.50 M sodium acetate

    33.4 mL H_2O
    _____
    100.0 mL

2. **By mixing the appropriate amounts of acetic acid and sodium hydroxide, and adding water to obtain a final volume of 100 mL.**

Sodium hydroxide is used to convert some of the acetic acid into sodium acetate. This means that acetic acid will be the source of both acetic acid and sodium acetate.

The concentrations needed are $[CH_3COOH] = 1.00$ M

$$[NaOH] = 0.23 \text{ M}$$

The amount of acetic acid needed is $[CH_3COOH] = 1.00$ M

$$\frac{x \text{ mmol}}{100 \text{ mL}} = 1.00 \text{ M}$$
$$x = 100 \text{ mmol}$$

To obtain 100 mmol of acetic acid from a 1.50 M solution of acetic acid,

$$\frac{100 \text{ mmol}}{y \text{ mL}} = 1.50 \text{ M}$$
$$y = 66.7 \text{ mL}$$

The amount of sodium hydroxide needed is $[NaOH] = 0.23$ M

$$\frac{x \text{ mmol}}{100 \text{ mL}} = 0.23 \text{ M}$$
$$x = 23 \text{ mmol}$$

To obtain 23 mmol of sodium hydroxide from a 1.50 M solution of NaOH,

$$\frac{23 \text{ mmol}}{y \text{ mL}} = 1.50 \text{ M}$$
$$y = 15.3 \text{ mL}$$

The desired buffer solution can be prepared using   66.7 mL 1.50 M acetic acid

15.3 mL 1.50 M NaOH

$$\underline{18.0 \text{ mL } H_2O}$$

100.0 mL

3. **By mixing the appropriate amounts of sodium acetate and hydrochloric acid, and adding water to obtain a final volume of 100 mL.**

Hydrochloric acid is used to convert some of the sodium acetate into acetic acid.

This means that sodium acetate will be the source of both acetic acid and sodium acetate.

The concentrations needed are $[CH_3COONa] = 1.00$ M

$$[HCl] = 0.77 \text{ M}$$

The amount of sodium acetate needed is $\left[\text{CH}_3\text{COONa}\right] = 1.00\text{ M}$

$$\frac{x\text{ mmol}}{100\text{ mL}} = 1.00\text{ M}$$

$$x = 100\text{ mmol}$$

To obtain 100 mmol of sodium acetate from a 1.50 M solution of sodium acetate,

$$\frac{100\text{ mmol}}{y\text{ mL}} = 1.50\text{ M}$$

$$y = 66.7\text{ mL}$$

The amount of hydrochloric acid needed is $[\text{HCl}] = 0.77\text{ M}$

$$\frac{x\text{ mmol}}{100\text{ mL}} = 0.77\text{ M}$$

$$x = 77\text{ mmol}$$

To obtain 77 mmol of hydrochloric acid from a 1.50 M solution of HCl,

$$\frac{77\text{ mmol}}{y\text{ mL}} = 1.50\text{ M}$$

$$y = 51.3\text{ mL}$$

100 mL of a 1.00 M acetic acid/acetate buffer cannot be made from these reagents, because the volume needed (66.7 mL + 51.3 mL) add up to more than 100 mL. To make this buffer using sodium acetate and hydrochloric acid, you would need to use a more concentrated solution (>1.50 M) of sodium acetate or more concentrated solution (>1.50 M) of HCl.

Problems on pH, pK_a, and Buffers

. Calculate the pH of each of the following solutions:

a. 1×10^{-3} M HCl

b. 0.60 M HCl

c. 1.40×10^{-2} M HCl

d. 1×10^{-3} M KOH

e. 3.70×10^{-4} M NaOH

f. a 1.20 M solution of an acid with a pK_a = 4.23

g. 1.60×10^{-2} M sodium acetate (pK_a of acetic acid = 4.76)

. Calculate the pH of each of the following buffer solutions:

a. a buffer prepared by mixing 20 mL of 0.10 M formic acid and 15 mL of 0.50 M sodium formate (pK_a of formic acid = 3.75)

b. a buffer prepared by mixing 10 mL of 0.50 M aniline and 15 mL of 0.10 M HCl (pK_a of the anilinium ion = 4.60)

c. a buffer prepared by mixing 15 mL of 1.00 M acetic acid and 10 mL of 0.50 M NaOH (pK_a of acetic acid = 4.76)

. What fraction of a carboxylic acid with pK_a = 5.23 would be ionized at pH = 4.98?

. What would be the concentration of formic acid and sodium formate in a 1.00 M buffer solution with a pH = 3.12 (pK_a of formic acid = 3.75)?

. You have found a bottle labeled 1.00 M RCOOH. You want to determine what carboxylic acid it is, so you decide to determine its pK_a value. How would you do this?

. a. How would you prepare 100 mL of a buffer solution that is 0.30 M in acetic acid and 0.20 M in sodium acetate using a 1.00 M acetic acid solution and a 2.00 M sodium acetate solution?

b. The pK_a of acetic acid is 4.76. Would the pH of the above solution be greater or less than 4.76?

. You have 100 mL of a 1.50 M acetic acid/sodium acetate buffer solution that has a pH = 4.90. How could you change the pH of the solution to 4.50?

. You have 100 mL of a 1.00 M solution of an acid with a pK_a = 5.62 to which you add 10 mL of 1.00 M sodium hydroxide. What fraction of the acid will be in the acidic form? How much more sodium hydroxide will you need to add in order to have 40% of the acid in its acidic form (that is, with its proton)?

9.   Describe three ways to prepare a 1.00 M acetic acid/sodium acetate buffer solution with a pH = 4.00.

10. You have available to you 1.50 M solutions of acetic acid, sodium acetate, sodium hydroxide, and hydrochloric acid. How would you make 50 mL of each of the buffers described in the preceding problem?

11. How would you make a 1.0 M buffer solution with a pH = 3.30?

12. You are planning to carry out a reaction that will produce protons. In order for the reaction to take place at constant pH, it will be carried in a solution buffered at pH = 4.2. Would it be better to use a formic acid/formate buffer or an acetic acid/acetate buffer?

Solutions to Problems on pH, pK_a, and Buffers

a. $pH = -\log(1 \times 10^{-3})$
$pH = 3$

b. $pH = -\log 0.60$
$pH = 0.22$

c. $pH = -\log(1.40 \times 10^{-2})$
$pH = 1.85$

d. $pOH = -\log(1 \times 10^{-3})$
$pOH = 3$
$pH = 14 - 3 = 11$

e. $pOH = -\log(3.70 \times 10^{-4})$
$pOH = 3.43$
$pH = 10.57$

f. $pK_a = 4.23, \quad K_a = 5.89 \times 10^{-5}$

$$K_a = \frac{[H^+][A^-]}{[HA]}$$

$$5.89 \times 10^{-5} = \frac{x^2}{1.20}$$

$$x^2 = 7.07 \times 10^{-5}$$

$$x = 8.41 \times 10^{-3}$$

$$pH = 2.08$$

g. $\dfrac{K_w}{K_a} = \dfrac{[HO^-][HA]}{[A^-]} \qquad (K_a = 10^{-4.76} = 1.74 \times 10^{-5})$

$$\frac{1.0 \times 10^{-14}}{1.74 \times 10^{-5}} = \frac{x^2}{1.60 \times 10^{-2}}$$

$$5.75 \times 10^{-10} = \frac{x^2}{1.60 \times 10^{-2}}$$

$$x^2 = 9.20 \times 10^{-12}$$

$$x = 3.03 \times 10^{-6}$$

$$pOH = 5.52$$

$$pH = 14.00 - 5.52 = 8.48$$

2.   **a.**   formic acid: 20 mL $\times$ 0.10 M = 2.0 mmol
sodium formate: 15 mL $\times$ 0.50 M = 7.5 mmol

$$pK_a = pH + \log\frac{[HA]}{[A^-]}$$

$$3.75 = pH + \log\frac{2.0}{7.5}$$

$$3.75 = pH + \log 0.27$$

$$3.75 = pH + (-0.57)$$

$$pH = 4.32$$

b.   aniline: 10 mL $\times$ 0.50 M = 5.0 mmol $\longrightarrow$ 3.5 mmol aniline (RNH_2)
HCl: 15 mL $\times$ 0.10 M = 1.5 mmol $\longrightarrow$ 1.5 mmol anilinium hydrochloride ($\overset{+}{R}NH_3$)

$$pK_a = pH + \log\frac{[HA]}{[A^-]}$$

$$4.60 = pH + \log\frac{1.5}{3.5}$$

$$4.60 = pH + \log 0.43$$

$$4.60 = pH + (-0.37)$$

$$pH = 4.97$$

c.   acetic acid: 15 mL $\times$ 1.00 M = 15 mmol $\longrightarrow$ 10 mmol acetic acid
NaOH: 10 mL $\times$ 0.50 M = 5.0 mmol $\longrightarrow$ 5.0 mmol sodium acetate

$$pK_a = pH + \log\frac{[HA]}{[A^-]}$$

$$4.76 = pH + \log\frac{10}{5.0}$$

$$4.76 = pH + \log 2$$

$$4.76 = pH + 0.30$$

$$pH = 4.46$$

3.   The ionized form is the basic form. Therefore, we need to use the equation that allows us to calculate the fraction present in the basic form.

$$\text{fraction of buffer in the basic form} = \frac{K_a}{K_a + [H^+]} = \frac{5.89 \times 10^{-6}}{(5.89 \times 10^{-6}) + (10.47 \times 10^{-6})} = \frac{5.89 \times 10^{-6}}{16.36 \times 10^{-6}}$$

$$= 0.36$$

fraction of buffer in the basic form = $\dfrac{K_a}{K_a + [H^+]} = \dfrac{1.78 \times 10^{-4}}{(1.78 \times 10^{-4}) + (7.59 \times 10^{-4})}$

$$= \dfrac{1.78 \times 10^{-4}}{9.37 \times 10^{-4}}$$

$$= 0.19$$

$$[\text{sodium formate}] = 0.19 \text{ M}$$
$$[\text{formic acid}] = 0.81 \text{ M}$$

From the Henderson–Hasselbalch equation, we see that the pH of the solution is the same as the pK_a of the species in the acidic form when the concentration of the species in the acidic form is the same as the concentration of the species in the basic form.

$$pK_a = pH + \log \dfrac{[HA]}{[A^-]}$$

$$\text{when } [HA] = [A^-],$$

$$pK_a = pH$$

Thus, in order to have a solution in which the pH will be the same as the pK_a, the number of mmol of acid must equal the number of mmol of conjugate base.

Preparing a solution of x mmol of RCOOH and $1/2\,x$ mmol NaOH will give a solution in which $[\text{RCOOH}] = [\text{RCOO}^-]$.

For example, 20 mL of 1.00 M RCOOH = 20 mmol
                    10 mL of 1.00 M NaOH = 10 mmol

This will give a solution that contains 10 mmol RCOOH and 10 mmol RCOO⁻.

The pH of this solution is the pK_a of RCOOH.

a. $\dfrac{x \text{ mmol}}{100 \text{ mL}} = 0.30 \text{ M}$ $\dfrac{x \text{ mmol}}{100 \text{ mL}} = 0.20 \text{ M}$

   x = 30 mmol of acetic acid x = 20 mmol of sodium acetate

   $\dfrac{30 \text{ mmol}}{y \text{ mL}} = 1.00 \text{ M}$ $\dfrac{20 \text{ mmol}}{y \text{ mL}} = 2.00 \text{ M}$

   y = 30 mL of 1.00 M acetic acid y = 10 mL of 2.00 M acetic acid

   The buffer solution could be prepared by mixing   30 mL of 1.00 M acetic acid

   $$\begin{array}{l} 10 \text{ mL of 2.00 M sodium acetate} \\ \underline{60 \text{ mL of water}} \\ 100 \text{ mL} \end{array}$$

b. Because the concentration of buffer in the acidic form (0.30 M) is greater than the concentration of buffer in the basic form (0.20 M), the pH of the solution will be less than 4.76.

7.   Original solution

fraction of buffer in the basic form = $\dfrac{K_a}{K_a + [H^+]}$ = $\dfrac{1.74 \times 10^{-5}}{(1.74 \times 10^{-5}) + (1.26 \times 10^{-5})}$

$$= \dfrac{1.74 \times 10^{-5}}{3.00 \times 10^{-5}}$$

$$= 0.58$$

$$0.58 \times 1.50 \text{ M} = 0.87 \text{ M}$$

$$[A^-] = 0.87 \text{ M}$$

$$[HA] = 0.63 \text{ M}$$

Desired solution

fraction of buffer in the basic form = $\dfrac{K_a}{K_a + [H^+]}$ = $\dfrac{1.74 \times 10^{-5}}{(1.74 \times 10^{-5}) + (3.16 \times 10^{-5})}$

$$= \dfrac{1.74 \times 10^{-5}}{4.90 \times 10^{-5}}$$

$$= 0.35$$

$$0.35 \times 1.50 \text{ M} = 0.53 \text{ M}$$

$$[A^-] = 0.53 \text{ M}$$

$$[HA] = 0.97 \text{ M}$$

The original solution contains 87 mmol of A^- (100 mL $\times$ 0.87 M).

The desired solution with a pH $= 4.50$ must contain 53 mmol of A^-.

Therefore, 34 mmol of A^- $(87 - 53 = 34)$ must be converted to HA.

This can be done by adding 34 mmol of HCl to the original solution.

If you have a 1.00 M HCl solution, you will need to add 34 mL to the original solution in order to change its pH from 4.90 to 4.50.

$$\dfrac{34 \text{ mmol}}{x \text{ mL}} = 1.00 \text{ M}$$

$$x = 34 \text{ mL}$$

Note that after adding HCl to the original solution, it will no longer be a 1.50 M buffer; it will be more dilute (150 mmol/134 mL $= 1.12$ M).

The change in the concentration of the buffer solution will be less if a more concentrated solution of HCl is used to change the pH. For example, if you have a 2.00 M HCl solution,

$$\dfrac{34 \text{ mmol}}{x \text{ mL}} = 2.00 \text{ M}$$

$$x = 17 \text{ mL}$$

You will need to add 17 mL to the original solution, and the concentration of the buffer will be 1.28 M (150 mmol/117 mL = 1.28 M).

8. acid: 100 mL × 1.00 M = 100 mmol HA $\longrightarrow$ 90 mmol HA

NaOH: 10 mL × 1.00 M = 10 mmol HO$^-$ $\longrightarrow$ 10 mmol A$^-$

Therefore, 90% is in the acidic form.

For 40% to be in the acidic form, you need   40 mmol HA
60 mmol A$^-$

You need to have 60 mmol rather than 10 mmol in the basic form. To get the additional 50 mmol in the basic form, you would need to add 50 mL of 1.0 M NaOH.

9. $$\frac{\text{fraction of buffer}}{\text{in the basic form}} = \frac{K_a}{K_a + [\text{H}^+]} = \frac{1.74 \times 10^{-5}}{(1.74 \times 10^{-5}) + (1.00 \times 10^{-4})} = \frac{1.74 \times 10^{-5}}{(1.74 \times 10^{-5}) + (10.00 \times 10^{-5})}$$

$$= \frac{1.74 \times 10^{-5}}{11.74 \times 10^{-5}}$$

$$= 0.15$$

$$[\text{A}^-] = 0.15 \text{ M}$$

$$[\text{HA}] = 0.85 \text{ M}$$

a. [acetic acid] = 0.85 M   **b.** [acetic acid] = 1.00 M   **c.** [sodium acetate] = 1.00 M
[sodium acetate] = 0.15 M   [NaOH] = 0.15 M   [HCl] = 0.85 M

10. a. $$\frac{x \text{ mmol}}{50 \text{ mL}} = 0.85 \text{ M}$$

$$x = 42.5 \text{ mmol of acetic acid}$$

$$\frac{42.5 \text{ mmol}}{y \text{ mL}} = 1.50 \text{ M}$$

$$y = 28.3 \text{ mL of 1.50 M acetic acid}$$

$$\frac{x \text{ mmol}}{50 \text{ mL}} = 0.15 \text{ M}$$

$$x = 7.5 \text{ mmol of sodium acetate}$$

$$\frac{7.5 \text{ mmol}}{y \text{ mL}} = 1.50 \text{ M}$$

$$y = 5.0 \text{ mL of 1.50 M sodium acetate}$$

28.3 mL of 1.50 M acetic acid

5.0 mL of 1.50 M sodium acetate

$$\frac{16.7 \text{ mL of H}_2\text{O}}{50.0 \text{ mL}}$$

b. $\dfrac{x \text{ mmol}}{50 \text{ mL}} = 1.00 \text{ M}$

$x = 50 \text{ mmol of acetic acid}$

$\dfrac{50 \text{ mmol}}{y \text{ mL}} = 1.50 \text{ M}$

$y = 33.3 \text{ mL of } 1.50 \text{ M acetic acid}$

$\dfrac{x \text{ mmol}}{50 \text{ mL}} = 0.15 \text{ M}$

$x = 7.5 \text{ mmol of NaOH}$

$\dfrac{7.5 \text{ mmol}}{y \text{ mL}} = 1.50 \text{ M}$

$y = 5.0 \text{ mL of } 1.50 \text{ M NaOH}$

$\ \ 33.3 \text{ mL of } 1.50 \text{ M acetic acid}$

$\ \ \ \ 5.0 \text{ mL of } 1.50 \text{ M NaOH}$

$\dfrac{11.7 \text{ mL of } H_2O}{50.0 \text{ mL}}$

c. $\dfrac{x \text{ mmol}}{50 \text{ mL}} = 1.00 \text{ M}$

$x = 50 \text{ mmol of sodium acetate}$

$\dfrac{50 \text{ mmol}}{y \text{ mL}} = 1.50 \text{ M}$

$y = 33.3 \text{ mL of } 1.50 \text{ M sodium acetate}$

$\dfrac{x \text{ mmol}}{50 \text{ mL}} = 0.85 \text{ M}$

$x = 42.5 \text{ mmol of HCl}$

$\dfrac{42.5 \text{ mmol}}{y \text{ mL}} = 1.5 \text{ M}$

$y = 28.3 \text{ mL of } 1.5 \text{ M HCl}$

We cannot make the required buffer with these solutions, because 33.3 mL + 28.3 mL > 50 mL.

1. Because formic acid has a pK_a = 3.75, a formic acid/formate buffer can be a buffer at pH = 3.30, since this pH is within one pH unit of the pK_a value.

fraction of buffer in the basic form $= \dfrac{K_a}{K_a + [H^+]} = \dfrac{1.78 \times 10^{-4}}{(1.78 \times 10^{-4}) + (5.01 \times 10^{-4})}$

$$= \dfrac{1.78 \times 10^{-4}}{6.79 \times 10^{-4}}$$

$$= 0.26$$

The solution must have [formic acid] = 0.74 M and [sodium formate] = 0.26 M.

2. The reaction to be carried out will generate protons that will react with the basic form of the buffer in order to keep the pH constant.

Therefore, the better buffer would be the one that has the larger percentage of the buffer in the basic form.

The pK_a of formic acid is 3.74. Because the pH of the solution (4.2) is more basic than the pK_a value of the compound, formic acid will exist primarily in its basic form.

The pK_a of acetic acid is 4.76. Because the pH of the solutions is more acidic than the pK_a value of the compound, acetic acid will exist primarily in its acidic form.

Therefore, the formate buffer is preferred, because it has a greater percentage of the buffer in the basic form.

CHAPTER 3
An Introduction to Organic Compounds:
Nomenclature, Physical Properties, and Representation of Structure

Important Terms

alcohol	a compound with an OH group in place of one of the hydrogens of an alkane (ROH).
alkane	a hydrocarbon that contains only single bonds.
alkyl halide	a compound with a halogen in place of one of the hydrogens of an alkane.
alkyl substituent	a substituent formed by removing a hydrogen from an alkane.
amine	a compound in which one or more of the hydrogens of NH_3 are replaced by an alkyl substituent (RNH_2, R_2NH, R_3N).
angle strain	the strain introduced into a molecule as a result of its bond angles being distorted from their ideal values.
anti conformer	the staggered conformer in which the largest substituents bonded to the two carbons are opposite each other. It is the most stable of the staggered conformers.
axial bond	a bond of the chair conformer of cyclohexane that points directly up or directly down.
boat conformer	a conformer of cyclohexane that roughly resembles a boat.
boiling point	the temperature at which the vapor pressure of a liquid equals the atmospheric pressure.
chair conformer	a conformer of cyclohexane that roughly resembles a chair. It is the most stable conformer of cyclohexane.
cis fused	two rings fused together in such a way that if the second ring were considered to be two substituents of the first ring, the two substituents would be on the same side of the first ring.
cis isomer (for a cyclic compound)	the isomer with two substituents on the same side of the ring.
cis–trans stereoisomers	see the definitions of "cis isomer" and "trans isomer."
common name	non systematic nomenclature.
conformation	the three-dimensional shape of a molecule at a given instant.
conformational analysis	the investigation of various conformers of a molecule and their relative stabilities.

conformers	different conformations of a molecule.
constitutional isomers (structural isomers)	molecules that have the same molecular formula but differ in the way the atoms are connected.
cycloalkane	an alkane with its carbon chain arranged in a closed ring.
1,3-diaxial interaction	the interaction between an axial substituent and one of the other two axial substituents on the same side of a cyclohexane ring.
dipole–dipole interaction	an interaction between the dipole of one molecule and the dipole of another.
eclipsed conformer	a conformer in which the bonds on adjacent carbons are parallel to each other when viewed looking down the carbon–carbon bond.
equatorial bond	a bond of the chair conformer of cyclohexane that juts out from the ring but does not point directly up or directly down.
ether	a compound in which an oxygen is bonded to two alkyl substituents (ROR).
functional group	the center of reactivity of a molecule.
gauche conformer	a staggered conformer in which the largest substituents bonded to the two carbons are gauche to each other—that is, their bonds have a dihedral angle of approximately 60°.

The substituents are gauche to each other.

gauche interaction	the interaction between two atoms or groups that are gauche to each other.
geometric isomers	cis–trans (or *E,Z*) isomers.
half-chair conformer	the least stable conformer of cyclohexane.
hydrocarbon	a compound that contains only carbon and hydrogen.
hydrogen bond	an unusually strong dipole–dipole attraction (5 kcal/mol) between a hydrogen bonded to O, N, or F and the lone pair of a different O, N, or F.
induced dipole-induced dipole interaction	an interaction between a temporary dipole in one molecule and the dipole that the temporary dipole induces in another molecule.
IUPAC nomenclature	systematic nomenclature developed by the International Union of Pure and Applied Chemistry.

| **melting point** | the temperature at which a solid becomes a liquid. |

methylene group   a CH_2 group.

Newman projection   a way to represent the three-dimensional spatial relationships of atoms by looking down the length of a particular carbon–carbon bond.

packing   a property that determines how well individual molecules fit into a crystal lattice.

parent hydrocarbon   the longest continuous carbon chain in a molecule; if the molecule has a functional group, it is the longest continuous carbon chain that contains the functional group.

perspective formula   a way to represent the three-dimensional spatial relationships of atoms using two adjacent solid lines, one solid wedge, and one hatched wedge.

primary alcohol   an alcohol in which the OH group is bonded to a primary carbon.

primary alkyl halide   an alkyl halide in which the halogen is bonded to a primary carbon.

primary amine   an amine with one alkyl group bonded to the nitrogen.

primary carbon   a carbon bonded to only one other carbon.

primary hydrogen   a hydrogen bonded to a primary carbon.

ring-flip (chair–chair interconversion)   the conversion of a chair conformer of cyclohexane into the other chair conformer; bonds that are axial in one chair conformer are equatorial in the other chair conformer.

secondary alcohol   an alcohol in which the OH group is bonded to a secondary carbon.

secondary alkyl halide   an alkyl halide in which the halogen is bonded to a secondary carbon.

secondary amine   an amine with two alkyl groups bonded to the nitrogen.

secondary carbon   a carbon bonded to two other carbons.

secondary hydrogen   a hydrogen bonded to a secondary carbon.

skeletal structure   a structure that shows the carbon–carbon bonds as lines and does not show the carbon–hydrogen bonds.

skew-boat conformer   one of the conformers of a cyclohexane ring.

solubility   the extent to which a compound dissolves in a solvent.

solvation   the interaction between a solvent and another molecule (or ion).

staggered conformer	a conformer in which the bonds on one carbon bisect the bond angles on the adjacent carbon when viewed looking down the carbon–carbon bond.

steric hindrance	hindrance due to groups occupying a volume of space.
steric strain	the repulsion between the electron cloud of an atom or group of atoms and the electron cloud of another atom or group of atoms.
straight-chain alkane	an alkane in which the carbons form a continuous chain with no branches.
structural isomers (constitutional isomers)	molecules that have the same molecular formula but differ in the way the atoms are connected.
systematic nomenclature	a system of nomenclature based on rules, such as IUPAC nomenclature.
tertiary alcohol	an alcohol in which the OH group is bonded to a tertiary carbon.
tertiary alkyl halide	an alkyl halide in which the halogen is bonded to a tertiary carbon.
tertiary amine	an amine with three alkyl groups bonded to the nitrogen.
tertiary carbon	a carbon bonded to three other carbons.
tertiary hydrogen	a hydrogen bonded to a tertiary carbon.
trans-fused	two rings fused together in such a way that if the second ring were considered to be two substituents of the first ring, the two substituents would be on opposite sides of the first ring.
trans isomer (for a cyclic compound)	the isomer with two substituents on opposite sides of the ring.
twist-boat conformer	one of the conformers of a cyclohexane ring.
van der Waals forces	induced-dipole–induced-dipole interactions.

Solutions to Problems

1. **a.** C_nH_{2n+2} If there are 17 carbons, then there are 36 hydrogens.
b. C_nH_{2n+2} If there are 74 hydrogens, then there are 36 carbons.

2. $CH_3CH_2CH_2CH_2CH_2CH_2CH_2CH_3$ $CH_3CHCH_2CH_2CH_2CH_2CH_3$

            octane CH_3 isooctane

3. **a.** propyl alcohol   **b.** butyl methyl ether   **c.** propylamine

4. Notice that each carbon forms four bonds, and each hydrogen and bromine forms one bond.

$$CH_3$$

$CH_3CH_2CH_2CH_2Br$ $CH_3CHCH_2CH_3$ CH_3CHCH_2Br CH_3CCH_3

                                 Br CH_3 Br

n-butyl bromide   *sec*-butyl bromide   isobutyl bromide   *tert*-butyl bromide
       or
butyl bromide

5. **a.** CH_3CHOH **c.** CH_3CH_2CHI **e.** CH_3CNH_2

            CH_3 CH_3 CH_3

$$CH_3$$

b. $CH_3CHCH_2CH_2F$ **d.** CH_3COH **f.** $CH_3CH_2CH_2CH_2CH_2CH_2CH_2CH_2B$

          CH_3 CH_2CH_3

6. **a.** ethyl methyl ether   **c.** *sec*-butylamine   **e.** isobutyl bromide
b. methylpropylamine   **d.** butyl alcohol or *n*-butyl alcohol   **f.** *sec*-butyl chloride

7. **a.** $CH_3CHCH_2CH_3$ **b.** CH_3CCH_3

            CH_3 CH_3

      2-methylbutane   2,2-dimethylpropane

8. You can draw substituents pointing up and/or pointing down from the chain.

a. Solved in the text.

$$\text{d.}\quad \underset{\underset{\underset{\underset{CH_3}{|}}{CHCH_3}}{|}}{\underset{CH_3}{\overset{CH_3}{\overset{|}{CH_3CHCH_2C}}}} \overset{CH_3}{\overset{|}{—}} \overset{CH_3}{\underset{|}{CHCH_2CH_3}}$$

b. $\underset{\underset{CH_3}{|}}{\overset{\overset{CH_3}{|}}{CH_3CHCHCH_2CH_2CH_3}}$

e. $\underset{\underset{\underset{CH_3CHCH_3}{|}}{CH_2}}{\overset{\overset{CH_3}{|}}{CH_3CHCH_2CHCHCH_2CH_2CH_3}}\overset{CH_3}{\overset{|}{}}$

c. $\underset{\underset{CH_2CH_3}{|}}{\overset{\overset{CH_2CH_3}{|}}{CH_3CH_2CH_2CCH_2CH_2CH_2CH_2CH_2CH_3}}$

f. $\underset{\underset{\underset{CH_3}{|}}{CH_3CCH_3}}{\overset{}{CH_3CH_2CH_2CHCH_2CH_2CH_2CH_3}}\overset{|}{}$

9.

a.

#1 $CH_3CH_2CH_2CH_2CH_2CH_2CH_2CH_3$
octane

#2 $\underset{\underset{CH_3}{|}}{CH_3CHCH_2CH_2CH_2CH_2CH_3}$

2-methylheptane

#3 $\underset{\underset{CH_3}{|}}{CH_3CH_2CHCH_2CH_2CH_2CH_3}$

3-methylheptane

#4 $\underset{\underset{CH_3}{|}}{CH_3CH_2CH_2CHCH_2CH_2CH_3}$

4-methylheptane

#5 $\underset{\underset{CH_3}{|}}{\overset{\overset{CH_3}{|}}{CH_3CCH_2CH_2CH_2CH_3}}$

2,2-dimethylhexane

#10 $\underset{\underset{}{}}{\overset{\overset{CH_3\ \ CH_3}{|\quad\ |}}{CH_3CH_2CH—CHCH_2CH_3}}$
3,4-dimethylhexane

#11 $\underset{\underset{CH_3}{|}}{\overset{\overset{CH_3\ \ CH_3}{|\quad\ |}}{CH_3C—CHCH_2CH_3}}$

2,2,3-trimethylpentane

#12 $\underset{\underset{CH_3}{|}}{\overset{\overset{CH_3\ \ CH_3}{|\quad\ |}}{CH_3CCH_2CHCH_3}}$

2,2,4-trimethylpentane

#13 $\underset{\underset{CH_3}{|}}{\overset{\overset{CH_3\ \ CH_3}{|\quad\ |}}{CH_3CH—CCH_2CH_3}}$

2,3,3-trimethylpentane

#14 $\underset{\underset{CH_3}{|}}{\overset{\overset{CH_3\qquad\ CH_3}{|\qquad\quad\ |}}{CH_3CH—CH—CHCH_3}}$

2,3,4-trimethylpentane

$$\#6 \quad \underset{\underset{CH_3}{|}}{\overset{\overset{CH_3}{|}}{CH_3CH_2CCH_2CH_2CH_3}}$$

3,3-dimethylhexane

$$\#7 \quad \underset{}{\overset{\overset{CH_3}{|} \quad \overset{CH_3}{|}}{CH_3CH - CHCH_2CH_2CH_3}}$$

2,3-dimethylhexane

$$\#8 \quad \overset{\overset{CH_3}{|} \quad \overset{CH_3}{|}}{CH_3CHCH_2CHCH_2CH_3}$$

2,4-dimethylhexane

$$\#9 \quad \overset{\overset{CH_3}{|} \qquad \overset{CH_3}{|}}{CH_3CHCH_2CH_2CHCH_3}$$

2,5-dimethylhexane

$$\#15 \quad \overset{\overset{CH_3}{|} \; \overset{CH_3}{|}}{\underset{\underset{CH_3}{|} \; \underset{CH_3}{|}}{CH_3C - CCH_3}}$$

2,2,3,3-tetramethylbutane

$$\#16 \quad \underset{\underset{CH_2CH_3}{|}}{CH_3CH_2CHCH_2CH_2CH_3}$$

3-ethylhexane

$$\#17 \quad \underset{\underset{CH_2CH_3}{|}}{\overset{\overset{CH_3}{|}}{CH_3CH_2CHCHCH_3}}$$

3-ethyl-2-methylpentane

$$\#18 \quad \underset{\underset{CH_2CH_3}{|}}{\overset{\overset{CH_3}{|}}{CH_3CH_2CCH_2CH_3}}$$

3-ethyl-3-methylpentane

b. The systematic name is under each structure.
c. Only **#1** (octane or *n*-octane) and **#2** (isooctane) have common names.
d. **#2, #7, #8, #9, #12, #13, #14, #17**
e. **#3, #8, #10, #11**
f. **#5, #11, #12, #15**

10. a. 2,2,4-trimethylhexane
  b. 2,2-dimethylbutane
  c. 2,2,5-trimethylhexane
  d. 5-ethyl-4,4-dimethyloctane
  e. 3,3-diethylhexane
  f. 2,5-dimethylheptane

11. a. $CH_3CH_2CH_2CH_2CH_3$

pentane

b. $\underset{\underset{CH_3}{|}}{\overset{\overset{CH_3}{|}}{CH_3CCH_3}}$

2,2-dimethylpropane

c. $\underset{}{\overset{\overset{CH_3}{|}}{CH_3CHCH_2CH_3}}$

2-methylbutane

d. $\underset{}{\overset{\overset{CH_3}{|}}{CH_3CHCH_2CH_3}}$

2-methylbutane

12.

13. **a.**  OH

b.

c.

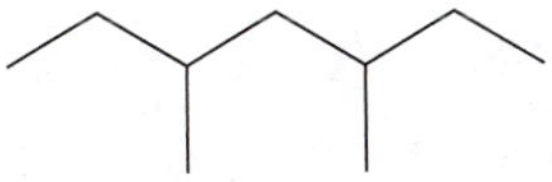

d.  O

e.  N H

f. 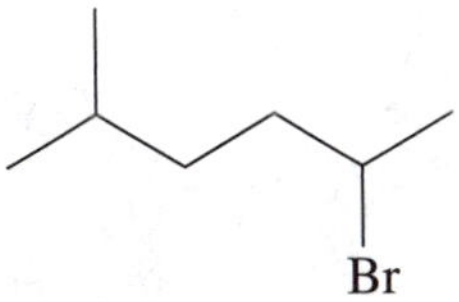 Br

14.   menthol = $C_{10}H_{20}O$ terpin hydrate = $C_{10}H_{20}O_2$

15. **a.** 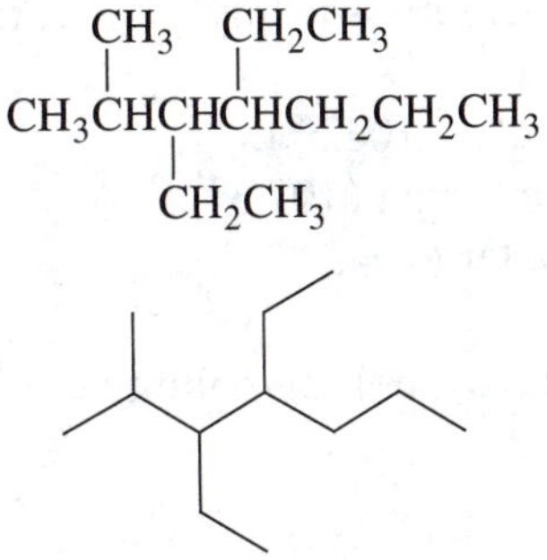 $\underset{\underset{\text{CH}_2\text{CH}_3}{|}}{\text{CH}_3\text{CHCHCHCH}_2\text{CH}_2\text{CH}_3}$ with CH_3 and CH_2CH_3 substituents

b. 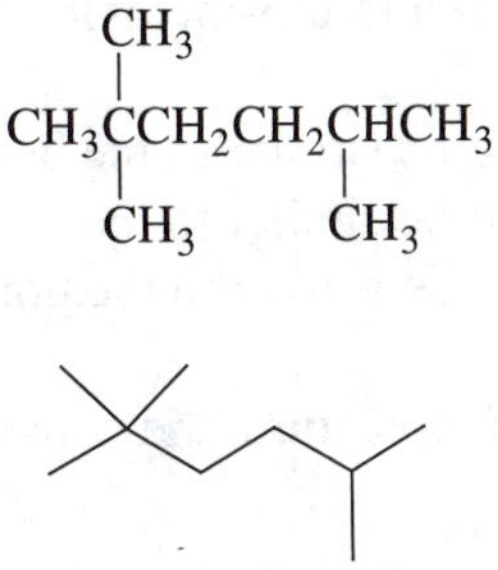 $CH_3CCH_2CH_2CHCH_3$ with CH_3 substituents

16. **a.** 1-ethyl-2-methylcyclopentane
b. ethylcyclobutane
c. 1-ethyl-3-methylcyclohexane
d. 3,6-dimethyldecane
e. heptane
f. 1-bromohexane

17. **a.** Both compounds have the same name (1-bromo-3-methylhexane), so they are the same compound.
b. Both compounds have the same name (1-iodo-2-methylcyclohexane), so they are the same compound.

18. **a.** *sec*-butyl chloride
2-chlorobutane
b. cyclohexyl bromide
bromocyclohexane
c. isohexyl chloride
1-chloro-4-methylpentane
d. isopropyl fluoride
2-fluoropropane

19. **a.** tertiary alkyl halide   **b.** tertiary alcohol   **c.** primary amine

20. **a.** methylpropylamine
secondary
b. trimethylamine
tertiary
c. diethylamine
secondary
d. butyldimethylamine
tertiary

21. **a.** CH₂Cl   Note that the name of a CH₂Cl substituent is "chloromethyl," because a Cl is in place of one of the Hs of a methyl substituent.

b. Cl   CH₃

chloromethylcyclohexane

1-chloro-1-methylcyclohexane

c.   CH₃   Cl   CH₃   Cl   CH₃   Cl

1-chloro-2-methylcyclohexane   1-chloro-3-methylcyclohexane   1-chloro-4-methylcyclohexane

22. **a.** The bond angle is predicted to be similar to the bond angle in water $(104.5°)$.
  b. The bond angle is predicted to be similar to the bond angle in ammonia $(107.3°)$.
  c. The bond angle is predicted to be similar to the bond angle in water $(104.5°)$.

23. Pentane. To be a liquid at room temperature, the compound must have a boiling point that is greater than room temperature.

24. **a.** An O — H hydrogen bond is longer.
  b. Because it is shorter, an O — H covalent bond is stronger.

25. **a.** 1, 4, and 5
  b. 1, 2, 4, 5, and 6

26. HO⟍⟋OH OH > ⟍⟋OH OH > ⟍⟋OH > ⟍⟋NH₂ >

⟍⟋⟍⟋ > ⟍⟋⟍

27. **a.** ⟍⟋⟍⟋Br > ⟍⟋⟍Br > ⟍⟋Br

  b. ⟍⟋⟍⟋⟍ > ⟍⟋⟍⟋⟍ > ⟍⟋⟍⟋ > ✕

  c. ⟍⟋⟍OH > ⟍⟋⟍OH > ⟍⟋⟍Cl > ⟍⟋⟍

28. **a.** ⟍⟋OH OH O > ⟍⟋⟍OH O > ⟍⟋⟍OH > ⟍⟋

  b. HOCH₂CH₂CH₂OH > CH₃CH₂CH₂OH > CH₃CH₂CH₂CH₂OH > CH₃CH₂CH₂CH₂Cl

29. Because cyclohexane is a nonpolar compound, it will have the lowest solubility in the most polar solvent, which, of the solvents given, is ethanol.

$CH_3CH_2CH_2CH_2CH_2OH$ $CH_3CH_2OCH_2CH_3$ CH_3CH_2OH $CH_3CH_2CH_2CH_2CH_2CH_3$
1-pentanol   diethyl ether   ethanol   hexane

30. Hexethal would be expected to be the more effective sedative, because hexethal has a hexyl group in place of the ethyl group of barbital and is, therefore, less polar than barbital. Being less polar, hexethal will be better able to penetrate the nonpolar membrane of the cell.

31. **a.** The Newman projection shows rotation about the C-2—C-3 bond.

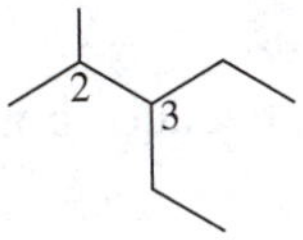

2-methyl-3-ethylpentane

   b. The Newman projection shows rotation about the C-2—C-3 bond.

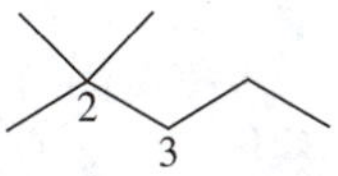

2,2-dimethylpentane

32. **a.**

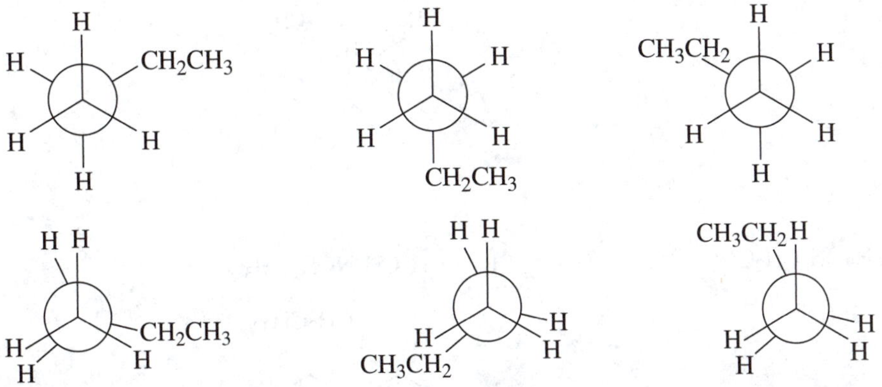

   b. yes   **c.** yes

33. **a.**

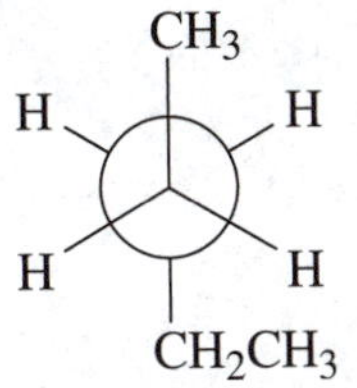

   b.

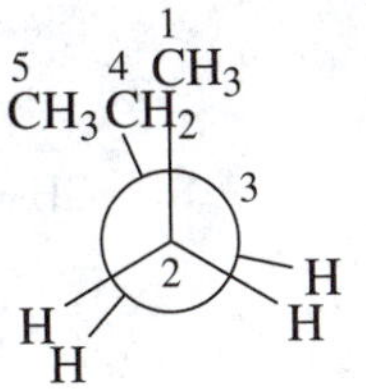

34. **a.**

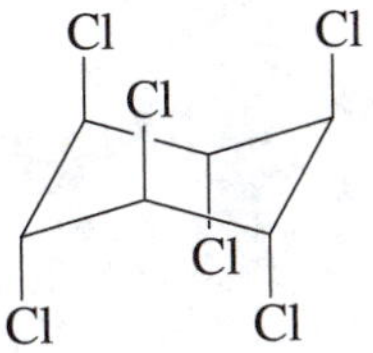

   b.

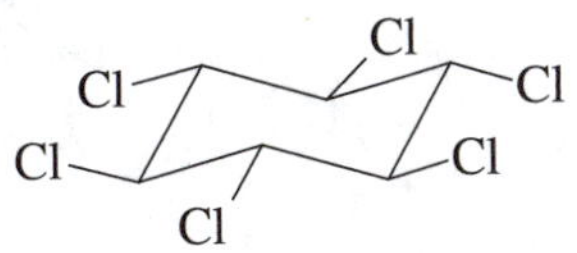

35. At any one time, there will be more molecules of isopropylcyclohexane with the substituent in the equatorial position, because the isopropyl substituent is larger than the ethyl substituent. Because the isopropyl substituent is larger, the axial conformer of isopropylcyclohexane has more steric strain than

the axial conformer of ethylcyclohexane, so the isopropyl group will have a greater preference for th
equatorial position.

36.   If both substituents point downward or both point upward, it is a cis isomer.
If one substituent points upward and the other downward, it is a trans isomer.

    a. cis   **b.** trans   **c.** cis   **d.** trans

37.   **a.**

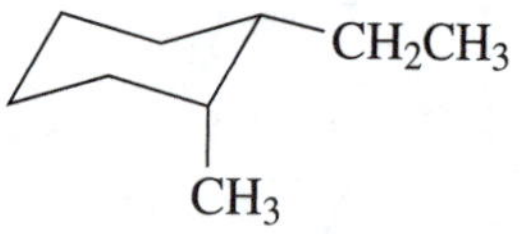

    b.

    c.   *trans*-1-Ethyl-2-methylcyclohexane is more stable, because both substituents can be in equatoria
positions.

38.   Both condensed and skeletal structures are shown.

a.

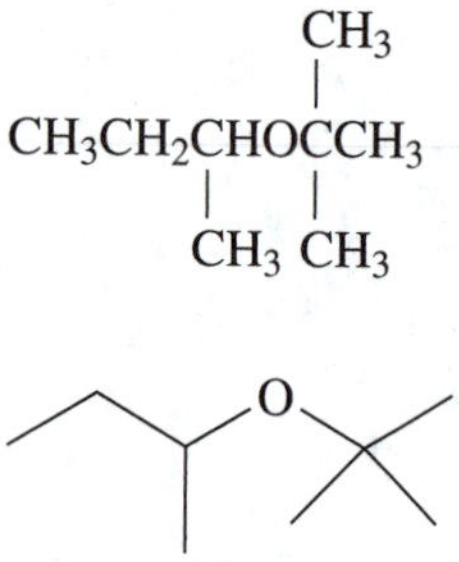

$$CH_3CH_2CHOCCH_3$$

with CH_3 above the fourth carbon and CH_3 CH_3 below.

d.

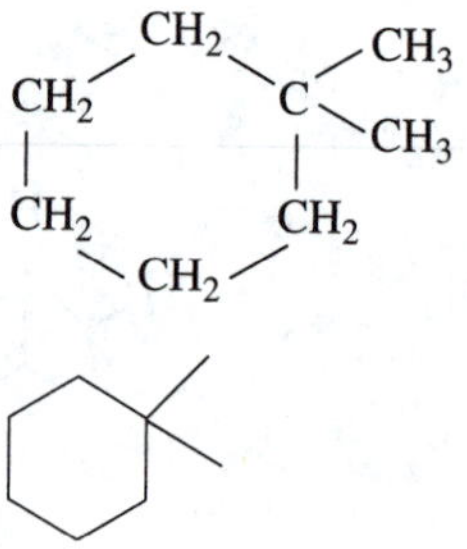

b. 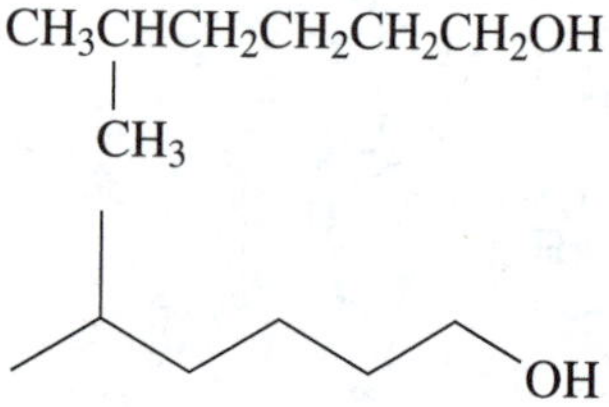 $CH_3CHCH_2CH_2CH_2CH_2OH$ with CH_3 below

e. 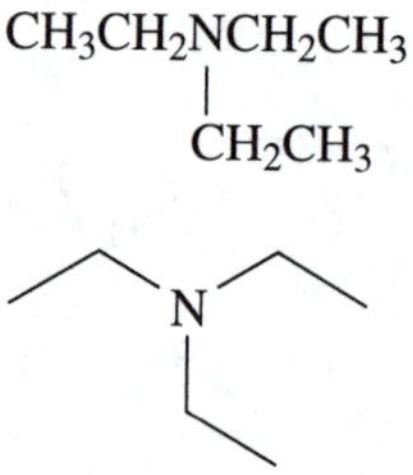 $CH_3CH_2NCH_2CH_3$ with CH_2CH_3 below

c. 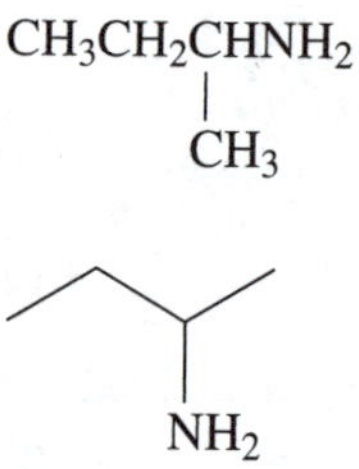 $CH_3CH_2CHNH_2$ with CH_3 below

f. 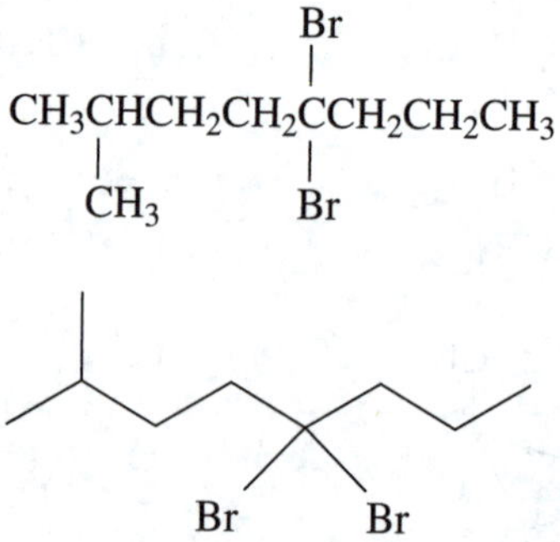 $CH_3CHCH_2CH_2CCH_2CH_2CH_3$ with Br above, CH_3 and Br below

39.

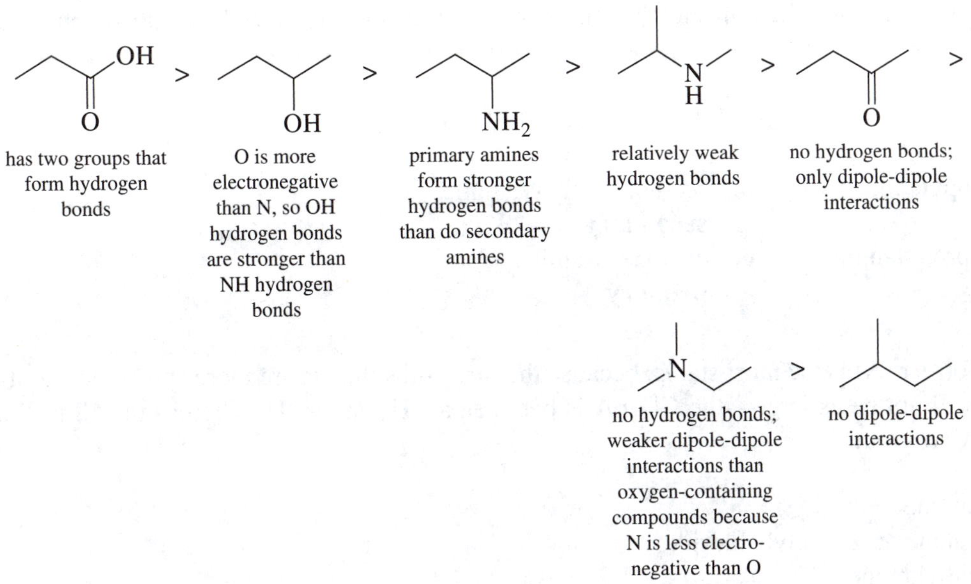

has two groups that form hydrogen bonds   >   O is more electronegative than N, so OH hydrogen bonds are stronger than NH hydrogen bonds   >   primary amines form stronger hydrogen bonds than do secondary amines   >   relatively weak hydrogen bonds   >   no hydrogen bonds; only dipole-dipole interactions   >

no hydrogen bonds; weaker dipole-dipole interactions than oxygen-containing compounds because N is less electro-negative than O   >   no dipole-dipole interactions

40.   **a.**
  1. 2,2,6-trimethylheptane
  2. 5-bromo-2-methyloctane
  3. 3,3-diethylpentane
  4. 2,3,5-trimethylhexane
  5. dimethylcyclohexylamine
  6. 1-bromo-4-methylcyclohexane
  7. 3-methylpentane

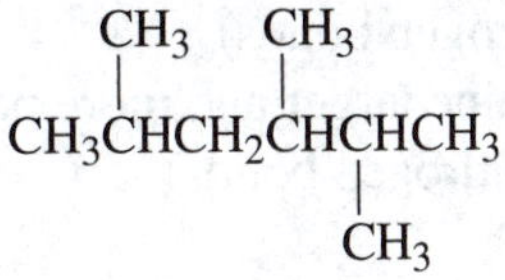

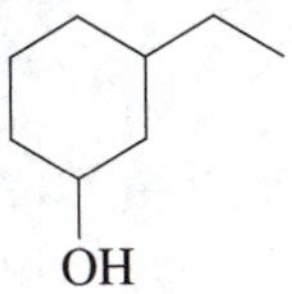

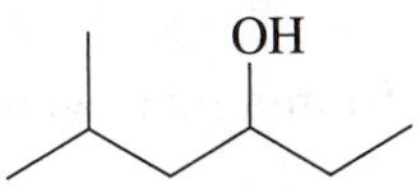

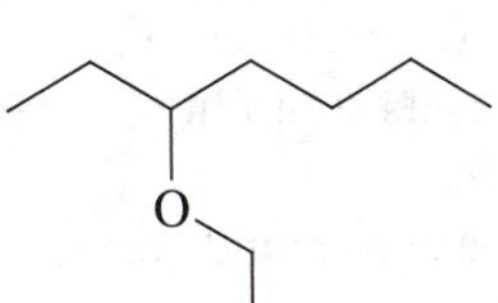

b.

1.

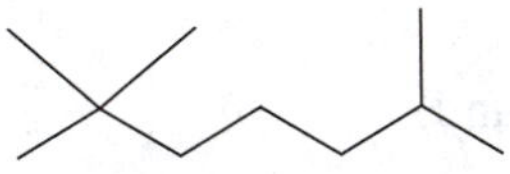

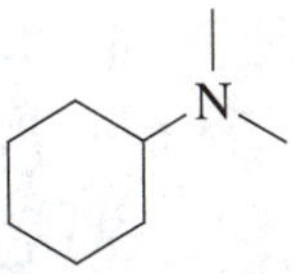

2. 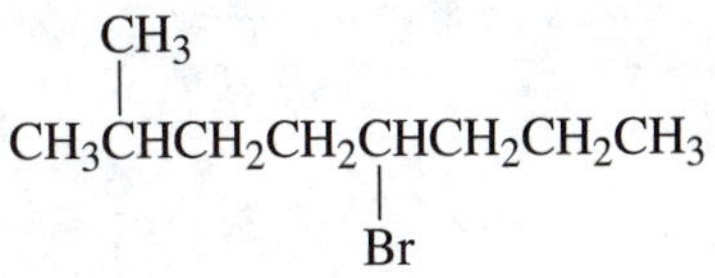 $CH_3CHCH_2CH_2CHCH_2CH_2CH_3$ with CH_3 and Br substituents

3.

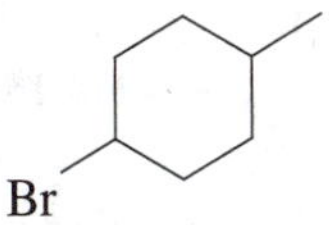

4.

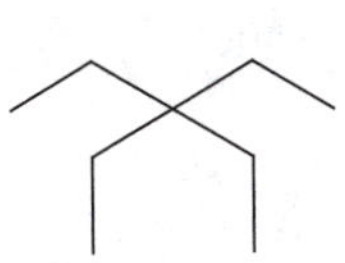

5.

6. $CH_3CHCH_2CHCHCH_3$ with CH_3, CH_3, and CH_3 substituents

7.

8.

9.

10.

11.

41.   **C** and **D** are cis isomers. (In **C,** both substituents are downward pointing; in **D,** both substituents are upward pointing.)

42.   **a.** **1.** 3   **2.** 4   **b.** **1.** 6   **2.** 5   **c.** **1.** 3   **3.** 4

43.   **a.**   diethylpropylamine   **b.**   *sec*-butylisobutylamine
                tertiary   **secondary**
           c.   isopentylpropylamine   **d.**   cyclohexylamine
                secondary   **primary**

44.   The first conformer (**A**) is the most stable, because the three substituents are more spread out, so its gauche interactions will not be as large—the Cl in **A** is between a CH_3 and an H, whereas the Cl in **B** and **C** is between two CH_3 groups.

45.   **a.**   *sec*-butylamine
           b.   2-chlorobutane or *sec*-butyl chloride
           c.   *sec*-butylethylamine
           d.   ethyl propyl ether
           e.   2-methylpentane or isohexane
           f.   isopropylamine
           g.   2-bromo-2-methylbutane or *tert*-pentylamine
           h.   bromocyclopentane or cyclopentyl bromide

46.   **a.**   **b.** $CH_3\overset{\overset{\displaystyle CH_3}{|}}{C}-\overset{\overset{\displaystyle CH_3}{|}}{\underset{\underset{\displaystyle CH_3}{|}}{C}}CH_3$ **c.** $CH_3\overset{}{C}HCH_2\overset{}{C}HCH_3$
                                                                                              $\quad\quad\; CH_3 \quad\quad CH_3$

47.   **a.**   1-bromohexane (larger, so greater surface area)
           b.   pentyl chloride (greater surface area than the branched compound)
           c.   1-butanol (fewer carbons)
           d.   1-hexanol (forms hydrogen bonds)
           e.   hexane (greater area of contact)
           f.   1-pentanol (forms hydrogen bonds)
           g.   1-bromopentane (bromine larger and more polarizable)
           h.   butyl alcohol (forms hydrogen bonds)
           i.   octane
           j.   isopentyl alcohol (forms stronger hydrogen bonds)
           k.   hexylamine (primary amines form stronger hydrogen bonds than do secondary amines)

48.   Ansaid is more soluble in water. It has a fluoro substituent that can form a hydrogen bond with water. Hydrogen bonding increases its solubility in water.

49.   The student named only one compound correctly.
           a.   4-ethyl-2,2-dimethylheptane   **d.**   2,5-dimethylheptane
           b.   1-bromo-3-methylbutane   **e.**   2,4-dimethylhexane
           c.   correct   **f.**   3-ethyl-2-methylhexane

0. All three compounds are diaxial-substituted cyclohexanes. **B** has the highest energy. Only **B** has a 1,3-diaxial interaction between CH_3 and Cl, which will be greater than a 1,3-diaxial interaction between CH_3 and **H** or between Cl and **H**.

1. The only one is 2,2,3-trimethylbutane.

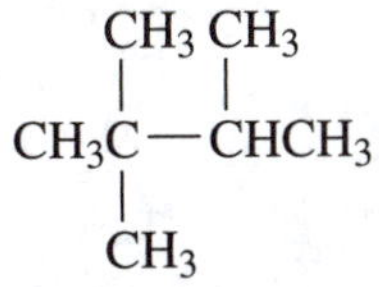

CH3 CH3
CH3C—CHCH3
CH3

or

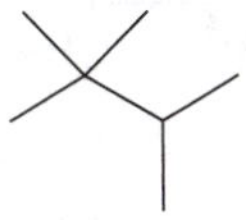

2. a.

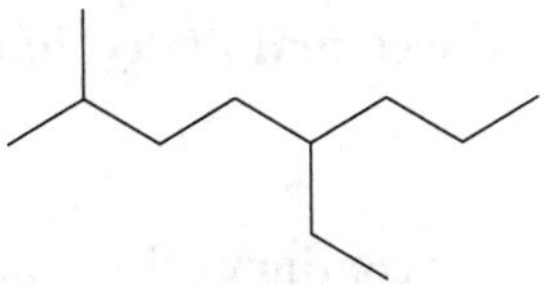

c.

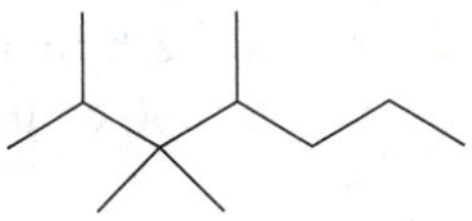

b.

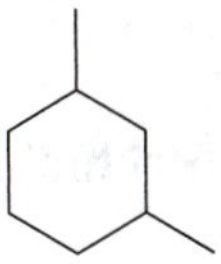

d.

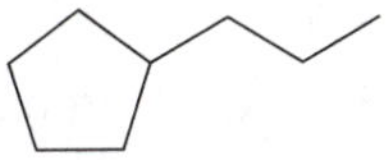

3. “Dibromomethane does not have constitutional isomers” proves that carbon is tetrahedral.

If carbon were flat rather than tetrahedral, dibromomethane would have constitutional isomers, because the two structures shown below would be different as a result of the bromines being 90° apart in one compound and 180° apart in the other compound. Only because carbon is tetrahedral, are the two structures identical.

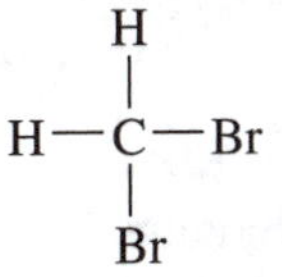

H
H—C—Br
Br

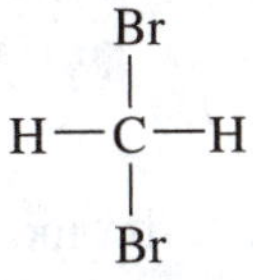

Br
H—C—H
Br

54. First draw the structure, so that you know what groups to put on the bonds in the Newman projections.

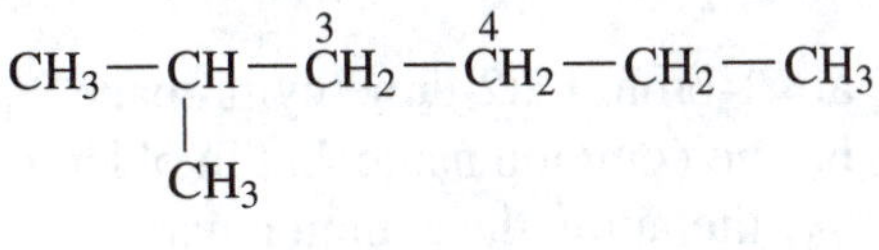

3
4
CH3—CH—CH2—CH2—CH2—CH3
CH3

a.

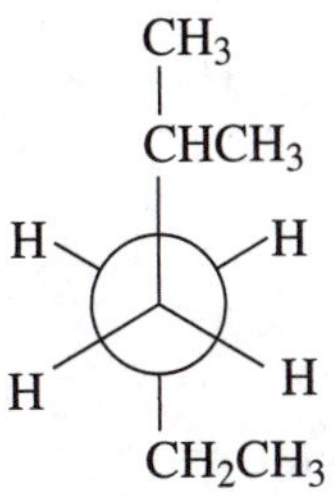

CH3
CHCH3
H
H
H
H
CH2CH3

most stable

b.

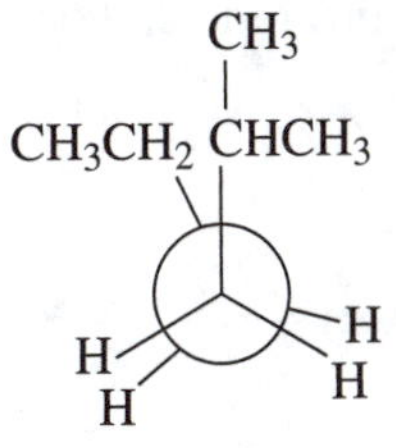

CH3
CH3CH2 CHCH3
H
H
H
H

least stable

c. Rotation can occur about all the C—C bonds. There are six carbon–carbon bonds in the compound, so there are five other carbon–carbon bonds, in addition to the C_3—C_4 bond, about which rotation can occur.

d. Three of the carbon–carbon bonds have staggered conformers that are equally stable, because each is bonded to a carbon with three identical substituents.

$$CH_3 \overset{\Updownarrow}{-} CH - CH_2 - CH_2 - CH_2 \overset{\Updownarrow}{-} CH_3$$
$$\underset{CH_3}{\overset{\Rightarrow}{|}}$$

55. $CH_3CH_2CH_2CH_2CH_2Br$

    a. 1-bromopentane
    b. pentyl bromide

    primary alkyl halide

$CH_3CH_2CH_2CHCH_3$
      |
      Br

    a. 2-bromopentane
    b. no common name

    secondary alkyl halide

$CH_3CH_2CHCH_2CH_3$
    |
    Br

    a. 3-bromopentane
    b. no common name

    secondary alkyl halide

    CH_3
     |
$CH_3CHCH_2CH_2Br$

    a. 1-bromo-3-methylbutane
    b. isopentyl bromide

    primary alkyl halide

    CH_3
      |
$CH_3CH_2CHCH_2Br$

    a. 1-bromo-2-methylbutane
    b. no common name

    primary alkyl halide

    Br
    |
$CH_3CH_2CCH_3$
    |
    CH_3

    a. 2-bromo-2-methylbutane
    b. *tert*-pentyl bromide

    tertiary alkyl halide

    Br
    |
$CH_3CHCHCH_3$
     |
    CH_3

    a. 2-bromo-3-methylbutane
    b. no common name

    secondary alkyl halide

    CH_3
     |
CH_3CCH_2Br
    |
    CH_3

    a. 1-bromo-2,2-dimethylpropane
    b. no common name, but in older literature, the common name neopentyl bromide can be found.

    primary alkyl halide

c. Four isomers are primary alkyl halides.
d. Three isomers are secondary alkyl halides.
e. One isomer is a tertiary alkyl halide.

56. a. butane
    b. 1-bromopropane
    c. 5-propyldecane
    d. 5-ethyl-2-methyloctane

    e. 6-chloro-4-ethyl-3-methyloctane
    f. 2,3-dimethyl-6-propyldecane
    g. 1-methyl-2-propylcyclohexane

57.

a.

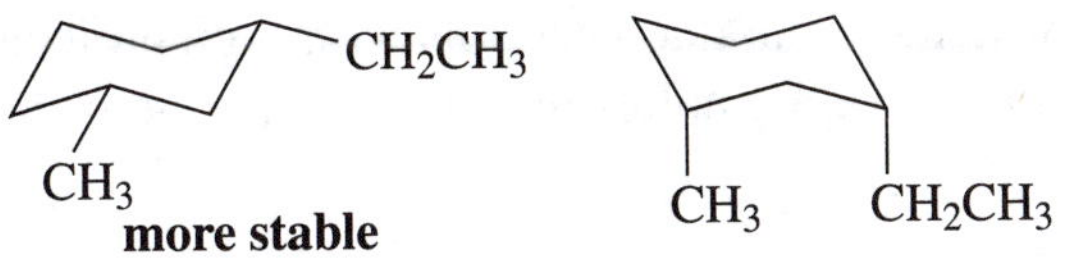

CH₃
more stable

CH₃   CH₂CH₃

b.

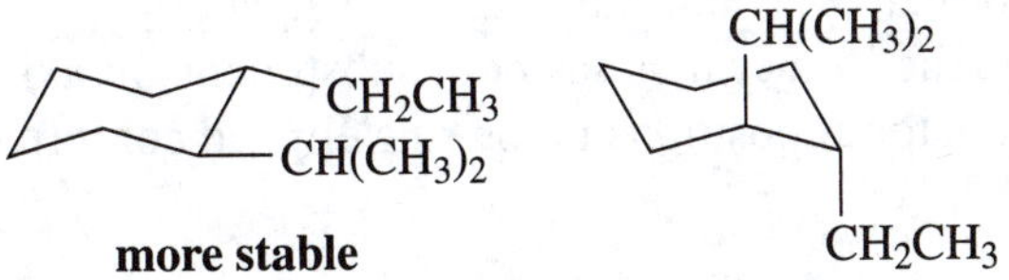

CH₂CH₃
CH(CH₃)₂
more stable

CH(CH₃)₂
CH₂CH₃

c.

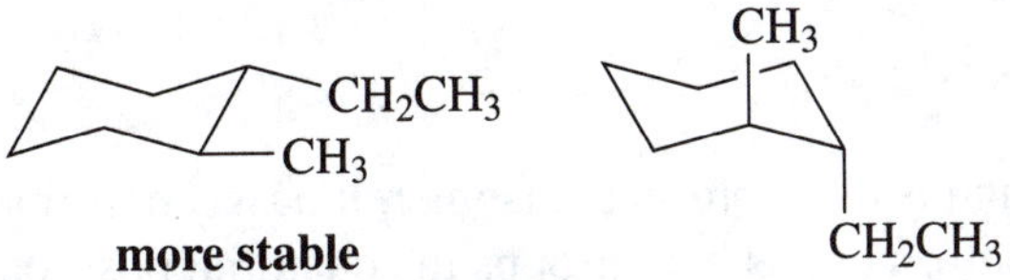

CH₂CH₃
CH₃
more stable

CH₃
CH₂CH₃

d.

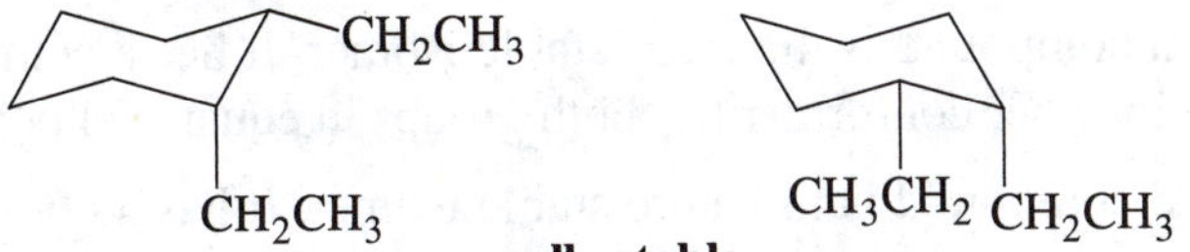

CH₂CH₃
CH₂CH₃

CH₃CH₂   CH₂CH₃
equally stable

e.

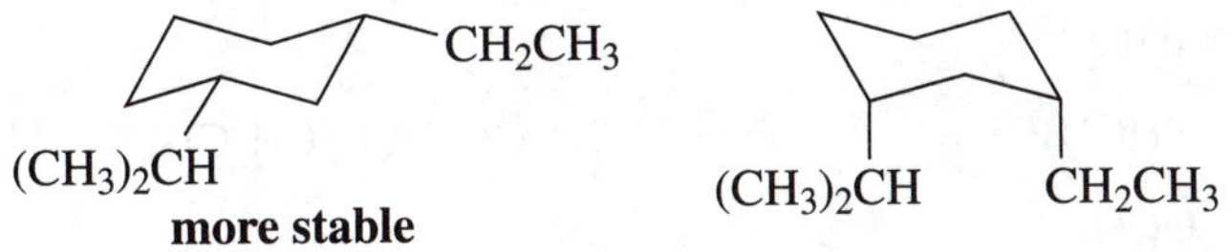

CH₂CH₃
(CH₃)₂CH
more stable

(CH₃)₂CH   CH₂CH₃

f.

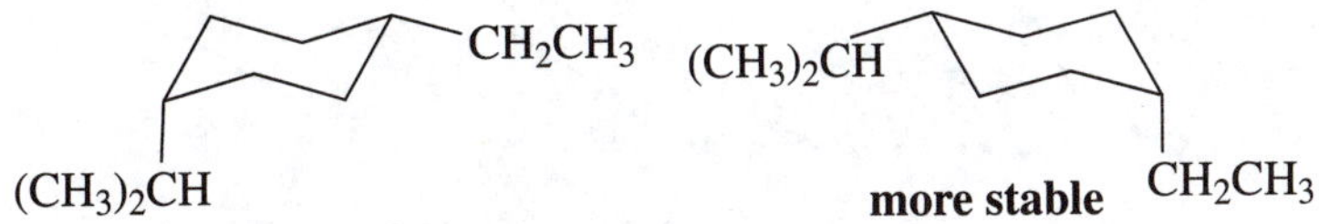

CH₂CH₃
(CH₃)₂CH

(CH₃)₂CH
more stable   CH₂CH₃

58.

#1 CH₃CH₂CH₂CH₂CH₂CH₂CH₃
heptane

#4
$$CH_3$$
$$CH_3CCH_2CH_2CH_3$$
$$CH_3$$
2,2-dimethylpentane

#2
$$CH_3$$
$$CH_3CHCH_2CH_2CH_2CH_3$$
2-methylhexane

#5
$$CH_3$$
$$CH_3CH_2CCH_2CH_3$$
$$CH_3$$
3,3-dimethylpentane

#3
$$CH_3$$
$$CH_3CH_2CHCH_2CH_2CH_3$$
3-methylhexane

#6
$$CH_3 \quad CH_3$$
$$CH_3CH-CHCH_2CH_3$$
2,3-dimethylpentane

59. Alcohols with low molecular weights are more water soluble than alcohols with high molecular weights because, as a result of having fewer carbons, they have a smaller nonpolar component that has to be dragged into water.

60. The more stable isomer is the one that has a conformer with both substituents in equatorial positions. Using the following structure, you can easily determine the isomer that has both substituents in axial positions. That will be the isomer that has both groups in equatorial positions in the ring-flipped conformer.

a. The cis isomer of a 1,3-disubstituted compound is the more stable isomer. It has a conformer with both substituents in axial positions, so its other conformer has both groups in equatorial positions.

b. The trans isomer of a 1,4-disubstituted compound is the more stable isomer. It has a conformer with both substituents in axial positions, so its other conformer has both groups in equatorial positions.

c. The trans isomer of a 1,2-disubstituted compound is the more stable isomer. It has a conformer with both substituents in axial positions, so its other conformer has both groups in equatorial positions.

61. Six ethers have the molecular formula $C_5H_{12}O$.

$CH_3OCH_2CH_2CH_2CH_3$ $CH_3CHCH_2CH_3$ $CH_3CH_2OCH_2CH_2CH_3$
                                                  |
                                                OCH_3

1-methoxybutane   2-methoxybutane   1-ethoxypropane
butyl methyl ether   *sec*-butyl methyl ether   ethyl propyl ether

                                       CH_3
                                        |
CH_3CHCH_3 CH_3COCH_3 $CH_3CHCH_2OCH_3$
     | | |
   OCH_2CH_3 CH_3 CH_3

2-ethoxypropane   2-methoxy-2-methylpropane   1-methoxy-2-methylpropane
ethyl isopropyl ether   *tert*-butyl methyl ether   isobutyl methyl ether

62.   **a.**   **b.**

63.   **a.** one equatorial and one axial   **d.** one equatorial and one axial
          b. both equatorial and both axial   **e.** one equatorial and one axial
          c. both equatorial and both axial   **f.** both equatorial and both axial

64. Both *trans*-1,4-dimethylcyclohexane and *cis*-1-*tert*-butyl-3-methylcyclohexane have a conformer with two substituents in the equatorial position and a conformer with two substituents in the axial position. *cis*-1-*tert*-Butyl-3-methylcyclohexane will have a higher percentage of the conformer with two substituents in the equatorial position, because the bulky *tert*-butyl substituent will have a greater preference for the equatorial position.

65. The most stable conformer has two CH_3 groups in equatorial positions and one in an axial position. (The other conformer would have two CH_3 groups in axial positions and one in an equatorial position.)

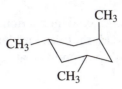

66.
a. 1-bromo-2-propylcyclopentane
b. 5-bromo-2-methyloctane
c. 1,2-dichloro-3-methylpentane
d. 2,3-dimethylpentane

e. 5-butylnonane
f. 7-butyl-3-methylundecane
(undecane is a straight-chain compound with eleven carbons)

67.

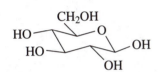

68. Because bromine has a larger diameter than chlorine, one would expect bromine to have a greater preference for the equatorial position as would be indicated by a larger $\Delta G°$. However, Table 9 shows that it has a smaller $\Delta G°$, indicating that it has less preference for the equatorial position than chlorine has. The C—Br bond is longer than the C—Cl bond, which causes bromine to be farther away than chlorine from the other axial substituents. Apparently, the longer bond more than offsets the larger diameter.

69. One of the chair conformers of *cis*-1,3-dimethylcyclohexane has both substituents in equatorial positions, so there are no unfavorable 1,3-diaxial interactions. The other chair conformer has three 1,3-diaxial interactions, two between a CH_3 and an H and one between two CH_3 groups.

   We know that a 1,3-diaxial interaction between a CH_3 and an H is 0.87 kcal/mol. Subtracting 1.7, for the two interactions between a CH_3 and an H, from 5.4 (the energy difference between the two conformers) results in a value of 3.7 kcal/mol for the 1,3-diaxial interaction between the two CH_3 groups.

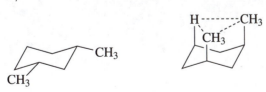

70.   The conformer on the left has two 1,3-diaxial interactions between a CH_3 and an H (2×0.87 kcal/mol for a total strain energy of 1.7 kcal/mol.

The conformer on the right has three 1,3-diaxial interactions, two between a CH_3 and an H (1.7 kcal/mol and one between two CH_3 groups (3.7 kcal/mol; see Problem 69) for a total strain energy of 5.4 kcal/mol Therefore, the conformer on the left will predominate at equilibrium.

Chapter 3 Practice Test

. Name the following compounds:

a.

b. Br

Cl

c.

2. Using Newman projections, draw the following conformers of hexane considering rotation about the C_3—C_4 bond:

a. the most stable of all the conformers

b. the least stable of all the conformers

c. a gauche conformer

3. What are the common and systematic names of the following compounds?

a. $CH_3CH_2CHCH_3$
   |
   Cl

b. $CH_3CHCH_2CH_2CH_2Br$
   |
   CH_3

c. (cyclopentane) Br

4. Label the three compounds in each set in order from highest boiling to lowest boiling.

a. $CH_3CH_2CH_2CH_2CH_2Br$ $CH_3CH_2CH_2Br$ $CH_3CH_2CH_2CH_2Br$

b. $CH_3CH_2CH_2CH_2CH_3$ $CH_3CH_2CH_2CH_2OH$ $CH_3CH_2CH_2CH_2Cl$

   $\quad\quad CH_3\ \ CH_3$
   $\quad\quad\ |\quad\ \ |$
c. CH_3C—CCH_3 $CH_3CH_2CH_2CH_2CH_2CH_2CH_2CH_3$ $CH_3CHCH_2CH_2CH_2CH_2CH_3$
   $\quad\quad\ |\quad\ \ |$ $\quad\quad\ |$
   $\quad\quad CH_3\ \ CH_3$ $\quad\quad CH_3$

5. Name each of the following compounds:

a. $CH_3CHCH_2CH_2CHCH_2CH_3$
   $\quad\ |\quad\quad\quad\ |$
   $\quad CH_3\quad\quad Cl$

b. $CH_3CH_2CHCH_2CH_3$
   $\quad\quad\quad |$
   $\quad\quad CH_2CH_2CH_2CH_3$

c. (cyclopentane) Br ... CH_3

d. $CH_3CHCHCH_2CH_2CH_2Cl$
   $\quad\ |\ \ |$
   $\quad\ \ \ Cl$
   $\quad\quad CH_2CH_3$

6.   Draw the other chair conformer for the following compound:

7.   Which is more stable: *cis*-1-isopropyl-3-methylcyclohexane or *trans*-1-isopropyl-3-methylcyclohexane?

8.   Which of the following has

    a.   the higher boiling point: diethyl ether or butyl alcohol?
    b.   the greater solubility in water: butyl alcohol or pentyl alcohol?
    c.   the higher boiling point: hexane or isohexane?
    d.   the higher boiling point: pentylamine or ethylmethylamine?
    e.   the greater solubility in water: ethyl alcohol or ethyl chloride?

9.   What is the name for each of the following compounds?

    a. $CH_3CHCH_2CH_2Br$ **b.** $CH_3CHCH_2CH_2OH$ **c.** $CH_3CHCH_2CH_2NH_2$
              CH_3 CH_3 CH_3

10.   Draw the more stable conformer of each of the following:

    a.   *cis*-1-*sec*-butyl-4-isopropylcyclohexane

    b.   *trans*-1-*sec*-butyl-4-isopropylcyclohexane

    c.   *trans*-1-*sec*-butyl-3-isopropylcyclohexane

11.   Draw the structure for each of the following:

    a.   a secondary alkyl bromide that has three carbons

    b.   a secondary amine that has three carbons

    c.   an alkane with no secondary hydrogens

    d.   a constitutional isomer of butane

    e.   three compounds with molecular formula C_3H_8O

12.   Name the following compounds:

    a. $CH_3CHCH_2CH_2CHCH_3$ **d.**
            CH_3 CH_2CH_3

    b. $CH_3CHCH_2CH_2CH_2CH_2CH_2Br$ **e.** $CH_3CH_2CH_2CHCH_2CH_2CH_2CH_2CH_3$
            CH_2CH_3 CH_3

    c. $CH_3CHCH_2CHCH_2CH_2CH_3$ **f.** $CH_3CHCH_2CHCH_2CHCH_2CH_3$
          Cl CH_3 CH_3 CH_3 CH_3

Answers to Chapter 3 Practice Test

1. **a.** 3-methyloctane **b.** 1-bromo-2-methylheptane **c.** 3-chlorooctane

2. **a.**
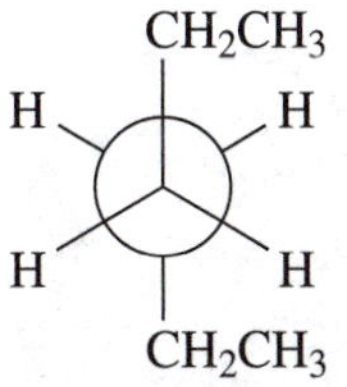

b.
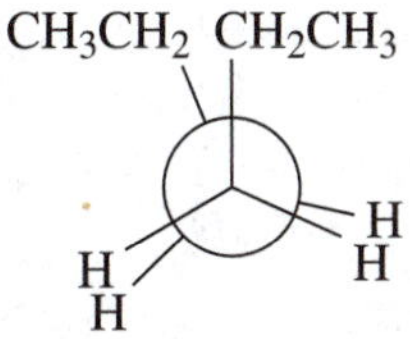

c.
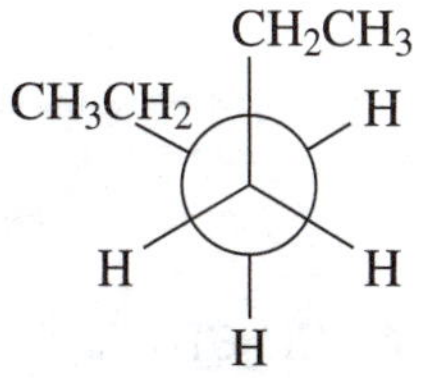

3. **a.** *sec*-butyl chloride, 2-chlorobutane **c.** cyclopentyl bromide, bromocyclopentane
   b. isohexyl bromide, 1-bromo-4-methylpentane

4. **a.** $CH_3CH_2CH_2CH_2CH_2Br$ **1** $CH_3CH_2CH_2Br$ **3** $CH_3CH_2CH_2CH_2Br$ **2**

   b. $CH_3CH_2CH_2CH_2CH_3$ **3** $CH_3CH_2CH_2CH_2OH$ **1** $CH_3CH_2CH_2CH_2Cl$ **2**

   c.
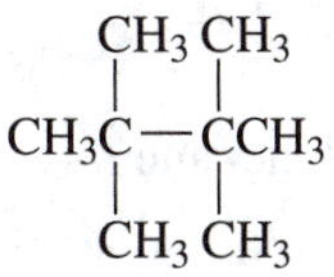

3 $CH_3CH_2CH_2CH_2CH_2CH_2CH_2CH_3$ **1**
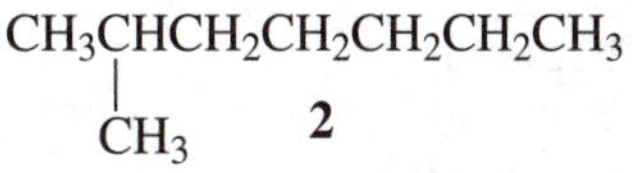

2

5. **a.** 5-chloro-2-methylheptane **c.** 1-bromo-3-methylcyclopentane
   b. 3-ethylheptane **d.** 1,4-dichloro-5-methylheptane

6.
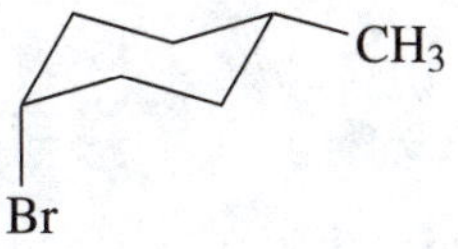

7. *cis*-1-isopropyl-3-methylcyclohexane

8. **a.** butyl alcohol **c.** hexane **e.** ethyl alcohol
   b. butyl alcohol **d.** pentylamine

9. **a.** isopentyl bromide or 1-bromo-3-methylbutane **b.** isopentyl alcohol **c.** isopentylamine

10. **a.**

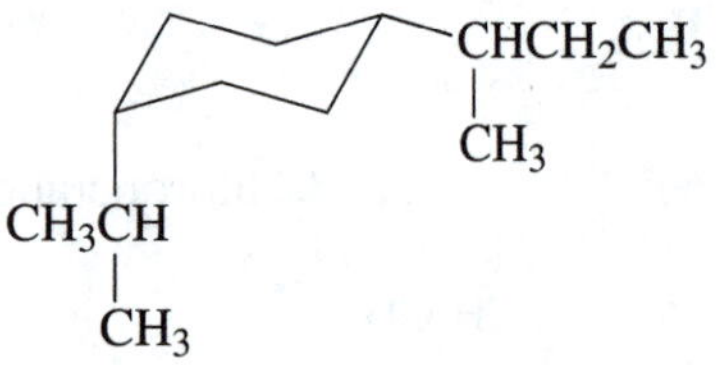

$CHCH_2CH_3$
CH_3
CH_3CH
CH_3

b.

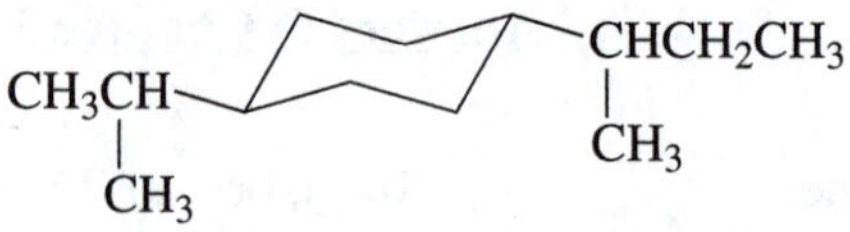

CH_3CH — $CHCH_2CH_3$
CH_3
CH_3

c.

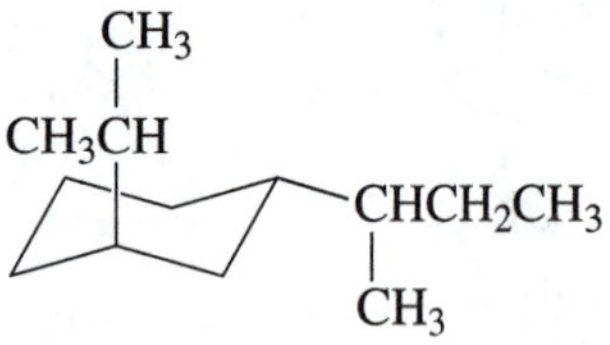

CH_3
CH_3CH

$CHCH_2CH_3$
CH_3

11. **a.** 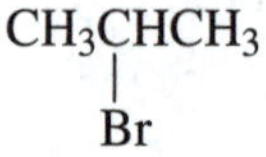 CH_3CHCH_3
Br

d. 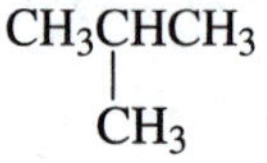 CH_3CHCH_3
CH_3

e. $CH_3CH_2CH_2OH$ 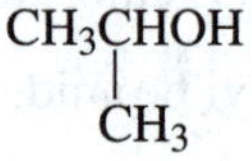 CH_3CHOH $CH_3CH_2OCH_3$
CH_3

b. $CH_3CH_2NHCH_3$

c. 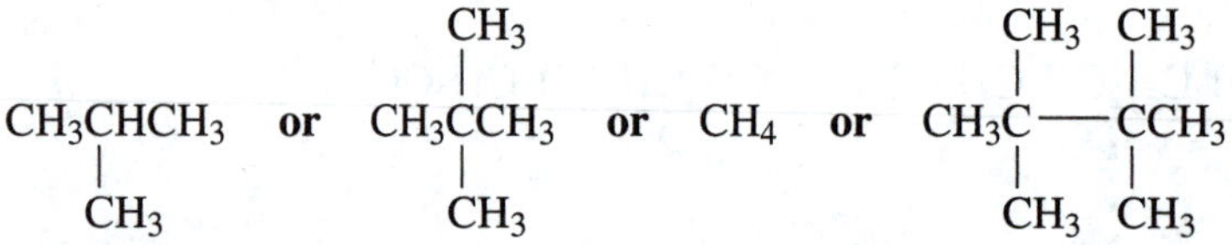 CH_3CHCH_3 **or** CH_3CCH_3 **or** CH_4 **or** CH_3C — CCH_3
CH_3 $\begin{matrix}CH_3\\CH_3\end{matrix}$ $\begin{matrix}CH_3 & CH_3\\CH_3 & CH_3\end{matrix}$

12. **a.** 2,5-dimethylheptane

b. 1-bromo-6-methyloctane

c. 2-chloro-4-methylheptane

d. 4-bromo-2-chloro-1-methylcyclohexane

e. 4-methylnonane

f. 2,4,6-trimethyloctane

Important Terms

achiral	a molecule or object that contains an element (a plane or a point) of symmetry; an achiral molecule is superimposable on its mirror image.
asymmetric center	an atom that is bonded to four different substituents.
chiral	a molecule that has a nonsuperimposable mirror image.
chiral probe	something capable of distinguishing between enantiomers.
chromatography	a separation technique in which the mixture to be separated is dissolved in a solvent and the solution is passed through a column packed with an adsorbent stationary phase.
cis isomer	the isomer with substituents on the same side of a cyclic structure, or the isomer with the hydrogens on the same side of a double bond.
cis–trans isomers	isomers that result from not being able to rotate about a carbon–carbon double bond.
configuration	the three-dimensional structure of a chiral compound. The configuration at a specific atom is designated by R or S.
configurational isomers	stereoisomers that cannot interconvert unless a covalent bond is broken. Cis–trans isomers and isomers with asymmetric centers are configurational isomers.
constitutional isomers (structural isomers)	molecules that have the same molecular formula but differ in the way the atoms are connected.
dextrorotatory	the enantiomer that rotates the plane of polarization of plane-polarized light in a clockwise direction $(+)$.
diastereomers	stereoisomers that are not enantiomers.
E isomer	the isomer with the high-priority groups on opposite sides of the double bond.
enantiomers	nonsuperimposable mirror-image molecules.
isomers	nonidentical compounds with the same molecular formula.
levorotatory	the enantiomer that rotates the plane of polarization of plane-polarized light in a counterclockwise direction $(-)$.
meso compound	a compound that possesses asymmetric centers and a plane of symmetry; it is achiral, because it has a plane of symmetry.

123

observed rotation	the amount of rotation observed in a polarimeter.
optically active	rotates the plane of polarization of plane-polarized light.
optically inactive	does not rotate the plane of polarization of plane-polarized light.
perspective formula	a method of representing the spatial arrangement of groups bonded to an asymmetric center. Two adjacent bonds are drawn in the plane of the paper; a solid wedge depicts a bond that projects out of the plane of the paper toward the viewer and a hatched wedge depicts a bond that projects back from the paper away from the viewer.
plane-polarized light	light that oscillates in a single plane.
plane of symmetry	an imaginary plane that bisects a molecule into pieces that are a pair of mirror images.
polarimeter	an instrument that measures the rotation of the plane of polarization of plane polarized light.
polarized light	light that oscillates in only one plane.
racemic mixture (racemate)	a mixture of equal amounts of a pair of enantiomers.
***R* configuration**	after assigning relative priorities to the four groups bonded to an asymmetric center, if the lowest-priority group is pointing away from the viewer in a perspective formula, an arrow drawn from the highest-priority group to the next highest-priority group and then to the next highest-priority group goes in a clockwise direction.
resolution of a racemic mixture	separation of a racemic mixture into the individual enantiomers.
***S* configuration**	after assigning relative priorities to the four groups bonded to an asymmetric center, if the lowest-priority group is pointing away from the viewer in a perspective formula, an arrow drawn from the highest-priority group to the next highest-priority group and then to the next highest-priority group goes in a counterclockwise direction.
specific rotation	the amount of rotation that will be observed for a compound with a concentration given in grams per 100 mL of solution (or g/mL if it is a pure liquid) in a sample tube 1.0 dm long.
stereocenter (stereogenic center)	an atom at which the interchange of two groups produces a stereoisomer.
stereoisomers	isomers that differ in the way the atoms are arranged in space.
trans isomer	the isomer with substituents on the opposite sides of a cyclic structure, or the isomer with the hydrogens on the opposite sides of a double bond.
***Z* isomer**	the isomer with the high-priority groups on the same side of the double bond.

olutions to Problems

a. $CH_3CH_2CH_2OH$ $CH_3\underset{\underset{\textstyle CH_3}{|}}{C}HOH$ $CH_3CH_2OCH_3$

b. There are seven constitutional isomers with molecular formula $C_4H_{10}O$.

$CH_3CH_2CH_2CH_2OH$ $CH_3\underset{\underset{\textstyle CH_3}{|}}{C}HCH_2OH$ $CH_3\underset{\underset{\textstyle CH_3}{|}}{\overset{\overset{\textstyle CH_3}{|}}{C}}OH$ $CH_3\underset{\underset{\textstyle OH}{|}}{C}HCH_2CH_3$

$CH_3CH_2OCH_2CH_3$ $CH_3OCH_2CH_2CH_3$ $CH_3O\underset{\underset{\textstyle CH_3}{|}}{C}HCH_3$

a. Br–⬡–Cl   Br–⬡····Cl   **b.**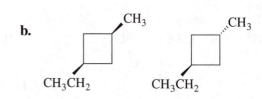

a. 5   **b.** 4   **c.** 4   **d.** 6

a. 1 and 3

b. **1.** $\underset{\text{cis}}{\underset{\displaystyle H_3C\quad\quad CH_2CH_2CH_3}{\underset{\displaystyle H\quad\quad\quad H}{C=C}}}$ $\underset{\text{trans}}{\underset{\displaystyle H_3C\quad\quad H}{\underset{\displaystyle H\quad\quad CH_2CH_2CH_3}{C=C}}}$ **3.** $\underset{\text{cis}}{\underset{\displaystyle H_3C\quad\quad CH_3}{\underset{\displaystyle H\quad\quad\quad H}{C=C}}}$ $\underset{\text{trans}}{\underset{\displaystyle H_3C\quad\quad H}{\underset{\displaystyle H\quad\quad CH_3}{C=C}}}$

1. ⟋⟍⟋⟍ trans   ⟋⟍⟋⟍ cis   **3.** ⟋⟍ trans   ⟋⟍ cis

2. ⟋⟍⟋

4. ⟋⟍

$CH_3CH_2CH_2CH=CH_2$ $CH_3\underset{\underset{\textstyle }{}}{\overset{\overset{\textstyle CH_3}{|}}{C}}=CHCH_3$ $CH_3CH_2\overset{\overset{\textstyle CH_3}{|}}{C}=CH_2$

a. —I > —Br > —OH > —CH$_3$

b. —OH > —CH$_2$Cl > —CH=CH$_2$ > —CH$_2$CH$_2$OH

8.   The high-priority groups are on the same side of the double bond, so tamoxifen has the Z configuration.

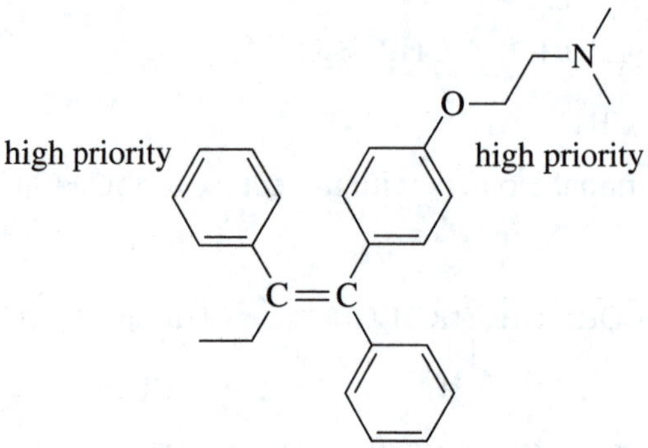

9.

a.

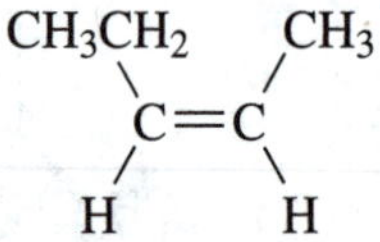

Z

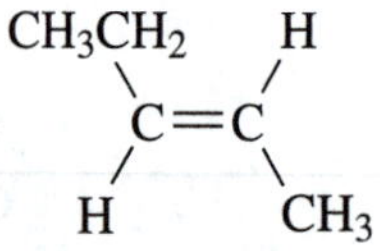

E

b.

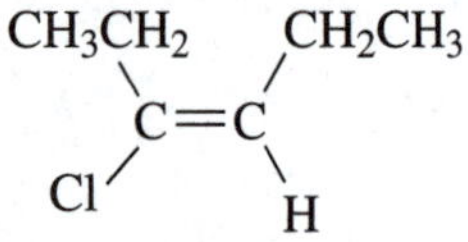

E

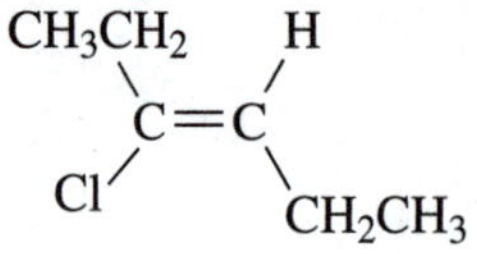

Z

c.

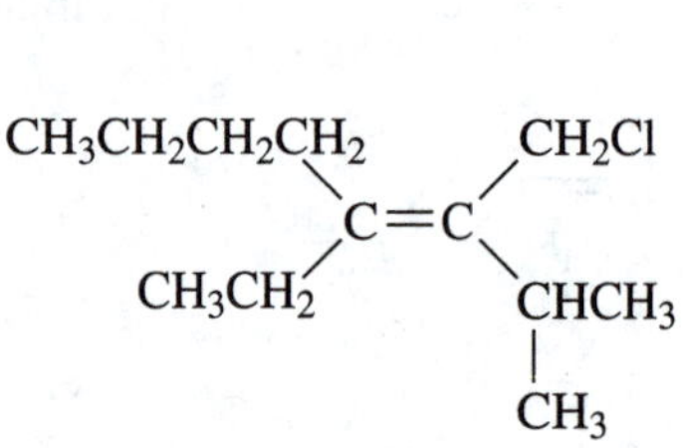

Z

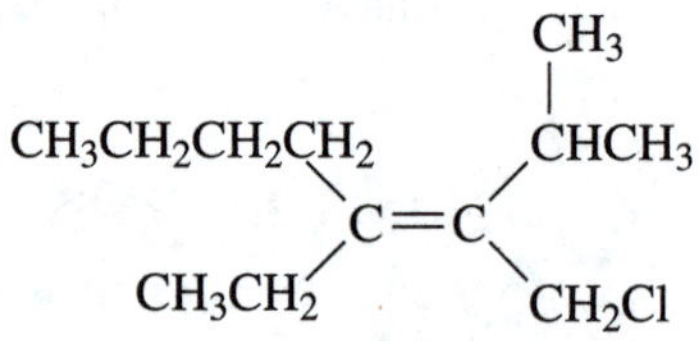

E

d.

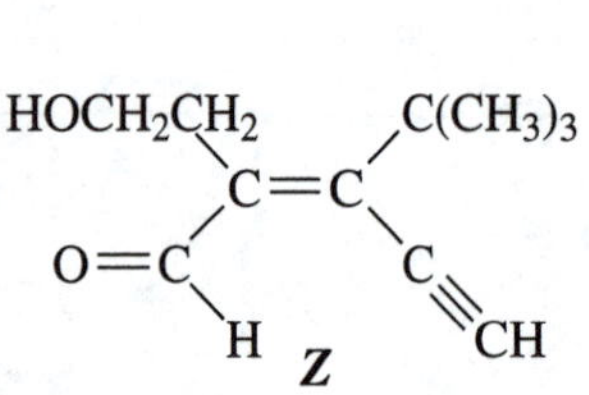

Z

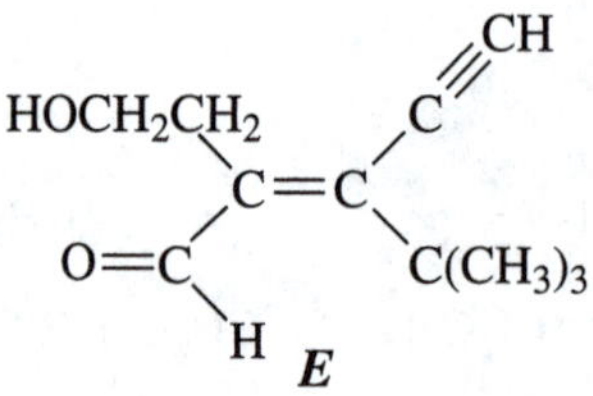

E

. **a.**

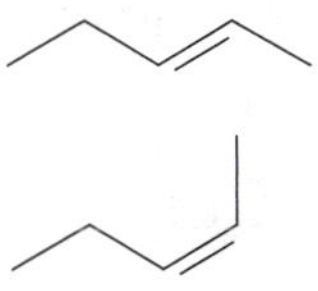

b.

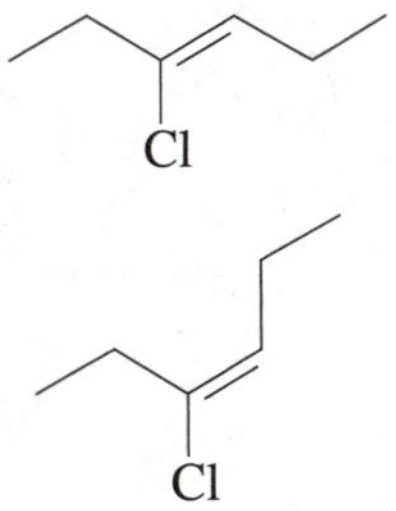
Cl
Cl

c.

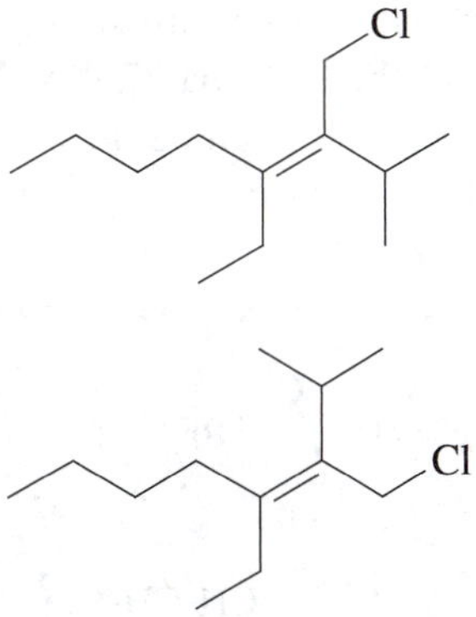
Cl
Cl

d.

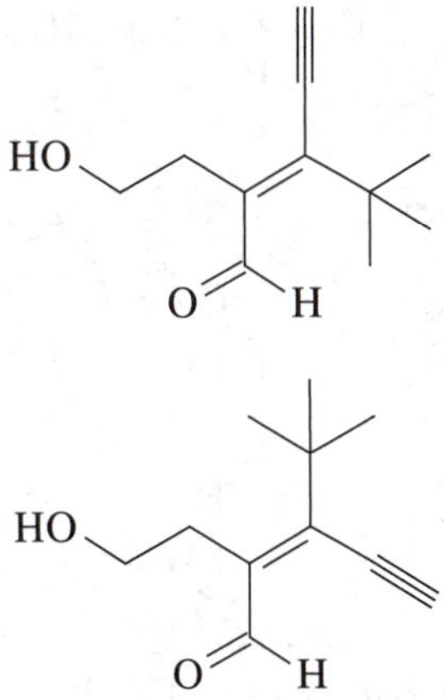
HO
O
H
HO
O
H

1. **a.** (*E*)-2-heptene **b.** (*Z*)-3,4-dimethyl-2-pentene **c.** (*Z*)-1-chloro-3-ethyl-4-methyl-3-hexene

2.

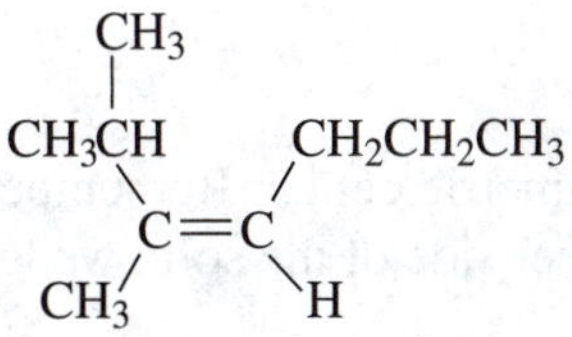
CH3
CH3CH
CH2CH2CH3
C=C
CH3
H

3. **b** and **d** are chiral.
a and **c** are each superimposable on its mirror image. These, therefore, are achiral.

4. **a, c,** and **f** have asymmetric centers.

5. Solved in the text.

6. **a, c,** and **f,** because in order to be able to exist as a pair of enantiomers, the compound must have an asymmetric center.

17. Draw the first enantiomer with the groups in any order you want. Then draw the second enantiomer b...
drawing the mirror image of the first enantiomer. Your answer might not look exactly like the ones show...
below, because the first enantiomer can be drawn with the four groups on any of the four bonds. The nex...
one is the mirror image of the first one.

a.

CH_3
Br—C—CH_2OH $HOCH_2$—C—Br
   H   H

c.

OH
$(CH_3)_2CH$—C—CH_3 CH_3—C—$CH(CH_3)_2$
          H   H
OH

b.

CH_2CH_2Cl
CH_3CH_2—C—CH_3 CH_3—C—CH_2CH_3
       H   H
CH_2CH_2Cl

18. Solved in the text.

19.   **a.** —CH_2OH ①   —CH_3 ③   —H ④   —CH_2CH_2OH ②

   b. —CH_2Br ②   —OH ①   —CH_3 ④   —CH_2OH ③

   c. —$CH(CH_3)_2$ ②   —CH_2CH_2Br ③   —Cl ①   —$CH_2CH_2CH_2Br$ ④

20.   **a.** (R)-2-bromobutane   **b.** (R)-1,3-dichlorobutane

21.   **a.** Solved in the text.
   b. R
   c. To determine the configuration, first add the fourth bond to the asymmetric center. Remember that
   cannot be drawn between the two solid bonds. (It can be drawn on either side of the solid wedge.)

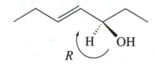

   d. R.

22. The easiest way to determine whether two compounds are identical or enantiomers is to determine thei...
configurations: If both are R (or both are S), they are identical. If one is R and the other is S, they ar...
enantiomers.
   a. identical   **b.** enantiomers   **c.** enantiomers

23.   **a.**

Cl
CH_3CH_2—C—H
       CH_3

   b.

Br
CH_3CH_2—C—H
       CH_2Br

4.   **a.** levorotatory   **b.** dextrorotatory

5.   **a.** *S*   **b.** *R*   **c.** *R*   **d.** *S*

6.   Solved in the text.

7.   We see that the (*R*)-alkyl halide reacts with HO⁻ to form the (*R*)-alcohol. We are told that the product (the (*R*)-alcohol) is (−). We can, therefore, conclude that the (+)-alcohol has the *S* configuration.

8. $$\text{specific rotation} = \frac{\text{observed rotation (degrees)}}{\text{concentration (g in 100 mL)} \times \text{length (dm)}}$$

$$[\alpha] = \frac{+13.4°}{\dfrac{1\text{ g}}{100\text{ mL}} \times 2\text{ dm}} = \frac{+13.4°}{2} = +6.7$$

9.   **a.** −24   **b.** 0

10.   From the data given, you cannot determine the configuration of naproxen.

11.   **a.** enantiomers
         b. identical compounds (Therefore, they are not stereoisomers.)
         c. diastereomers

12.   **a.** Find the sp^3 carbons that are bonded to four different substituents; these are the asymmetric centers. Cholesterol has eight asymmetric centers. They are indicated by arrows.

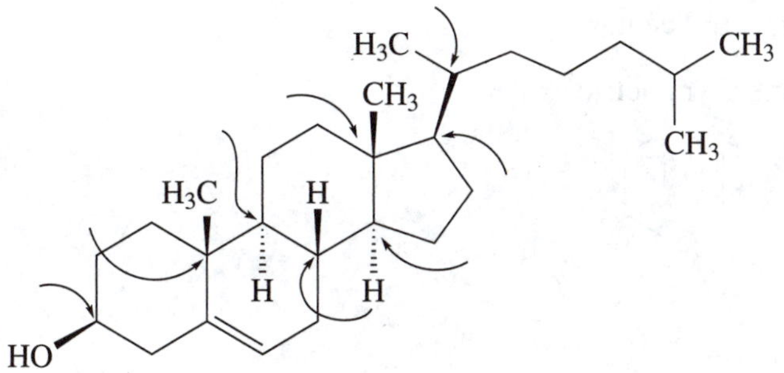

         b. $2^8 = 256$
         Only the stereoisomer shown above is found in nature.

13.   Your perspective formulas may not look exactly like the ones drawn here, because you can draw the first one with the groups attached to any bonds you want. Just make certain that the second one is a mirror image of the first one.

         a. Leucine has one asymmetric center, so it will have two stereoisomers.

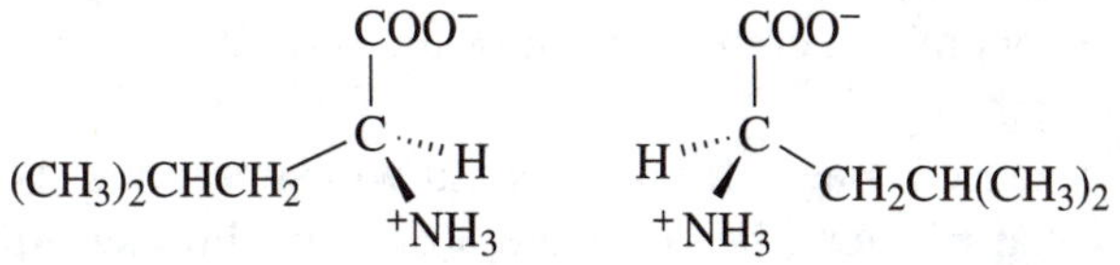

b. Isoleucine has two asymmetric centers, so it has four stereoisomers. To draw a perspective **formula** with two asymmetric centers, first draw the three solid lines of the perspective formula. **Add** a solid and a hatched wedge at each carbon, making sure that the solid wedge is below the hatched wedge.

Again your perspective formulas may not look exactly like the ones drawn here.

To make sure you have all four, determine the configuration of each of the asymmetric centers. You should have *R,R*, *S,S*, *R,S*, and *S,R*. Now put the substituents on the bonds in any order. Then draw the mirror image of the structure you just drew. To get the other pair of enantiomers, switch one pair on one of the asymmetric centers and then draw its mirror image.

34.   **B** and **D** have no symmetric centers.

A and **C** each have one asymmetric center.

A   **C**

E has two asymmetric centers.

E

35.   **a.** One asymmetric center has the same configuration in both compounds and the other asymmetric center has the opposite configuration in both, so the compounds are diastereomers.
b. Both asymmetric centers in one compound have the opposite configuration in the other, so the compounds are enantiomers.
c. They are identical, because if one is flipped over, it will superimpose on the other.
d. They are constitutional isomers, because the atoms are hooked up differently; one compound is 1-chloro-2-methylcyclopentane and the other is 1-chloro-3-methylcyclopentane.

36. a.

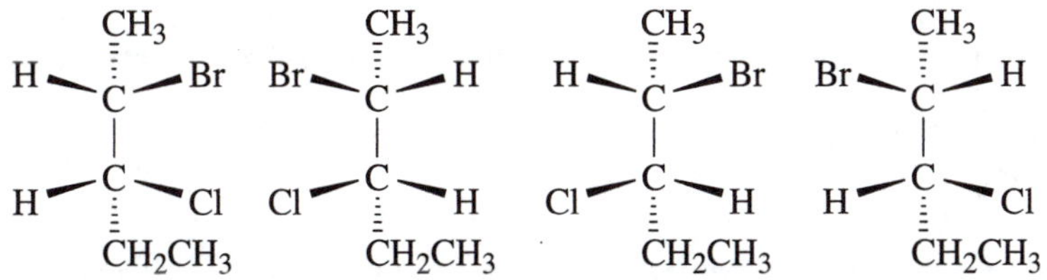

CH₃ — H—C—Br, H—C—Cl, CH₂CH₃ | CH₃ — Br—C—H, Cl—C—H, CH₂CH₃ | CH₃ — H—C—Br, Cl—C—H, CH₂CH₃ | CH₃ — Br—C—H, H—C—Cl, CH₂CH₃

b.

(cyclopentane rings with Br and Cl substituents)

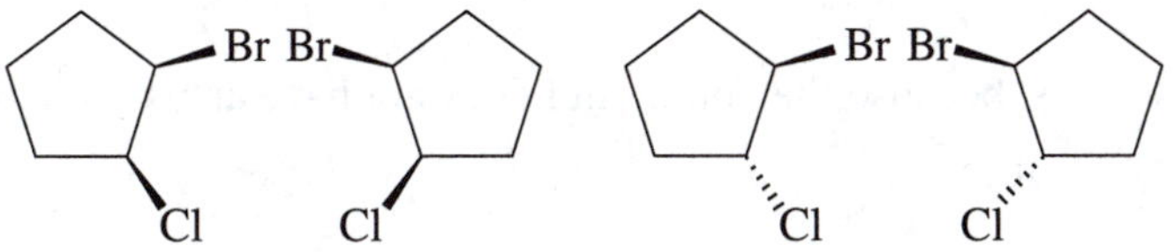

37.   **a, b,** and **d** have more than one diastereomer; **c** has only one diastereomer.

To draw a diastereomer, switch any one pair of substituents bonded to one of the asymmetric centers. Because any one pair can be switched, your diastereomer may not be the one drawn here, unless you happened to switch the same pair that is switched here.

a.

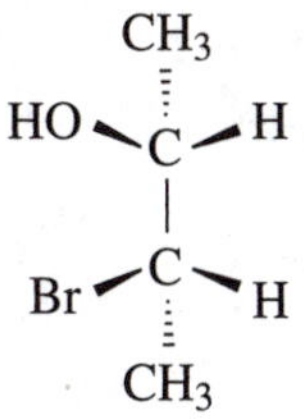

CH₃ — HO—C—H, Br—C—H, CH₃

b.

HO (cyclopentane ring) CH₃

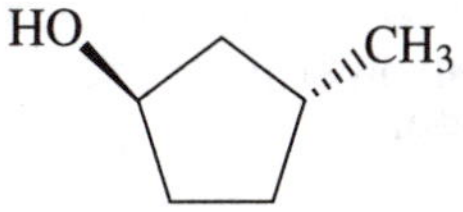

38.   **A** = identical   **B** = enantiomer   **C** = diastereomer   **D** = identical

39.   **B, D,** and **F,** because each has two asymmetric centers and the same four groups bonded to each of the asymmetric centers.
A has two asymmetric centers, but it does not have a stereoisomer that is a meso compound because it does not have the same four groups bonded to each of the asymmetric centers.
C and E do not have a stereoisomer that is a meso compound, because they do not have asymmetric centers.

40.   Solved in the text.

41. **a.**

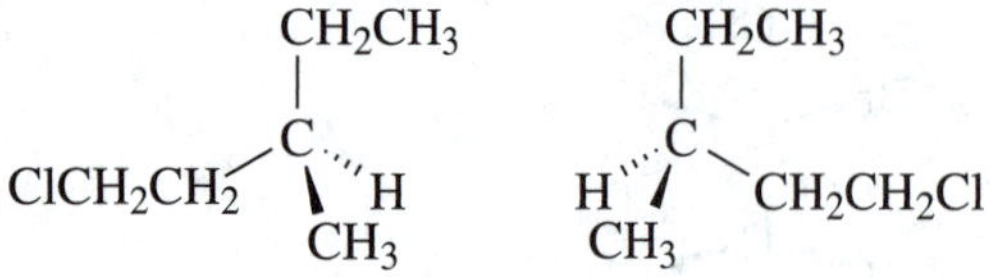

R   S

b.

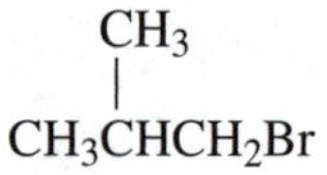

CH_3CHCH_2Br No stereoisomers, because the compound does not have an asymmetric center.

c.

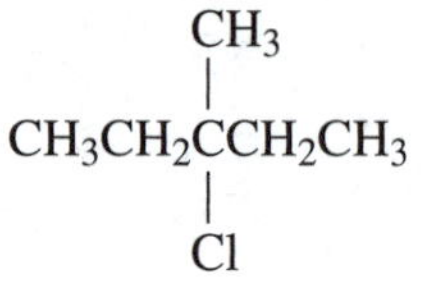

$CH_3CH_2CCH_2CH_3$ No stereoisomers, because the compound does not have an asymmetric center.

d.

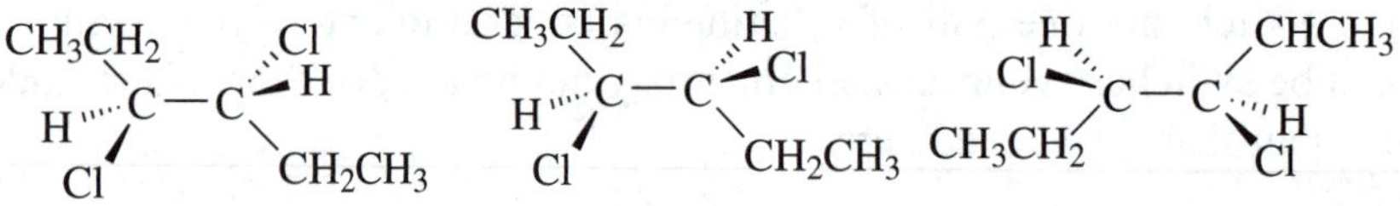

S   R   S   S   R   R

a meso compound

e.

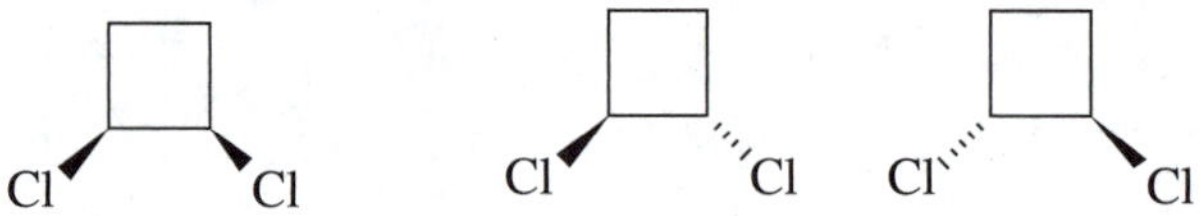

The cis stereoisomer is
a meso compound.

f.

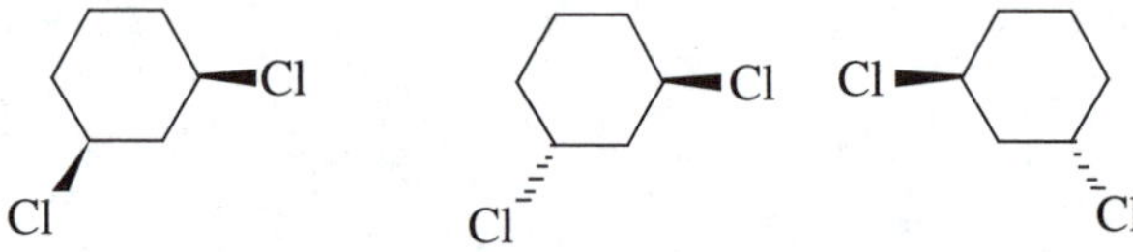

The cis stereoisomer is
a meso compound.

g.

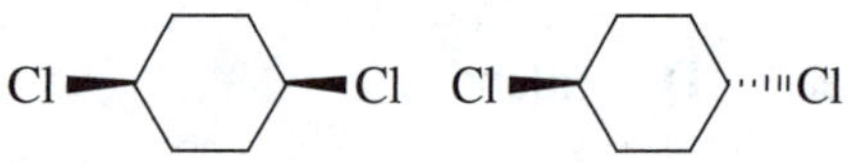

This compound does not have any asymmetric centers,
so it has only cis–trans isomers.

h.

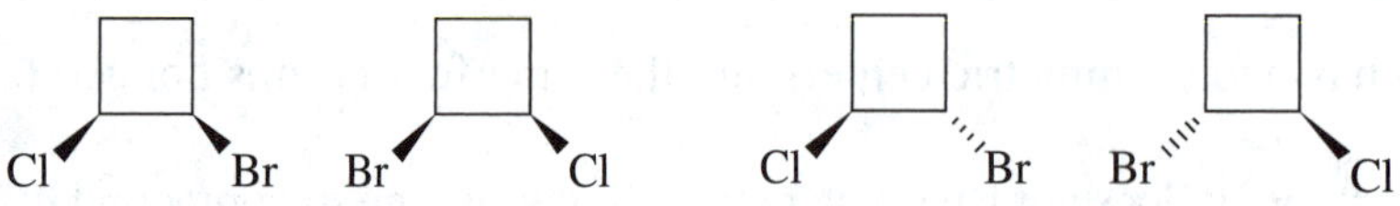

42.   (+)-Limonene has the *R* configuration, so it is the enantiomer found in oranges and lemons.

43.   An asymmetric center is an atom attached to four different atoms (or groups). CHFBrCl is the only one
with an asymmetric center.

4.

a.

Br
|
CH$_3$CHCH$_2$—C⸻H   +   H⸻C—CH$_2$CHCH$_3$
| \   / |
CH$_3$ CH$_3$ CH$_3$ CH$_3$

b.

CH$_3$ CH$_3$ CH$_3$ CH$_3$
| | | |
H—C—Br   Br—C—H   H—C—Br   Br—C—H
| | | |
CH$_2$ CH$_2$ CH$_2$ CH$_2$
| | | |
H—C—Cl   Cl—C—H   Cl—C—H   H—C—Cl
| | | |
CH$_3$ CH$_3$ CH$_3$ CH$_3$

c.

CH$_3$CH$_2$ CH$_2$CH$_2$CH$_3$ CH$_3$CH$_2$ H
\   /   \   /
C=C   C=C
/   \   /   \
H   H   H   CH$_2$CH$_2$CH$_3$

d.

e.

f.

CH$_2$CH$_2$CH$_3$ CH$_2$CH$_2$CH$_3$
| |
C   C
/   \   /   \
CH$_3$ H   H   CH$_3$
I   I

g.

CH
‖
CH$_3$CH$_2$CCH$_2$CH   no stereoisomers
|
CH$_3$

h.

CH$_3$ CH$_3$
| |
CH$_2$=CH—C⸻H   H⸻C—CH=CH$_2$
\   /
Cl   Cl

45.   CH$_3$CH=CHCH$_2$CH$_3$ CH$_2$=CHCH$_2$CH$_2$CH$_3$

CH$_3$
|
CH$_3$C=CHCH$_3$

CH$_3$
|
CH$_3$CHCH=CH$_2$

CH$_3$
|
CH$_3$CH$_2$C=CH$_2$

2 stereoisomers   no stereoisomers   no stereoisomers   no stereoisomers   no stereoisomers
[cis and trans]

46. **a.** (*E*)-1-bromo-2-chloro-2-fluoro-1-iodoethene   **c.** (*Z*)-2-bromo-1-chloro-1-fluoroethene
  b. (*S*)-2,3-dimethylpentane   **d.** (*S*)-1-chloro-3-methylpentane

47.

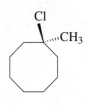

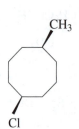

1-chloro-1-methylcyclooctane   *cis*-1-chloro-5-methylcyclooctane   *trans*-1-chloro-5-methylcyclooctane

48.   Mevacor has eight asymmetric centers, which are indicated by the arrows.

49.   Only the compound on the far right is optically active, because it is the only one that has one or mor
  asymmetric centers.

50. **a.** diastereomers   **c.** diastereomers   **e.** enantiomers
  b. identical   **d.** enantiomers   **f.** identical

  Diastereomers are stereoisomers that are not enantiomers. Therefore, cis–trans isomers are diastereomers.

51. **a.** $-CH=CH_2$ > $-CH(CH_3)_2$ > $-CH_2CH_2CH_3$ > $-CH_3$

  b. $-OH$ > $-NH_2$ > $-CH_2OH$ > $-CH_2NH_2$

  c. $-Cl$ > $-C(=O)CH_3$ > $-C\equiv N$ > $-CH=CH_2$

52.   Compounds **a** and **c** have a stereoisomer that is achiral.

53.

enantiomers = 1 and 2; 3 and 4
diastereomers = 1 and 3; 1 and 4; 2 and 3; 2 and 4

54.   **a.** *Z*   **b.** *E*   **c.** *E*   **d.** *Z*   **e.** *E*   **f.** *E*

55. Switch the CH₃ and H so that the H is on the hatched wedge. An arrow drawn from the highest priority to the next highest priority is clockwise, so the compound with the switched pair has the *R* configuration. Therefore, the original molecule is (*S*)-naproxen.

(*S*)-naproxen

(*R*)-naproxen

56. $[\alpha] = \dfrac{\alpha}{l \times c} = \dfrac{+180°}{[2.0 \text{ dm}][15 \text{ g}/100 \text{ mL}]} = -6.0$

57. The only way that *R* and *S* are related to (+) and (−) is that if one configuration (say, *R*) is (+), the other one is (−).

Because some compounds with the *R* configuration are (+) and some are (−), there is no way to determine whether a particular *R* enantiomer is (+) or (−) without putting the compound in a polarimeter or finding out whether someone else has previously determined how the compound rotates the plane of polarization of plane-polarized light.

58.   **a.** identical   **c.** diastereomers   **e.** enantiomers
          b. constitutional isomers   **d.** constitutional isomers   **f.** identical

59.   **a.** *R*   **b.** [structures] = *R*   **c.** [structures] = *S*

60.   **a.** If you rotate one of the structures 180° (keeping it in the same plane), you can see that they are identical.
          b. Because the two molecules have the opposite configuration at both asymmetric centers, they are enantiomers.

61.

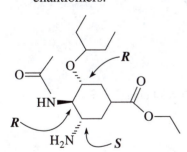

62.   **a.** diastereomers (the configuration of all the symmetric centers is not the same in both and not opposite in both)
          b. identical (by rotating one compound, you can see that it is superimposable on the other)
          c. constitutional isomers
          d. diastereomers (the configuration of all the stereoisomers is not the same in both and not opposite in both)

63. a.

Br
|
C
/ \
Cl   H
CH₂CH₂CH₃

c.

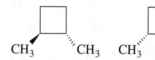

Br   Br   or   Br   Br

b.

CH₃   CH₃

64.   Because of the double bond, the compound has cis and trans isomers. Because of the asymmetric center the cis and trans isomers each one has a pair of enantiomers.

CH₃   CH₃   CH₃   CH₃
| | | |
H₃C   C‴H   H‴C   CH₃   H   C‴H   H‴C   H
\   OH   HO   /   \   OH   HO   /
C=C   C=C   C=C   C=C
/   \   /   \   /   \   /   \
H   H   H   H   CH₃   H   H   CH₃

enantiomers for the cis stereoisomer   enantiomers for the trans stereoisomer

65.   a. and b.

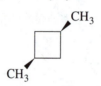

CH₂CH₃

ethylcyclobutane

CH₃
CH₃

1,1-dimethylcyclobutane

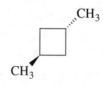

CH₃   CH₃

cis-1,2-dimethylcyclobutane

CH₃   CH₃   CH₃   CH₃

trans-1,2-dimethylcyclobutane

CH₃

CH₃

cis-1,3-dimethylcyclobutane

CH₃

CH₃

trans-1,3-dimethylcyclobutane

c. 1.   ethylcyclobutane
1,1-dimethylcyclobutane
1,2-dimethylcyclobutane
1,3-dimethylcyclobutane

2.   the three isomers of 1,2-dimethylcyclobutane
the two isomers of 1,3-dimethylcyclobutane

3. *cis*- and *trans*-1,2-dimethylcyclobutane
    cis- and *trans*-1,3-dimethylcyclobutane

4. the two trans stereoisomers of 1,2-dimethylcyclobutane

5. all the isomers except the two trans stereoisomers of 1,2-dimethylcyclobutane

6. *cis*-1,2-dimethylcyclobutane
    (Note: *cis*-1,3-dimethylcyclobutane is not a meso compound, because it does not have any asymmetric centers.)

7. the two trans stereoisomers of 1,2-dimethylcyclobutane

8. *cis*-1,3-dimethylcyclobutane and *trans*-1,3-dimethylcyclobutane
    cis-1,2-dimethylcyclobutane and either of the enantiomers of *trans*-1,2-dimethylcyclobutane

66.   **a.** 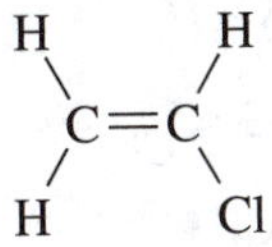   **b.**

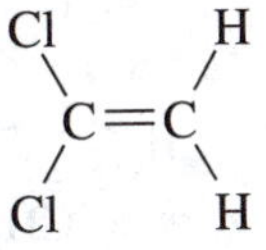

67.   **a.**   Because there are two asymmetric centers, there are four possible stereoisomers.

b.

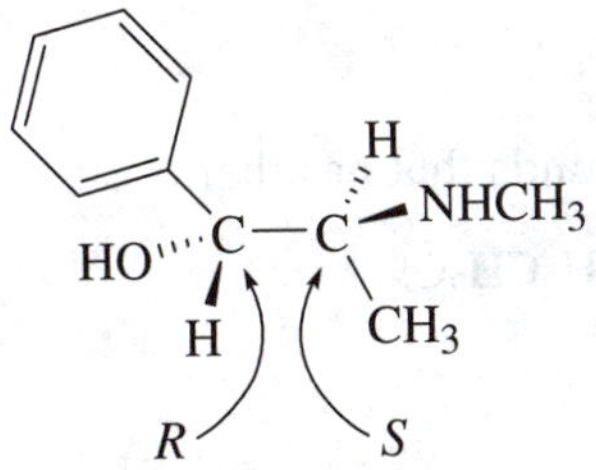

68.   **a.**

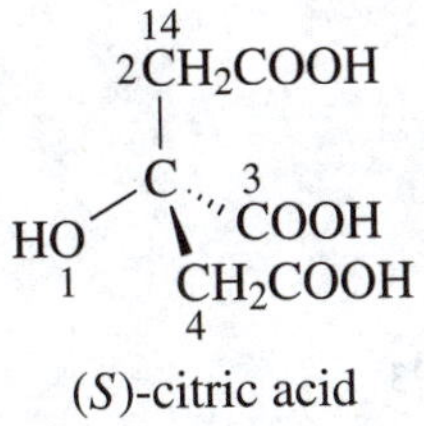

(*S*)-citric acid

b.   The product of the reaction will be achiral, because if it does not have a ^{14}C label, the two CH_2COOH groups will be identical, so it will not have an asymmetric center.

69.

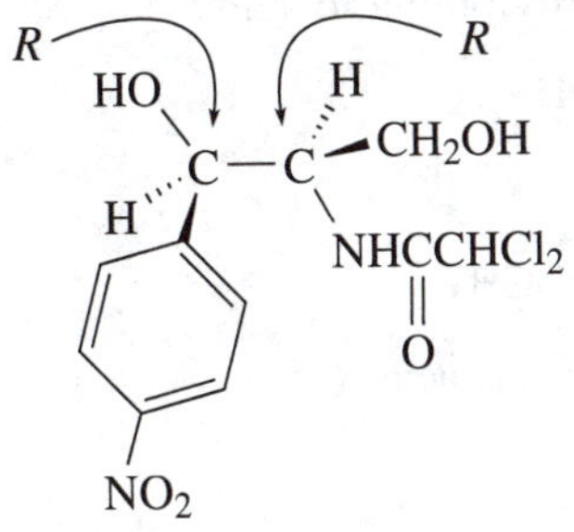

Chapter 4 Practice Test

1. Are the following compounds identical or a pair of enantiomers?

   a.

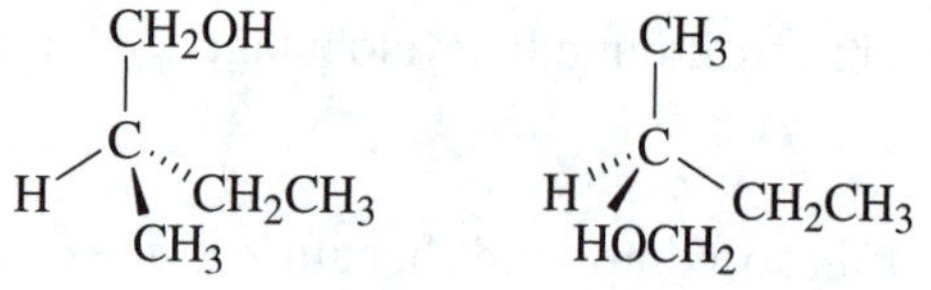

   b.

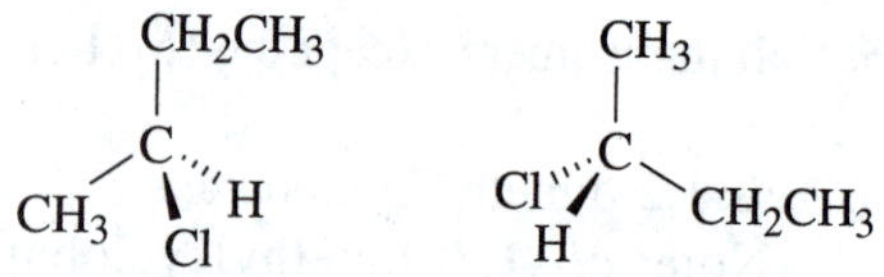

2. 100 mL of a solution containing 0.80 g of a compound rotates the plane of polarized light $-4.8°$ in a polarimeter with a 2-dm sample tube. What is the specific rotation of the compound?

3. Which are meso compounds?

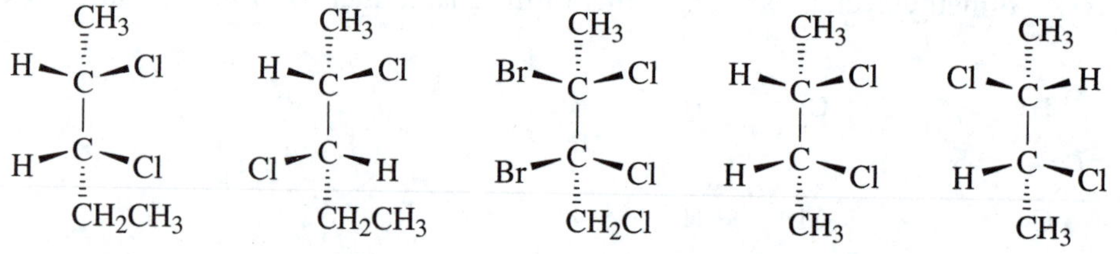

   A   B   C   D   E

4. Draw all the constitutional isomers with molecular formula C_4H_9Cl.

5. Draw all the possible stereoisomers for each of the following compounds that has them.

   a.

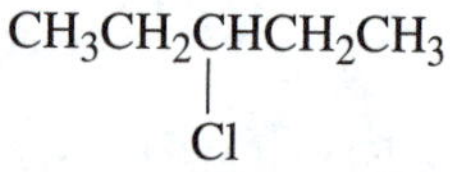

   d. $CH_3CH_2CHCH_2CH_2Cl$
                |
                Cl

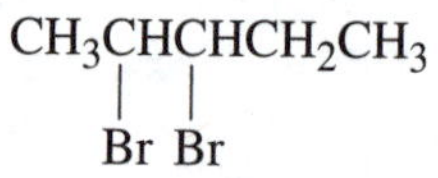

   b. $CH_3CHCHCH_2CH_3$
              | |
              Br Br

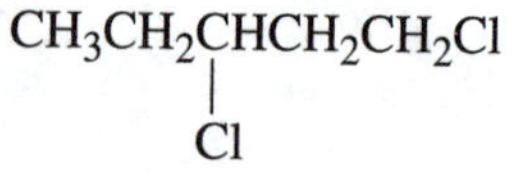

   e.

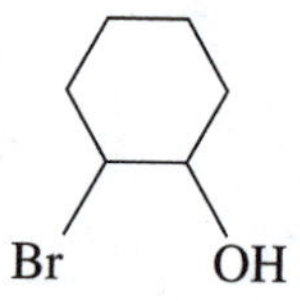

   c. $CH_3CH_2CHCH_2CH_3$
                |
                Cl

   f. HO—⬡—CH_3

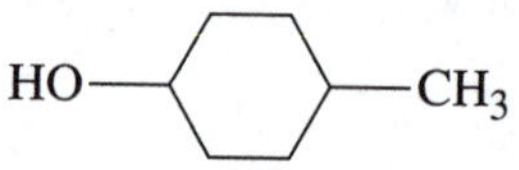

6. (R)-$(-)$-2-Methyl-1-butanol can be oxidized to $(+)$-2-methylbutanoic acid without breaking any of the bonds to the asymmetric center. What is the configuration of $(-)$-2-methylbutanoic acid?

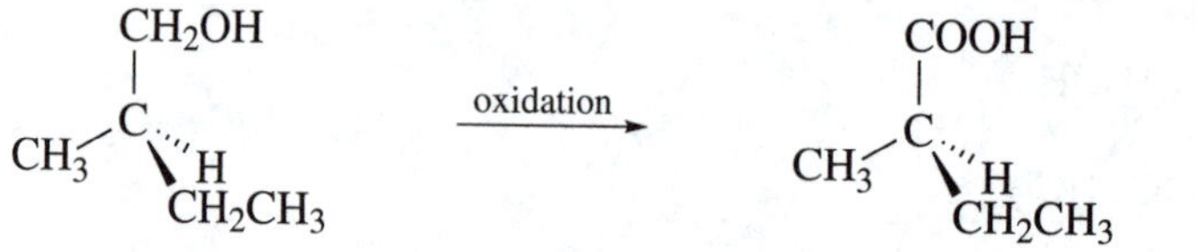

   (R)-$(-)$-2-methyl-1-butanol $(+)$-2-methylbutanoic acid

7. (−)-Cholesterol has a specific rotation of −32. What would be the observed specific rotation of a solution that contains 25%(+)-cholesterol and 75% (−)-cholesterol?

8. Which of the following have the *R* configuration?

9. Answer the following:

a. Put the remaining groups on the structure so that it represents (*R*)-2-butanol.

b. Draw a diastereomer for the following:

10. Indicate whether each of the following statements is true or false:

		T	F
a.	Diastereomers have the same melting points.	T	F
b.	3-Chloro-2,3-dimethylpentane has two asymmetric centers.	T	F
c.	Meso compounds do not rotate the plane of polarization of plane-polarized light.	T	F
d.	2,3-Dichloropentane has a stereoisomer that is a meso compound.	T	F
e	All chiral compounds with the *R* configuration are dextrorotatory.	T	F
f	A compound with three asymmetric centers can have a maximum of nine stereoisomers.	T	F

11. Which of the following have cis–trans isomers?

1-pentene   4-methyl-2-hexene   2-bromo-3-hexene   2-methyl-2-hexene

12. Label the following substituents in order from highest priority to lowest priority in the *E,Z* system of nomenclature.

$$\underset{\text{—CCH}_3}{\overset{\text{O}}{\overset{\|}{}}} \qquad \text{—CH=CH}_2 \qquad \text{—Cl} \qquad \text{—C}\equiv\text{N}$$

13.   Do the following compounds have the *E* or the *Z* configuration?

a.
$$\begin{array}{c}
CH_3 \\
|\\
CH_3CHCH_2 \quad CH_2CH_2CH_2Br \\
C=C \\
CH_3CH \quad CH_2OH \\
|\\
CH_3
\end{array}$$

b.
$$\begin{array}{c}
Cl \quad\quad O \\
|\quad\quad\; || \\
CH_3CH \quad\quad C \\
C=C \quad CH_3 \\
CH_3CH \quad CH_2OH \\
|\\
CH_3
\end{array}$$

14.   Draw and label the *E* and *Z* stereoisomers of the following:

a.   1-bromo-2,3-dimethyl-2-pentene

b.   2,3,4-trimethyl-3-hexene

Answers to Chapter 4 Practice Test

1. **a.** a pair of enantiomers   **b.** a pair of enantiomers

2. -3.0 **3.** D

4. 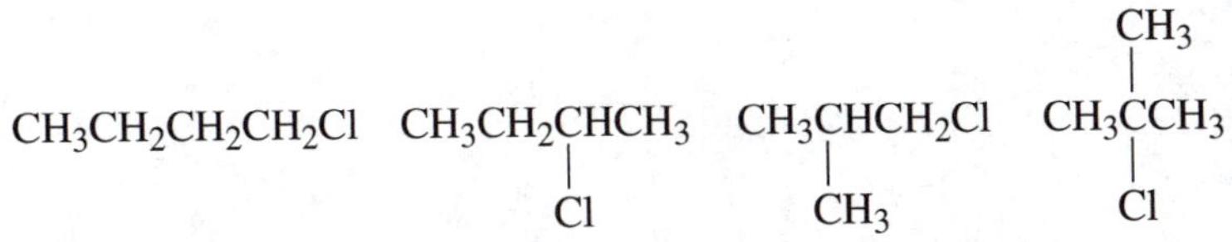 $CH_3CH_2CH_2CH_2Cl$ $CH_3CH_2CHCH_3$ CH_3CHCH_2Cl CH_3CCH_3

5. **a.** 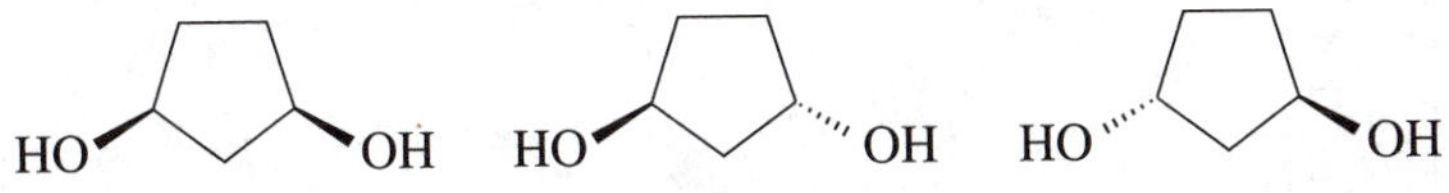 HO   OH   HO   OH   HO   OH

b.

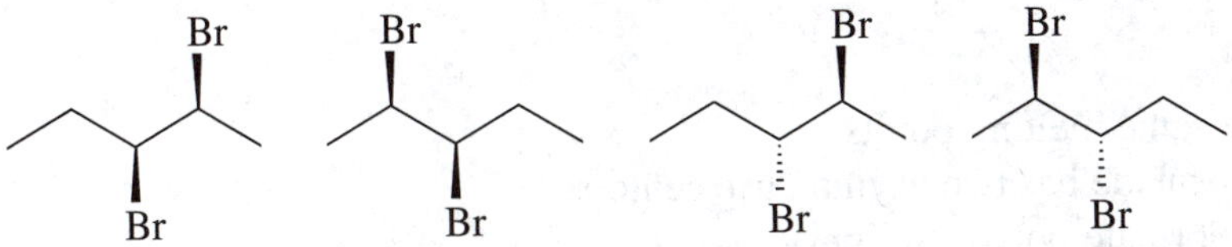

or

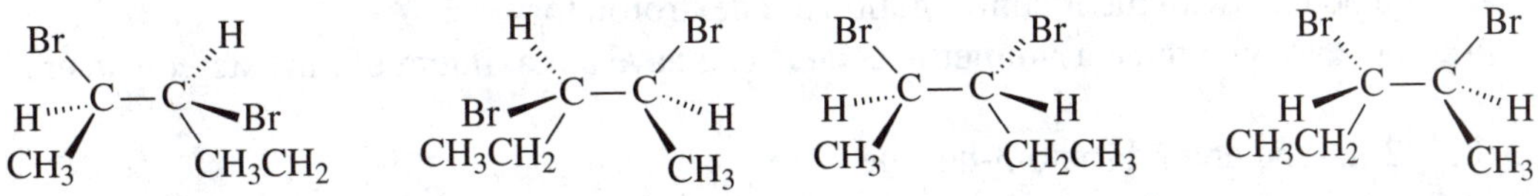

c. no stereoisomers

d.

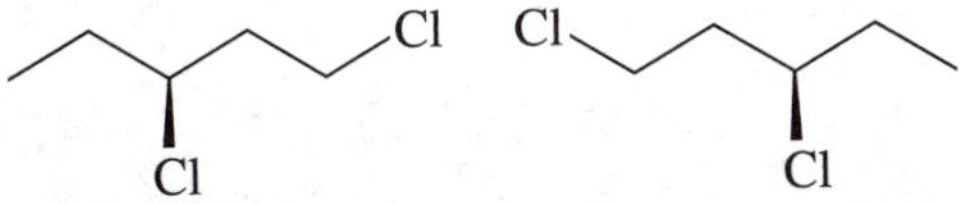

or

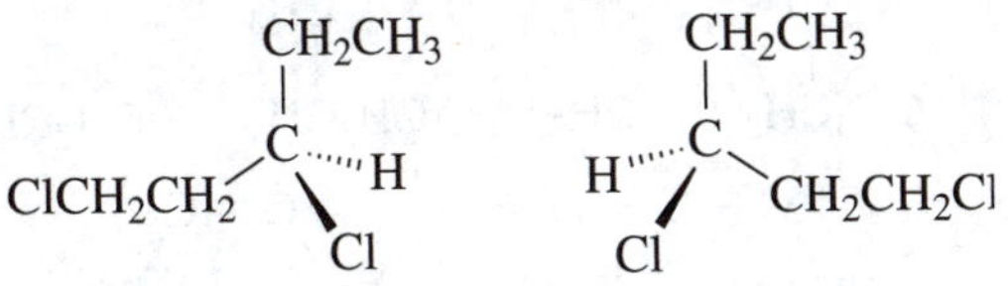

e.

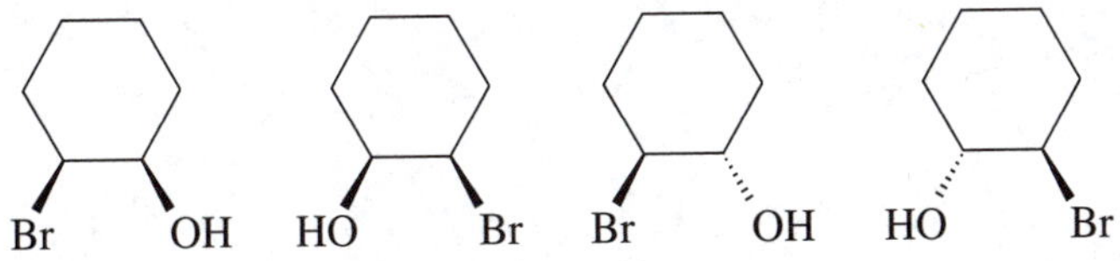

f. 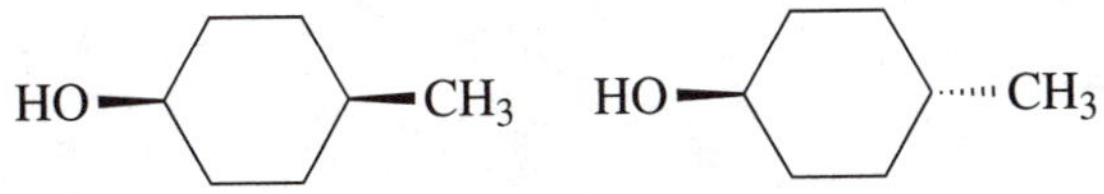 HO—◯—CH$_3$ HO—◯····CH$_3$

6. (−)-2-Methylbutanoic acid has the *S* configuration.

7. −16

8. 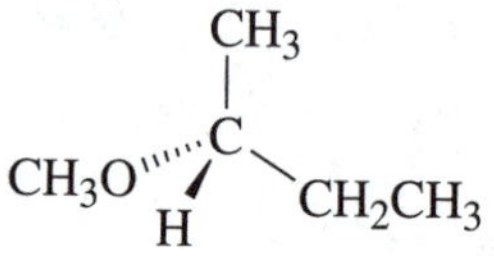 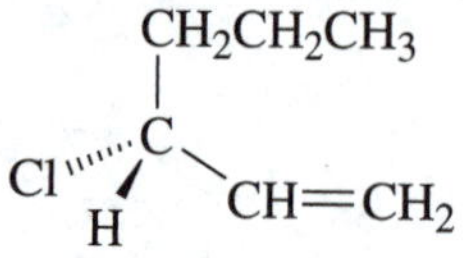

9. **a.** 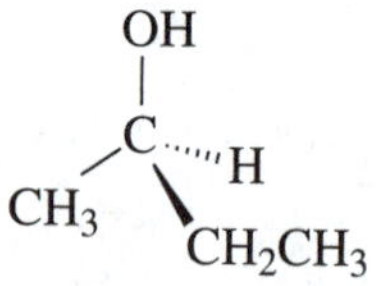

b. 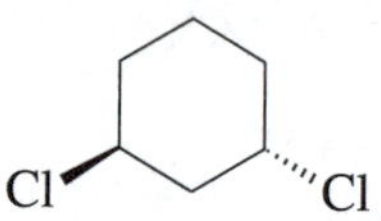 **or** 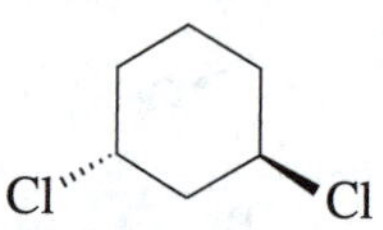

10.
- **a.** Diastereomers have the same melting points. F
- **b.** 3-Chloro-2,3-dimethylpentane has two asymmetric centers. F
- **c.** Meso compounds do not rotate polarized light. T
- **d.** 2,3-Dichloropentane has a stereoisomer that is a meso compound. F
- **e.** All compounds with the *R* configuration are dextrorotatory. F
- **f.** A compound with three asymmetric centers can have a maximum of nine stereoisomers. F

11. 4-methyl-2-hexene and 2-bromo-3-hexene

12. 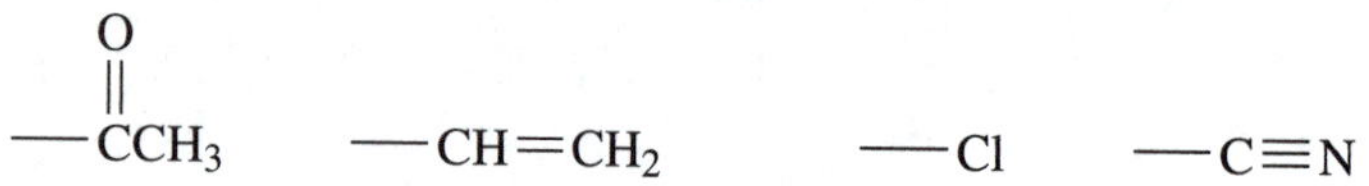

2 **4** **1** **3**

13. **a.** *Z* **b.** *Z*

14. **a.** 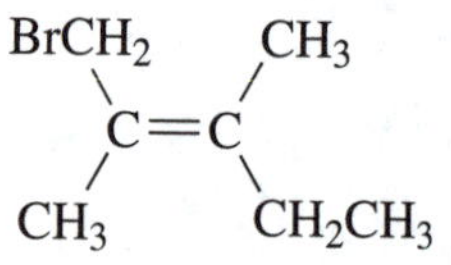

E

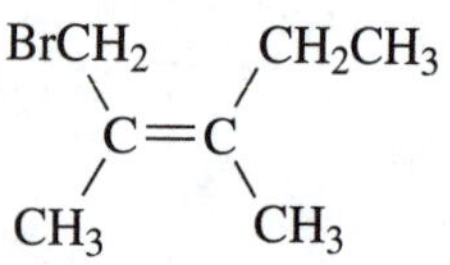

Z

b. 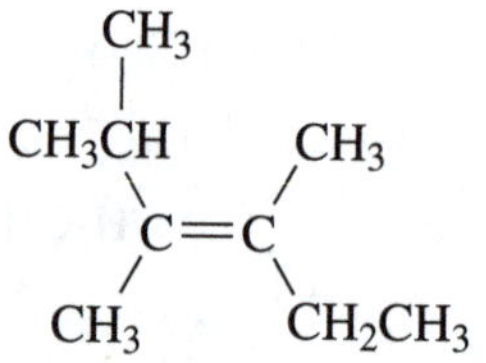

E

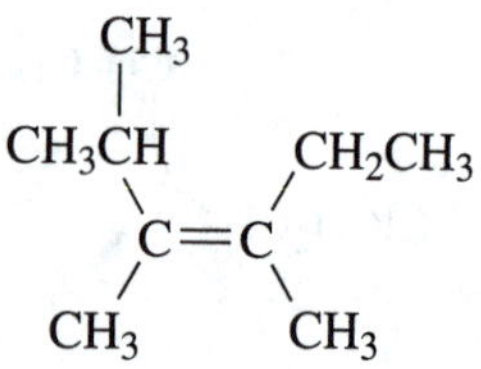

Z

CHAPTER 5
Alkenes: Structure, Nomenclature, Stability, and an Introduction to Reactivity • Thermodynamics and Kinetics

Important Terms

active site	the pocket of an enzyme where all the bond-making and bond-breaking steps of an enzyme-catalyzed reaction occur.
addition reaction	a reaction in which atoms or groups are added to the reactant.
alkene	a hydrocarbon that contains a double bond.
allyl group	$CH_2\!=\!CHCH_2\!-$
allylic carbon	an sp^3 carbon adjacent to a vinyl carbon.
allylic hydrogen	a hydrogen bonded to an allylic carbon.
catalyst	a species that increases the rate at which a reaction occurs without being consumed in the reaction.
catalytic hydrogenation	a hydrogenation reaction that takes place with a catalyst.
coupled reactions	an endergonic reaction followed by an exergonic reaction.
electrophile	an electron-deficient atom or molecule.
electrophilic addition reaction	an addition reaction in which the first species that adds to the reactant is an electrophile.
endergonic reaction	a reaction with a positive $\Delta G°$; it consumes more energy than it releases.
endothermic reaction	a reaction with a positive $\Delta H°$.
enthalpy	the heat given off $(\text{if } \Delta H° < 0)$ or the heat absorbed $(\text{if } \Delta H° > 0)$ during the course of a reaction.
entropy	a measure of the freedom of motion in a system.
enzyme	a protein that is a biological catalyst.
exergonic reaction	a reaction with a negative $\Delta G°$; it releases more energy than it consumes.
exothermic reaction	a reaction with a negative $\Delta H°$.
free energy of activation $(\Delta G^{\ddagger})$	the true energy barrier to a reaction.

functional group	the center of reactivity of a molecule.
Gibbs free energy change ($\Delta G°$)	the difference between the free energy content of the products and the free energy content of the reactants at equilibrium under standard conditions (1 M, 25 °C, and 1 atm).
heat of hydrogenation	the heat $(\Delta H°)$ released in a hydrogenation reaction.
hydrogenation	addition of hydrogen.
intermediate	a species formed during a reaction that is not the final product of the reaction.
kinetics	the field of chemistry that deals with the rates of chemical reactions.
kinetic stability	kinetic stability is indicated by $\Delta G^{\ddagger}$. If $\Delta G^{\ddagger}$ is large, the compound is kinetically stable (is not very reactive). If $\Delta G^{\ddagger}$ is small, the compound is kinetically unstable (is very reactive).
Le Châtelier's principle	a principle that states that if an equilibrium is disturbed, the components of the equilibrium will adjust in a way that will offset the disturbance.
mechanism of the reaction	a description of the step-by-step process by which reactants are changed into products.
metabolic pathway	a series of reactions that convert complex nutrient molecules into simple molecules.
molecular recognition	the ability of one molecule to recognize another as a result of intermolecular interactions.
nucleophile	an electron-rich atom or molecule.
pheromone	a chemical substance used for the purpose of communication.
rate of a reaction	depends on the number of collisions per unit of time times the fraction with sufficient energy to get over the energy barrier times the fraction with the proper orientation.
rate-determining step or rate-limiting step	the step in a reaction that has the transition state with the highest energy.
reaction coordinate diagram	a diagram that describes the energy changes that take place during the course of a reaction.
reduction reaction	a reaction that increases the number of C—C bonds.
saturated hydrocarbon	a hydrocarbon that is completely saturated with hydrogen (contains no double or triple bonds).

substrate	the reactant of an enzyme-catalyzed reaction.
thermodynamic stability	thermodynamic stability is indicated by $\Delta G°$. If $\Delta G°$ is negative, the products are more stable than the reactants. If $\Delta G°$ is positive, the reactants are more stable than the products.
thermodynamics	the field of chemistry that describes the properties of a system at equilibrium.
transition state	the energy maximum in a reaction step on a reaction coordinate diagram. In the transition state, bonds in the reactant that will break are partially broken and bonds in the product that will form are partially formed.
unsaturated hydrocarbon	a hydrocarbon that contains one or more double or triple bonds.
vinyl group	$CH_2=CH-$
vinylic carbon	a carbon that is doubly bonded to another carbon.
vinylic hydrogen	a hydrogen bonded to a vinylic carbon.

Solutions to Problems

1. **a.** It has two vinylic hydrogens.
    b. It has four allylic hydrogens.

2. **a.**

    c. $CH_3CH_2OCH{=}CH_2$

    b. $CH_3C{=}CCH_2CH_2CH_2Br$ (with CH₃ groups on both alkene carbons)

    d. $CH_2{=}CHCH_2OH$

3. **a.** 4-methyl-2-pentene
    b. 2-chloro-3,4-dimethyl-3-hexene
    c. 1-bromocyclopentene
    d. 1,5-dimethylcyclohexene
    e. 1-bromo-4-methyl-3-hexene
    f. 1-bromo-2-methyl-1,3-pentadiene

4. nucleophiles: H^- CH_3O^- $CH_3C{\equiv}CH$ NH_3

    electrophiles: $CH_3\overset{+}{C}HCH_3$

5. **1.**

    Drawing the arrows incorrectly leads to a bromine with an incomplete octet and a positive charge as well as an oxygen with 10 valence electrons and a charge of minus 2.

    Drawing the arrows incorrectly leads to an oxygen with an incomplete octet and a charge of plus 2.

    2. This one cannot be drawn, because the arrow is supposed to show where the electrons move to, but there are no electrons on the H to go anywhere.

    3.

    CH_3COCH_3 + $H\ddot{O}:^-$ ⟶ ?

    The product cannot be drawn, because the destination of the electrons in the breaking π bond is not clear.

    4. This one cannot be drawn, because the arrow is supposed to show where the electrons move to, but there are no electrons on the C to go anywhere.

6. **a.**

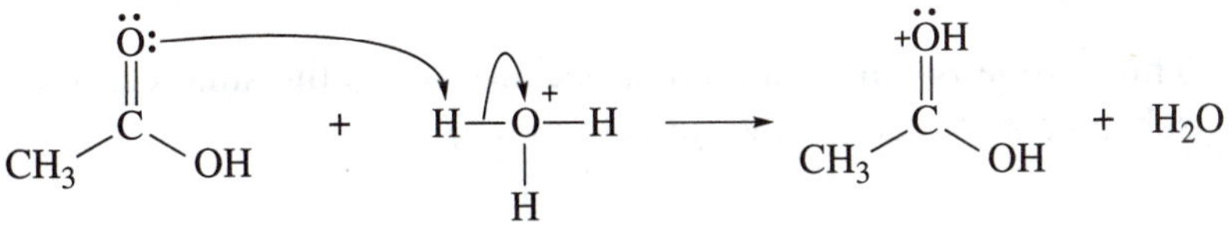

nucleophile electrophile

b.

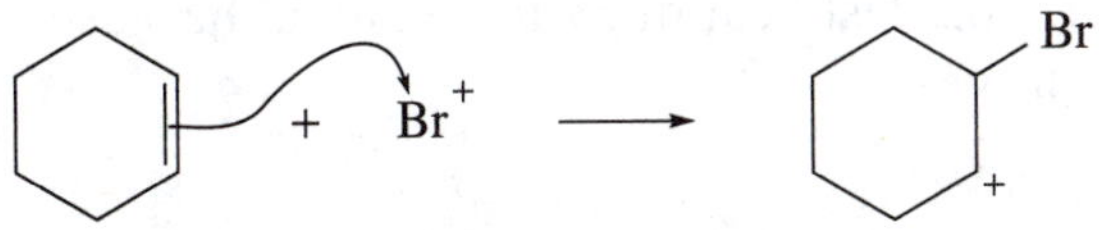

nucleophile electrophile

c.

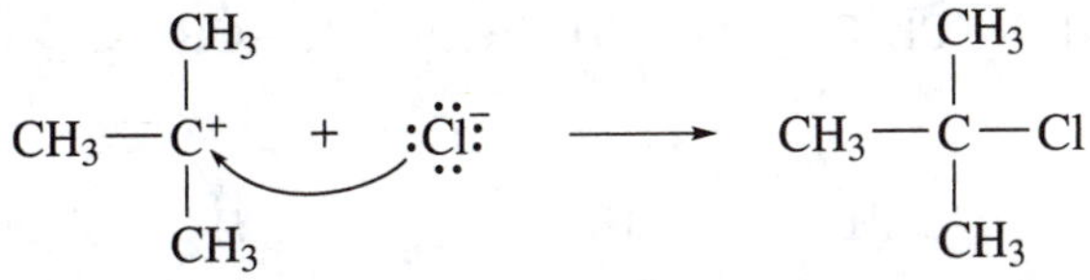

electrophile nucleophile

7. The labels are under the structures in Problem 6.

8. $\Delta S°$ is more significant in reactions in which the number of reactant molecules and the number of product molecules are not the same.

a. **1.** $A + B \rightleftharpoons C$
2. $A + B \rightleftharpoons C$

b. None of the four reactions has a positive $\Delta S°$.
In order to have a positive $\Delta S°$, the products must have greater freedom of motion than the reactants. (In other words, there should be more molecules of products than molecules of reactant.)

9. **a.** $CH_2{=}CHCH_2CH_2CH_3$ **or** $CH_3CH{=}CHCH_2CH_3$

b.

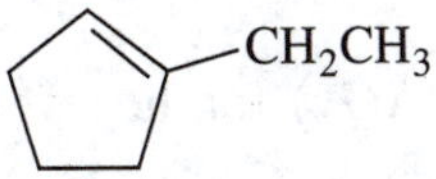

or

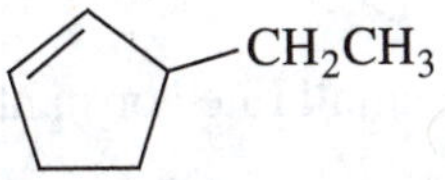

or

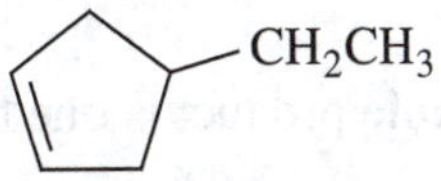

or

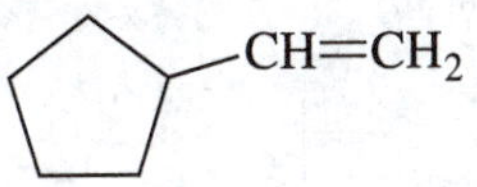

or

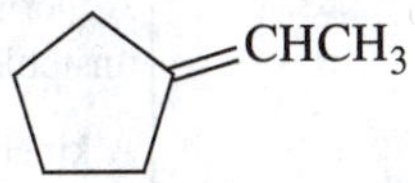

10. **a.** three alkenes: 1-butene, *cis*-2-butene, *trans*-2-butene
b. four alkenes: 3-methyl-1-pentene, (*E*)-3-methyl-2-pentene, (*Z*)-3-methyl-2-pentene, 2-ethyl-1-butene
c. five alkenes: 1-hexene, *cis*-2-hexene, *trans*-2-hexene, *cis*-3-hexene, *trans*-3-hexene

11. Because alkene **A** has the smaller heat of hydrogenation, it is more stable.

12. **a.**

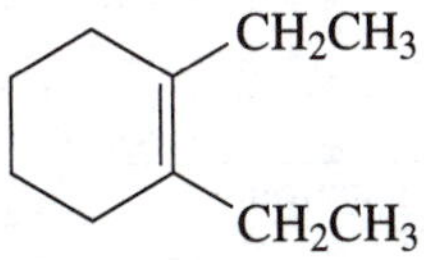

This alkene is the most stable, because it has the greatest number of alkyl substituents bonded to the sp^2 carbons.

b.

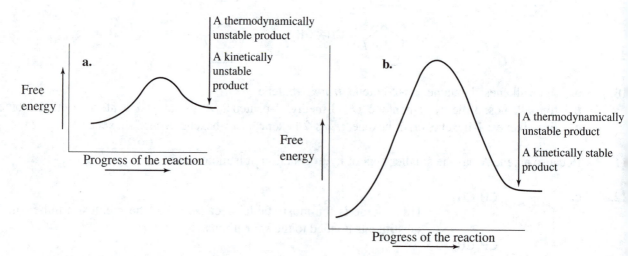

This alkene is the least stable, because it has the smallest number of alkyl substituents bonded to the sp^2 carbons.

c.

This alkene has the smallest heat of hydrogenation, because it is the most stable of the three alkenes.

13.

4 alkyl substituents   2 trans alkyl substituents   2 cis alkyl substituents   2 cis alkyl substituents that cause greater steric strain than those in *cis*-3-hexene

14.   The rate constant for a reaction can be increased by **decreasing** the stability of the reactant (increasing its energy) or by **increasing** the stability of the transition state (decreasing its energy).

15. **a.** **a** and **b**, because the product is more stable than the reactant.
   b. **b** is the most kinetically stable product, because it has the smallest rate constant (greatest $\Delta G^{\ddagger}$) leading from the product to the transition state.
   c. **c** is the least kinetically stable product, because it has the largest rate constant (smallest $\Delta G^{\ddagger}$) leading from the product to the transition state.

16. **a.** A thermodynamically **unstable** product is one that is less stable than the reactant.
   A kinetically **unstable** product is one that has a large rate constant (small $\Delta G^{\ddagger}$) for the reverse reaction.
   b. A kinetically **stable** product is one that has a small rate constant (large $\Delta G^{\ddagger}$) for the reverse reaction.

7.

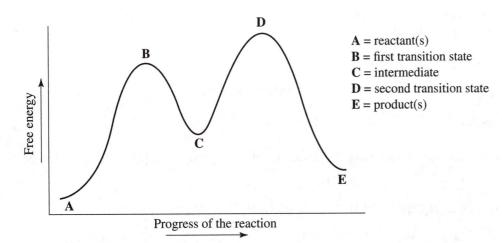

A = reactant(s)
B = first transition state
C = intermediate
D = second transition state
E = product(s)

18.   **a.**   The first step in the forward direction $(A \rightarrow B)$ has the greatest free energy of activation.

b.   The first-formed intermediate (B) is more apt to revert to reactants, because the free energy of activation for **B** to form **A** (the reactants) is less than the free energy of activation for **B** to form **C**.

c.   The second step $(B \rightarrow C)$ is the rate-determining step, because it has the transition state with the highest energy.

Notice that the second step is rate determining even though the first step has the greater energy of activation (steeper hill to climb). That is because it is easier for the intermediate that is formed in the first step to go back to starting material than to undergo the second step of the reaction. So the second step is the rate-limiting step.

19.

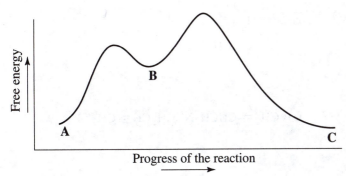

a.   one (B)

b.   two

c.   the second step (k_2), B to form C (In this particular diagram, $k_2 > k_1$: if you had made the transition state for the second step a lot higher, you could have had a diagram in which $k_1 > k_2$.)

d.   the second step in the reverse direction (k_{-1}) **f.**   B to C

e.   the second step in the reverse direction (k_{-1}), B to form A   **g.**   C to B

20.   A catalyst will change the energy difference between the reactants and the transition state, but it will not change the energy difference between the reactants and the products.

Therefore, $\Delta G^{\ddagger}$ is the only parameter that would be different when carried out in the presence of a catalyst.

21. **a.** 1,5-dimethylcyclopentene
b. 4-methylcyclohexene
c. 4-ethyl-5-methylcyclohexene

22. **a.** $CH_2{=}CHCH_2CH_3$

b. 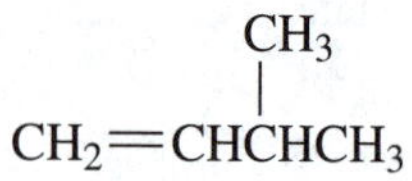

c. 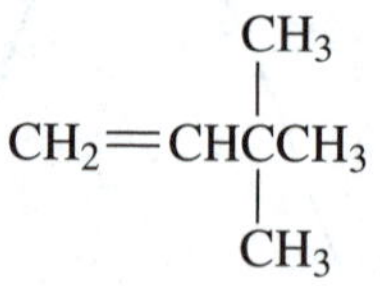

23. First draw the structures so that you can see the number of alkyl groups bonded to the sp^2 carbons:

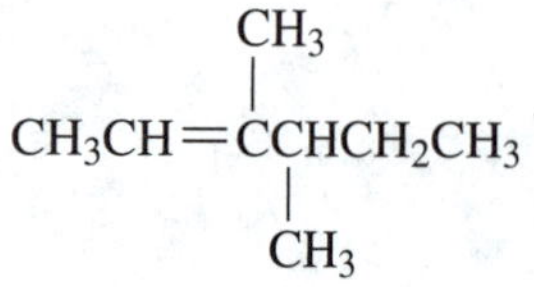
3,4-dimethyl-2-hexene

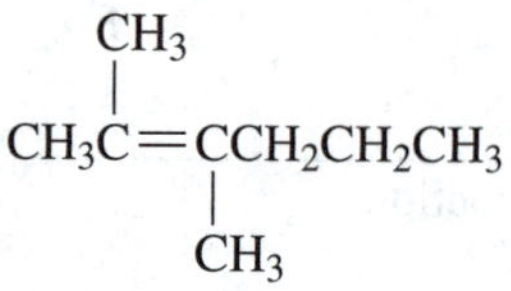
2,3-dimethyl-2-hexene

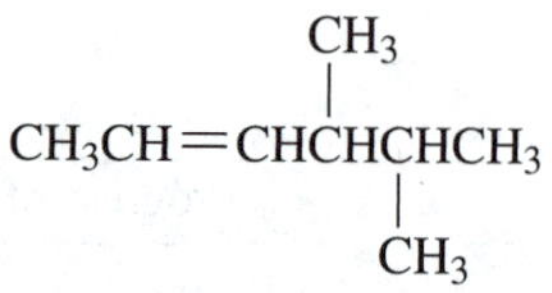
4,5-dimethyl-2-hexene

2,3-Dimethyl-2-hexene is the most stable of the three alkenes, because it has the greatest number of alkyl substituents bonded to the sp^2 carbons.
4,5-Dimethyl-2-hexene has the fewest alkyl substituents bonded to the sp^2 carbons, making it the least stable of the three alkenes.

24. **a.** 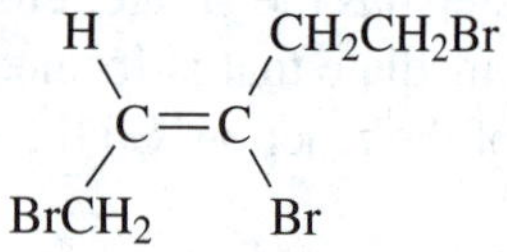

b. 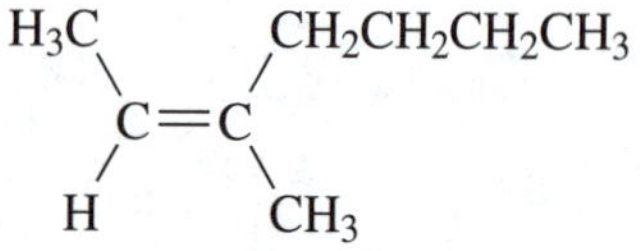

c. 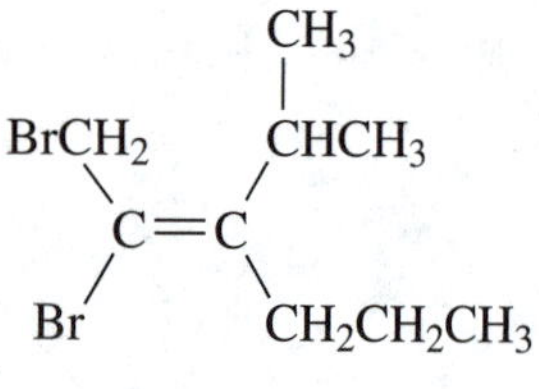

d. $CH_2{=}CHBr$

e. 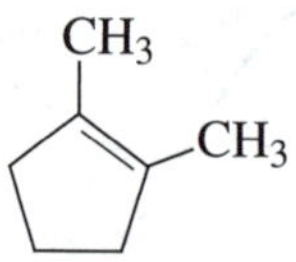

f. $CH_2{=}CHCH_2NHCH_2CH{=}CH_2$

25. **a.** 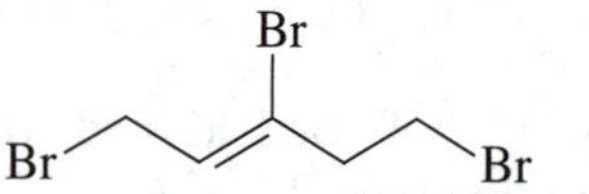

b. 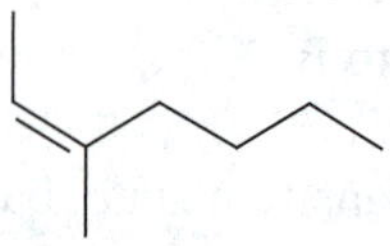

c. 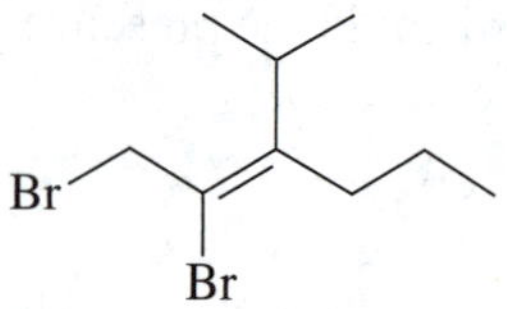

d. 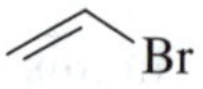

e. 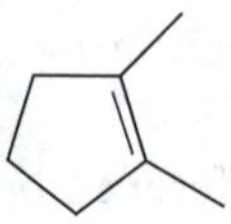

f. 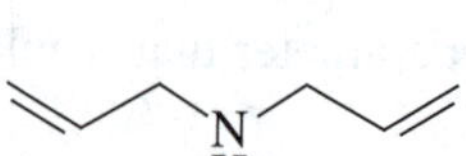

6. a. $CH_2\!\!=\!\!CHCH_2CH_2CH_2CH_3$ $CH_3CH\!\!=\!\!CHCH_2CH_2CH_3$ $CH_3CH_2CH\!\!=\!\!CHCH_2CH_3$

1-hexene   2-hexene   3-hexene

$CH_2\!\!=\!\!CCH_2CH_2CH_3$ $CH_2\!\!=\!\!CHCHCH_2CH_3$ $CH_2\!\!=\!\!CHCH_2CHCH_3$
    | | |
   CH_3 CH_3 CH_3

2-methyl-1-pentene   3-methyl-1-pentene   4-methyl-1-pentene

$CH_3C\!\!=\!\!CHCH_2CH_3$ $CH_3CH\!\!=\!\!CCH_2CH_3$ $CH_3CH\!\!=\!\!CHCHCH_3$
 | | |
CH_3 CH_3 CH_3

2-methyl-2-pentene   3-methyl-2-pentene   4-methyl-2-pentene

     CH_3 CH_3
     | |
$CH_2\!\!=\!\!CCHCH_3$ $CH_3CCH\!\!=\!\!CH_2$ $CH_3CH_2C\!\!=\!\!CH_2$
     | | |
    CH_3 CH_3 CH_2CH_3

2,3-dimethyl-1-butene   3,3-dimethyl-1-butene   2-ethyl-1-butene

     CH_3
     |
$CH_3C\!\!=\!\!CCH_3$
       |
      CH_3

2,3-dimethyl-2-butene

b. Of the compounds shown in part **a**, the following can have E and Z isomers:

2-hexene, 3-hexene, 3-methyl-2-pentene, 4-methyl-2-pentene

c. 2,3-dimethyl-2-butene

27. $H\!-\!\ddot{O}\!:^- \;+\; H\!-\!\overset{H}{\underset{H}{C}}\!-\!\overset{H}{\underset{Br}{C}}\!-\!H \;\longrightarrow\; \overset{H}{\underset{H}{}}C\!\!=\!\!C\overset{H}{\underset{H}{}} \;+\; H_2O \;+\; Br^-$

28. a. (E)-3-methyl-3-hexene
    b. *trans*-8-methyl-4-nonene or (E)-8-methyl-4-nonene
    c. *trans*-9-bromo-2-nonene or (E)-9-bromo-2-nonene
    d. 2,4-dimethyl-1-pentene
    e. 2-ethyl-1-pentene
    f. *cis*-2-pentene or (Z)-2-pentene

29.   **a.** $\underset{\displaystyle \overset{|}{\mathrm{CH_3C}}}{\overset{\displaystyle \mathrm{CH_3}}{}}=\mathrm{CHCH_2CH_3}$

has the more substituted
double bond

b.

has the more substituted
double bond

c.

has the more substituted
double bond

30.   **a.** $\mathrm{CH_3CH_2}-\ddot{\mathrm{B}}\mathrm{r}\colon$
electrophile
$\colon\mathrm{NH_3}$
nucleophile

b.

$\overset{\displaystyle :\ddot{\mathrm{O}}:}{\underset{\displaystyle \mathrm{CH_3}\qquad\mathrm{CH_3}}{\overset{\displaystyle \|}{\mathrm{C}}}}$
electrophile
$\mathrm{H\ddot{O}}\colon^{-}$
nucleophile

c. $\underset{\displaystyle \mathrm{CH_3C}=\mathrm{CCH_3}}{\overset{\displaystyle \overset{|}{\mathrm{CH_3}}\;\overset{|}{\mathrm{CH_3}}}{}}$
nucleophile
$\mathrm{H}-\ddot{\mathrm{C}}\mathrm{l}\colon$
electrophile

31.   Only one name is correct.
   a.   2-pentene
   b.   correct
   c.   3-methyl-1-hexene (The parent hydrocarbon is the longest chain that contains the functional group.)
   d.   3-heptene
   e.   4-ethylcyclohexene
   f.   2-chloro-3-hexene
   g.   3-methyl-2-pentene
   h.   2-methyl-1-hexene (It does not have E and Z isomers.)
   i.   1-methylcyclopentene

32.

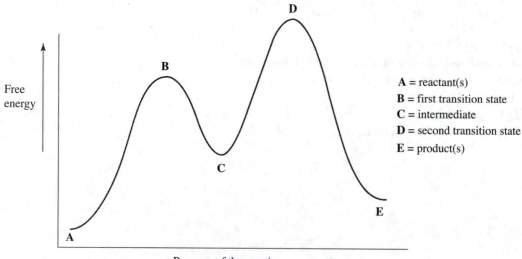

A = reactant(s)
B = first transition state
C = intermediate
D = second transition state
E = product(s)

3.

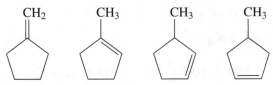

4.   a. 4 alkenes

b. 1-Methylcyclopentene is the most stable.
c. Because 1-methylcyclopentene is the most stable, it has the smallest heat of hydrogenation.

5.   a. 2
b. B, D, F
c. E to G (The fastest step has the smallest energy of activation to overcome.)
d. G
e. A
f. C
g. endergonic
h. exergonic
i. C
j. E to G (The largest rate constant corresponds to the lowest energy of activation.)
k. G to E (The smallest rate constant corresponds to the highest energy of activation.)

6.   a.   **b.**   **c.**   **d.**

7.   a. (Z)-1-chloro-3-ethyl-4-methyl-3-hexene
b. (Z)-2-bromo-3-propyl-2-heptene
c. (E)-1,3-dibromo-4,7-diethyl-2-octene

8.   a. **B** will have the larger $\Delta S°$ value because, unlike **A**, the number of reactants is not the same as the number of products.

b. $\Delta S° = $ (the freedom of motion of the products) $-$ (the freedom of motion of the reactants).
Because the three products have a greater freedom of motion than the two reactants, $\Delta S°$ will be positive.

9.
$$\Delta G° = -RT \ln K_{eq}$$
$$\ln K_{eq} = -\Delta G°/RT$$
$$\ln K_{eq} = -\Delta G°/0.59 \text{ kcal/mol}$$
$$\ln K_{eq} = -5.3/0.59 \text{ kcal/mol}$$
$$\ln K_{eq} = -9.0$$
$$K_{eq} = [B]/[A] = 0.00013$$
$$[B]/[A] = 0.00013/1 = 0.13/1000$$

The calculation shows that for every 1000 molecules in the chair conformation there is 0.13 molecule in a twist-boat conformation.

This agrees with the statement that for every 10,000 chair conformers of cyclohexane, there is no more than one twist-boat conformer.

40.

a.

$$\text{fluorocyclohexane} \rightleftharpoons \text{fluorocyclohexane}$$
$$\text{axial} \qquad\qquad \text{equatorial}$$

$$\Delta G^\circ = -0.25 \text{ kcal/mol at } 25°C$$

$$\Delta G^\circ = -RT \ln K_{eq}$$

$$-0.25 \frac{\text{kcal}}{\text{mol}} = -1.986 \times 10^{-3} \frac{\text{kcal}}{\text{mol K}} \times 298 \text{ K} \times \ln K_{eq}$$

$$\ln K_{eq} = 0.422$$

$$K_{eq} = 1.53 = \frac{[\text{fluorocyclohexane}]_{equatorial}}{[\text{fluorocyclohexane}]_{axial}} = \frac{1.53}{1}$$

Now we must determine the percentage of the total that is equatorial, as follows:

$$\frac{[\text{fluorocyclohexane}]_{equatorial}}{[\text{fluorocyclohexane}]_{equatorial} + [\text{fluorocyclohexane}]_{axial}} = \frac{1.53}{1.53 + 1} = \frac{1.53}{2.53} = 0.60 \text{ or } 60\%$$

b.

$$\Delta G^\circ = -RT \ln K_{eq}$$

$$-2.1 = -1.986 \times 10^{-3} \times 298 \times \ln K_{eq}$$

$$\ln K_{eq} = 3.56$$

$$K_{eq} = 35$$

$$K_{eq} = \frac{[\text{isopropylcyclohexane}]_{equatorial}}{[\text{isopropylcyclohexane}]_{axial}} = \frac{35}{1}$$

$$\begin{array}{l} \% \text{ of equatorial} \\ \text{isopropylcyclohexane} \end{array} = \frac{[\text{isopropylcyclohexane}]_{equatorial}}{[\text{isopropylcyclohexane}]_{equatorial} + [\text{isopropylcyclohexane}]_{axial}} \times 100$$

$$= \frac{35}{35 + 1} \times 100$$

$$= \frac{35}{36} \times 100$$

$$= 97\%$$

c. Isopropylcyclohexane has a greater percentage of the conformer with the substituent in the equatorial position because an isopropyl substituent is larger than a fluoro substituent. The larger the substituent, the less stable is the conformer in which the substituent is in the axial position because of the 1,3-diaxial interactions.

Chapter 5 Practice Test

. Name each of the following compounds:

a. CH₃CH₂CHCH₂CH=CH₂
    |
    CH₃

c. CH₃CH₂CH=CHCH₂CH₂CHCH₃
                |
                CH₂CH₃

b.

d.

. Which member of each pair is more stable?

a.
    or

b.
    or

. Correct the incorrect names.

a. 3-pentene
b. 2-vinylpentane
c. 2-ethyl-2-butene
d. 2-methylcyclohexene

. Indicate whether each of the following statements is true or false:

a. Increasing the energy of activation increases the rate of the reaction.   T   F

b. Decreasing the entropy of the products compared to the entropy of the reactants makes the equilibrium constant more favorable.   T   F

c. An exothermic reaction is one with a $-\Delta G°$.   T   F

d. An alkene is an electrophile.   T   F

e. The higher the energy of activation, the more slowly the reaction will take place.   T   F

f. 2,3-Dimethyl-2-pentene is more stable than 3,4-dimethyl-2-pentene.   T   F

g. A reaction with a negative $\Delta G°$ has an equilibrium constant greater than one.   T   F

h. Increasing the free energy of the reactants increases the rate of the reaction.   T   F

i. Increasing the free energy of the products increases the rate of the reaction.   T   F

j. The magnitude of a rate constant is not dependent on the concentration of the reactants.   T   F

k. A catalyst increases the equilibrium constant of a reaction.   T   F

5. The addition of H_2 in the presence of Pd/C to alkenes **A** and **B** results in the formation of the same alkane. The addition of H_2 to alkene **A** has a heat of hydrogenation $(-\Delta H°)$ of 29.7 kcal/mol, whereas the addition of H_2 to alkene **B** has a heat of hydrogenation of 27.3 kcal/mol. Which is the more stable alkene: **A** or **B**?

6. Draw structures for each of the following:

    a. allyl alcohol   **b.** 3-methylcyclohexene   **c.** *cis*-3-heptene   **d.** vinyl bromide

7. Using curved arrows, show the movement of electrons in the following reaction mechanism:

$$CH_3CH{=}CH_2 + H{-}\ddot{C}l\colon \rightleftharpoons CH_3\underset{+}{C}H{-}CH_3 + \colon\!\ddot{C}l\!\colon^{\!-} \longrightarrow CH_3CH{-}CH_3$$
$$\colon\!\!\overset{|}{\underset{\colon}{C}l}\!\colon$$

8. A favorable (negative) $\Delta G°$ is given by
(a positive or negative $\Delta H°$), (a positive or negative $\Delta S°$), and (a high or low temperature).

9. Which of the following has a more favorable equilibrium constant (that is, which reaction favors product more)?

    a. a reaction with a $\Delta H°$ of 4 kcal/mol or a reaction with a $\Delta H°$ of 7 kcal/mol? (assume a constant $\Delta S°$ value)
    b. a reaction with a positive $\Delta S°$ value that takes place at 25 °C or the same reaction that takes place at 35 °C?
    c. a reaction in which two reactants form one product or a reaction in which one reactant forms two products? (assume a constant $\Delta H°$ value)

10. Draw a reaction coordinate diagram for a one-step reaction with a product that is thermodynamically unstable but kinetically stable.

Answers to Chapter 5 Practice Test

1. **a.** 4-methyl-1-hexene   **c.** 7-methyl-3-nonene

    b. 4-bromocyclopentene   **d.** 4-chloro-3-methylcyclohexene

2. **a.**   **b.**

3. **a.** 2-pentene   **c.** 3-methyl-2-pentene

    b. 3-methyl-1-hexene   **d.** 1-methylcyclohexene

4. **a.** Increasing the energy of activation increases the rate of the reaction.   F

    b. Decreasing the entropy of the products compared to the entropy of the reactants
        makes the equilibrium constant more favorable.   F

    c. An exothermic reaction is one with a $-\Delta G°$.   F

    d. An alkene is an electrophile.   F

    e. The higher the energy of activation, the more slowly the reaction will take place.   T

    f. 2,3-Dimethyl-2-pentene is more stable than 3,4-dimethyl-2-pentene.   T

    g. A reaction with a negative $\Delta G°$ has an equilibrium constant greater than one.   T

    h. Increasing the energy of the reactants increases the rate of the reaction.   T

    i. Increasing the energy of the products increases the rate of the reaction.   F

    j. The magnitude of a rate constant is not dependent on the concentration of the reactants.   T

    k. A catalyst increases the equilibrium constant of a reaction.   F

5. **B**

6. **a.** $CH_2{=}CHCH_2OH$ **c.**

    b.   **d.** $CH_2{=}CHBr$

7.

8. a negative $\Delta H°$, a positive $\Delta S°$, and a high temperature

9.   **a.** 4 kcal/mol   **b.**   35 °C   **c.**   one reactant forms two products

10.

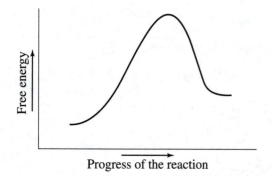

Important Terms

acid-catalyzed reaction	a reaction catalyzed by an acid.
aldehyde	a compound with a carbonyl group that is bonded to an alkyl group and to a hydrogen (or bonded to two hydrogens).

alkyne	a hydrocarbon that contains a triple bond.
biochemistry	the chemistry associated with living organisms.
carbocation rearrangement	the rearrangement of a carbocation to a more stable carbocation.
carbonyl group	a carbon doubly bonded to an oxygen.

$$\underset{\diagup\;\diagdown}{\overset{\overset{O}{\|}}{C}}$$

catalyst	a species that increases the rate at which a reaction occurs without being consumed in the reaction. Because it does not change the equilibrium constant of the reaction, it does not change the amount of product that is formed.
electrophilic addition reaction	an addition reaction in which the first species that adds to the reactant is an electrophile.
enol	an alkene with an OH group bonded to one of the sp^2 carbons.

hydration	addition of water to a compound.
1,2-hydride shift	the movement of a hydride ion (a hydrogen with its pair of bonding electrons) from one carbon to an adjacent carbon.
hyperconjugation	delocalization of electrons by overlap of carbon–hydrogen or carbon–carbon σ bonds with an empty orbital on an adjacent carbon.
internal alkyne	an alkyne with its triple bond not at the end of the carbon chain.

keto–enol tautomers   a ketone or aldehyde and its isomeric enol. The keto and enol tautomers differ only in the location of a double bond and a hydrogen.

$$\underset{\text{keto}}{\overset{\displaystyle\overset{O}{\|}}{RCH_2CR}} \quad\rightleftharpoons\quad \underset{\text{enol}}{\overset{\displaystyle\overset{OH}{|}}{RCH=CR}}$$

ketone   a compound with a carbonyl group that is bonded to two alkyl groups.

$$\overset{\displaystyle\overset{O}{\|}}{\underset{R\qquad\quad R}{C}}$$

mechanism of the reaction   a description of the step-by-step process by which reactants are changed into products.

1,2-methyl shift   the movement of a methyl group with its bonding electrons from one carbon to an adjacent carbon.

primary carbocation   a carbocation with the positive charge on a primary carbon.

regioselective reaction   a reaction that leads to the preferential formation of one constitutional isomer over another.

secondary carbocation   a carbocation with the positive charge on a secondary carbon.

stereoisomers   isomers that differ in the way the atoms are arranged in space.

tautomerization   interconversion of tautomers.

tautomers   constitutional isomers that are in rapid equilibrium—for example, keto and enol tautomers. The keto and enol tautomers differ only in the location of a double bond and a hydrogen.

terminal alkyne   an alkyne with its triple bond at the end of the carbon chain.

tertiary carbocation   a carbocation with the positive charge on a tertiary carbon.

Solutions to Problems

. The σ bond orbitals of the carbon adjacent to the positively charged carbon are available for overlap

a. The σ bond orbitals of the carbon adjacent to the positively charged carbon are available for overlap with the vacant p orbital. Because the methyl cation does not have a carbon adjacent to the positively charged carbon, there are no σ bond orbitals available for overlap with the vacant p orbital.

b. An ethyl cation is more stable, because the carbon adjacent to the positively charged carbon has three σ bond orbitals available for overlap with the vacant p orbital, whereas a methyl cation does not have any σ bond orbitals available for overlap with the vacant p orbital.

a.

$$CH_3-\overset{H}{\underset{CH_3}{\overset{|}{\underset{|}{C}}}}CH_2 \qquad CH_3CH_2-\overset{H}{\underset{H}{\overset{|}{\underset{|}{C}}}}CH_2 \qquad CH_3-\overset{H}{\underset{H}{\overset{|}{\underset{|}{C}}}}\overset{H}{\underset{H}{\overset{|}{\underset{|}{C}}}}HC-H$$

$$\qquad 3 \qquad\qquad\qquad 3 \qquad\qquad\qquad 6$$

b. *sec*-butyl cation

a.

$$CH_3CH_2\overset{CH_3}{\underset{+}{\overset{|}{C}}}CH_3 \;>\; CH_3CH_2\overset{}{\underset{+}{C}}HCH_3 \;>\; CH_3CH_2CH_2\overset{}{\underset{+}{C}}H_2$$

b. A halogen atom decreases the stability of the carbocation because, since it is an electronegative atom, it withdraws electrons away from the positively charged carbon. This increases the concentration of positive charge on the carbocation, which makes it less stable.

$$CH_3\overset{}{C}HCH_2CH_2 \;>\; CH_3\overset{}{C}HCH_2CH_2 \;>\; CH_3\overset{}{C}HCH_2CH_2$$
$$\underset{CH_3}{|}\;\overset{+}{} \qquad\qquad \underset{Cl}{|}\;\overset{+}{} \qquad\qquad \underset{F}{|}\;\overset{+}{}$$

Because fluorine is more electronegative than chlorine and therefore withdraws electrons more strongly, the fluorine-substituted carbocation is less stable than the chlorine-substituted carbocation.

a.

$$CH_3CH_2\overset{CH_3}{\underset{}{\overset{|}{C}}}=CH_2$$

In both **a** and **b,** the compound that is more highly regioselective is the one in which the choice is between forming a tertiary carbocation and forming a primary carbocation.

b.

In both **a** and **b,** the less regioselective compound is the one in which the choice is between forming a tertiary carbocation and forming a secondary carbocation, because the difference in the stability of the two possible carbocations—and, therefore, the difference in the amount of product formed—is not as great as it is when the choice is between a tertiary and a primary carbocation.

6. **a.** 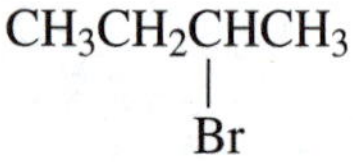 **c.** 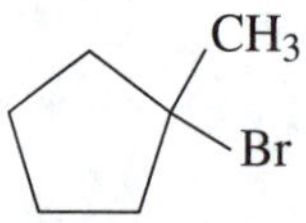 **e.** 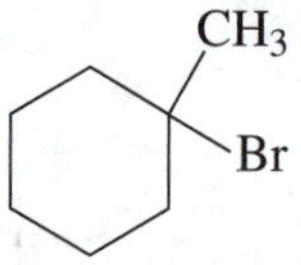

b. 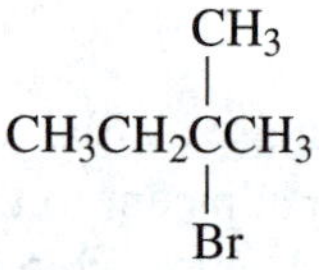 **d.** 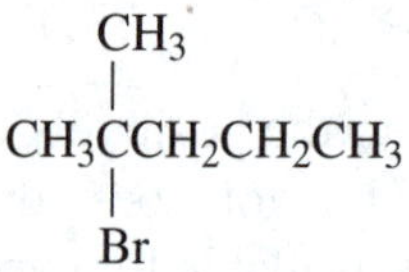 **f.** 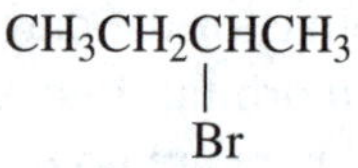

7. **a.** 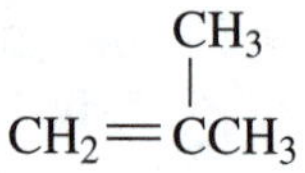 **c.** 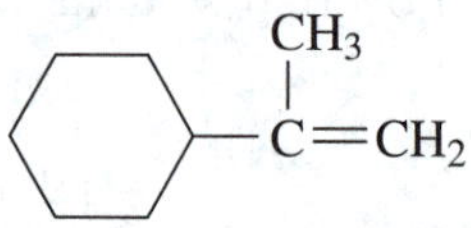

b. 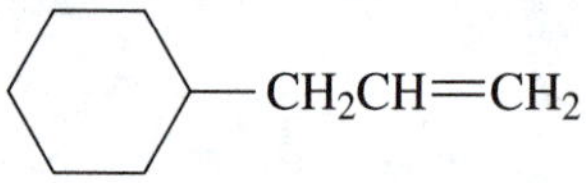 **d.** 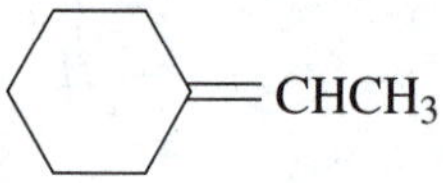 or 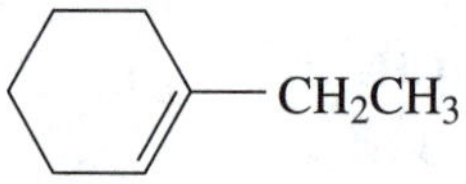

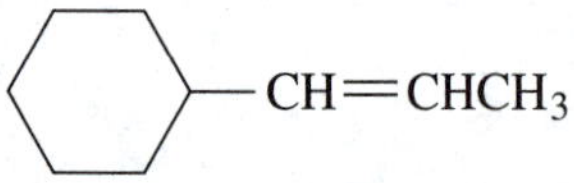

This would not be a good choice, because it would lead to a mixture of two products.

8. Solved in the text.

9. **a.** three transition states **b.** two intermediates

c. The first step is the *slowest step,* so it has the *smallest rate constant;* the second step is fast because n bonds are being broken; the third step is fast because transfer of a proton from or to an O or an N i always a fast reaction.

10. **a.** 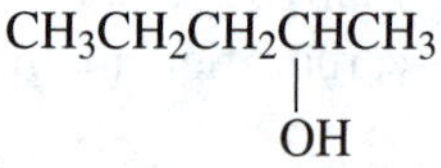 **c.** 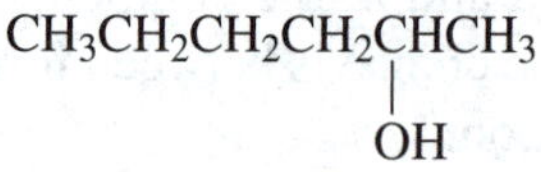 and 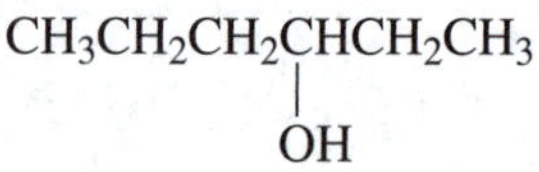

formed in equal amounts

b. 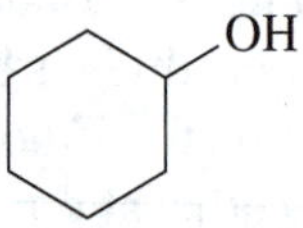 **d.** 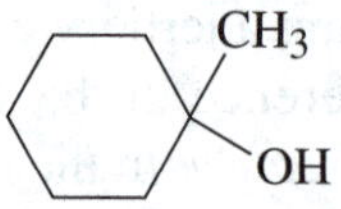

11. a. 1.
$$CH_3\underset{\underset{Cl}{|}}{\overset{\overset{CH_3}{|}}{C}}CH_3$$
2.
$$CH_3\underset{\underset{Br}{|}}{\overset{\overset{CH_3}{|}}{C}}CH_3$$
3.
$$CH_3\underset{\underset{OH}{|}}{\overset{\overset{CH_3}{|}}{C}}CH_3$$
4.
$$CH_3\underset{\underset{OH}{|}}{\overset{\overset{CH_3}{|}}{C}}CH_3$$

b. 1. The first step in all the reactions is addition of an electrophilic proton (H^+) to the carbon of the CH_2 group.

  2. A *tert*-butyl carbocation is formed as an intermediate in each of the reactions.

c. 1. The nucleophile that adds to the *tert*-butyl carbocation is different in each reaction.

  2. In reactions #3 and #4, there is a third step—a proton is lost from the group that acted as the nucleophile in the second step of the reaction.

3.
$$CH_3\overset{\overset{CH_3}{|}}{\underset{\underset{\overset{+}{O}H}{|}}{C}}CH_3 \longrightarrow CH_3\overset{\overset{CH_3}{|}}{\underset{\underset{OH}{|}}{C}}CH_3 + H_3O^+$$
$$H_2\ddot{O}\quad H$$

4.
$$CH_3\overset{\overset{CH_3}{|}}{\underset{\underset{\overset{+}{O}CH_3}{|}}{C}}CH_3 \longrightarrow CH_3\overset{\overset{CH_3}{|}}{\underset{\underset{OCH_3}{|}}{C}}CH_3 + CH_3\overset{+}{O}H_2$$
$$CH_3\ddot{O}H\quad H$$

12. a.

⬡ (cyclohexene) $+ H_2O \xrightarrow{H_2SO_4}$ ⬡—OH (cyclohexanol)

b.
$$CH_3\overset{\overset{CH_3}{|}}{C}{=}CH_2 + CH_3OH \xrightarrow{H_2SO_4} CH_3\underset{\underset{OH}{|}}{\overset{\overset{CH_3}{|}}{C}}CH_3$$

c.

⬠ (cyclopentene) $+\quad H_2O \xrightarrow{H_2SO_4}$ ⬠—OH (cyclopentanol)

d. $CH_3CH{=}CHCH_3 + H_2O \xrightarrow{H_2SO_4} CH_3\underset{\underset{OH}{|}}{CH}CH_2CH_3$

or

$CH_2{=}CHCH_2CH_3 + H_2O \xrightarrow{H_2SO_4} CH_3\underset{\underset{OH}{|}}{CH}CH_2CH_3$

13. Only the stereoisomers of the major product of each reaction are shown.

a.
$$\underset{Cl}{\overset{CH_2CH_2CH_3}{\underset{CH_3}{\overset{S}{C}}}}H \quad \text{racemic}\atop\text{mixture} \quad H\underset{Cl}{\overset{CH_2CH_2CH_3}{\overset{R}{C}}}CH_3$$

b.
$$CH_3CH_2\underset{OH}{\overset{CH_2CH_2CH_3}{\overset{S}{C}}}H \quad \text{racemic}\atop\text{mixture} \quad H\underset{HO}{\overset{CH_2CH_2CH_3}{\overset{R}{C}}}CH_2CH_3$$

c.

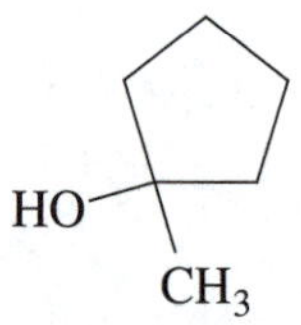

This compound does not have any stereoisomers, because it does not have any asymmetric centers.

d.

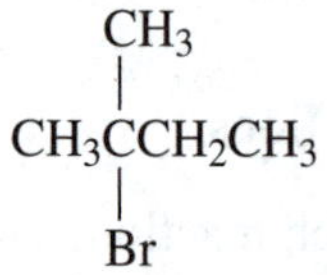

$$CH_3CCH_2CH_3$$

with CH$_3$ above and Br below the central carbon

This compound does not have any stereoisomers, because it does not have an asymmetric center.

14.   a.

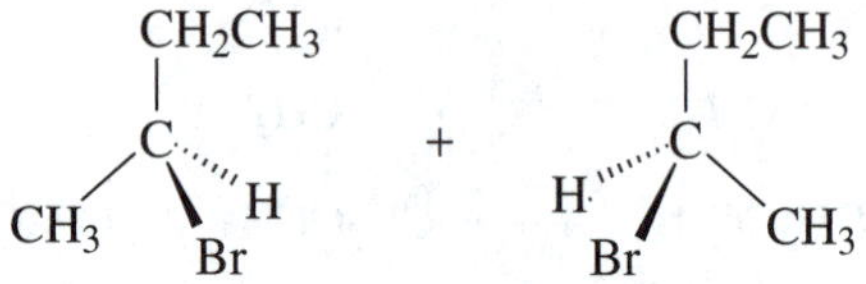

S + R

b. 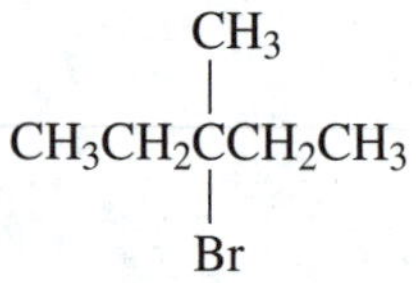 $CH_3CH_2CCH_2CH_3$ This compound does not have an asymmetric center.

with CH$_3$ above and Br below the central carbon

c.

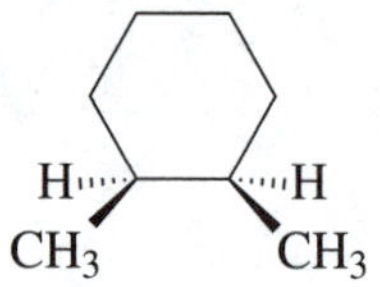

d.

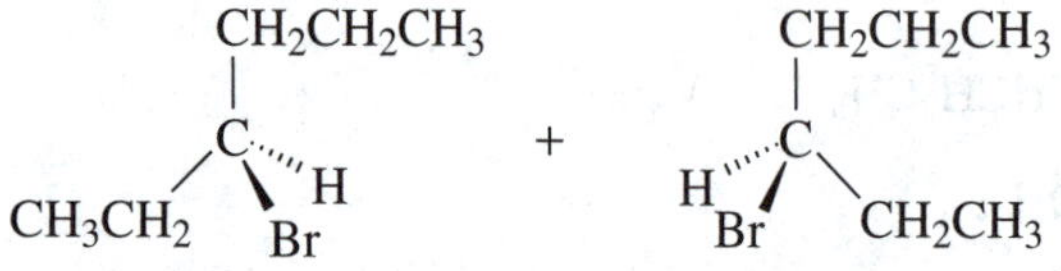

S + R

e. 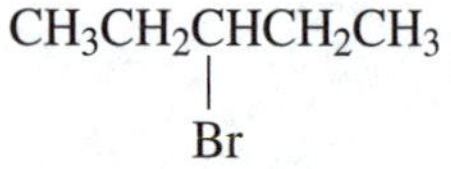 $CH_3CH_2CHCH_2CH_3$ This compound does not have an asymmetric center.

with Br below

f.

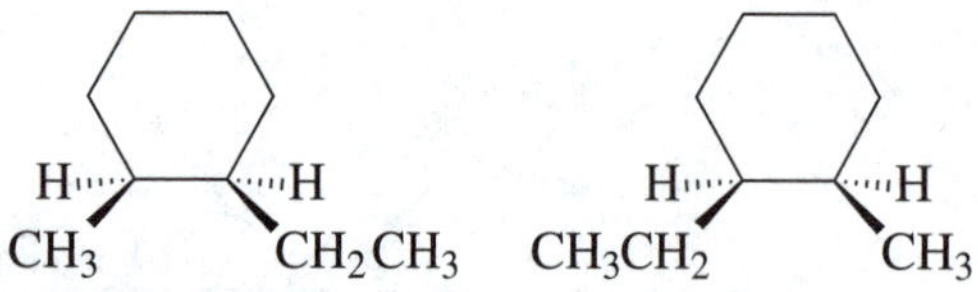

15.   a.   A racemic mixture of (R)-malate and (S)-malate would be obtained. (A product with one asymmetric center would be formed from a reactant with no asymmetric centers.)

b.   A racemic mixture of (R)-malate and (S)-malate would again be obtained. In the absence of an enzyme, the reactions are neither stereoselective (part **a**) nor stereospecific (part **b**).

16.   a.   4-methyl-1-pentyne   **b.**   2-hexyne

17.   **a.** $ClCH_2CH_2C \equiv CCH_2CH_3$

b.

c.   CH₃CCH₂C≡CH

$$\underset{\underset{CH_3}{|}}{\overset{\overset{CH_3}{|}}{CH_3\!-\!C\!-\!CH_2\!-\!C}}\equiv CH$$

18.

$HC \equiv CCH_2CH_2CH_2CH_3$	$CH_3C \equiv CCH_2CH_2CH_3$	$CH_3CH_2C \equiv CCH_2CH_3$
1-hexyne	2-hexyne	3-hexyne
butylacetylene	methylpropylacetylene	diethylacetylene

$$CH_3CH_2\underset{\underset{CH_3}{|}}{CH}C \equiv CH \qquad CH_3\underset{\underset{CH_3}{|}}{CH}CH_2C \equiv CH \qquad CH_3\underset{\underset{CH_3}{|}}{CH}C \equiv CCH_3$$

3-methyl-1-pentyne	4-methyl-1-pentyne	4-methyl-2-pentyne
sec-butylacetylene	isobutylacetylene	isopropylmethylacetylene

$$CH_3\underset{\underset{CH_3}{|}}{\overset{\overset{CH_3}{|}}{C}}C \equiv CH$$

3,3-dimethyl-1-butyne
tert-butylacetylene

19.   **a.** 5-bromo-2-pentyne   **b.** 6-bromo-2-chloro-4-octyne   **c.** 3-ethyl-1-hexyne

20.   **a.** 6-methyl-2-octyne   **b.** 5-ethyl-4-methyl-1-heptyne   **c.** 2-bromo-4-octyne

21.   **a.** sp^2-sp^2 **d.** $sp-sp^3$ **g.** sp^2-sp^3
   b. sp^2-sp^3 **e.** $sp-sp$ **h.** $sp-sp^3$
   c. $sp-sp^2$ **f.** sp^2-sp^2 **i.** sp^2-sp

22.   **a.** $CH_2 = \underset{\underset{Br}{|}}{C}CH_3$ **c.** $CH_3CH_2\underset{\underset{Br}{|}}{\overset{\overset{Br}{|}}{C}}CH_3$

b. $CH_3\underset{\underset{Br}{|}}{\overset{\overset{Br}{|}}{C}}CH_3$ **d.** $CH_3\underset{\underset{Br}{|}}{\overset{\overset{Br}{|}}{C}}CH_2CH_2CH_3$ + $CH_3CH_2\underset{\underset{Br}{|}}{\overset{\overset{Br}{|}}{C}}CH_2CH_3$

23.   Because the alkyne is not symmetrical, two ketones will be obtained.

$$CH_3CH_2\overset{\overset{\displaystyle O}{\|}}{C}CH_2CH_2CH_2CH_3 \quad \text{and} \quad CH_3CH_2CH_2\overset{\overset{\displaystyle O}{\|}}{C}CH_2CH_2CH_3$$

24.   **a.** $CH_3C\equiv CH$ **b.** $CH_3CH_2C\equiv CCH_2CH_3$ **c** $HC\equiv C-\bigcirc$

The best answer for **b** is 3-hexyne, because it would form only the desired ketone. 2-Hexyne would form two different ketones, so only half of the product would be the desired ketone.

25.   Note that both of the enols can exist as *E* and *Z* isomers.

$$\underset{\displaystyle CH_3}{CH_3CH=\overset{\overset{\displaystyle OH}{|}}{C}CH_2CH_2\underset{|}{C}HCH_3} \quad \text{and} \quad CH_3CH_2\overset{\overset{\displaystyle OH}{|}}{C}=CHCH_2\underset{\underset{\displaystyle CH_3}{|}}{C}HCH_3$$

26.   **a.** $CH_2=\overset{\overset{\displaystyle OH}{|}}{C}CH_3$ Because the ketone has identical substituents bonded to the carbonyl carbon, it has only one enol tautomer.

b. $CH_3CH=\overset{\overset{\displaystyle OH}{|}}{C}CH_2CH_2CH_3$ and $CH_3CH_2\overset{\overset{\displaystyle OH}{|}}{C}=CHCH_2CH_3$

E and *Z* isomers are possible for each of these enols.

c. $CH_2=\overset{\overset{\displaystyle OH}{|}}{C}-\bigcirc$ and $CH_3\overset{\overset{\displaystyle OH}{|}}{C}=\bigcirc$ Because each enol has identical groups bonded to one of its sp^2 carbons, *E* and *Z* isomers are not possible for either enol.

27.   **a.** $CH_3CH_2CH_2C\equiv CH$ **or** $CH_3CH_2C\equiv CCH_3$ $\xrightarrow[\text{Pd/C}]{H_2}$ $CH_3CH_2CH_2CH_2CH_3$
       1-pentyne   2-pentyne

b. $CH_3C\equiv CCH_3$ $\xrightarrow[\substack{\text{Lindlar} \\ \text{catalyst}}]{H_2}$ $\underset{\displaystyle H}{\overset{\displaystyle H_3C}{\diagdown}}C=C\underset{\displaystyle H}{\overset{\displaystyle CH_3}{\diagup}}$
       2-butyne

c. $CH_3CH_2CH_2CH_2C\equiv CH$ $\xrightarrow[\substack{\text{Lindlar} \\ \text{catalyst}}]{H_2}$ $CH_3CH_2CH_2CH_2CH=CH_2$
       1-hexyne

28. **a.**

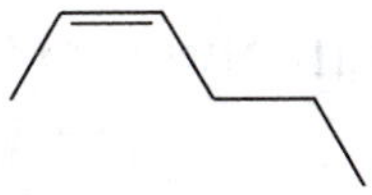

c.

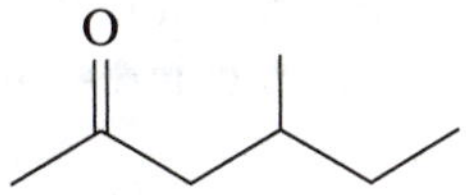

b.

d.

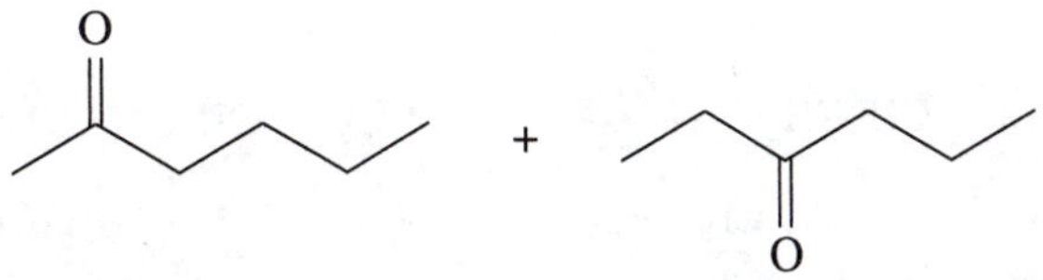

electrophile   nucleophile

29. **a.** 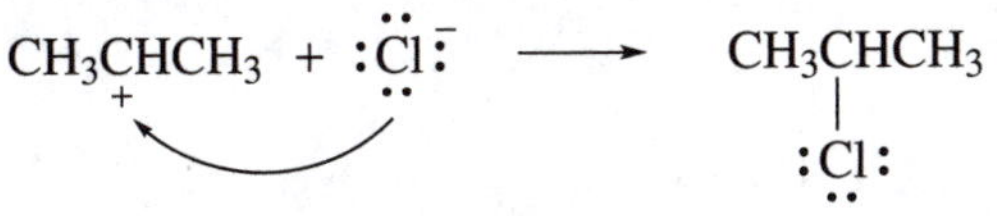 $CH_3\overset{+}{C}HCH_3$ + $:\overset{..}{\underset{..}{Cl}}:^-$ ⟶ $CH_3\overset{\displaystyle CHCH_3}{}$

$$:\overset{..}{\underset{..}{Cl}}:$$

b.   nucleophile   electrophile

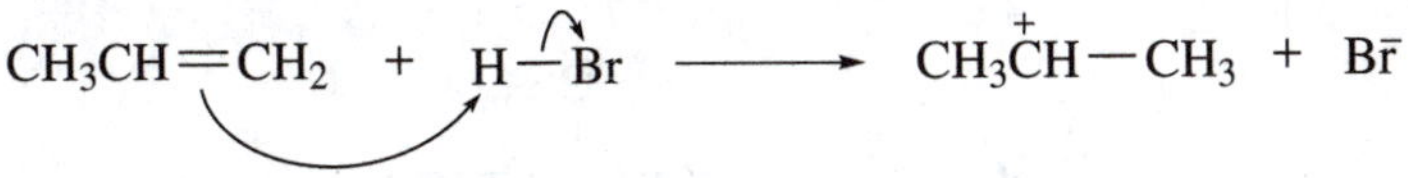

$CH_3CH{=}CH_2$ + H—Br ⟶ $CH_3\overset{+}{C}H{-}CH_3$ + $\bar{Br}$

30. **a.**

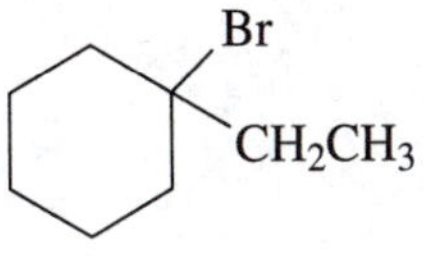

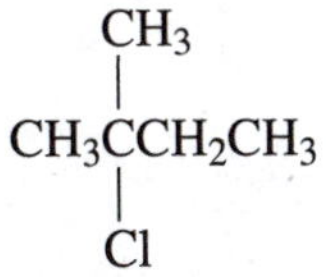

c. 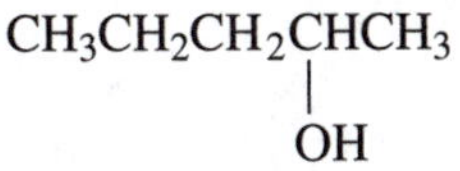 $CH_3CH_2CH_2\underset{\underset{OH}{|}}{C}HCH_3$

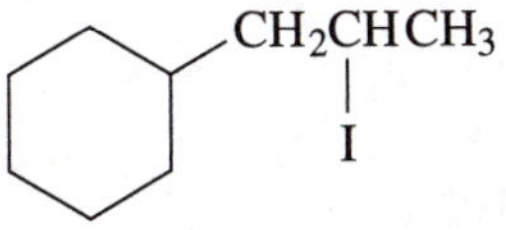

e.

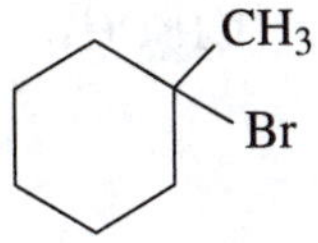

b. $CH_3\underset{\underset{Cl}{|}}{\overset{\overset{CH_3}{|}}{C}}CH_2CH_3$

d.

f.

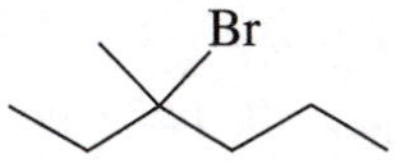

31. **a.** 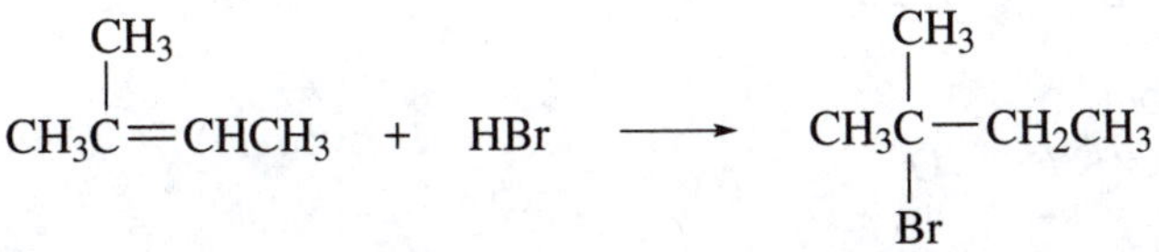 $CH_3\overset{\overset{CH_3}{|}}{C}{=}CHCH_3$ + HBr ⟶ $CH_3\underset{\underset{Br}{|}}{\overset{\overset{CH_3}{|}}{C}}{-}CH_2CH_3$

b. $CH_3\overset{\overset{CH_3}{|}}{C}{=}CHCH_3$ + HI ⟶ $CH_3\underset{\underset{I}{|}}{\overset{\overset{CH_3}{|}}{C}}{-}CH_2CH_3$

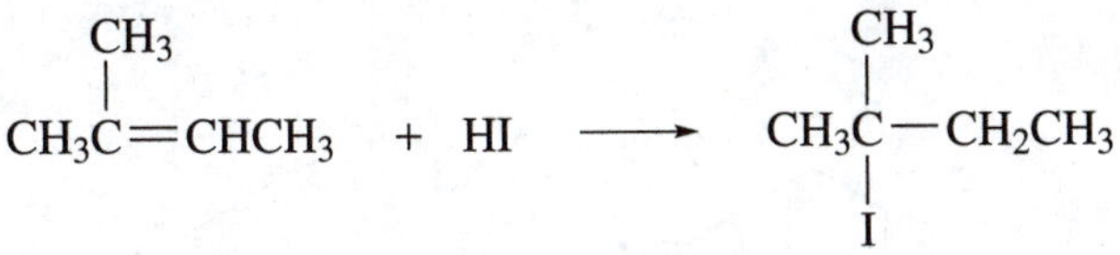

c. $CH_3\overset{\overset{CH_3}{|}}{C}{=}CHCH_3$ $\xrightarrow[\text{Pd/C}]{H_2}$ $CH_3\overset{\overset{CH_3}{|}}{C}H{-}CH_2CH_3$

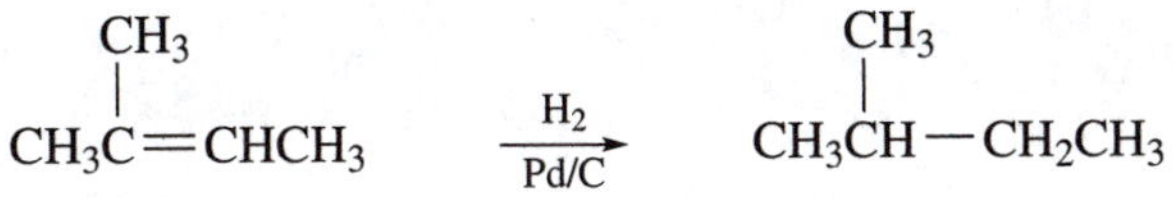

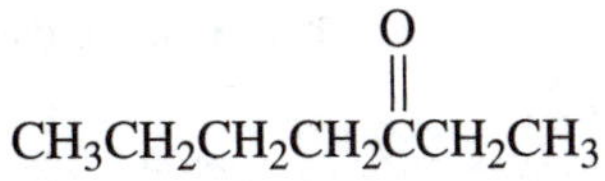

d. 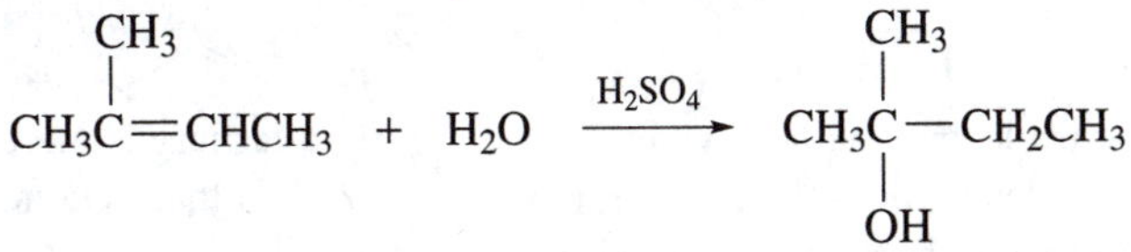 $CH_3\overset{\overset{CH_3}{|}}{C}{=}CHCH_3$ + H_2O $\xrightarrow{H_2SO_4}$ $CH_3\underset{\underset{OH}{|}}{\overset{\overset{CH_3}{|}}{C}}{-}CH_2CH_3$

32. 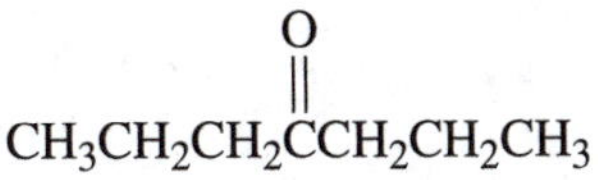 $CH_3CH_2CH_2\overset{\overset{O}{\|}}{C}CH_2CH_2CH_3$ and $CH_3CH_2CH_2CH_2\overset{\overset{O}{\|}}{C}CH_2CH_3$

33. **a.** 5-bromo-2-hexyne

     b. 5-methyl-2-octyne

     c. 5,5-dimethyl-2-hexyne

     d. 6-chloro-2-methyl-3-heptyne

34. **a.** 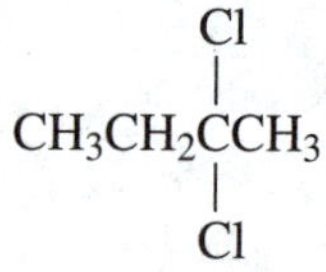 $CH_3CH_2CCH_3$ with Cl above and Cl below the third carbon

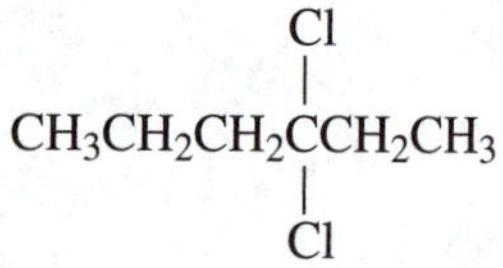

     c. 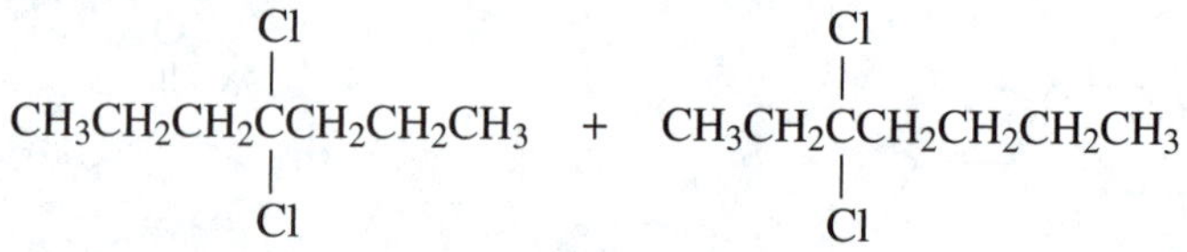 $CH_3CH_2CH_2CCH_2CH_2CH_3$ (Cl above and Cl below the fourth carbon) + $CH_3CH_2CCH_2CH_2CH_2CH_3$ (Cl above and Cl below the third carbon)

         equal amounts

     b. $CH_3CH_2CH_2CCH_2CH_3$ with Cl above and Cl below the fourth carbon

35. **a.** $CH_3C{\equiv}CCH_2CH_2CH_3$

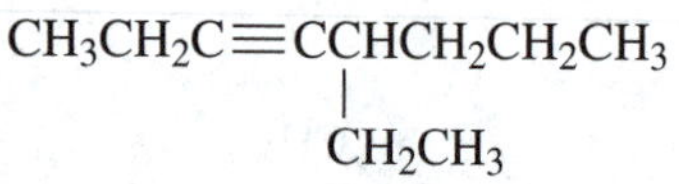

     c. $BrC{\equiv}CCH_2CH_2CH_3$

     b. $CH_3CH_2C{\equiv}CCHCH_2CH_2CH_3$ with CH_2CH_3 branch

     d. 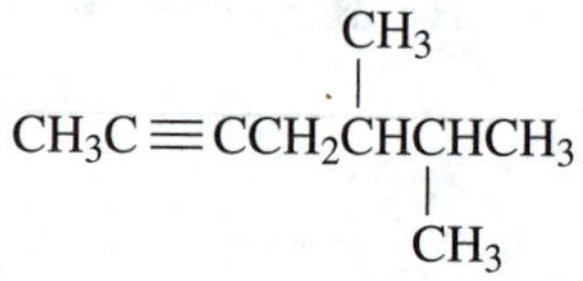 $CH_3C{\equiv}CCH_2CHCHCH_3$ with CH_3 above and CH_3 below

36.

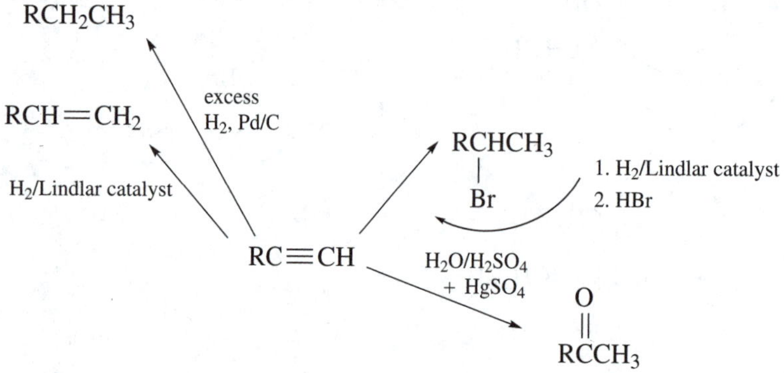

37. **a.** 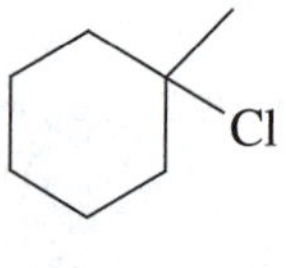 cyclohexane ring with methyl and Cl substituents

     b. no reaction without an acid catalyst

     c. 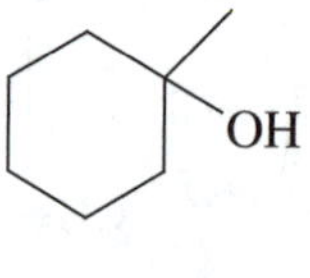 cyclohexane ring with methyl and OH substituents

     d. 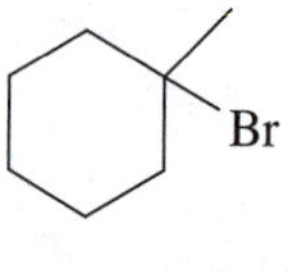 cyclohexane ring with methyl and Br substituents

38. **a.** 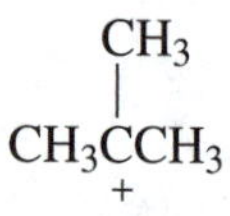 $CH_3\overset{+}{C}CH_3$ with CH_3 above

         tertiary is more stable than secondary

     b. 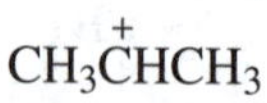 $CH_3\overset{+}{C}HCH_3$

         the electron withdrawing chlorine destabilizes the carbocation by increasing the amount of positive charge on the carbon

     c. 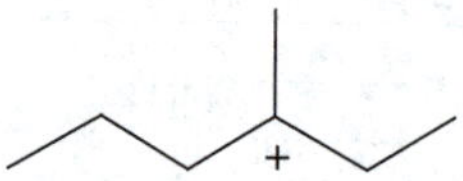 branched carbon chain with positive charge

         tertiary is more stable than secondary

9. The student named only one correctly.
   a. 4-methyl-2-hexyne   **b.** 7-bromo-3-heptyne   **c.** correct   **d.** 2-pentyne

0. First draw the straight-chain compounds with seven carbons; then draw the straight-chain compounds with six carbons and one methyl group; then draw the straight-chain compounds with five carbons and two methyl groups (or with one ethyl group). Naming them will tell you whether you have drawn one compound more than once, because if two compounds have the same name, they are the same compound.

$HC\equiv CCH_2CH_2CH_2CH_2CH_3$
1-heptyne
pentylacetylene

$CH_3C\equiv CCH_2CH_2CH_2CH_3$
2-heptyne
butylmethylacetylene

$CH_3CH_2C\equiv CCH_2CH_2CH_3$
3-heptyne
ethylpropylacetylene

$CH_3CH_2CH_2CHC\equiv CH$
|
CH_3
3-methyl-1-hexyne

$CH_3CH_2CHCH_2C\equiv CH$
|
CH_3
4-methyl-1-hexyne

$CH_3CHCH_2CH_2C\equiv CH$
|
CH_3
5-methyl-1-hexyne
isopentylacetylene

$CH_3CH_2CHC\equiv CCH_3$
|
CH_3
4-methyl-2-hexyne
sec-butylmethylacetylene

$CH_3CHCH_2C\equiv CCH_3$
|
CH_3
5-methyl-2-hexyne
isobutylmethylacetylene

$CH_3CHC\equiv CCH_2CH_3$
|
CH_3
2-methyl-3-hexyne
ethylisopropylacetylene

CH_3
|
$CH_3CC\equiv CCH_3$
|
CH_3
4,4-dimethyl-2-pentyne
tert-butylmethylacetylene

CH_3
|
$CH_3CCH_2C\equiv CH$
|
CH_3
4,4-dimethyl-1-pentyne

CH_3
|
$CH_3CH_2CC\equiv CH$
|
CH_3
3,3-dimethyl-1-pentyne
tert-pentylacetylene

41.

42. **a.** $CH_2=CCH_3$
              |
              Br

   c. $CH_3\overset{\displaystyle O}{\overset{\|}{C}}CH_3$

   e. $CH_3CH=CH_2$

   b. $\overset{Br}{\underset{Br}{CH_3CCH_3}}$

   d. $CH_3CH_2CH_3$

   f. $CH_3C\equiv C^-$

   Recall that the hydrogen attached to an *sp* carbon can be removed by a strong base (Sections 2.5 and 2.6).

43.   **a.** 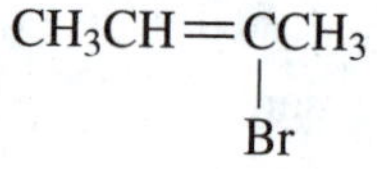 $CH_3CH{=}CCH_3$ **d.** $CH_3CH_2CH_2CH_3$
                       |
                      Br

               Br
b. 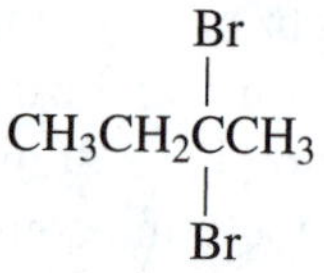 $CH_3CH_2CCH_3$ **e.**

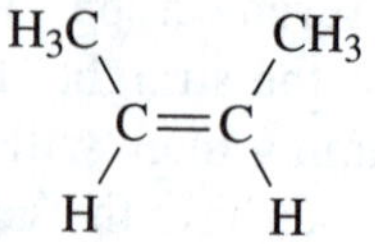

               Br

          O
          ||
c. 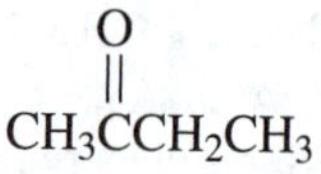 $CH_3CCH_2CH_3$ **f.**   no reaction

44.   **a.**   1,5-octadiyne   **c.**   5-chloro-1,3-cyclohexadiene
     b.   1,6-dimethyl-1,3-cyclohexadiene   **d.**   1-methyl-1,3,5-cycloheptatriene

45.   **a.** 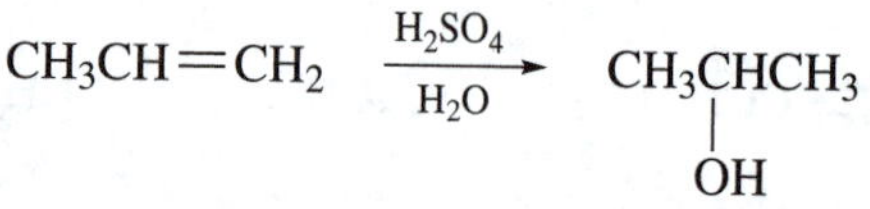 $CH_3CH{=}CH_2 \xrightarrow[\text{H}_2\text{O}]{\text{H}_2\text{SO}_4} CH_3CHCH_3$
                                  |
                                 OH

b.

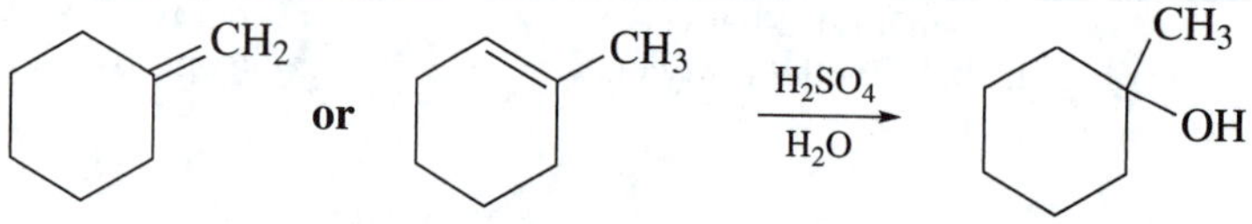

c.

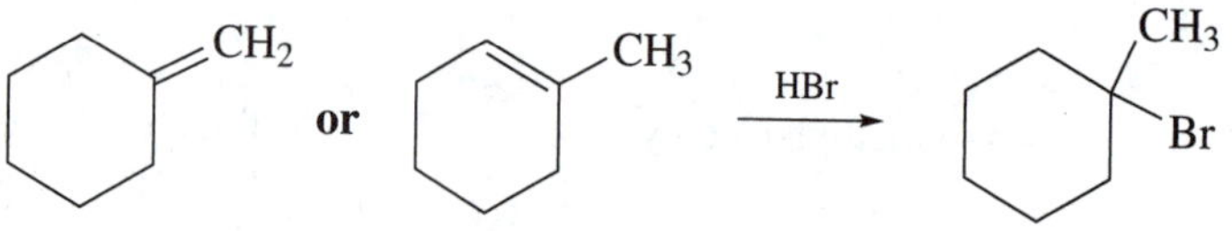

d.

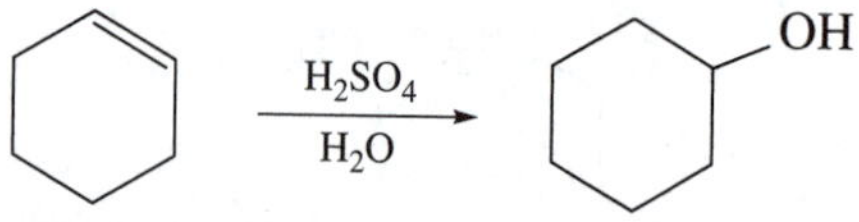

46.   **a.**   Both *cis*- and *trans*-2-butene give these products; in each case, a product with one asymmetric center is formed, so the product is a racemic mixture.

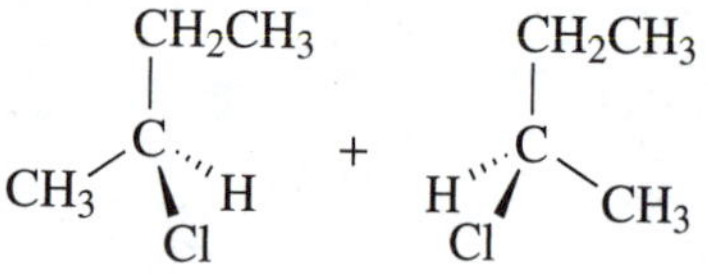

     b.   Both *cis*- and *trans*-2-butene form this product; a product with no asymmetric centers is formed.

$$CH_3CH_2CH_2CH_3$$

c. Both *cis-* and *trans*-2-butene form these products; a product with one asymmetric center is formed.

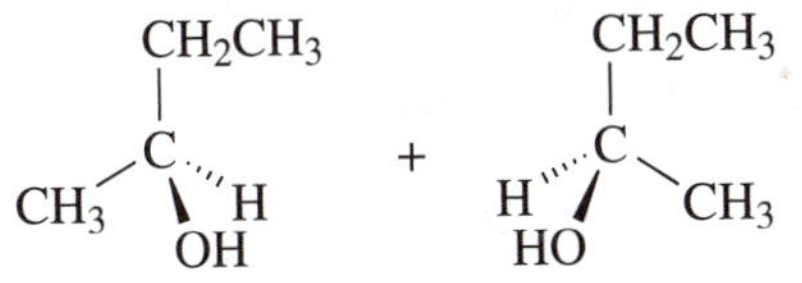

47. Only **C** and **D** are keto–enol tautomers. Notice that an enol tautomer has an OH group bonded to an *sp*^{2} carbon. The structures in **D** are not enol tautomers, because they do not have the oxygen on the same carbon in each structure.

48. Three of the names are correct.

A 3-heptyne	**C** correct	**E** correct
B 5-methyl-3-heptyne	**D** 6,7-dimethyl-3-octyne	**F** correct

49. **a.**

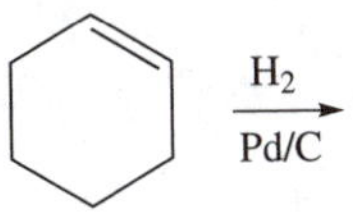

b.

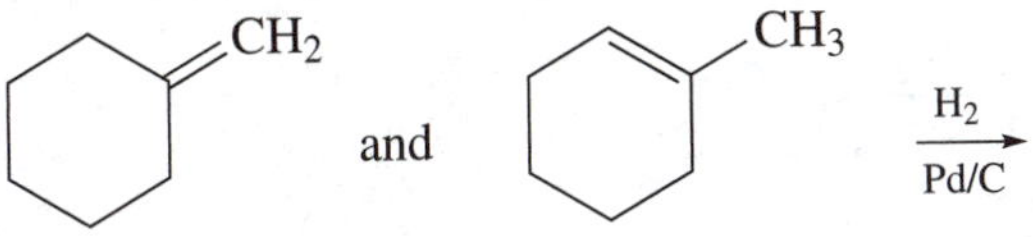

c.

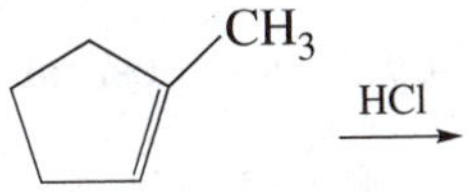

50. **a.**

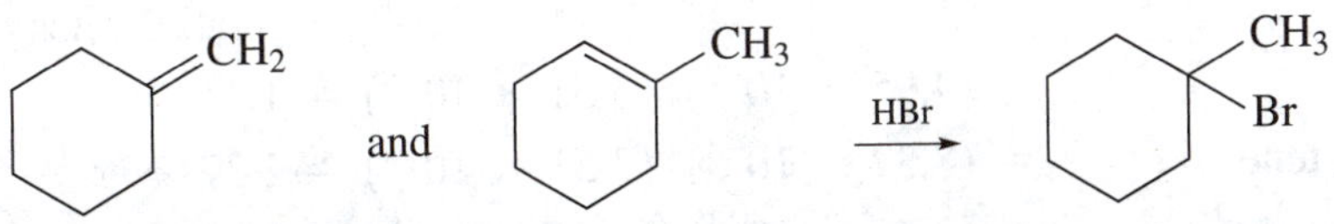

b. (Note: D stands for deuterium, an isotope of hydrogen; DBr reacts in a manner very similar to that of HBr.) While HBr forms the same product when it reacts with the two alkenes, DBr would form different products. They are shown here.

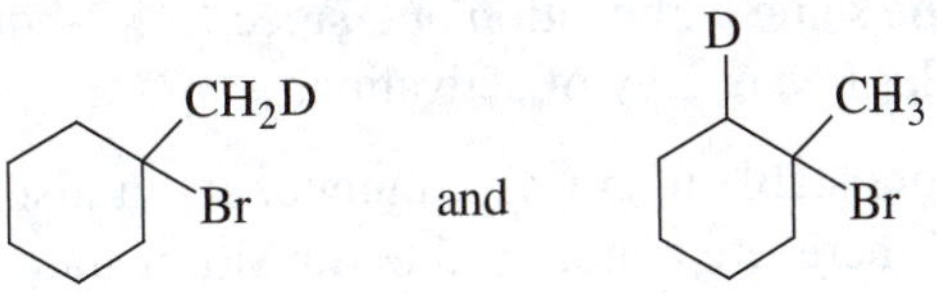

51. **a.** $CH_3CH_2\overset{\overset{\displaystyle O}{\|}}{C}CH_3$ **b.** $CH_3CH_2CH_2\overset{\overset{\displaystyle O}{\|}}{C}CH_3$ **c.**

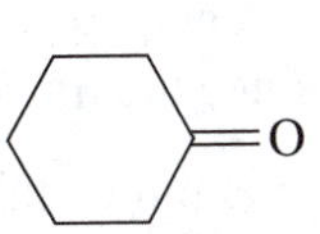

d.

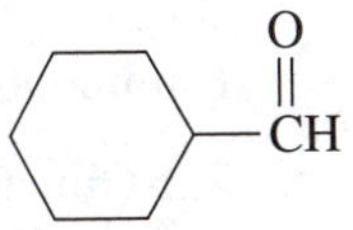

52. **a.**

b. the first step (see the answer to Problem 9)
c. H⁺
d. 1-butene
e. the *sec*-butyl cation
f. methanol

53. **a.**

c.

b.

d.

54. **a.** To determine relative rates, the rate constant of each alkene is divided by the smallest rate constant of the series (3.51×10^{-8}).

relative rates

propene	$= (4.95 \times 10^{-8})/(3.51 \times 10^{-8}) = 1.41$
(Z)-2-butene	$= (8.32 \times 10^{-8})/(3.51 \times 10^{-8}) = 2.37$
(E)-2-butene	$= (3.51 \times 10^{-8})/(3.51 \times 10^{-8}) = 1$
2-methyl-2-butene	$= (2.15 \times 10^{-4})/(3.51 \times 10^{-8}) = 6.12 \times 10^{3}$
2,3-dimethyl-2-butene	$= (3.42 \times 10^{-4})/(3.51 \times 10^{-8}) = 9.74 \times 10^{3}$

b. Both compounds form the same carbocation but, since (Z)-2-butene is less stable than (E)-2-butene, (Z)-2-butene has a smaller free energy of activation.

c. 2-Methyl-2-butene is more stable than (Z)-2-butene, and it also forms a more stable carbocation intermediate (tertiary) and therefore a more stable transition state than does (Z)-2-butene (secondary). Knowing that 2-methyl-2-butene reacts faster tells us that the energy difference between the transition states is greater than the energy difference between the alkenes. This is what we would expect from the Hammond postulate, since the transition states look more like the carbocations than like the alkenes.

d. 2,3-Dimethyl-2-butene is more stable than 2-methylbutene, and both compounds form a tertiary carbocation intermediate. On this basis, you would predict that 2,3-dimethyl-2-butene would react more slowly than 2-methylbutane. However, 2,3-dimethyl-2-butene has two sp^2 carbons that can react with a proton to form the tertiary carbocation, whereas 2-methyl-2-butene has only one sp^2 carbon that can react with a proton to form the tertiary carbocation. The fact that 2,3-dimethyl-2-butene reacts faster in spite of being more stable tells us that the more important factor is the greater number of collisions with the proper orientation that lead to a productive reaction in the case of 2,3-dimethyl-2-butene.

5.

a. 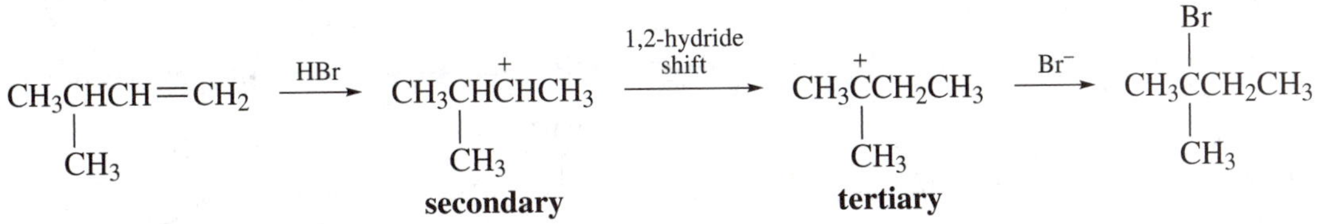 CH₃CHCH=CH₂ →(HBr)→ CH₃CHCHCH₃ →(1,2-hydride shift)→ CH₃CCH₂CH₃ →(Br⁻)→ CH₃CCH₂CH₃

$$ \text{a. } CH_3\underset{CH_3}{CH}CH=CH_2 \xrightarrow{HBr} CH_3\underset{CH_3}{CH}\overset{+}{C}HCH_3 \xrightarrow[\text{shift}]{\text{1,2-hydride}} CH_3\overset{+}{C}\underset{CH_3}{C}H_2CH_3 \xrightarrow{Br^-} CH_3\underset{CH_3}{\overset{Br}{C}}CH_2CH_3 $$

secondary   tertiary

$$ \text{b. } CH_3\underset{CH_3}{CH}CH_2CH=CH_2 \xrightarrow{HBr} CH_3\underset{CH_3}{CH}CH_2\overset{+}{C}HCH_3 \xrightarrow{Br^-} CH_3\underset{CH_3}{CH}CH_2\underset{Br}{CH}CH_3 $$

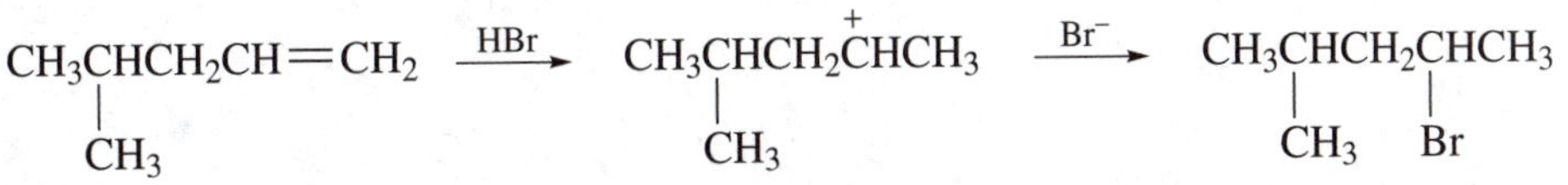

c.

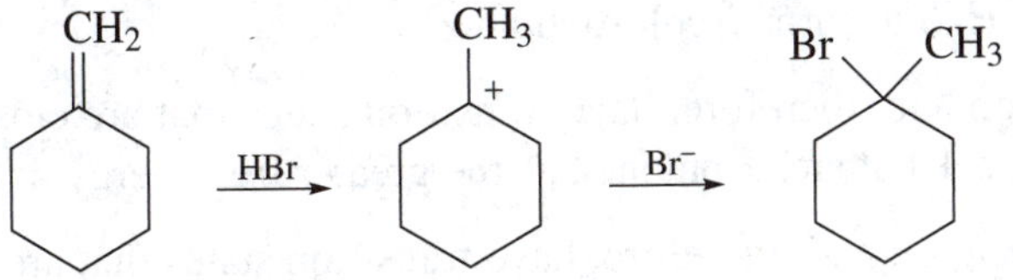

d.

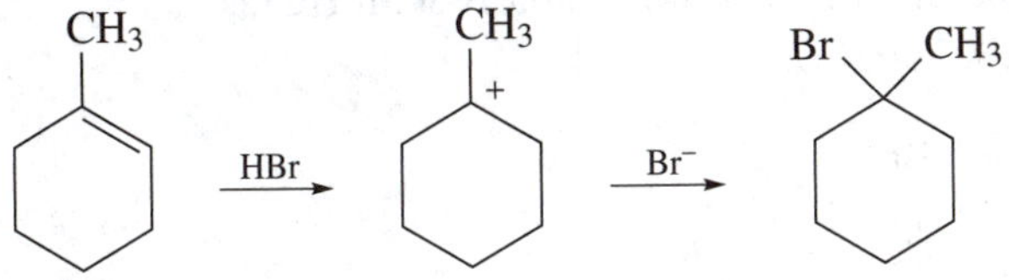

$$ \text{e. } CH_2=CH\underset{CH_3}{\overset{CH_3}{C}}CH_3 \xrightarrow{HBr} CH_3\overset{+}{C}H\underset{CH_3}{\overset{CH_3}{C}}CH_3 \xrightarrow[\text{shift}]{\text{1,2-methyl}} CH_3\underset{CH_3}{CH}\overset{CH_3}{\overset{+}{C}}CH_3 \xrightarrow{Br^-} CH_3CH-\underset{CH_3}{\overset{CH_3\ Br}{C}}CH_3 $$

secondary carbocation   tertiary carbocation

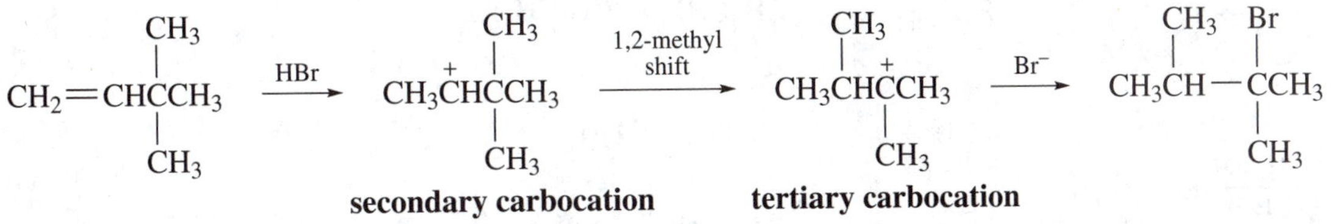

f.

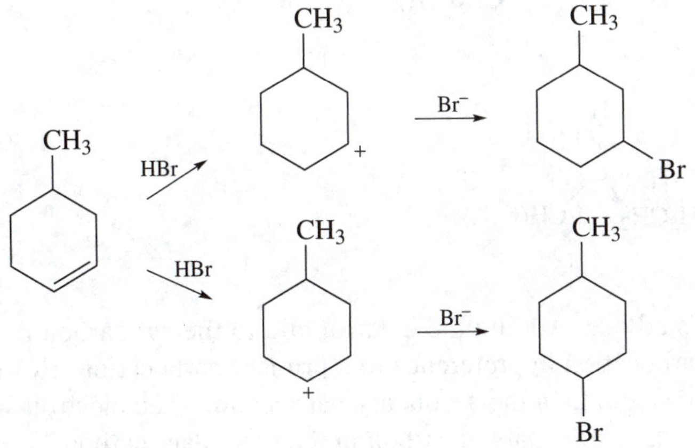

1-Bromo-3-methylcyclohexane and 1-bromo-4-methylcyclohexane will be obtained in approximately equal amounts because, in each case, the intermediate is a secondary carbocation, so the two compounds will be formed at about the same rate. A carbocation rearrangement will not occur, because it would just form another secondary carbocation.

56.

57. **a.** Both 1-butene and 2-butene react with HCl to form 2-chlorobutane.

b. Both alkenes form the same carbocation and, therefore, have transition states that are close in energy but because 2-butene is more stable than 1-butene, 2-butene has the greater free energy of activation.

c. Both compounds form the same carbocation and, therefore, have transition states that are close in energy, but since (Z)-2-butene is less stable, it will react more rapidly with HCl.

58.

59. No, he should not follow his friend's advice. Adding the electrophile to the sp^2 carbon bonded to the most hydrogens forms a secondary carbocation in preference to a primary carbocation. However, in this case, the primary carbocation is more stable than the secondary carbocation. The electron-withdrawing fluorine substituents are closer to the positively charged carbon in the secondary carbon. Therefore, they will destabilize the secondary carbocation more than the primary carbocation. So, the major product will be 1,1,1-trifluoro-3-iodopropane and not 1,1,1-trifluoro-2-iodopropane, the compound that would be predicted to be the major product by following the rule.

$$F_3CCH_2 \overset{+}{-} CH_2 \qquad\qquad F_3C\overset{+}{C}H - CH_3$$

more stable   less stable because of the nearby
                                    electron-withdrawing fluorines

50. 2-Methylpropene will be hydrated more rapidly.

1. It is more reactive than the chloro-substituted alkene, because the electron-withdrawing chlorine makes the alkene less nucleophilic.

2. The carbocation intermediate that 2-methylpropene forms (and therefore the transition state leading to its formation) is more stable, because the electron-withdrawing chlorine increases the amount of positive charge on the carbon.

51. It tells us that the first step of the mechanism is the slow step. If the first step is slow, the carbocation will react with water in a subsequent fast step, which means that the carbocation will not have time to lose a proton to re-form the alkene, so all the deuterium atoms (D) will be retained in the unreacted alkene.

If the first step were not the slow step, an equilibrium would be set up between the alkene and the carbocation and, because the carbocation could lose either H^+ or D^+ when it re-forms the alkene, all the deuterium atoms would not be retained in the unreacted alkene.

62. When forming a cyclic compound, start by numbering the atoms in the reactant and the product. You will then be able to see the atoms that become attached to each other in the cyclic product.

We see that the oxygen forms a bond with C-4. Adding the electrophile (H^+ from HCl) to the sp^2 carbon of the alkene that is bonded to the most hydrogens forms a carbocation (with the positive charge on C-4) that will react with a nucleophile.

There are two nucleophiles that could add to the carbocation, OH and Cl^-. The nucleophile present in greatest concentration is the OH group, so it adds to the carbocation to from the five-membered ring. (There is little Cl^- present because HCl is a catalyst, so only a small amount is needed because it is regenerated.)

Chapter 6 Practice Test

1.   Which member of each pair is more stable?

a. 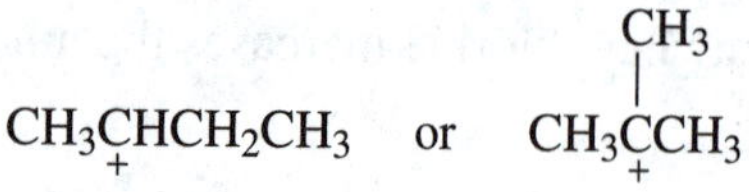 $CH_3\overset{+}{C}HCH_2CH_3$ or $CH_3\overset{+}{\underset{\displaystyle CH_3}{C}}CH_3$

b. $CH_3CH_2CH_2\overset{+}{C}H_2$ or $CH_3CH_2\overset{+}{C}HCH_3$

c. $CH_3CH_2\overset{+}{C}H_2$ or $ClCH_2CH_2\overset{+}{C}H_2$

2.   Which would be a better compound to use as a starting material for the synthesis of 2-bromopentane?

$$CH_3CH_2CH_2\,CH = CH_2 \quad \text{or} \quad CH_3CH_2CH = CHCH_3$$

3.   What is the major product of each of the following reactions?

a. 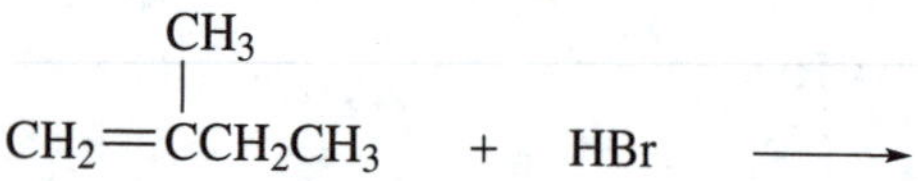 $CH_2{=}\overset{\displaystyle CH_3}{\overset{|}{C}}CH_2CH_3$ +   HBr $\longrightarrow$

b. $CH_3\overset{\displaystyle CH_3}{\overset{|}{C}}HCH{=}CH_2$ +   HCl $\longrightarrow$

c. $CH_3\overset{\displaystyle CH_3}{\underset{\displaystyle CH_3}{\overset{|}{\underset{|}{C}}}}CH{=}CH_2$ +   HBr $\longrightarrow$

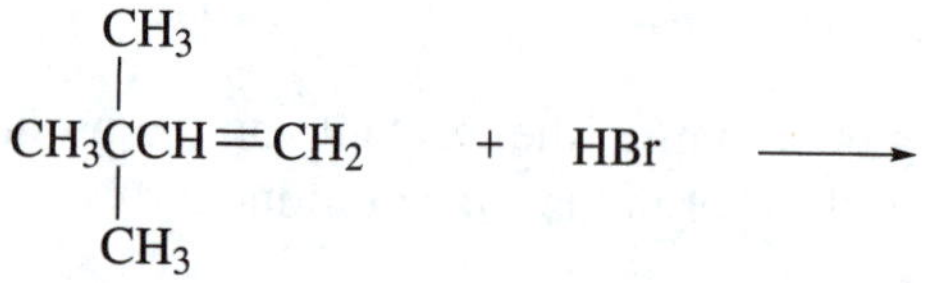

4.   Indicate how each of the following compounds could be synthesized using an alkene as one of the starting materials:

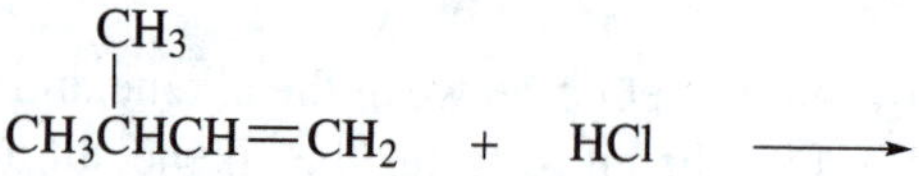

a. 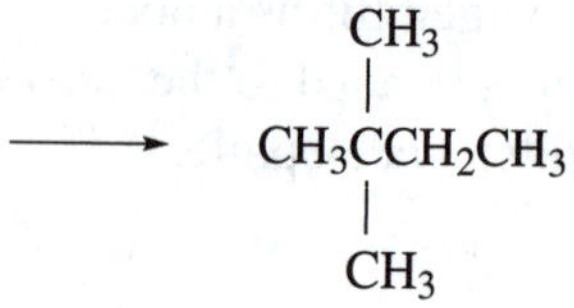 $\longrightarrow$ $CH_3\overset{\displaystyle CH_3}{\underset{\displaystyle CH_3}{\overset{|}{\underset{|}{C}}}}CH_2CH_3$

b. 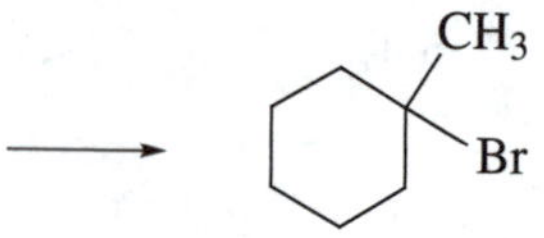 $\longrightarrow$ (cyclohexane ring with CH_3 and Br on same carbon)

5.   Indicate the carbocations that you would expect to rearrange to give a more stable carbocation.

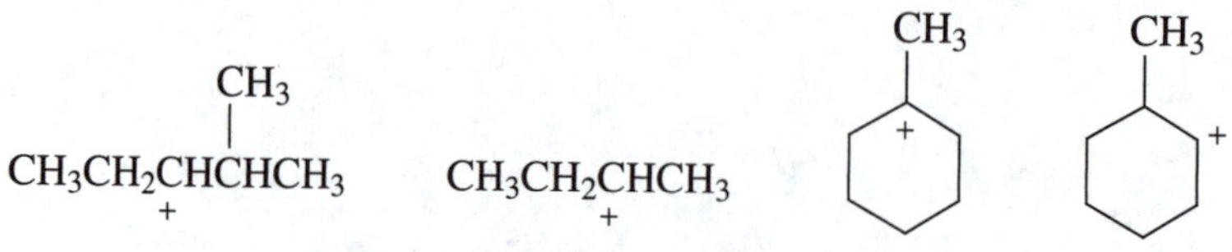

$CH_3CH_2\overset{\displaystyle CH_3}{\underset{+}{\overset{|}{C}}}HCHCH_3$ $CH_3CH_2\underset{+}{C}HCH_3$ (cyclohexane ring with CH_3 and $+$)   (cyclohexane ring with CH_3 and $+$)

Indicate whether each of the following statements is true or false:

a. The reaction of 1-butene with HCl will form 1-chlorobutane as the major product.   T   F

b. The reaction of HBr with 3-methylcyclohexene is more highly
    regioselective than is the reaction of HBr with 1-methylcyclohexene.   T   F

c. The addition of HBr to 3-methyl-2-pentene is a regioselective reaction.   T   F

Draw all the products that would be obtained from each of the following reactions, indicating which isomers are formed:

a. 1-butene + HCl

b. 2-pentene + HBr

Draw the stereoisomers that would be obtained from each of the following reactions:

a. 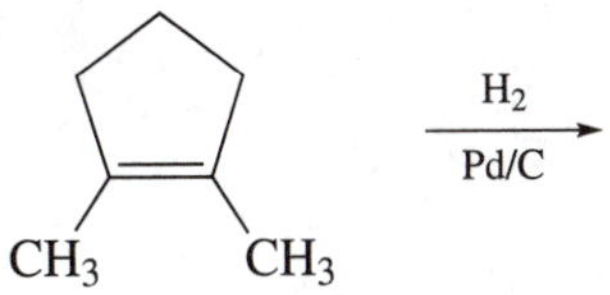 $\xrightarrow[\text{Pd/C}]{\text{H}_2}$ b. 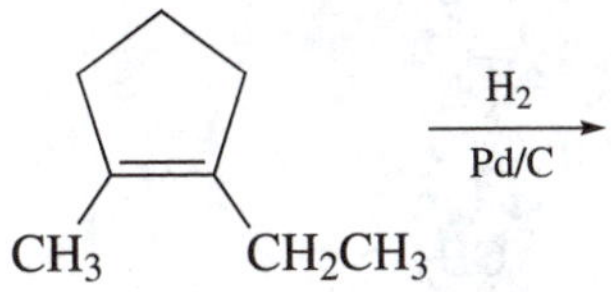 $\xrightarrow[\text{Pd/C}]{\text{H}_2}$

What reagents could be used to convert the given starting material into the desired product?

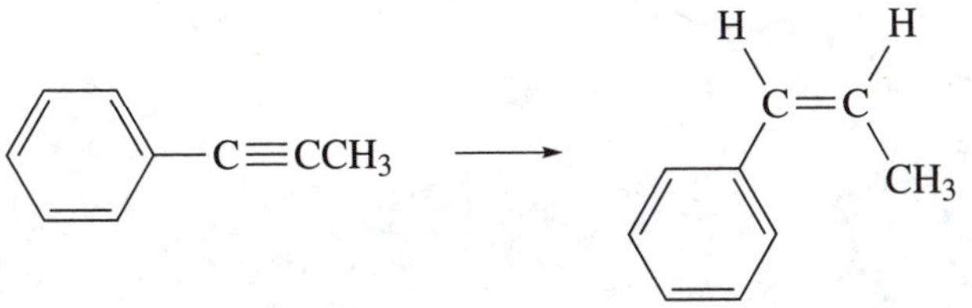

10. Draw the enol tautomer(s) of the following ketones:

a. 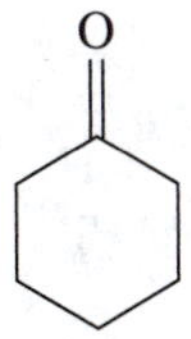   b.

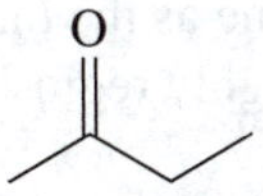

11. Draw the structure for 2-methyl-1,3-cyclohexadiene.

12. What is the compound's systematic name?

$CH_3CHC\equiv CCH_2CH_2Br$
  |
  CH_3

13. Which alkyne would be the best reagent to use for the synthesis of each of the following ketones?

a. $CH_3CH_2CH_2\overset{\overset{\displaystyle O}{\|}}{C}CH_3$ b. $CH_3CH_2\overset{\overset{\displaystyle O}{\|}}{C}CH_2CH_2CH_3$

14. Give an example of a ketone that has two enol tautomers.

Answers to Chapter 6 Practice Test

1. **a.** CH₃ĊCH₃ (with CH₃ substituent on the + carbon)   **b.** CH₃CH₂ĊHCH₃   **c.** CH₃CH₂ĊH₂

1.

a.

$$\overset{\displaystyle CH_3}{\underset{\displaystyle +}{\underset{|}{CH_3\overset{}{C}CH_3}}}$$

b. $CH_3CH_2\overset{+}{C}HCH_3$

c. $CH_3CH_2\overset{+}{C}H_2$

2. $CH_3CH_2CH_2CH{=}CH_2$

3. **a.**

$$\overset{\displaystyle CH_3}{\underset{\displaystyle Br}{\underset{|}{CH_3\overset{|}{C}CH_2CH_3}}}$$

c.

$$\overset{\displaystyle CH_3}{\underset{\displaystyle Br\ \ CH_3}{\underset{|\ \ \ \ |}{CH_3\overset{|}{C}{-}CHCH_3}}}$$

b.

$$\overset{\displaystyle CH_3}{\underset{\displaystyle Cl}{\underset{|}{CH_3\overset{|}{C}CH_2CH_3}}}$$

4. **a.**

$$\overset{\displaystyle CH_3}{\underset{\displaystyle CH_3}{\underset{|}{CH_3\overset{|}{C}CH{=}CH_2}}} \xrightarrow[\text{Pd/C}]{H_2} \overset{\displaystyle CH_3}{\underset{\displaystyle CH_3}{\underset{|}{CH_3\overset{|}{C}CH_2CH_3}}}$$

b. (cyclohexene with CH₃) $\xrightarrow{HBr}$ (cyclohexane with CH₃ and Br)

5. $CH_3CH_2\overset{+}{C}H\overset{\underset{|}{CH_3}}{C}HCH_3$ (methylcyclohexyl cation with +)

6. **a.** The reaction of 1-butene with HCl will form 1-chlorobutane as the major product.   F

   b. The reaction of HBr with 3-methylcyclohexene is more highly regioselective
   than is the reaction of HBr with 1-methylcyclohexene.   F

   c. The addition of HBr to 3-methyl-2-pentene is regioselective.   T

7. **a.** $CH_3CH_2\overset{\underset{\displaystyle Cl}{\underset{|}{}}}{C}HCH_3$ (both R and S)   **b.** $CH_3CH_2CH_2\overset{\underset{\displaystyle Br}{\underset{|}{}}}{C}HCH_3$ (both R and S)   and $CH_3CH_2\overset{\underset{\displaystyle Br}{\underset{|}{}}}{C}HCH_2CH_3$

(no stereoisomers)

8. **a.** (cyclopentane ring with H, H on top carbons and CH₃, CH₃ below)

   b. (cyclopentane with H, H and CH₃, CH₂CH₃) + (cyclopentane with H, H and CH₃CH₂, CH₃)

9.   H₂/Lindlar catalyst

0. **a.** 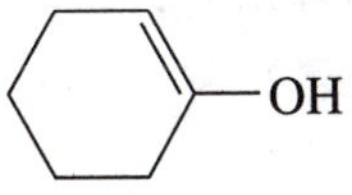 **b.** 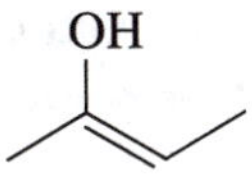 and

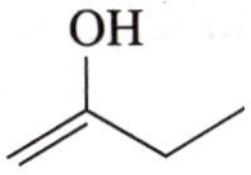

1.

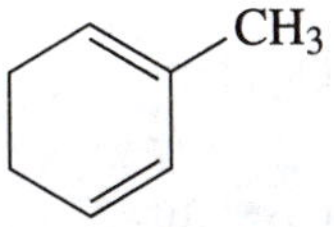

2. 1-bromo-5-methyl-3-hexyne

3. **a.** $CH_3CH_2CH_2C{\equiv}CH$ **b.** $CH_3CH_2C{\equiv}CCH_2CH_3$

4. $CH_3CH_2CH_2\overset{\overset{\displaystyle O}{\|}}{C}CH_2CH_3$

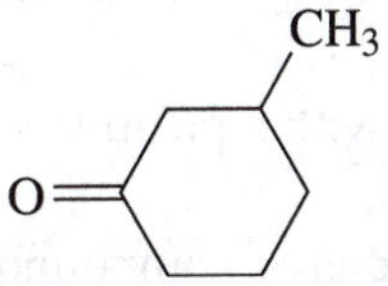

CHAPTER 7
Delocalized Electrons and Their Effect on Stability, pK_a, and the Products of a Reaction • Aromaticity and the Reactions of Benzene

Important Terms

1,2-addition (direct addition)	addition to the 1- and 2-positions of a conjugated system.
1,4-addition (conjugate addition)	addition to the 1- and 4-positions of a conjugated system.
aliphatic compound	a saturated or unsaturated organic compound that does not contain an aromatic ring.
allylic carbon	a carbon adjacent to an sp^2 carbon of a carbon–carbon double bond.
allylic cation	a compound with a positive charge on an allylic carbon.
aromatic compound	a cyclic and planar compound with an uninterrupted cloud of electrons containing an odd number of pairs of π electrons.
benzylic carbon	a carbon, joined to other atoms by single bonds, that is bonded to a benzene ring.
benzylic cation	a compound with a positive charge on a benzylic carbon.
conjugate addition	addition to the 1- and 4-positions of a conjugated system.
conjugated diene	a compound with two conjugated double bonds.
conjugated double bonds	double bonds separated by one single bond.
contributing resonance structure	a structure with localized electrons that approximates the true structure of a compound with delocalized electrons.
delocalization energy (resonance energy)	the extra stability associated with a compound as a result of having delocalized electrons.
delocalized electrons	electrons that are not localized on a single atom or between two atoms.
Diels–Alder reaction	a $[4+2]$ cycloaddition reaction.
diene	a hydrocarbon with two double bonds.
dienophile	an alkene that reacts with a diene in a Diels–Alder reaction.
direct addition (1,2-addition)	addition to the 1- and 2-positions of a conjugated system.
donation of electrons by resonance	donation of electrons through π bonds.
electron delocalization	the sharing of electrons by more than two atoms.

electrophilic aromatic substitution reaction	a reaction in which an electrophile substitutes for a hydrogen of an aromatic ring.
Friedel–Crafts acylation	an electrophilic aromatic substitution reaction that puts an acyl group on an aromatic ring.
Friedel–Crafts alkylation	an electrophilic aromatic substitution reaction that puts an alkyl group on an aromatic ring.
halogenation	reaction with a halogen (Br_2, Cl_2, I_2).
inductive electron withdrawal	withdrawal of electrons through a σ bond.
isolated double bonds	double bonds separated from one another by more than one single bond.
localized electrons	electrons that are restricted to a particular locality.
nitration	substitution of a nitro group (NO_2) for a hydrogen of an aromatic ring.
phenyl group	C_6H_5—
resonance	electron delocalization.
resonance contributor (resonance structure)	a structure with localized electrons that approximates the true structure of a compound with delocalized electrons.
resonance electron donation	donation of electrons through π bonds.
resonance electron withdrawal	withdrawal of electrons through π bonds.
resonance energy (delocalization energy)	the extra stability a compound possesses as a result of having delocalized electrons.
resonance hybrid	the actual structure of a compound with delocalized electrons; it is represented by two or more resonance contributors with localized electrons.
separated charges	a positive and a negative charge that can both be neutralized by the movement of electrons.
sulfonation	substitution of a hydrogen of an aromatic ring with a sulfonic acid group (SO_3H).
withdrawal of electrons by resonance	withdrawal of electrons through π bonds.

Solutions to Problems

1.   **A** and **C** have delocalized electrons.

B, D, and **E** do not have delocalized electrons, because lone-pair electrons cannot be moved toward an sp^3 carbon.

A   1. $CH_3CH\!=\!CH\!-\!CH\!=\!CH\!-\!\overset{+}{C}H_2$ $CH_3\overset{+}{C}H\!-\!CH\!=\!CH\!-\!CH\!=\!CH_2$

$CH_3CH\!=\!CH\!-\!\overset{+}{C}H\!-\!CH\!=\!CH_2$

C $\longleftrightarrow$

2.   The resonance contributor that makes the greatest contribution to the hybrid is labeled **A**. **B** contributes less to the hybrid than **A**, and **C** contributes less to the hybrid than **B**.

a.   Solved in the text.

b.

A is more stable than **B**, because **B** has separated charges and has a positive charge on an oxygen.

c.

A is more stable than **B**, because the negative charge in **A** is on an oxygen, whereas the negative charge in **B** is on carbon, which is less electronegative than oxygen.

d.

A is more stable than **B**, because **A** does not have separated charges. **B** is more stable than **C**, because the electronegative oxygen atom is closer to the positive charge in **C**. **C** is so unstable that it can be neglected.

e.

A is more stable than **B**, because the positive charge in **A** is on a less electronegative atom. (N is less electronegative than O.)

f. $CH_3\overset{+}{C}H\!-\!CH\!=\!CHCH_3$ $\longleftrightarrow$ $CH_3CH\!=\!CH\!-\!\overset{+}{C}HCH_3$

      A   **B**

A and **B** are equally stable.

a. 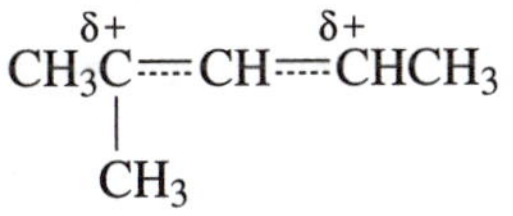 $\overset{\delta+}{CH_3C} =\!=\!= CH =\!=\!= \overset{\delta+}{CHCH_3}$
$\quad\quad\quad\;\; |$
$\quad\quad\quad\; CH_3$

c.

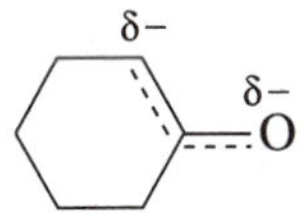

e.

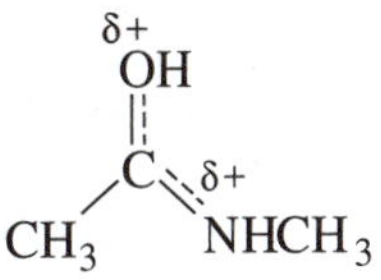

b.

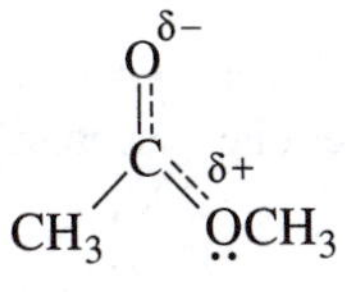

d.

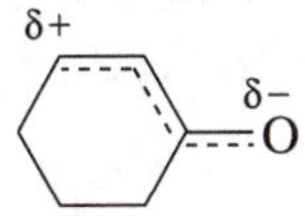

f. 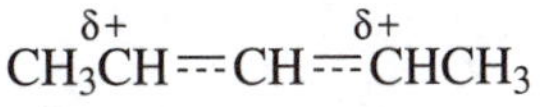 $\overset{\delta+}{CH_3CH} =\!=\!= CH =\!=\!= \overset{\delta+}{CHCH_3}$

a. All the carbon–oxygen bonds in the carbonate ion should be the same length, because each carbon–oxygen bond is represented in one resonance contributor by a double bond and in two resonance contributors by a single bond.

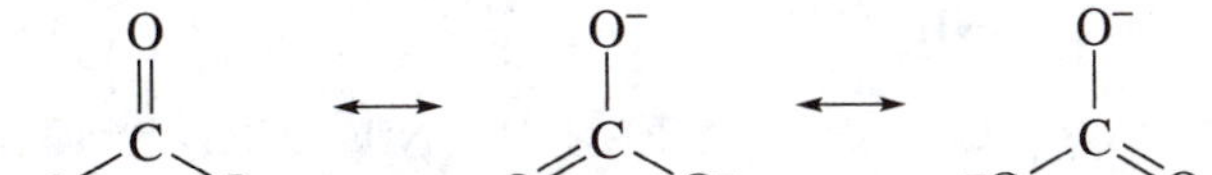

b. Because the two negative charges are shared equally by three oxygens, each oxygen will have two-thirds of a negative charge.

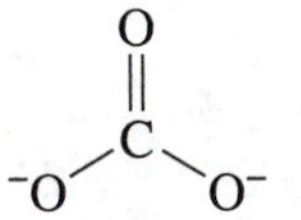

has the greatest delocalization energy, because it has three equivalent resonance contributors

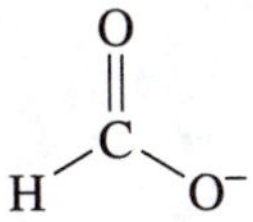

has two equivalent resonance contributors

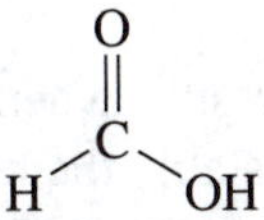

has the least delocalization energy, because its two resonance contributors have very different predicted stabilities

The acetate ion has the greater delocalization energy, because it has two equivalent resonance contributors.

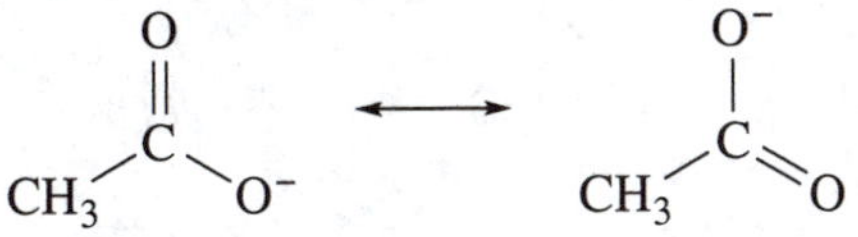

2,4-Heptadiene is more stable, because it has conjugated double bonds.

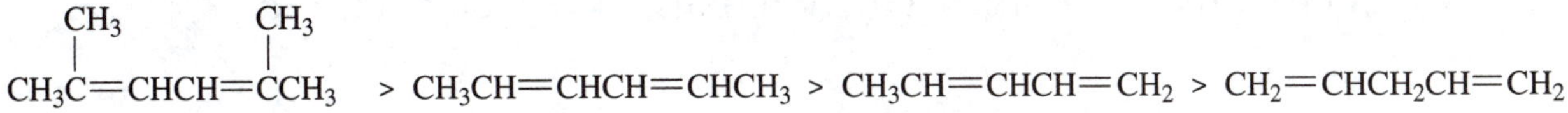

$CH_3\overset{CH_3}{\underset{|}{C}}=CHCH=\overset{CH_3}{\underset{|}{C}}CH_3$ > $CH_3CH=CHCH=CHCH_3$ > $CH_3CH=CHCH=CH_2$ > $CH_2=CHCH_2CH=CH_2$

2,5-dimethyl-2,4-hexadiene   2,4-hexadiene   1,3-pentadiene   1,4-pentadiene

9.

a. $\overset{+}{CH_3CH=CHCHCH_3}$

secondary allylic
(the other one is primary allylic)

b. [structure: cyclohexene ring with $\overset{+}{CHCH_3}$ substituent]

secondary allylic
(the other one is secondary)

c. [benzene ring with $-\overset{+}{C}CH_3$ and CH_3 below]

tertiary benzylic
(the other one is secondary benzylic)

10.

a.

$$\overset{\overset{+}{}OH}{CH_3-\underset{}{\overset{\|}{C}}-\ddot{N}H_2} \longleftrightarrow \overset{OH}{CH_3-\underset{}{C}=\overset{+}{N}H_2} \quad CH_3\overset{\overset{+}{O}H}{\diagdown}\overset{\|}{C}\diagdown NH_2 \quad CH_3\diagdown\overset{OH}{C}\underset{}{\overset{+}{\lessgtr}}NH_2$$

Less stable, because the positive charge is shared by an O and an N.

$$\overset{\overset{+}{}NH_2}{CH_3-\underset{}{\overset{\|}{C}}-\ddot{N}H_2} \longleftrightarrow \overset{NH_2}{CH_3-\underset{}{C}=\overset{+}{N}H_2} \quad CH_3\overset{\overset{+}{N}H_2}{\diagdown}\overset{\|}{C}\diagdown NH_2 \quad CH_3\diagdown\overset{NH_2}{C}\underset{}{\overset{+}{\lessgtr}}NH_2$$

More stable, because the positive charge is shared by two nitrogens. Nitrogen is less electronegative than oxygen, so nitrogen is more comfortable with a positive charge. In addition, this species is more stable because it has two equivalent resonance contributors.

b. $\overset{\overset{O^-}{|}}{CH_3CH-CH=CH_2}$

Less stable, because the negative charge is not delocalized.

$$\overset{\ddot{\overset{..}{O}}\overline{}}{CH_3C=CHCH_3} \longleftrightarrow \overset{O}{CH_3\overset{\|}{C}-\overset{..}{\overline{C}}HCH_3}$$

More stable, because the negative charge is delocalized.

11. In each case, the compound shown is the stronger acid, because the negative charge that results when it loses a proton can be delocalized. Electron delocalization stabilizes the base, and the more stable the base, the more acidic its conjugate acid. Electron delocalization is not possible for the other compound in each pair.

a. $CH_3\overset{\overset{O}{\|}}{\underset{}{C}}OH \longrightarrow H^+ + CH_3\overset{\overset{O}{\|}}{\underset{}{C}}O^- \longleftrightarrow CH_3\overset{\overset{O^-}{|}}{\underset{}{C}}=O$

b. $CH_3CH=CHOH \longrightarrow H^+ + CH_3CH=CHO^- \longleftrightarrow CH_3\overline{C}HCH=O$

c. $CH_3CH=CH\overset{+}{N}H_3 \longrightarrow H^+ + CH_3CH=CHNH_2 \longleftrightarrow CH_3\overline{C}HCH=\overset{+}{N}H_2$

2.   **a.**   Ethylamine is a stronger base, because when the lone pair on the nitrogen in aniline is protonated, it can no longer be delocalized into the benzene ring. Therefore, aniline is less apt to share its electrons with a proton.

   b.   Ethoxide ion is a stronger base, because a negatively charged oxygen is a stronger base than a neutral nitrogen.

   c.   Ethoxide ion is a stronger base, because when the phenolate ion is protonated, the pair of electrons that is protonated can no longer be delocalized into the benzene ring. Therefore, the phenolate ion is less apt to share its electrons with a proton.

3.   **a.**   donates electrons by resonance and withdraws electrons inductively
   b.   donates electrons by hyperconjugation
   c.   withdraws electrons by resonance and withdraws electrons inductively
   d.   donates electrons by resonance and withdraws electrons inductively
   e.   donates electrons by resonance and withdraws electrons inductively
   f.   withdraws electrons inductively

4.   For **a, b,** and **c,** the closer the electron-withdrawing substituent is to the COOH group, the stronger the acid.

   a. $ClCH_2COOH$

   c.   COOH
   (benzene ring with $CH_3C{=}O$ at para position)

   e.   HCOOH
   A hydrogen is electron-withdrawing compared to a methyl group, since a methyl group can donate electrons by hyperconjugation.

   b. O_2NCH_2COOH

   d. $H_3\overset{+}{N}CH_2COOH$

   f.   COOH
   (benzene ring with Cl at para position)

5.   Solved in the text.

6.   When *para*-nitrophenol loses a proton, the negative charge in the conjugate base can be delocalized onto the nitro substituent. Therefore, the *para*-nitro substituent decreases the pK_a both by resonance electron withdrawal and by inductive electron withdrawal.

When *meta*-nitrophenol loses a proton, the negative charge in the conjugate base cannot be delocalized onto the nitro substituent. Therefore, the *meta*-nitro substituent can decrease the pK_a only by inductive electron withdrawal. Therefore, the para isomer has a lower pK_a.

17.   a. Solved in the text.

b. The contributing resonance structures show that there are two sites that can be protonated, the lone pair on nitrogen or the lone pair on carbon.

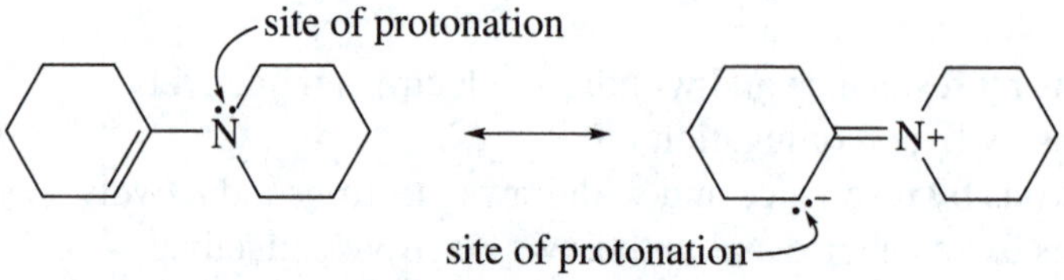

18.   a. The more reactive double bond is the one that forms a tertiary carbocation.

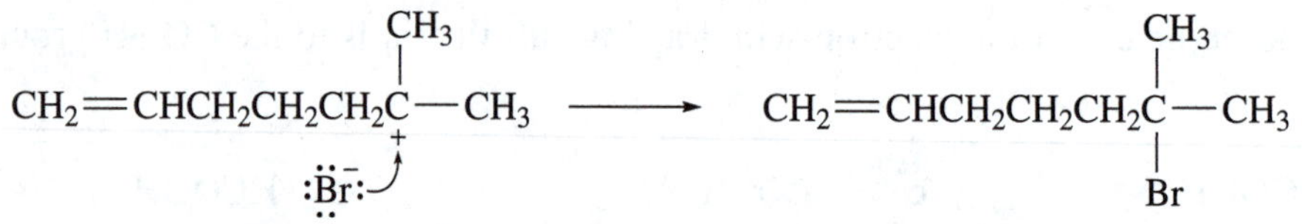

Instead of Br^- being the nucleophile that adds to the tertiary carbon, the π bond can be the nucleophile. In that case, a stable six-membered ring will be formed. This is expected to be a minor product because, unlike the above reaction of the carbocation with Br^-, bond breaking is required to form the product.

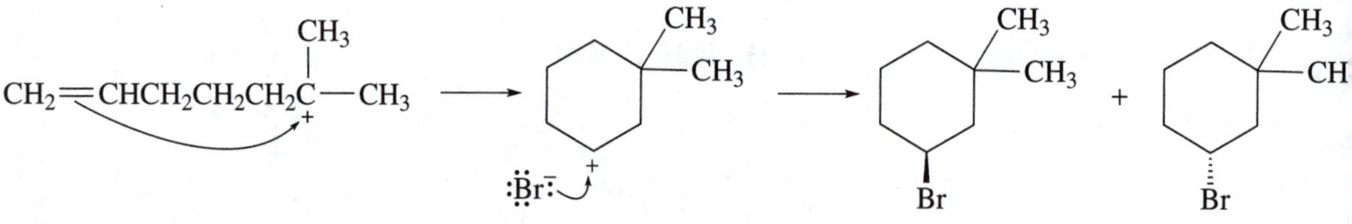

b. The more reactive double bond is the one that forms a tertiary carbocation.

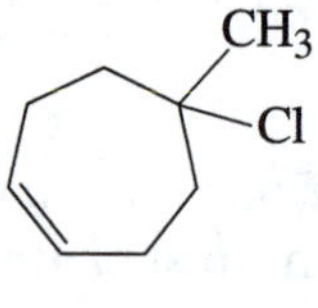

9. **a.** $CH_3CH=CH-CH=CHCH_3$

$\downarrow H^+$

$CH_3CH_2-\overset{+}{C}H-CH=CHCH_3$ $\longleftrightarrow$ $CH_3CH_2-CH=CH-\overset{+}{C}HCH_3$

$\downarrow Cl^-$ $\downarrow Cl^-$

$\overset{\displaystyle Cl}{\underset{\displaystyle |}{CH_3C_2HCHCH=CHCH_3}}$ + $\overset{\displaystyle Cl}{\underset{\displaystyle |}{CH_3CH_2CH=CHCHCH_3}}$

1,2-addition product   1,4-addition product

b. $CH_3CH_2\overset{\displaystyle Br}{\underset{\displaystyle |}{C}}-\overset{}{\underset{\displaystyle |}{C}}=CHCH_3$ + $CH_3CH_2\overset{\displaystyle Br}{\underset{\displaystyle |}{C}}=\overset{}{\underset{\displaystyle |}{C}}CHCH_3$

$\quad\quad\overset{}{\underset{\displaystyle CH_3\ CH_3}{}}$ $\overset{}{\underset{\displaystyle CH_3\ CH_3}{}}$

1,2-addition product   1,4-addition product

c. [structure with OH]   +   [structure with HO]   Only one product is
obtained, because
the 1,2- and 1,4-
products are
identical.

1,2-addition product   1,4-addition product

d. [structure with Cl]   Only one product is
obtained, because
the double bonds in
the reactant are not
conjugated.

10. The indicated double bond is the most reactive in an electrophilic addition reaction with HBr, because addition of an electrophile to this double bond forms the most stable carbocation (a tertiary allylic cation).

21.   first reaction:

$$CH_2-\overset{*}{C}H-CH=CH_2$$
$$\quad\;|\qquad|$$
$$\quad Cl\quad\; Cl$$

$$CH_2-CH=CH-CH_2$$
$$\;\;|\qquad\qquad\quad|$$
$$\;\;Cl\qquad\qquad\quad Cl$$

This compound has an asymmetric center, so both the *R* and *S* stereoisomers will be obtained. (Note that *E* and *Z* stereoisomers are not possbile for this double bond.)

This compound has a double bond, so both the *E* and *Z* stereoisomers will be obtained.

$$CH_2=CH-CH=CH-CH=CH_2 \xrightarrow{\text{HBr}} CH_3-CH-CH=CH-CH=CH_2$$
$$\qquad\qquad\qquad\qquad\qquad\qquad\qquad\qquad\qquad\quad\;|$$
$$\qquad\qquad\qquad\qquad\qquad\qquad\qquad\qquad\qquad\; Br$$

22.

$$CH_3-CH=CH-CH-CH=CH_2$$
$$\qquad\qquad\qquad\quad\;|$$
$$\qquad\qquad\qquad\quad Br$$

$$CH_3-CH=CH-CH=CH-CH_2$$
$$\qquad\qquad\qquad\qquad\qquad\quad\;|$$
$$\qquad\qquad\qquad\qquad\qquad\; Br$$

23.   **a.**   **b.**   **c.**

24.   **a.** 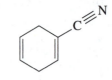   **b.**   **c.**

25.   This is the only one that is aromatic; it is cyclic and planar, every ring atom has a *p* orbital, and it has three pairs of π electrons.

The first compound is not aromatic, because one of the atoms is sp^3 hybridized and, therefore does not have a *p* orbital.

The third compound is not aromatic, because it has four pairs of π electrons.

26.   Solved in the text.

7.   Only **C** is aromatic.

C is aromatic, because it is cyclic and planar, every atom in the ring has a *p* orbital, and it has seven pairs of π electrons.

A and **B** are not aromatic, because each compound has two pairs of π electrons and every atom in the ring does not have a *p* orbital.

D is not aromatic, because it is not cyclic.

8.   Solved in the text.

9.   **a.**

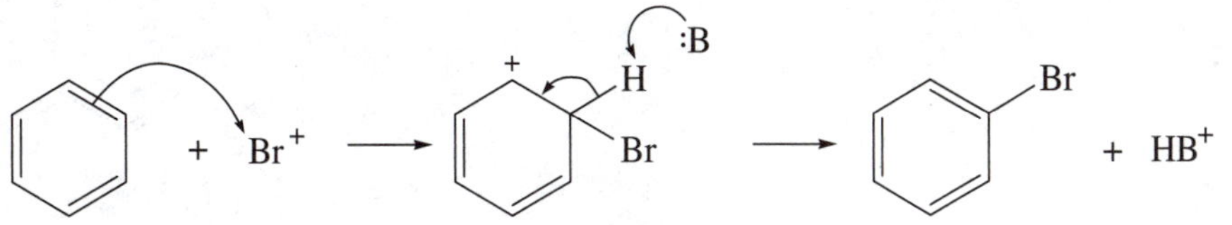

b.

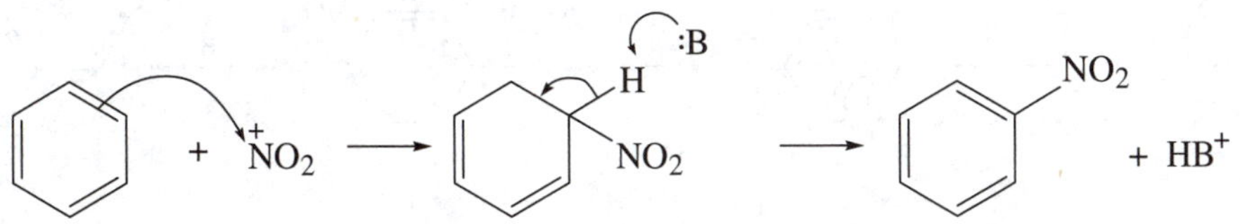

c.

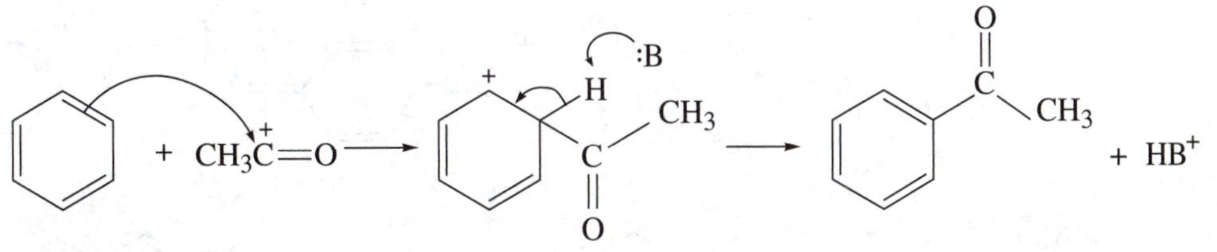

d.

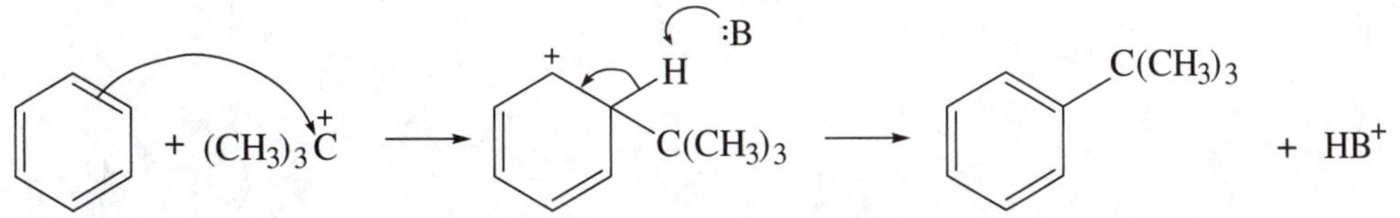

30.   A carbocation rearrangement occurs in **b** and **e**.

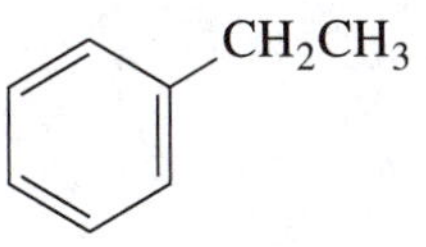

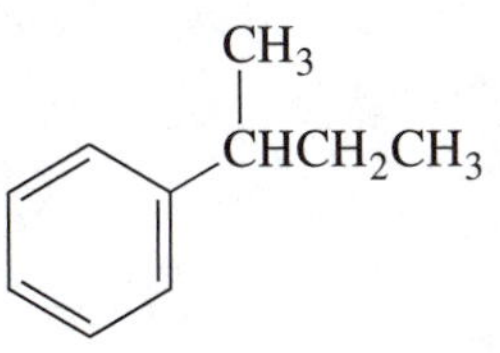

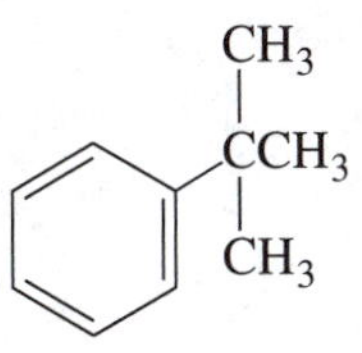

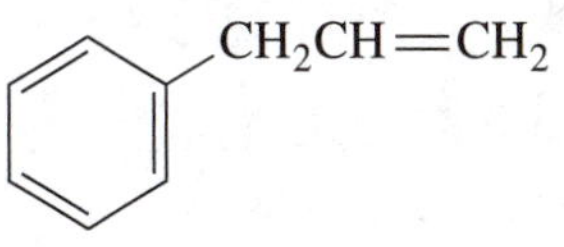

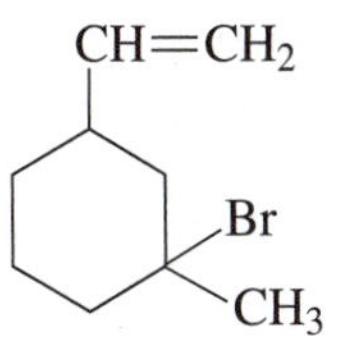

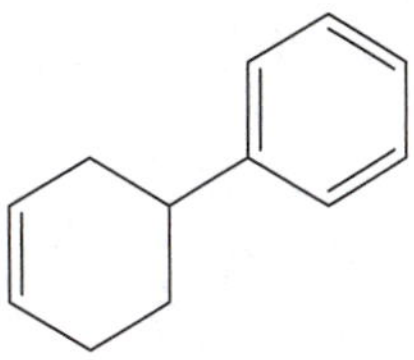

31.   **A, B, C, E, J, K, L**

32.   **a.**

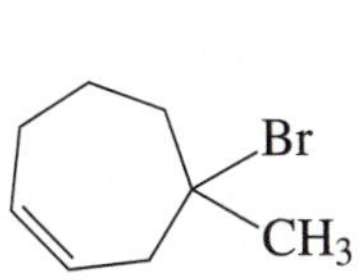

33. **a.** different compounds   **d.** resonance contributors
        b. different compounds   **e.** different compounds
        c. resonance contributors

Notice that in the structures that are different compounds, both atoms and electrons have changed their locations. In contrast, in the structures that are resonance contributors, only the electrons have moved.

34. **a.**

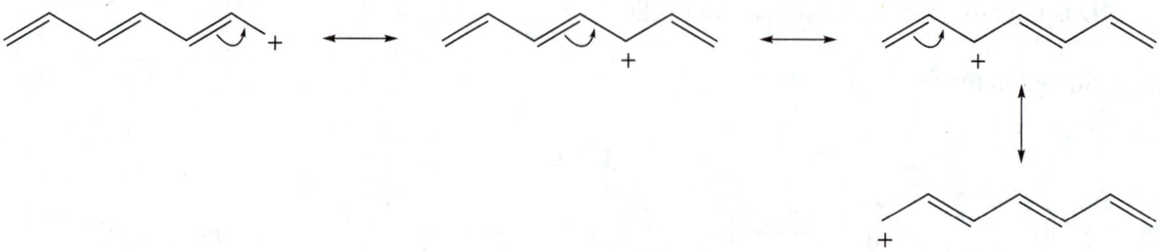

b.

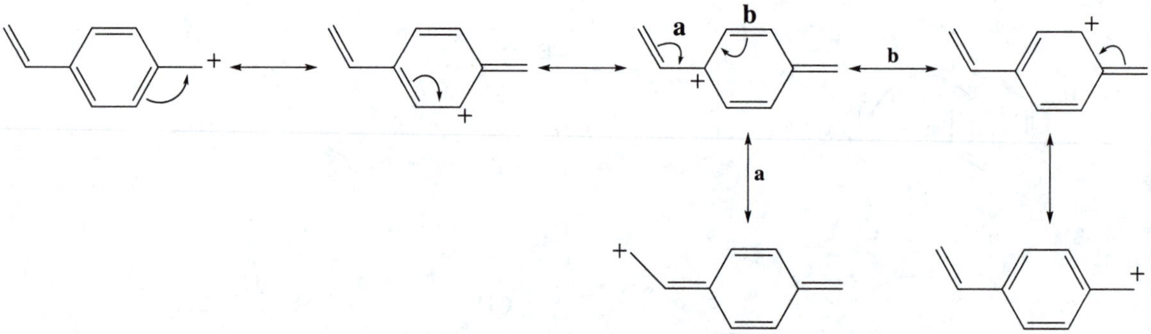

c.

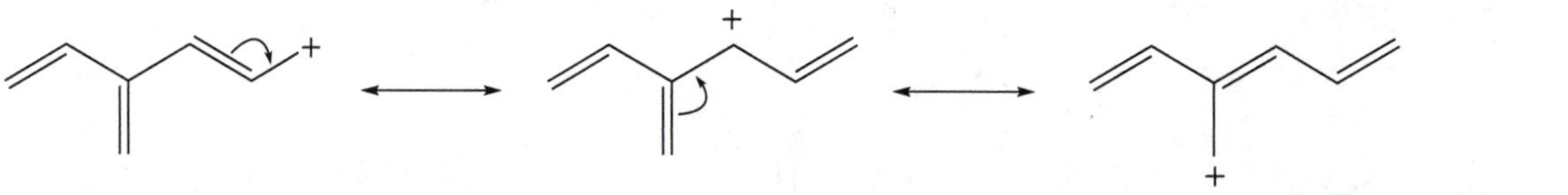

35. **a.**
        CH₃
        |
        CHCH₃

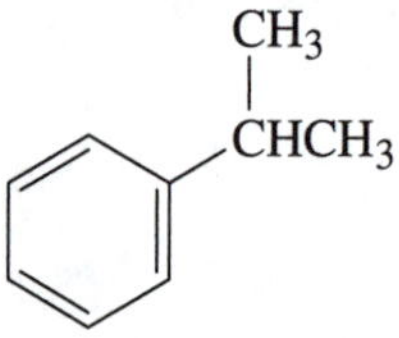

        b.   CH₂CH=CH₂

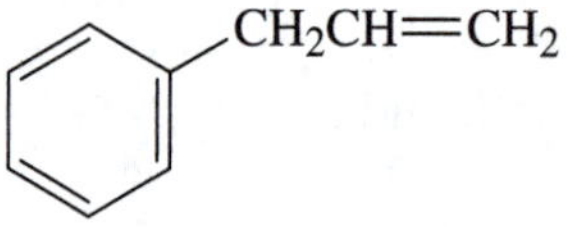

        c.
        O 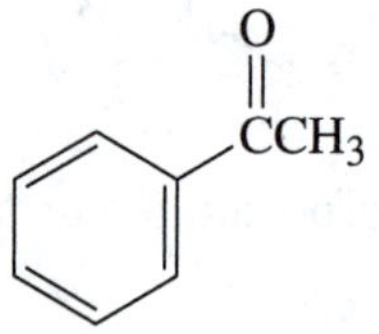
        ‖
        CCH₃

6. Draw the resonance contributors. Then have the Cl pointing up in all the products, with the OH pointing up in one product and down in the other. Notice that the products in row 4 are the mirror images of the products in row 1, and that the products in row 3 are the mirror images of the products in row 2. (The mirror images would also have been obtained if the Cl pointed down in all the products with the OH pointing up in one structure and down in the other.)

7.

a. CH$_2$CH$_3$ donates electrons inductively, but it does not donate or withdraw electrons by resonance.

b. NO$_2$ withdraws electrons inductively and withdraws electrons by resonance.

c. Br withdraws electrons inductively and donates electrons by resonance, but it is better at withdrawing inductively.

d. OH withdraws electrons inductively, donates electrons by resonance, but is better at donating by resonance.

e. $^+$NH$_3$ withdraws electrons inductively and does not donate or withdraw electrons by resonance.

8.   a.   b.

9. The compound with the methoxy substituent is the more reactive, because it forms the more stable carbocation intermediate. The carbocation intermediate is stabilized by resonance electron donation.

40.   Both compounds form the same product when they are hydrogenated, so the difference in the heats of hy
drogenation will depend only on the difference in the stabilities of the reactants. Because 1,2-pentadiene
has cumulated double bonds and 1,4-pentadiene has isolated double bonds, 1,2-pentadiene is less stable
and, therefore, will have a greater heat of hydrogenation (a more negative $\Delta H°$).

$$CH_2{=}C{=}CHCH_2CH_3 \xrightarrow[Pd/C]{H_2} CH_3CH_2CH_2CH_2CH_3$$

1,2-pentadiene

$$CH_2{=}CHCH_2CH{=}CH_2 \xrightarrow[Pd/C]{H_2} CH_3CH_2CH_2CH_2CH_3$$

41.   aromatic

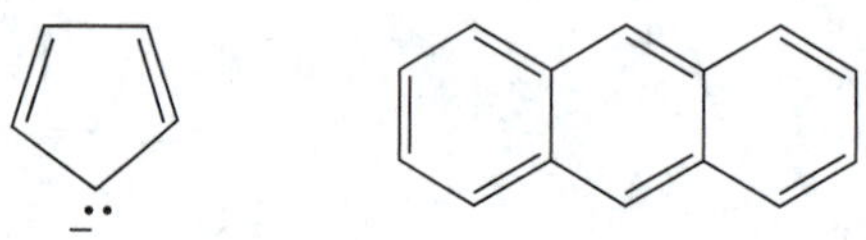

42.   **a.** **1.**

major   minor

2.

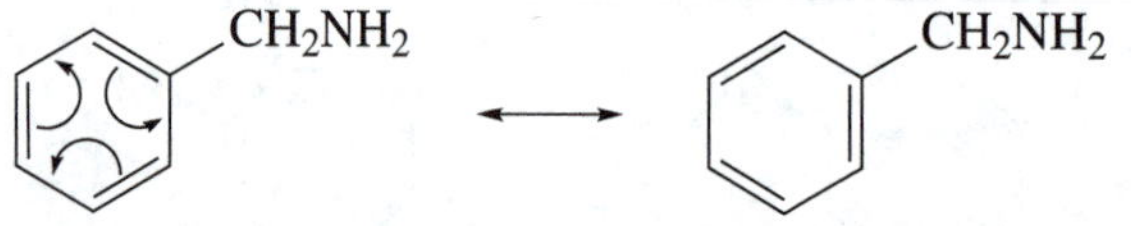

The two resonance contributors have the same stability and, therefore, contribute equally to the
resonance hybrid.

3.

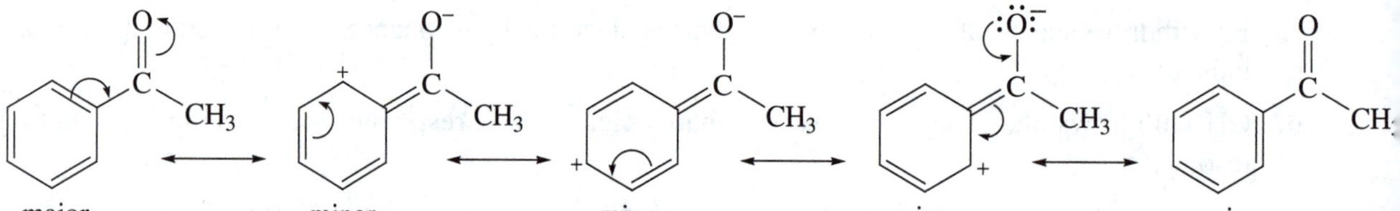

major   minor   minor   minor   major

4.

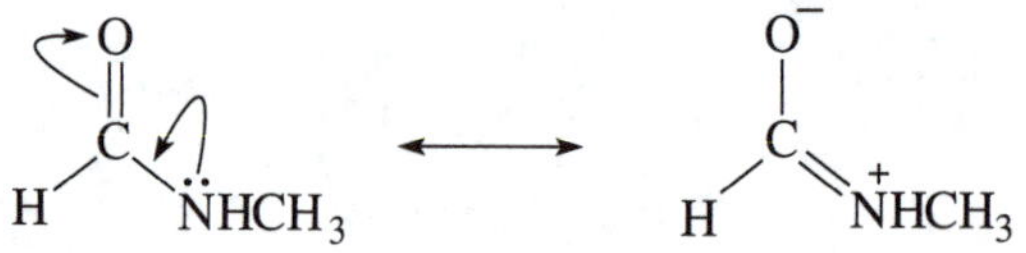

H   NHCH$_3$ H   NHCH$_3$
major   minor

5.

minor   major

6.

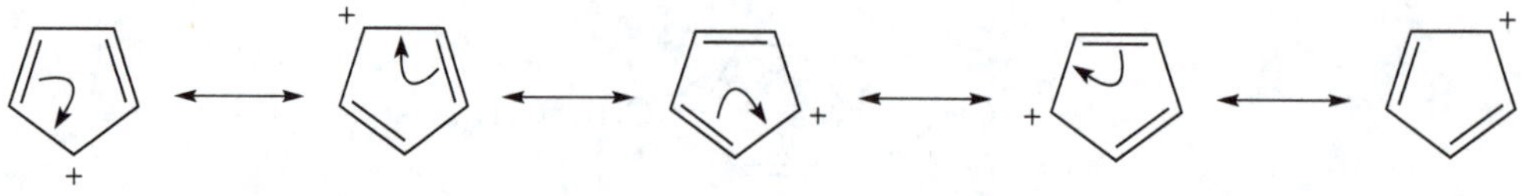

The five contributors are equally stable and therefore contribute equally to the resonance hybrid.

7.

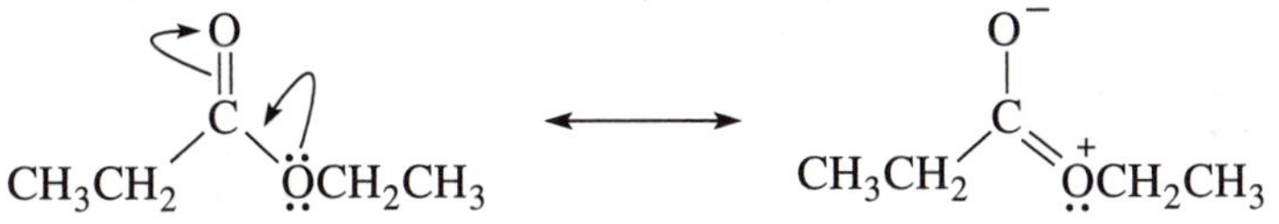

major   minor

8.

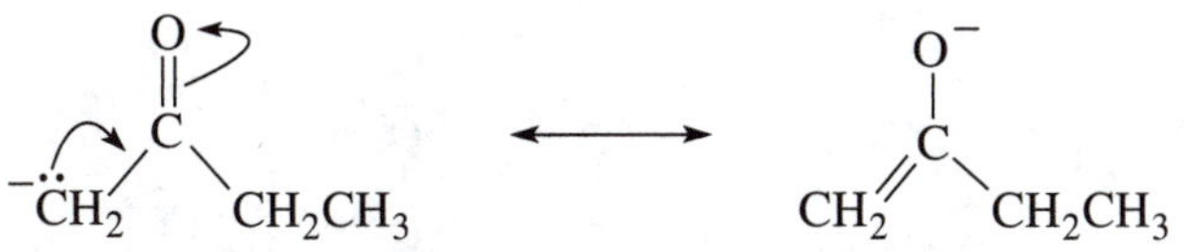

minor   major

The major contributor has a negative charge on oxygen, which is more stable than a contributor with a negative charge on carbon.

9. $CH_3CH{=}CHCH{=}CH\overset{+}{C}H_2$ $CH_3\overset{+}{C}HCH{=}CHCH{=}CH_2$

minor
(the positive
charge is on
a primary
allylic carbon)

major
(the positive
charge is on
a secondary
allylic carbon)

$CH_3CH{=}CH\overset{+}{C}HCH{=}CH_2$
major (the positive charge is on
a secondary allylic carbon)

10.

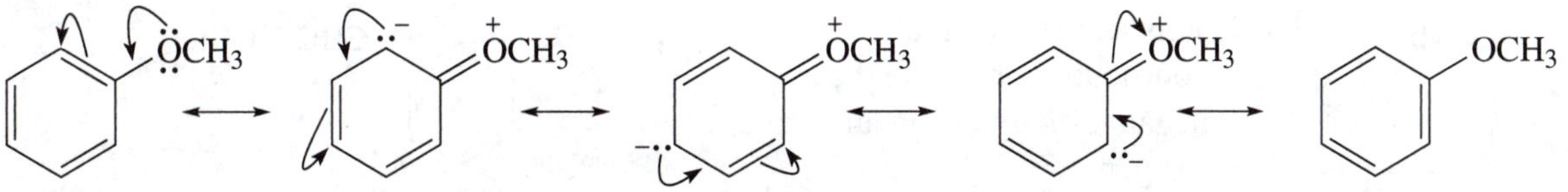

major   minor   minor   minor   major

11. 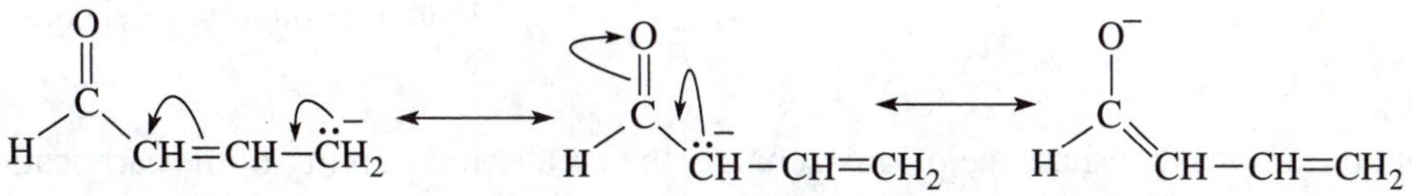   H   H   H

minor   minor   major

12. Notice that the electrons on the center carbon can be delocalized onto both of the carbonyl oxygen

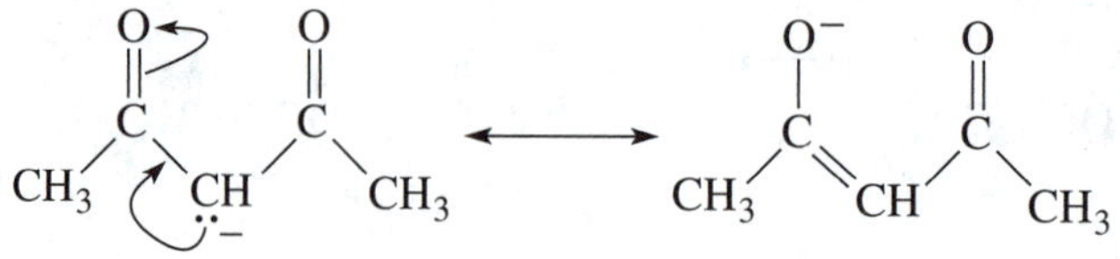

minor   major

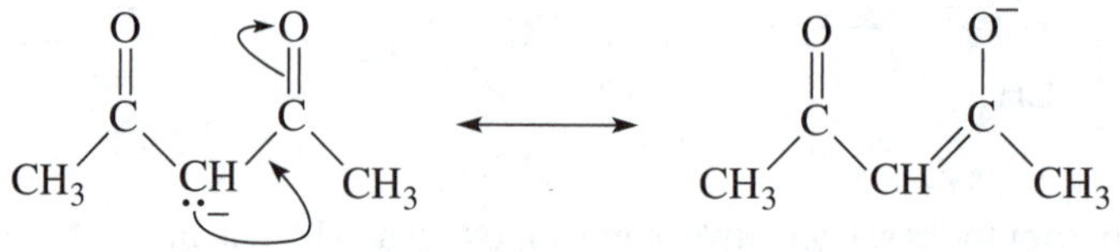

major

b. **2,** and **6**

43. **a.** $CH_3\overset{+}{C}HCH=CH_2$

This makes the greater contribution, because the positive charge is on a secondary allylic carbon.

c.

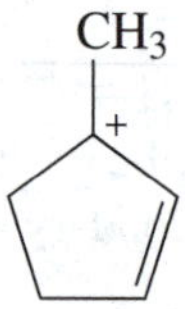

This makes the greater contribution, because the positive charge is on a tertiary allylic carbon.

b.

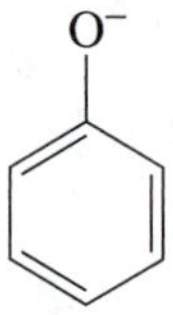

This makes the greater contribution, because the negative charge is on an oxygen.

d.

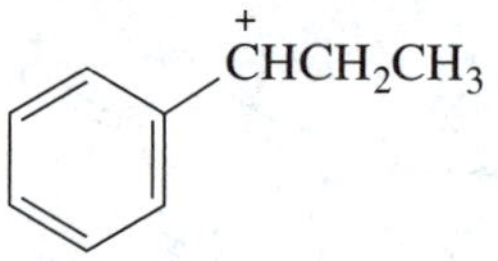

This makes the greater contribution, because a secondary benzylic cation is more stable than a secondary alkyl cation.

44.   The more carbons that share the positive charge, the greater the stability of the carbocation.

The positive charge on the carbocation on the left can be shared by two other carbons as a result of electro delocalization; the positive charge on the carbocation in the middle can be shared by one other carbon; th positive charge on the carbocation on the right cannot be shared by other carbons. Therefore, the carboca tions have the relative stabilities shown here:

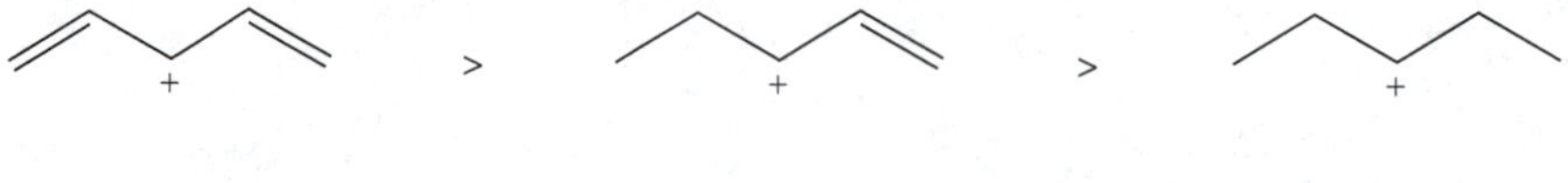

45.

a.   The negative charge is shared by two oxygens.

b.   The negative charge is shared by a carbon and an oxygen.

c.   The negative charge is shared by a carbon and two oxygens.

d.   The negative charge is shared by a nitrogen and two oxygens.

46.   The stronger base is the less stable base of each pair in Problem 45.

a.   Less stable because the negative charge cannot be delocalized.

b.   Less stable because the negative charge cannot be delocalized.

c.   Less stable because the negative charge can be delocalized onto only one oxygen.

d.   Less stable because the negative charge can be delocalized onto only one oxygen.

47.   The more the substituent can donate electrons into the ring, the more basic will be the oxyanion.

48.   **a.**   The resonance contributors show that the carbonyl oxygen has the greater electron density.

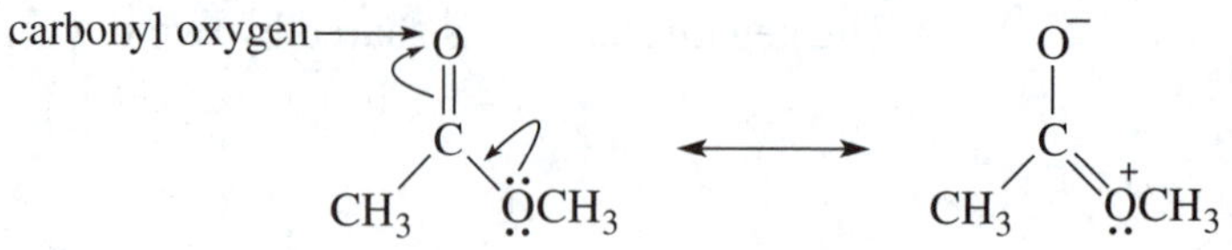

   b.   The compound on the right has the greater electron density on its nitrogen, because the compoun
   on the left has a resonance contributor with a positive charge on the nitrogen as a result of electro
   delocalization.

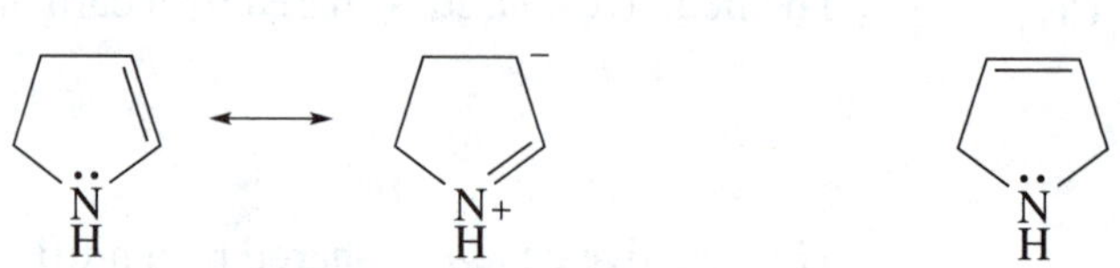

   c.   The compound with the cyclohexane ring has the greater electron density on its oxygen, because th
   lone pair on the nitrogen can be delocalized only onto the oxygen.

   d.   The first compound is the more stable (weaker) base, because it has a resonance contributor that is aro
   matic. Therefore, the other base (the one on the right) is the stronger base.

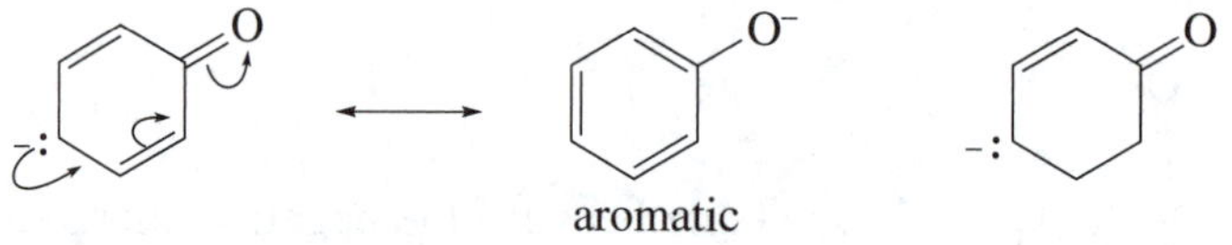

aromatic

49.   The methyl group on benzene can lose a proton easier than the methyl group on cyclohexane, because th
   electrons left behind on the carbon in the former can be delocalized into the benzene ring. In contrast, th
   electrons left behind in the other compound cannot be delocalized.

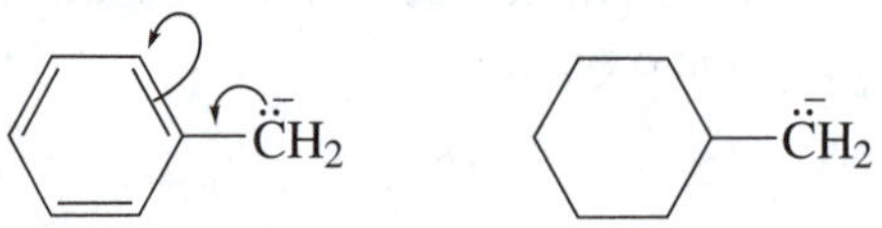

50.   The compound shown here is the strongest base. The lone-pair electrons on the nitrogens of the other tw
   compounds are delocalized, so they are less available for protonation.

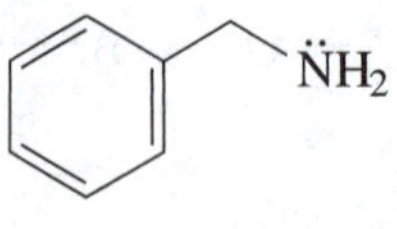

1.   a.

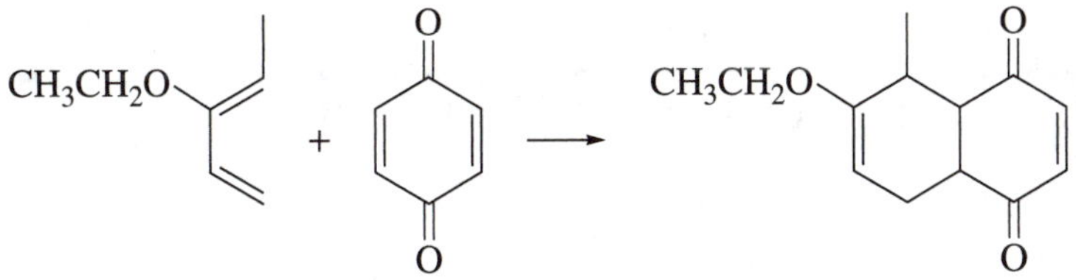

b.

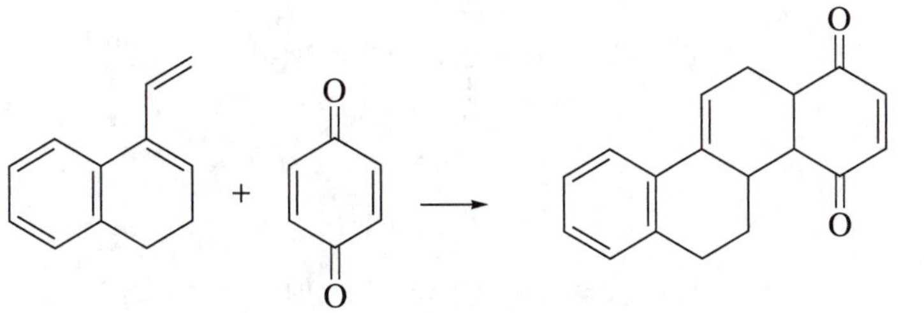

2.   **A** is the most acidic, because the electrons left behind when the proton is removed can be delocalized onto two oxygens. **B** is the next most acidic, because the electrons left behind when the proton is removed can be delocalized onto one oxygen. **C** is the least acidic, because the electrons left behind when the proton is removed cannot be delocalized.

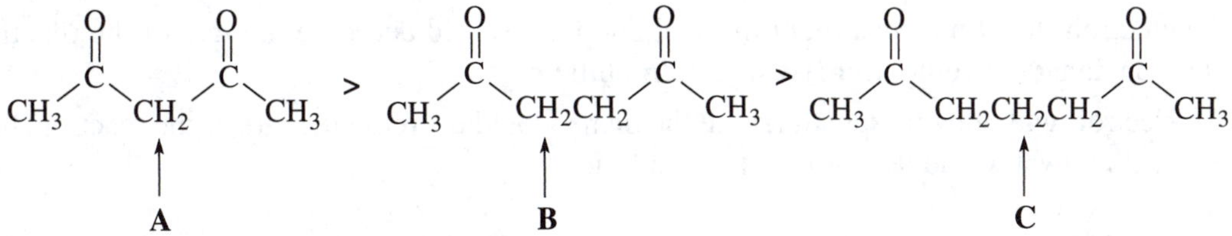

3.   a.

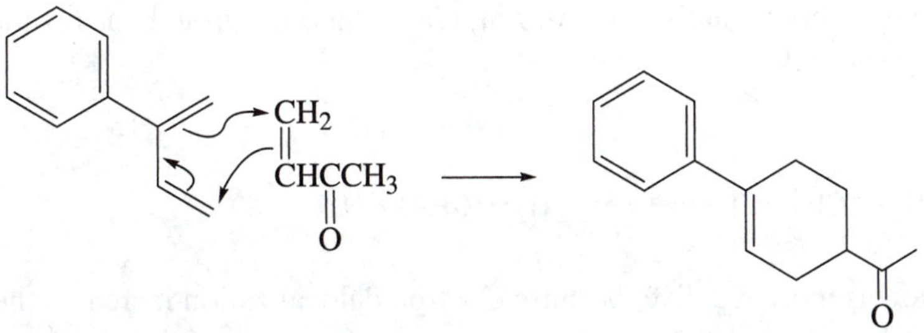

b.

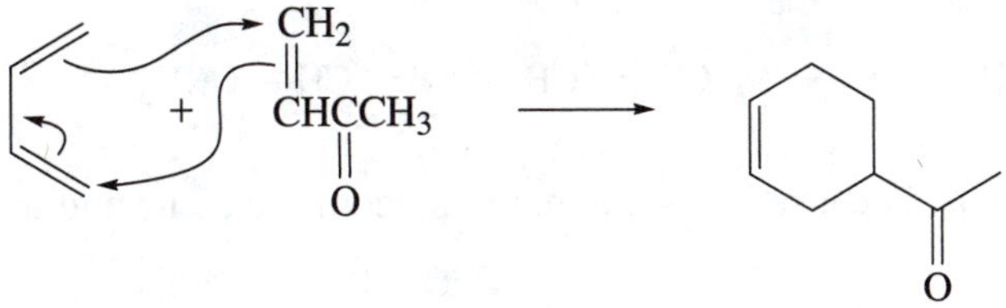

c.

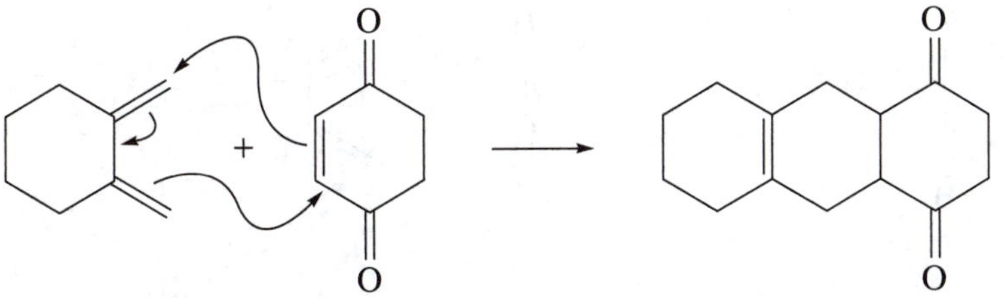

54.

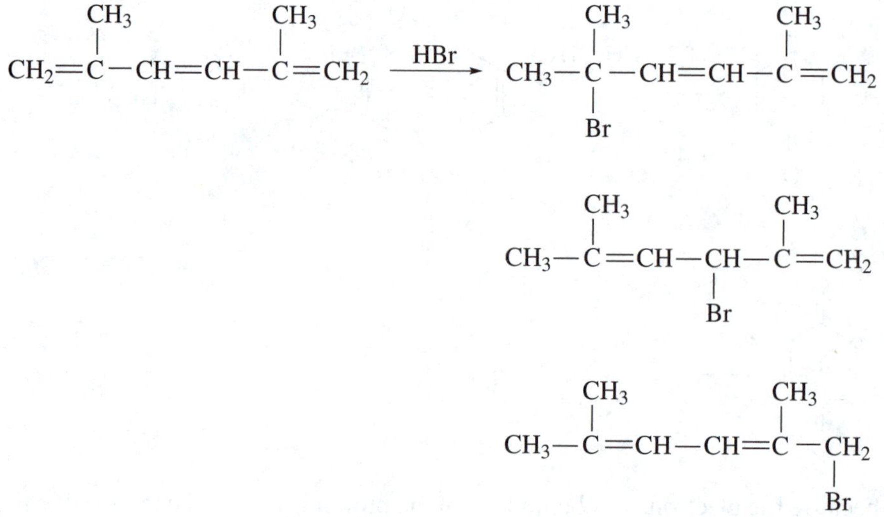

$$CH_2=\overset{CH_3}{\underset{}{C}}-CH=CH-\overset{CH_3}{\underset{}{C}}=CH_2 \xrightarrow{HBr} CH_3-\overset{CH_3}{\underset{Br}{C}}-CH=CH-\overset{CH_3}{\underset{}{C}}=CH_2$$

$$CH_3-\overset{CH_3}{\underset{}{C}}=CH-\overset{}{\underset{Br}{C}H}-\overset{CH_3}{\underset{}{C}}=CH_2$$

$$CH_3-\overset{CH_3}{\underset{}{C}}=CH-CH=\overset{CH_3}{\underset{Br}{C}}-CH_2$$

55. The diene is the nucleophile, and the dienophile is the electrophile in a Diels–Alder reaction.

a. An electron-donating substituent in the diene would increase the rate of the reaction, because electron donation would increase its nucleophilicity.

b. An electron-donating substituent in the dienophile would decrease the rate of the reaction, because electron donation would decrease its electrophilicity.

c. An electron-withdrawing substituent in the diene would decrease the rate of the reaction, because electron withdrawal would decrease its nucleophilicity.

56. A Diels–Alder reaction is a reaction between a nucleophilic diene and an electrophilic dienophile.

a. The compound shown below is more reactive in both **1** and **2**, because electron delocalization increases the electrophilicity of the dienophile.

$$CH_2=CH-\overset{O}{\overset{\|}{CH}} \longleftrightarrow \overset{+}{C}H_2-CH=CH-\overset{O^-}{\underset{}{}}$$

b. The compound shown below is more reactive, because electron delocalization increases the nucleophilicity of the diene.

$$CH_2=CH-CH=CH-\overset{..}{O}CH_3 \longleftrightarrow \overset{-}{:}CH_2-CH=CH-CH=\overset{+}{O}CH_3$$

57. The first pair is the preferred set of reagents, because it has the more nucleophilic diene and the more electrophilic dienophile.

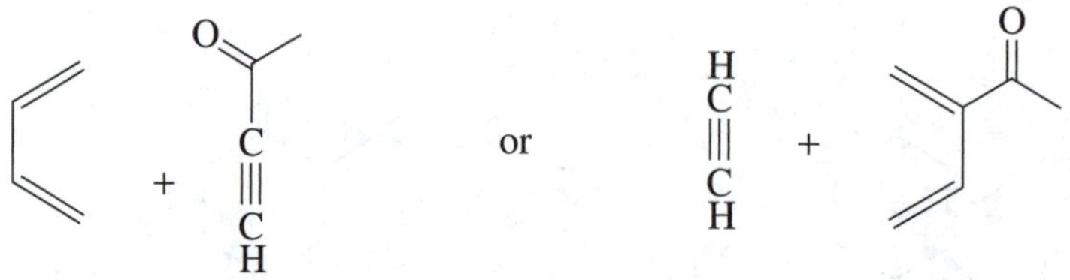

58. The reactions in part **a** and part **b** are intramolecular Friedel–Crafts acylation reactions.

a.

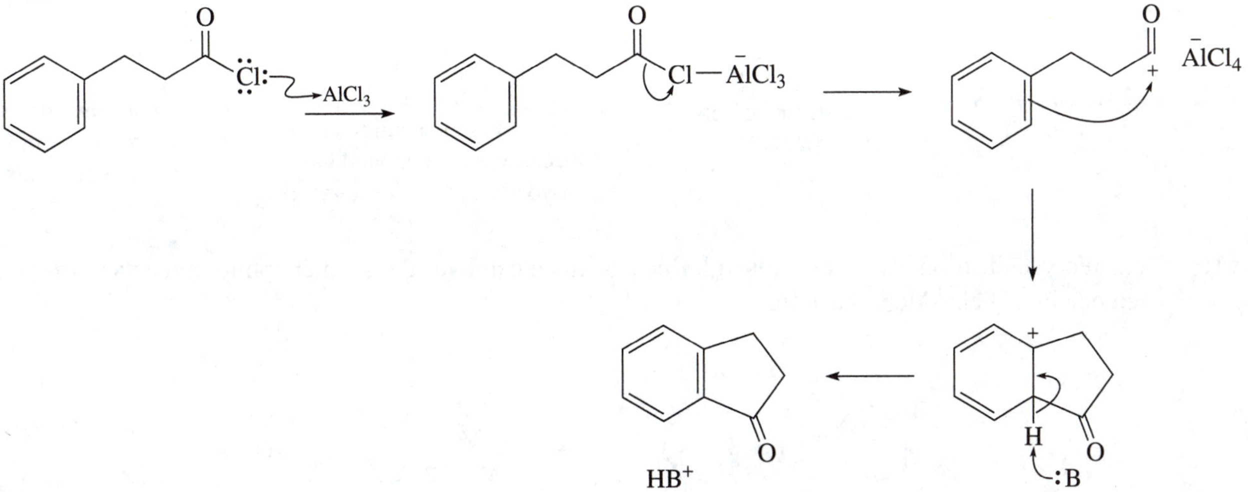

b.

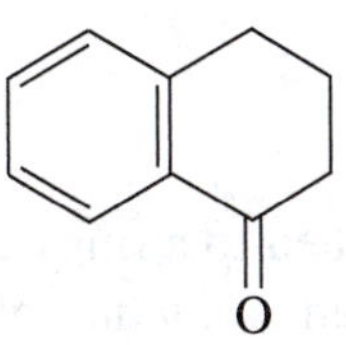

59. These are intramolecular Friedel–Crafts alkylation reactions. Notice that a carbocation rearrangement occurs in part **a.**

a.

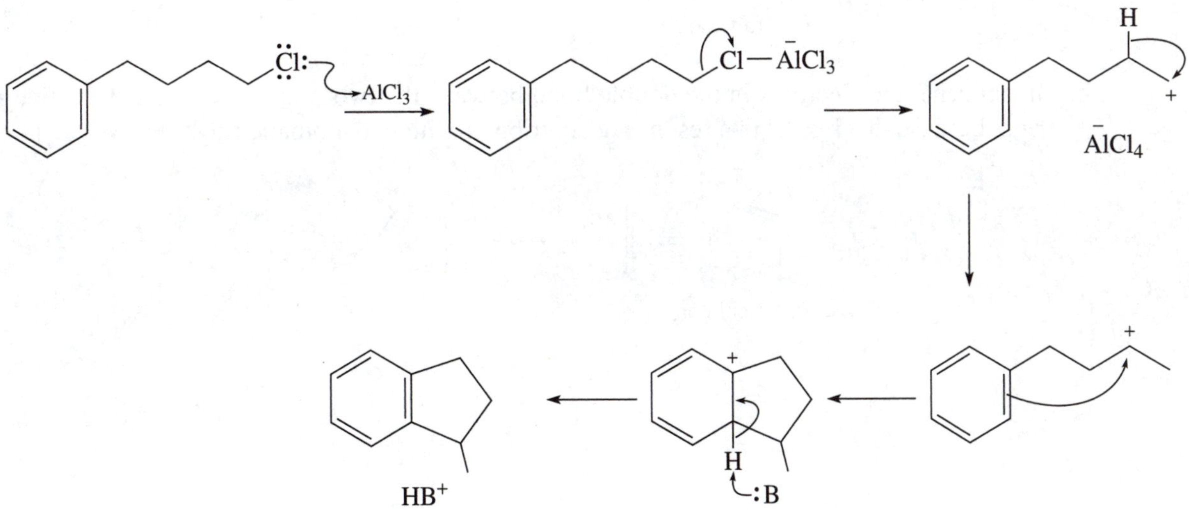

b.

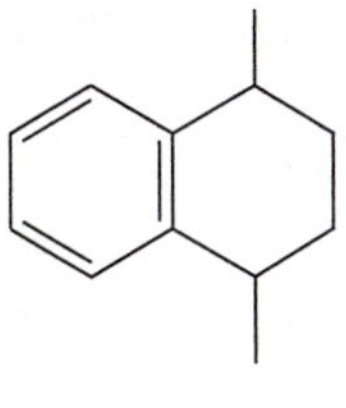

60.

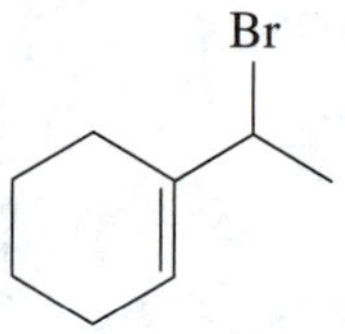

2 stereoisomers (R and S)	2 stereoisomers (R and S)	4 stereoisomers (two pairs of enantiomers, because the compound has two asymmetric centers)	no asymmetric center, therefore no stereoisomers

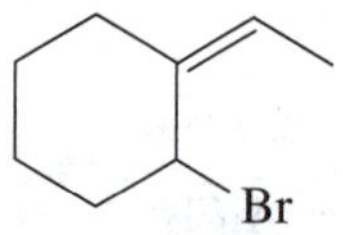

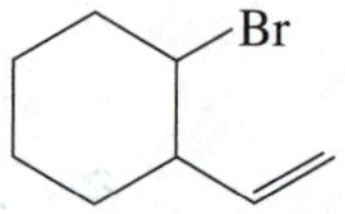

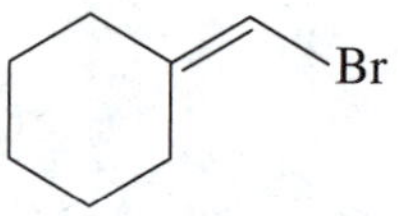

61.   His recrystallization was not successful. Because maleic anhydride is a dienophile, it reacts with cyclopentadiene in a Diels–Alder reaction.

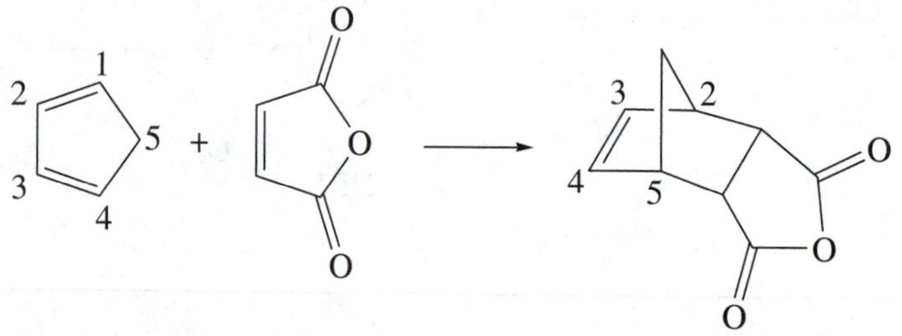

62.   **a.**   In fulvene, the electrons in the exocyclic double bond (a double bond attached to a ring) move toward the five-membered ring, because the resonance contributor that results has an aromatic ring.

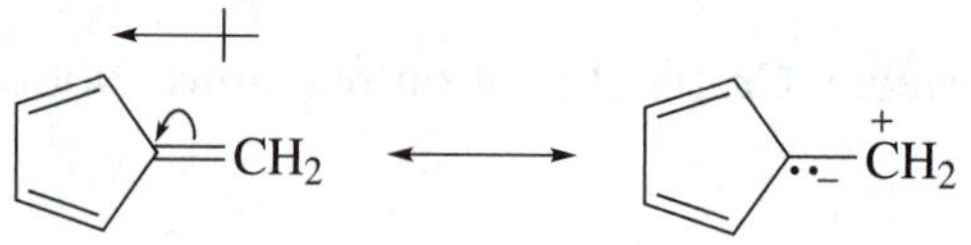

fulvene

b.   In calicene, the electrons in the double bond between the two rings move toward the five-membered ring, because that results in a resonance contributor with two aromatic rings.

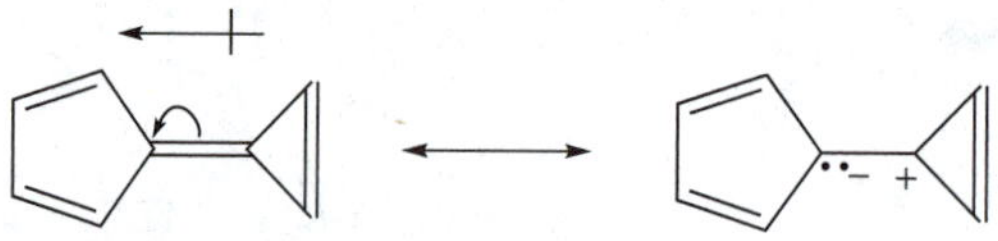

calicene

63.

a. HCl adds to the alkene, forming a secondary carbocation that undergoes a 1,2-hydride shift to form a tertiary carbocation. The tertiary carbocation is an electrophile that can either add to the double bond in a second molecule of the reactant (in an intermolecular reaction) or add to the benzene ring in the same molecule of the reactant (in an intramolecular reaction).

The intramolecular reaction is favored, because it forms a stable five-membered ring. After the electrophile adds to the benzene ring, a base (B:) in the reaction mixture removes a proton and the aromaticity of the benzene ring is restored.

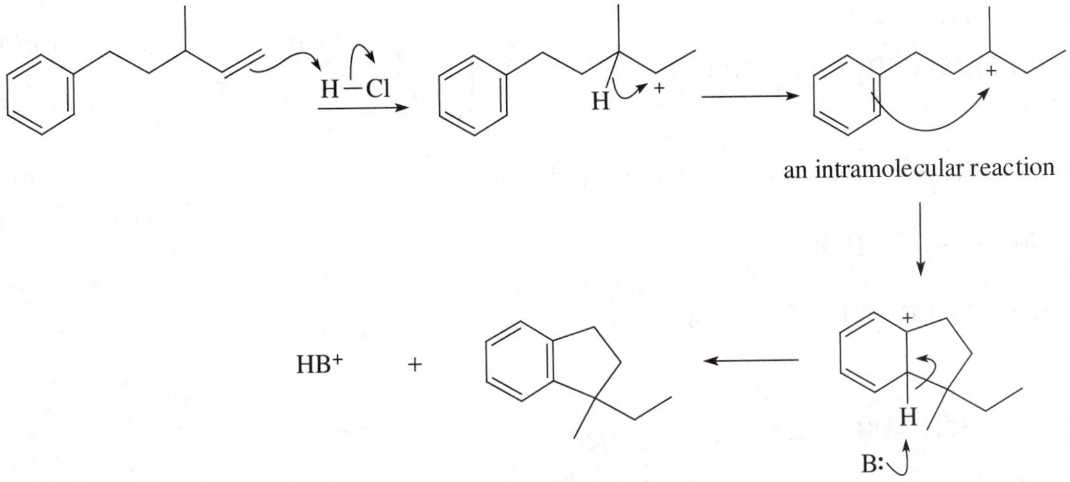

an intramolecular reaction

b. As in part **a,** an electrophile is formed that can react in either an intermolecular reaction or an intramolecular reaction. Seeing that the product of the reaction has two benzene rings and that there are twice as many carbons in the product as in the reactant indicates that two reactant molecules react in an intermolecular reaction. In this case, the intermolecular reaction is favored, because the intramolecular reaction would lead to a highly strained three-membered ring. The electrophile that is formed in the intermolecular reaction can add to the benzene ring in an intramolecular reaction to form a stable five-membered ring.

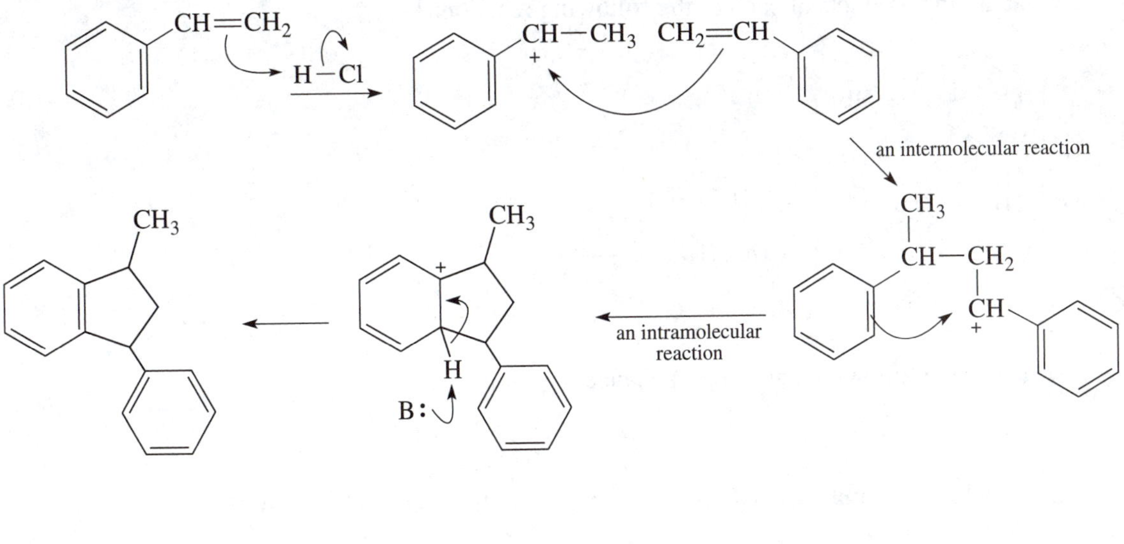

an intermolecular reaction

an intramolecular reaction

Chapter 7 Practice Test

1.   For each of the following pairs of compounds, indicate the one that is the more stable:

a. 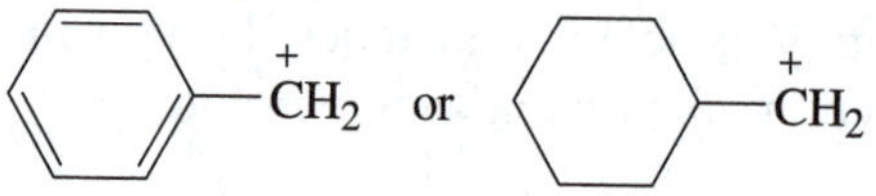 (benzene ring)–$\overset{+}{C}H_2$ or (cyclohexane ring)–$\overset{+}{C}H_2$ c. $CH_2\!=\!CH\overset{+}{C}H_2$ or $CH_2\!=\!CHCH_2\overset{+}{C}H_2$

b. $CH_3\overset{-}{C}HCH_2\overset{O}{\overset{||}{C}}CH_3$ or $CH_3\overset{-}{C}H\overset{O}{\overset{||}{C}}CH_3$ d. 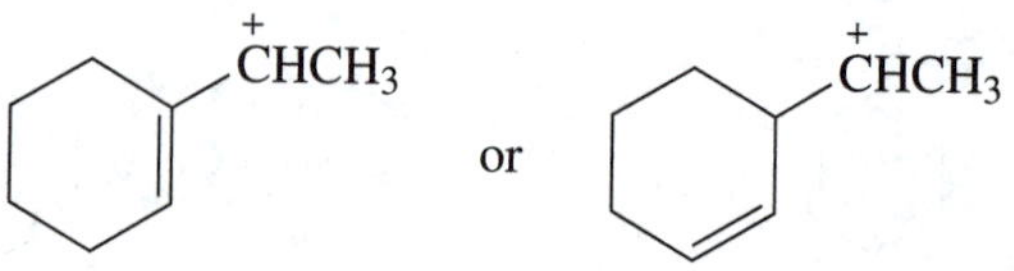 (cyclohexene ring)–$\overset{+}{C}HCH_3$ or (cyclohexene ring)–$\overset{+}{C}HCH_3$

2.   Draw resonance contributors for each of the following:

a. $CH_3CH\!=\!CH\!-\!\overset{..}{\overset{..}{O}}CH_3$

b. $CH_3CH\!=\!CH\!-\!CH\!=\!CH\!-\!\overset{+}{C}H_2$

c. $\overset{-..}{C}H_2\!-\!CH\!=\!CH\!-\!\overset{O}{\overset{||}{C}}H$

3.   Which compounds do not have delocalized electrons?

$CH_3CH_2NHCH\!=\!CHCH_3$ $CH_3\overset{CH_3}{\overset{|}{\underset{+}{C}}}CH_2CH\!=\!CH_2$ $CH_2\!=\!CHCH_2CH\!=\!CH_2$

$CH_2\!=\!CH\overset{O}{\overset{||}{C}}CH_3$ $CH_3CH_2NHCH_2CH\!=\!CHCH_3$ (cyclohexadiene ring)

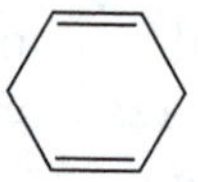

4.   What are the products of each of the following reactions?

a.  (cyclohexadiene ring) + HBr $\longrightarrow$

b. 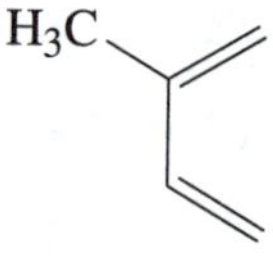 H_3C–(diene structure) + $\overset{O}{\overset{||}{C}H\overset{||}{\underset{CH_2}{C}}CH_3}$ $\longrightarrow$

5.   Which of the following pairs are resonance contributors?

a. CH_3CH_2OH and CH_3OCH_3 c. $CH_3\overset{O}{\overset{||}{C}}OH$ and $CH_3\overset{O^-}{\overset{|}{C}}\!=\!\overset{+}{O}H$

b. $CH_3\overset{O}{\overset{||}{C}}OH$ and $CH_3\overset{O}{\overset{||}{C}}O^-$ d. $CH_3CH_2\overset{O}{\overset{||}{C}}H$ and $CH_3CH\!=\!\overset{OH}{\overset{|}{C}}H$

6. Which is a stronger base?

a. 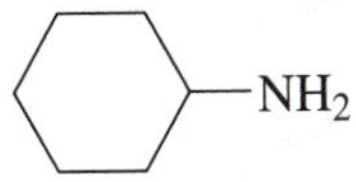 (cyclohexyl)—NH$_2$ or   (phenyl)—NH$_2$

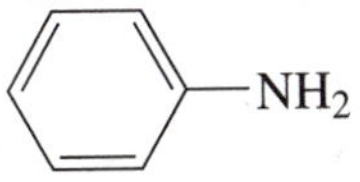

b. 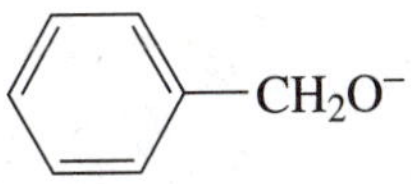 (phenyl)—CH$_2$O$^-$ or   (phenyl)—O$^-$

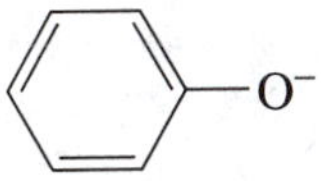

7. Draw resonance contributors for each of the following:

a. 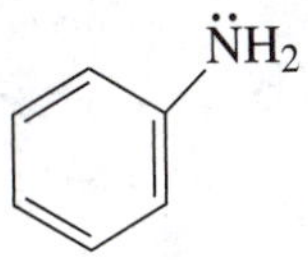 (benzene ring with $\ddot{N}H_2$)

b. 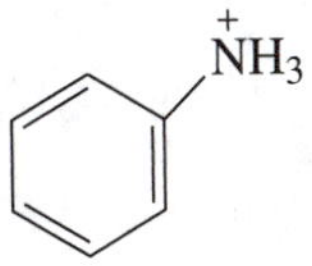 (benzene ring with $\overset{+}{N}H_3$)

c. 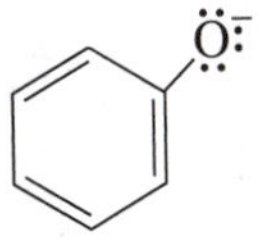 (benzene ring with $\ddot{\ddot{O}}{:}^-$)

8. Which resonance contributor makes a greater contribution to the resonance hybrid?

a. 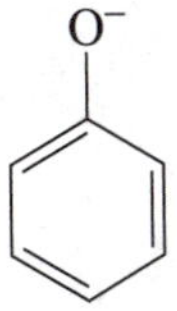 (benzene ring with O$^-$)   or   (cyclohexadienone with O and $^-$)

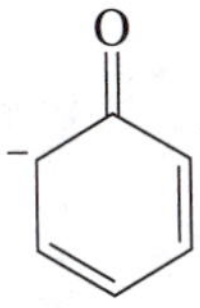

b. 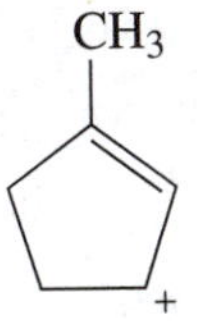 (cyclopentene ring with CH$_3$ and $+$)   or   (cyclopentadiene ring with CH$_3$ and $+$)

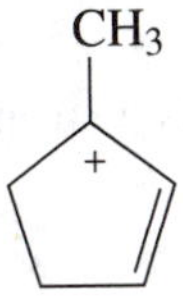

9. Indicate whether each of the following statements is true or false:

a. A conjugated diene is more stable than an isomeric-isolated diene.   T   F

b. A single bond formed by an sp^2—sp^2 overlap is longer than a single bond formed by an sp^2—sp^3 overlap.   T   F

c. 1,3-Hexadiene is more stable than 1,4-hexadiene.   T   F

10. Draw the four products that would be obtained from the following reaction. Ignore stereoisomers.

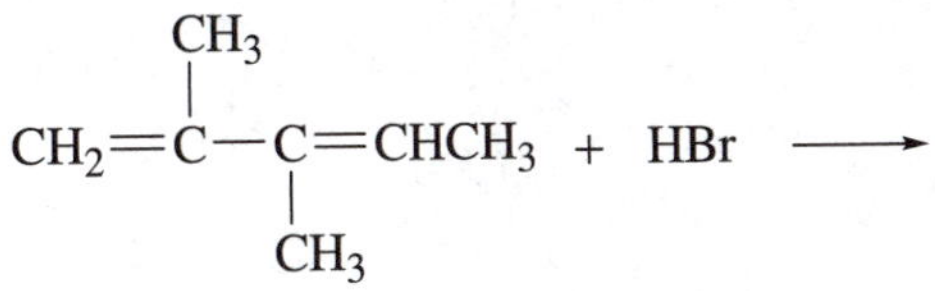

$$CH_2{=}\underset{\underset{CH_3}{|}}{\overset{\overset{CH_3}{|}}{C}}{-}C{=}CHCH_3 \ + \ HBr \longrightarrow$$

11. What reactants are necessary for the synthesis of the following compound via a Diels–Alder reaction?

12. Rank the following carbocations in order from most stable to least stable:

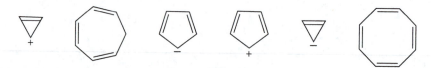

13. Which are aromatic compounds?

14. Which compound has the greater delocalization energy?

or

15. Which of the following is a stronger acid?

a. or

b. or

c. or

16. For each of the following pairs of compounds, circle the one that is the stronger acid:

a.

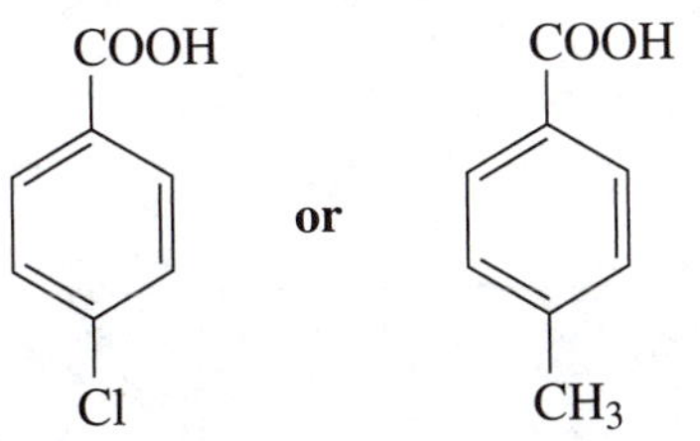

c.

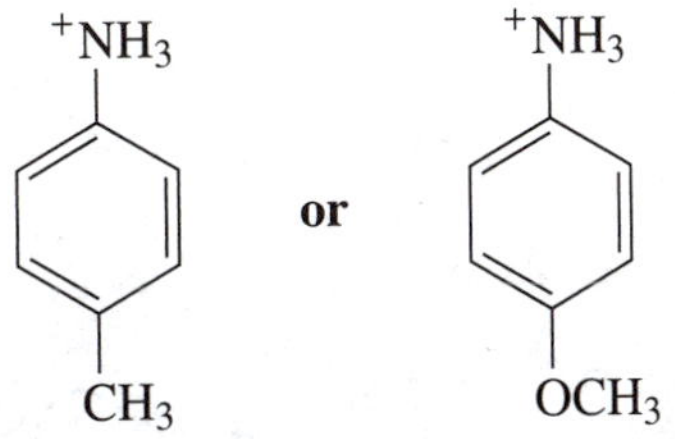

b.

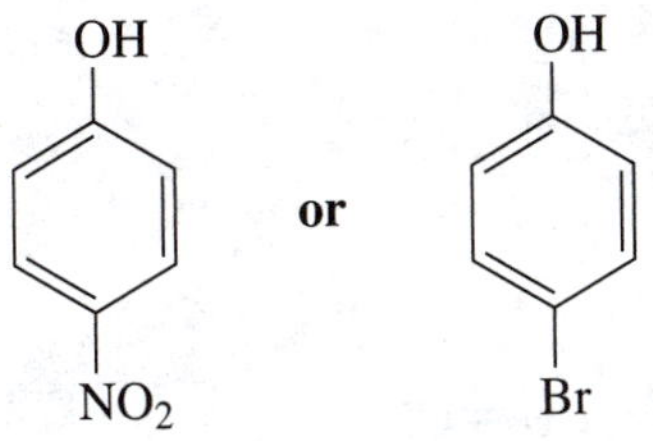

d.

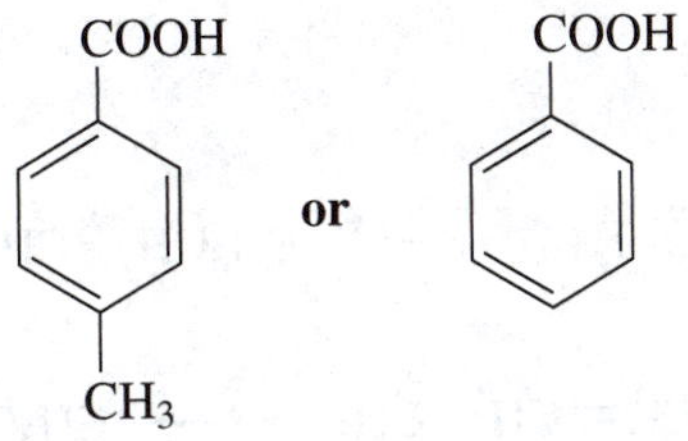

17. Draw the mechanism for the formation of the nitronium ion from nitric acid and sulfuric acid.

Answers to Chapter 7 Practice Test

1.   **a.** (phenyl)$-\overset{+}{C}H_2$

c. $CH_2{=}CH\overset{+}{C}H_2$

b. $CH_3\overset{-}{C}H\overset{O}{\overset{\|}{C}}CH_3$

d. (cyclohexenyl)$\overset{+}{C}HCH_3$

2.   **a.** $CH_3CH{=}CH{-}\overset{..}{\underset{..}{O}}CH_3 \longleftrightarrow CH_3\overset{-}{C}H{-}CH{=}\overset{+}{O}CH_3$

b. $CH_3CH{=}CH{-}CH{=}CH{-}\overset{+}{C}H_2 \longleftrightarrow CH_3CH{=}CH{-}\overset{+}{C}H{-}CH{=}CH_2$

$CH_3\overset{+}{C}H_2{-}CH{=}CH{-}CH{=}CH_2$

c. $\overset{-}{C}H_2{-}CH{=}CH{-}\overset{O}{\overset{\|}{C}}H \longleftrightarrow CH_2{=}CH{-}\overset{-}{C}H{-}\overset{O}{\overset{\|}{C}}H \longleftrightarrow CH_2{=}CH{-}CH{=}\overset{O^-}{\overset{|}{C}}H$

3. $CH_3\overset{+}{\underset{+}{C}}CH_2CH{=}CH_2$ (with CH_3 substituent) $CH_2{=}CHCH_2CH{=}CH_2$ $CH_3CH_2NHCH_2CH{=}CHCH_3$ (cyclohexadiene ring)

4.   **a.**   (cyclohexadiene) $+$ HBr $\longrightarrow$ (cyclohexene with Br)

b. H_3C (diene) $+$ $\overset{O}{\overset{\|}{C}}\underset{\overset{\|}{CH_2}}{CHCCH_3}$ $\longrightarrow$ H_3C (cyclohexene ring with $\overset{O}{\overset{\|}{C}}CH_3$) $+$ H_3C (cyclohexene ring with $\overset{\|}{\underset{O}{C}}CH_3$)

5. $CH_3\overset{O}{\overset{\|}{C}}OH$ and $CH_3\overset{O^-}{\overset{|}{C}}{=}\overset{+}{O}H$

5. **a.** (cyclohexane with $-NH_2$)   **b.** (benzene with $-CH_2O^-$)

7. **a.** (aniline resonance structures) $\ddot{N}H_2 \longleftrightarrow {}^{+}\!NH_2 \longleftrightarrow {}^{+}\!NH_2 \longleftrightarrow {}^{+}\!NH_2 \longleftrightarrow \ddot{N}H_2$

b. ${}^{+}\!NH_3 \longleftrightarrow {}^{+}\!NH_3$

c. (phenoxide resonance structures) $\ddot{O}^{:-} \longleftrightarrow O \longleftrightarrow O \longleftrightarrow O \longleftrightarrow \ddot{O}^{:-}$

8. **a.** (benzene with O^-)   **b.** (CH_3 on cyclopentenyl cation)

9. **a.** A conjugated diene is more stable than an isomeric isolated diene.   T

b. A single bond formed by an sp^2—sp^2 overlap is longer than a single bond formed by an sp^2—sp^3 overlap.   F

c. 1,3-Hexadiene is more stable than 1,4-hexadiene.   T

10.

$$\underset{\underset{\displaystyle Br}{|}}{\overset{\overset{\displaystyle CH_3}{|}}{CH_3C}}-\overset{\overset{\displaystyle }{|}}{\underset{\underset{\displaystyle CH_3}{|}}{C}}=CHCH_3 \qquad \underset{\underset{\displaystyle CH_3}{|}}{\overset{\overset{\displaystyle CH_3}{|}}{CH_3C}}=\overset{\overset{\displaystyle Br}{|}}{C}-CHCH_3 \qquad CH_2=\overset{\overset{\displaystyle CH_3}{|}}{\underset{\underset{\displaystyle CH_3}{|}}{C}}-\overset{\overset{\displaystyle Br}{|}}{C}CH_2CH_3 \qquad BrCH_2\overset{\overset{\displaystyle CH_3}{|}}{C}=\overset{}{\underset{\underset{\displaystyle CH_3}{|}}{C}}CH_2CH_3$$

11.

(2,4-hexadiene with CH_3 groups) $+$ (cyclopent-2-enone)

12. $CH_3CH=CH\overset{+}{\underset{\underset{\displaystyle CH_3}{|}}{C}}CH_3 > CH_3CH=CH\overset{+}{C}HCH_3 > CH_3CH=CH\overset{+}{C}H_2 > CH_3CH=CHCH_2\overset{+}{C}H_2$

13.

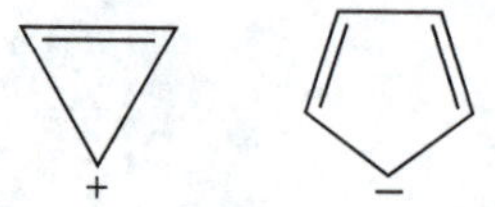

14.

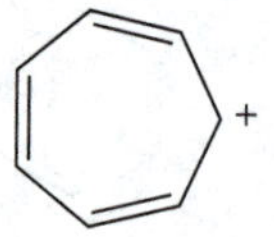

15.   **a.** 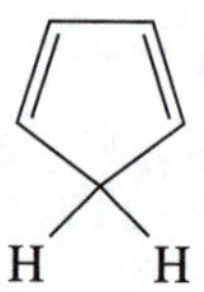   **b.** 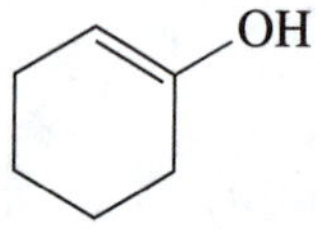   **c.**

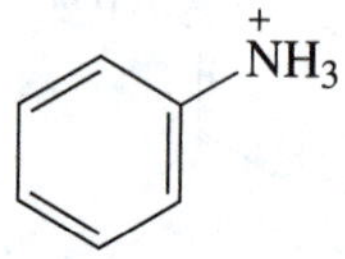

16.   **a.** 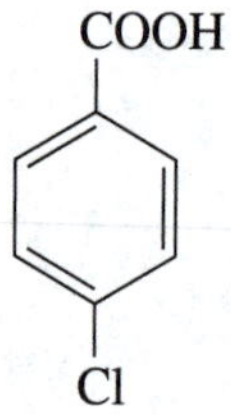   **b.** 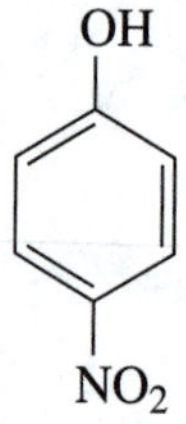   **c.**    **d.**

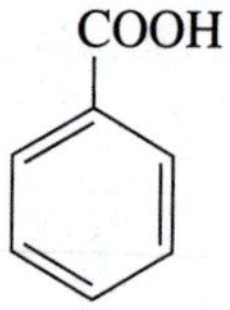

17.

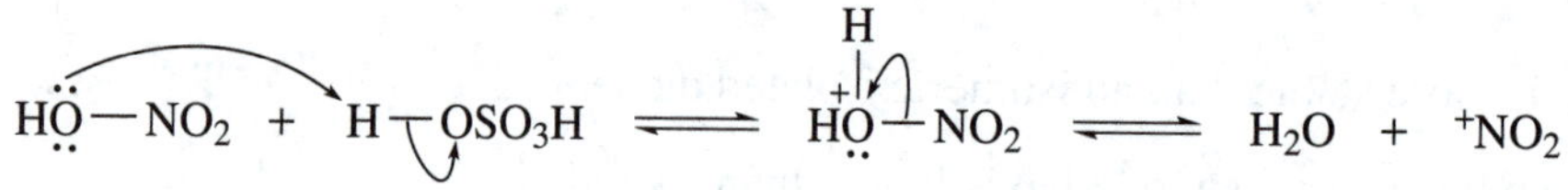

Substitution and Elimination Reactions of Alkyl Halides

Important Terms

backside attack	nucleophilic attack on the side of the carbon opposite to the side bonded to the leaving group.
base	a substance that accepts a proton.
basicity	the tendency of a compound to share its electrons with a proton.
bifunctional molecule	a molecule with two functional groups.
bimolecular reaction	a reaction in which two molecules are involved in the transition state of the rate-determining step.
dielectric constant	a measure of how well a solvent can insulate opposite charges from one another.
elimination reaction	a reaction that removes atoms or groups from the reactant to form a π bond.
E1 reaction	a unimolecular elimination reaction.
E2 reaction	a bimolecular elimination reaction.
intermolecular reaction	a reaction that takes place between two molecules.
intramolecular reaction	a reaction that takes place within a molecule.
inversion of configuration	turning the carbon inside out like an umbrella so that the resulting product has a configuration opposite to that of the reactant.
ion–dipole interaction	the interaction between an ion and the dipole of a molecule.
kinetics	the field of chemistry that deals with the rates of chemical reactions.
leaving group	the group that is displaced in a substitution reaction.
nucleophile	an electron-rich atom or molecule.
nucleophilicity	a measure of how readily an atom or molecule with a lone pair attacks another atom.
nucleophilic substitution reaction	a reaction in which a nucleophile substitutes for an atom or group.
racemization	formation of a pair of enantiomers.

rate constant	the constant of proportionality in the rate law for a reaction; it describes how difficult it is to overcome the energy barrier of a reaction.
rate law	the equation that shows the relationship between the rate of a reaction and the concentration of the reactants.
regioselectivity	the preferential formation of a constitutional isomer.
second-order reaction	a reaction whose rate is dependent on the concentration of two reactants, or on the square of the concentration of two reactants.
S_N1 reaction	a unimolecular nucleophilic substitution reaction.
S_N2 reaction	a bimolecular nucleophilic substitution reaction.
solvolysis	reaction with a solvent.
steric effects	effects due to the fact that groups occupy a certain volume of space.
steric hindrance	caused by bulky groups at the site of a reaction that make it difficult for the reactants to approach each other.
substitution reaction	a reaction that exchanges one substituent of a reactant for another.
unimolecular reaction	a reaction in which only one molecule is involved in the transition state of the rate-determining step.
Williamson ether synthesis	formation of an ether from the reaction of an alkoxide ion with an alkyl halide.

olutions to Problems

Methoxychlor has methoxy groups in place of the chlorines on the benzene rings of DDT. These methoxy groups can form hydrogen bonds with water, making methoxychlor more soluble in water and, therefore, less soluble in fatty tissues.

rate $= k \,[\,\text{alkyl halide}\,]\,[\,\text{nucleophile}\,]$

original: rate $= k\,[\,1.0\,]\,[\,1.0\,]$

a.   rate $= k\,[\,1.0\,]\,[\,3.0\,] \;=\; 3.0$ The rate is tripled.

b.   rate $= k\,[\,0.50\,]\,[\,1.0\,] \;=\; 0.50$ The rate is cut in half.

Increasing the height of the energy barrier decreases the magnitude of the rate constant; this causes the reaction to be slower.

The closer the methyl group is to the site of nucleophilic attack, the greater the steric hindrance to nucleophilic attack and the slower the rate of the reaction.

$$\underset{}{CH_3CH_2CH_2CH_2CH_2Br} \;>\; \overset{\overset{\textstyle CH_3}{\textstyle |}}{CH_3CHCH_2CH_2Br} \;>\; \overset{\overset{\textstyle CH_3}{\textstyle |}}{CH_3CH_2CHCH_2Br} \;>\; \overset{\overset{\textstyle CH_3}{\textstyle |}}{\underset{\underset{\textstyle CH_3}{\textstyle |}}{CH_3CH_2CBr}}$$

a.   Solved in the text.

b.
(R)-2-bromobutane   (S)-2-methoxybutane

c.
(S)-3-chlorohexane   (R)-3-hexanol

d. $CH_3CH_2\overset{\overset{\textstyle |}{\textstyle I}}{CH}CH_2CH_3 \xrightarrow{\;HO^-\;}$ $CH_3CH_2\overset{\overset{\textstyle |}{\textstyle OH}}{CH}CH_2CH_3$

3-iodopentane   3-pentanol

a.   Solved in the text.

b.

c.

7. a. 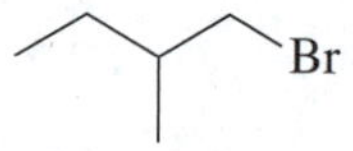 The primary alkyl halide is less sterically hindered than the secondary alkyl halid (the CH_3 group is farther away from the backside of the carbon attached to the Br)

b. 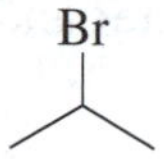 Br is a weaker base; therefore, it is a better leaving group.

c. 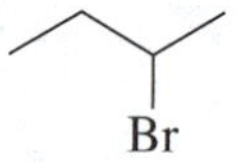 With one methyl and one ethyl group, this alkyl halide is less sterically hindere than the other alkyl halide that has two ethyl groups.

d. 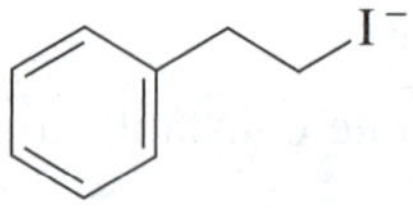 An iodide ion is a weaker base and, therefore, a better leaving group than a bromide ion.

8. a. $CH_3CH_2Br + HO^-$ HO^- is a better nucleophile than H_2O.

b. $CH_3CH_2Cl + CH_3O^-$ CH_3O^- is a better nucleophile than CH_3OH.

9. a. CH_3NH_2 b. CH_3NH_2

10. Solved in the text.

11. These are all S_N2 reactions.

a. $CH_3CH_2OCH_2CH_2CH_3$

b. $CH_3CH_2C{\equiv}CCH_3$

c. $CH_3CH_2\overset{+}{N}(CH_3)_3\ Br^-$

d. $CH_3CH_2SCH_2CH_3$

12. a. 

R and *S*, because the leaving group was attached to an asymmetric center.

b. 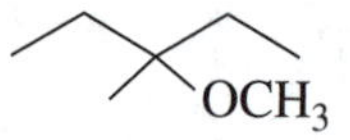

One product, because the leaving group was not attached to an asymmetric center.

13. 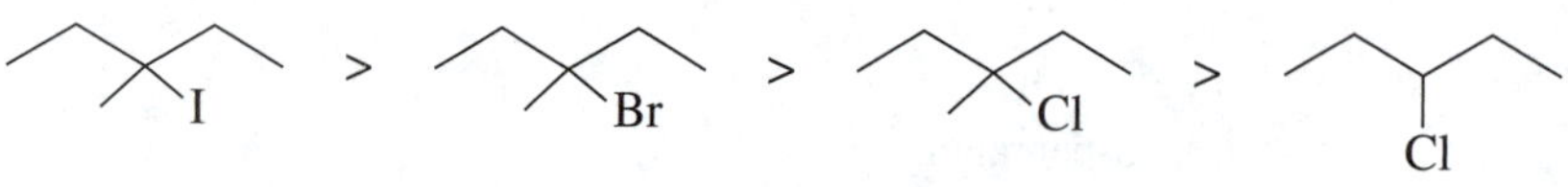

14. a. 1. 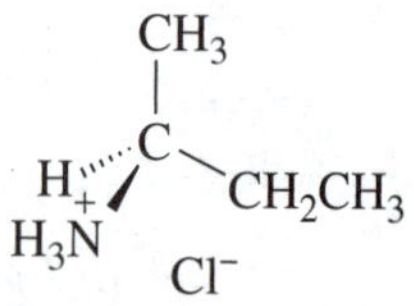 NH_3 is a good nucleophile, so this is an S_N2 reaction. The product has the inverted configuration compared to that of the reactant.

2. 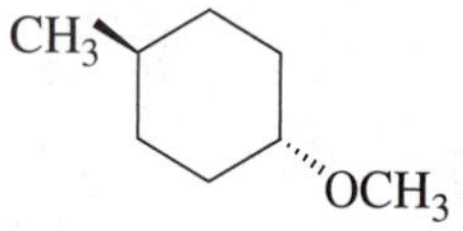 The S_N2 reaction takes place with inversion of configuration.

b. 1.

The product does not have an asymmetric center, so only a single product is formed.

2. (CH₃)₂CH $\longrightarrow$ ... OCH₃ + (CH₃)₂CH $\longrightarrow$ ... CH₃

After the tertiary carbocation forms, methanol can attack the sp^2 carbon from the top or the bottom of the planar carbocation.

15. The rate of an S_N1 reaction is not affected by increasing the concentration of the nucleophile, whereas the rate of an S_N2 reaction is increased when the concentration of the nucleophile is increased. Therefore, we first have to determine whether the reactions are S_N1 or S_N2 reactions.

A is an S_N2 reaction, because the configuration of the product is inverted compared with that of the reactant.

B is an S_N2 reaction, because the reactant is a primary alkyl halide.

C is an S_N1 reaction, because the reactant is a tertiary alkyl halide. (Recall that tertiary alkyl halides do not undergo S_N2 reactions.)

Because they are S_N2 reactions, the rate of **A** and **B** will increase if the concentration of the nucleophile is increased.
Because it is an S_N1 reaction, the rate of **C** will not change if the concentration of the nucleophile is increased.

16. **a.** HO ~~~~~ Br

because it forms a six-membered ring, whereas the other compound would form a seven-membered ring. A seven-membered ring is more strained than a six-membered ring, so the six-membered ring is formed more easily.

b. HO ~~~ Br

because it forms a five-membered ring, whereas the other compound would form a four-membered ring. A four-membered ring is more strained than a five-membered ring, so the five-membered ring is formed more easily.

c. HO ~~~~~ Br

because it forms a seven-membered ring, whereas the other compound would form an eight-membered ring. An eight-membered ring is more strained than a seven-membered ring, so the seven-membered ring is formed more easily; also, the Br and OH in the compound that leads to the eight-membered ring are less likely to be in the proper position relative to one another for reaction, because there are more bonds around which rotation to an unfavorable conformation can occur.

17.   **a.**   Hydride ion removes a proton from the OH group more rapidly than it attacks the alkyl chloride. Once the alkoxide ion is formed, it attacks the backside of the alkyl chloride, forming an epoxide.

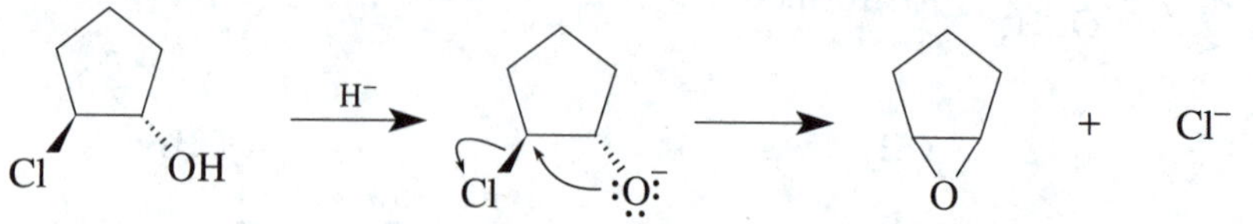

   b.   When hydride ion removes a proton from the OH group, the alkoxide ion cannot react in an intramolecular reaction with the alkyl chloride to form an epoxide, because it cannot reach the backside of the carbon attached to the chlorine. Thus, the major product will result from an intermolecular reaction.

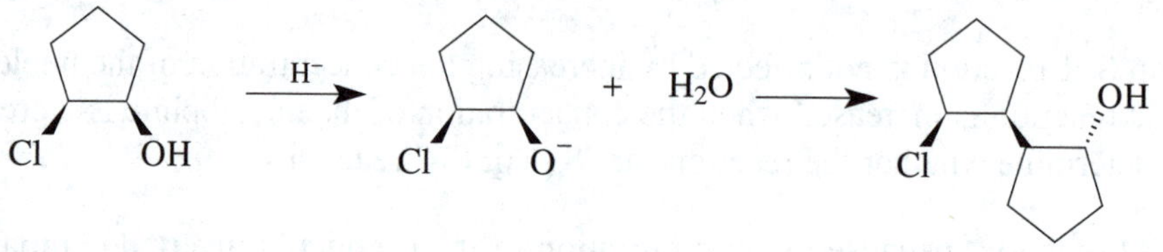

   c.   Hydride ion removes a proton from the OH group, and the alkoxide ion attacks the backside of the carbon attached to the bromine, forming a six-membered ring ether.

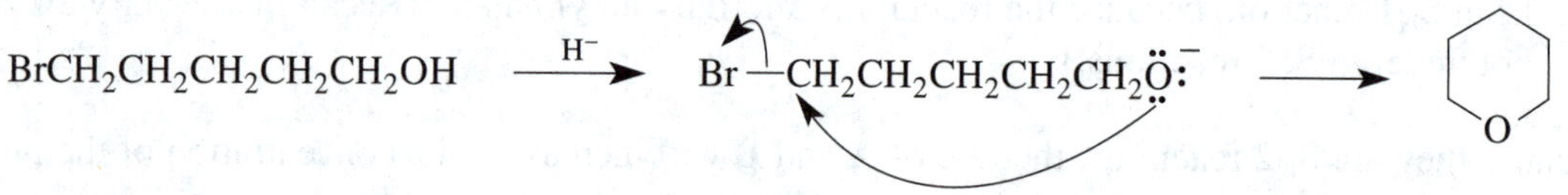

$$BrCH_2CH_2CH_2CH_2CH_2OH \xrightarrow{\text{H}^-} Br-CH_2CH_2CH_2CH_2CH_2\ddot{\underset{\cdot\cdot}{O}}\text{:}^- \longrightarrow$$

   d.   Hydride ion removes a proton from the OH group, and the alkoxide ion attacks the backside of the carbon attached to the chlorine, forming an epoxide.

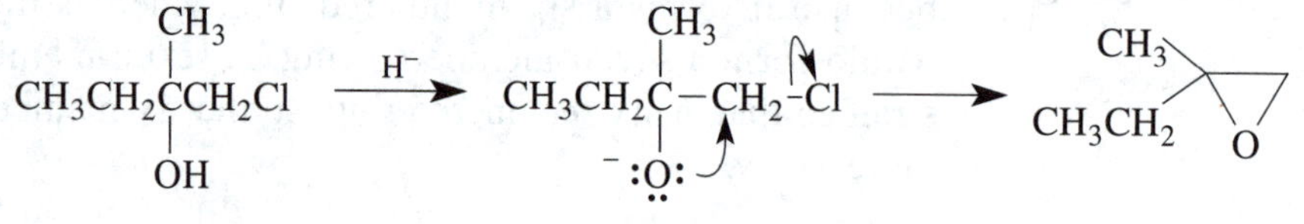

18.   **a.**   Solved in the text.

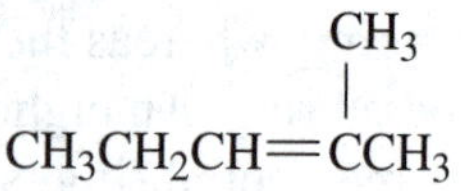

   c. 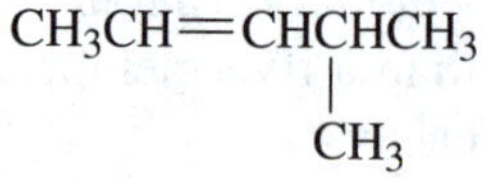 $CH_3CH{=}CHCHCH_3$
          $|$
          CH_3

   b. $CH_3CH_2CH{=}\overset{\overset{\textstyle CH_3}{|}}{C}CH_3$

   d. 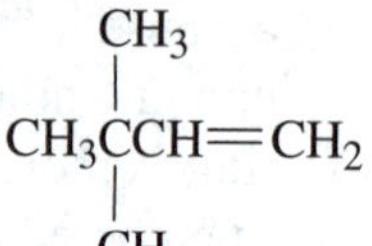 $CH_3\overset{\overset{\textstyle CH_3}{|}}{\underset{\underset{\textstyle CH_3}{|}}{C}}CH{=}CH_2$

9. **a.** **1.** Solved in the text.

2. $\underset{\displaystyle CH_3CH_2CH}{}\!\!=\!\!\underset{\displaystyle CCH_3}{\overset{\displaystyle CH_3}{|}}$

3.   The larger substituent attached to one sp^2 carbon and the larger substituent attached to the other sp^2 carbon are on opposite sides of the double bond.

4.

b. **1.** Same as the major product formed in an E1 reaction.

2. Same as the major product formed in an E1 reaction.

3. Same as the major product formed in an E1 reaction.

4. Same as the major product formed in an E1 reaction.

20. **a.** *tert*-Butyl bromide would be more reactive in an E2 reaction, because a bromide ion is a better leaving group than a chloride ion.
  b. *tert*-Butyl bromide would be more reactive in an E1 reaction, because a bromide ion is a better leaving group than a chloride ion.

21. **a.** B, because it forms the more stable carbocation.
  b. B, because it forms the more stable alkene.
  c. B, because it forms the more stable carbocation.
  d. A, because it is less sterically hindered.

22. The substitution product does not require a bond to be broken in the carbocation, but the elimination product does.

23.   **a.** $CH_3CH_2CH_2Br$ This compound has less steric hindrance.

b. I^- is a better leaving group (weaker base) than Br^-.

c. $CH_3CH_2CH_2\overset{\underset{\displaystyle |}{CH_3}}{\underset{\underset{\displaystyle Br}{|}}{C}}CH_3$ The tertiary alkyl halide, because a secondary alkyl halide doe
not undergo S_N1 reactions.

24.   **a.**   primarily substitution
b.   substitution and elimination
c.   substitution and elimination
d.   elimination

25.   Because both reactants in the rate-limiting step are neutral, the reaction will be faster if the polarity of the
solvent is increased.

26.   **a.**   Increasing the polarity will decrease the rate of the reaction, because the concentration of charge on the
reactants is greater (the reactants are charged) than the concentration of charge on the transition state.

b.   Increasing the polarity will decrease the rate of the reaction, because the concentration of charge on the
reactants is greater (the reactants are charged) than the concentration of charge on the transition state.

c.   Increasing the polarity will increase the rate of the reaction, because the concentration of charge on the
reactants is less (the reactants are not charged) than the concentration of charge on the transition state.

27.   **a.** $CH_3Br \ + \ HO^- \ \longrightarrow \ CH_3OH \ + \ Br^-$

HO^- is a better nucleophile than H_2O.

b. $CH_3I \ + \ HO^- \ \longrightarrow \ CH_3OH \ + \ I^-$

I^- is a better leaving group than Cl^-.

c. $CH_3Br \ + \ NH_3 \ \longrightarrow \ CH_3\overset{+}{N}H_3 \ + \ Br^-$

NH_3 is a better nucleophile than H_2O.

d. $CH_3Br \ + \ HO^- \ \xrightarrow{\text{DMSO}} \ CH_3OH \ + \ Br^-$

Unlike ethyl alcohol, DMSO will not stabilize the nucleophile (and therefore decrease the rate of the
reaction) by hydrogen bonding.

e. $CH_3Br \ + \ NH_3 \ \xrightarrow{\text{EtOH}} \ CH_3\overset{+}{N}H_3 \ + \ Br^-$

A more polar solvent is able to stabilize the transition state. (EtOH is ethanol.)

28.   Solved in the text.

9.

a. Because an S_N1 reaction is favored by a polar solvent, the reaction will be faster in 50% water/50% ethanol, the more polar of the two solvents.

b. In 50% water and 50% ethanol, the product will be 50% *tert*-butyl alcohol and 50% *tert*-butyl ethyl ether. In 100% ethanol, the product will be 100% *tert*-butyl ethyl ether.

0. In order to maximize the amount of ether formed in the S_N2 reaction, make sure the less hindered group is provided by the alkyl halide. In order to convert the alcohol ($pK_a \sim 15$) to an alkoxide ion in a reaction that favors products (parts **a, b,** and **c**), a strong base (H^-) is needed.

a.
$$\underset{\overset{|}{CH_3}}{CH_3CH_2CHOH} \xrightarrow{\text{NaH}} \underset{\overset{|}{CH_3}}{CH_3CH_2CHO^-} \xrightarrow{CH_3CH_2CH_2Br} \underset{\overset{|}{CH_3}}{CH_3CH_2CHOCH_2CH_2CH_3}$$

b.
$$\underset{\overset{|}{CH_3}}{CH_3CH_2CH_2CHCH_2OH} \xrightarrow{\text{Na}} \underset{\overset{|}{CH_3}}{CH_3CH_2CH_2CHCH_2O^-} \xrightarrow{CH_3CH_2Br} \underset{\overset{|}{CH_3}}{CH_3CH_2CH_2CHCH_2OCH_2CH_3}$$

c.

1.

a. $CH_3Br + CH_3O^- \longrightarrow CH_3OCH_3 + Br^-$

CH_3O^- is a better nucleophile than CH_3OH.

b. $CH_3I + NH_3 \longrightarrow CH_3\overset{+}{N}H_3 + I^-$

I^- is a better leaving group than Cl^-.

c. $CH_3Br + CH_3NH_2 \longrightarrow CH_3\overset{+}{N}H_2CH_3 + Br^-$

CH_3NH_2 is a better nucleophile than CH_3OH.

2.

a. CH_3OH **c.** CH_3SH **e.** $CH_3OCH_2CH_3$

b. CH_3NH_2 **d.** CH_3SH **f.** $CH_3\overset{+}{N}H_2CH_3$

(Notice that the product in **c** is not protonated, because its pK_a is ~ -7; the product in **f** is protonated, because its pK_a is ~ 11.)

3. The stronger base is the better nucleophile.

a. HO^- **b.** $^-NH_2$ **c.** $CH_3CH_2O^-$ **d.**

4. The weaker base is the better leaving group.

a. H_2O **b.** NH_3 **c.** $CH_3\overset{\overset{O}{\|}}{C}O^-$ **d.**

35. **a.** HO^- **c.** $CH_3CH_2S^-$ **e.** $CH_3\overset{\displaystyle O}{\overset{\|}{C}}O^-$

b. CH_3O^- **d.** $^-C\equiv N$ **f.** $CH_3C\equiv C^-$

36. **a.** $\underset{\overset{\displaystyle |}{CH_3}}{CH_3CH_2CHBr}$ This compound has less steric hindrance.

b. $\underset{\overset{\displaystyle |}{I}}{CH_3CH_2CHCH_3}$ I^- is a weaker base than Br^-, so I^- is a better leaving group.

c. $\underset{\overset{\displaystyle |}{CH_3}}{CH_3CH_2CHCH_2Br}$ A primary carbon is less sterically hindered than a secondary carbon.

d. —CH_2CH_2Br A primary carbon is less sterically hindered than a secondary carbon.

37. **a.** The rate will be increased ninefold.
b. The reaction will be slower, because of the more polar solvent.
c. The reaction will be slower, because it will be an S_N2 reaction with a poor nucleophile.
d. The reaction will be slower, because the leaving group will be poorer.
e. The reaction will be slower, because there will be more steric hindrance.

38. **a.** The reaction will be slower, because the leaving group will be poorer.
b. There would be no change in the rate, because the nucleophile does not participate in the rate-limiting step of an S_N1 reaction.

39. **a.**
$$\underset{H}{\overset{H_3C}{>}}C=C\underset{CH_3}{\overset{H}{<}}$$

c.
$$\underset{H}{\overset{H_3C}{>}}C=C\underset{CH_3}{\overset{H}{<}}$$

e.
$$\underset{H}{\overset{H_3C}{>}}C=C\underset{CH_2CH_3}{\overset{H}{<}}$$

b.

d.

f.

40.   Only part **f** can undergo an E1 reaction.

a. An S$_N$2 reaction forms the isomer with the inverted configuration.

CH$_2$CH$_2$CH$_3$

$\xrightarrow{\text{CH}_3\text{O}^-}$

Br—C$_{\cdots}$H
CH$_3$

(*R*)-2-bromopentane

CH$_2$CH$_2$CH$_3$

H—C—OCH$_3$
CH$_3$

(*S*)-2-methoxypentane

b. An S$_N$2 reaction forms the isomer with the inverted configuration (that is, the one that results from backside attack).

Br — ⬡ — CH$_3$ $\xrightarrow{\text{CH}_3\text{O}^-}$ CH$_3$O — ⬡ — CH$_3$

c. 3-Bromo-3-methylpentane does not have an asymmetric center, so there are no stereoisomers.

CH$_3$
|
CH$_3$CH$_2$CCH$_2$CH$_3$
|
Br

$\xrightarrow{\text{CH}_3\text{OH}}$

CH$_3$
|
CH$_3$CH$_2$CCH$_2$CH$_3$
|
OCH$_3$

d. An S$_N$1 reaction forms a pair of enantiomers.

CH$_2$CH$_2$CH$_3$
|
Br—C$_{\cdots}$CH$_3$
CH$_2$CH$_3$

(*R*)-2-bromopentane

$\xrightarrow{\text{CH}_3\text{OH}}$

CH$_2$CH$_2$CH$_3$
|
CH$_3$—C—OCH$_3$
CH$_3$CH$_2$

(*S*)-2-methoxypentane

+

CH$_2$CH$_2$CH$_3$
|
CH$_3$O—C$_{\cdots}$CH$_3$
CH$_2$CH$_3$

(*R*)-2-methoxypentane

+   HBr

e. Only one product is formed, because the reactant does not have an asymmetric center.

⬡ Br / CH$_3$ $\xrightarrow{\text{CH}_3\text{OH}}$ ⬡ Br / OCH$_3$

2. Alkyl chlorides and alkyl iodides could also be used. It is best to avoid alkyl fluorides, because they have the poorest leaving groups.

a. CH$_3$CHCH$_2$CH$_3$ $\xrightarrow{\text{CH}_3\text{O}^-}$ CH$_3$CHCH$_2$CH$_3$
        | |
        Br   OCH$_3$

b. CH$_3$CH$_2$CH$_2$CH$_2$Br $\xrightarrow{\text{CH}_3\text{O}^-}$ CH$_3$CH$_2$CH$_2$CH$_2$OCH$_3$

c. ⬡—Br $\xrightarrow{\text{⬡—O}^-}$ ⬡—O—⬡

43.   **a.** $(CH_3)_3CI$ $\xrightarrow[\text{H}_2\text{O}]{\text{HO}^-}$ $\underset{\underset{\text{H}_3\text{C}}{}}{\overset{\overset{\text{H}_3\text{C}}{}}{}}C{=}CH_2$ $+$ I^-

because I^- is a better leaving group than CI^-

b. $(CH_3)_3CBr$ $\xrightarrow[\text{H}_2\text{O}]{\text{HO}^-}$ $CH_3\overset{\overset{\text{CH}}{|}}{C}{=}CH_2$

because the reaction forms a more stable alkene

44.   **a.**   ⬡(with double bond)   **b.**   ⬡—CH=CH₃   **c.**   ⬡(with double bond)—CH₃

45.   **a.** $\underset{\text{H}}{\overset{\text{H}_3\text{C}}{}}C{=}C\underset{\text{CH}_3}{\overset{\text{H}}{}}$ **b.** $\underset{\text{H}}{\overset{\text{H}_3\text{C}}{}}C{=}C\underset{\text{CH}_3}{\overset{\text{H}}{}}$ **c.** $\underset{\text{H}}{\overset{\text{H}_3\text{C}}{}}C{=}C\underset{\text{CH}_2\text{CH}_3}{\overset{\text{H}}{}}$

46.   Notice that the major product in both E2 and E1 reactions is the isomer that has the larger of the two substituents on one sp^2 carbon trans to the larger of the two substituents on the other sp^2 carbon.

a. $\underset{\text{H}}{\overset{\text{CH}_3\text{CH}_2\text{CH}_2}{}}C{=}C\underset{\text{CH}_3}{\overset{\text{H}}{}}$ $+$ $\underset{\text{H}}{\overset{\text{CH}_3\text{CH}_2\text{CH}_2}{}}C{=}C\underset{\text{H}}{\overset{\text{CH}_3}{}}$

major

b. $\underset{\text{CH}_3}{\overset{\text{CH}_3}{}}C{=}C\underset{\text{CH}_2\text{CH}_3}{\overset{\text{CH}_3}{}}$

only product

c. $\underset{\text{CH}_3}{\overset{\text{CH}_3\text{CH}_2}{}}C{=}C\underset{\text{CH}_3}{\overset{\text{H}}{}}$ $+$ $\underset{\text{CH}_3}{\overset{\text{CH}_3\text{CH}_2}{}}C{=}C\underset{\text{H}}{\overset{\text{CH}_3}{}}$

major

d. $\underset{\text{CH}_3}{\overset{\text{CH}_3\text{CH}_2}{}}C{=}C\underset{\text{CH}_3}{\overset{\text{H}}{}}$ $+$ $\underset{\text{CH}_3}{\overset{\text{CH}_3\text{CH}_2}{}}C{=}C\underset{\text{H}}{\overset{\text{CH}_3}{}}$

major

7.   **a.**

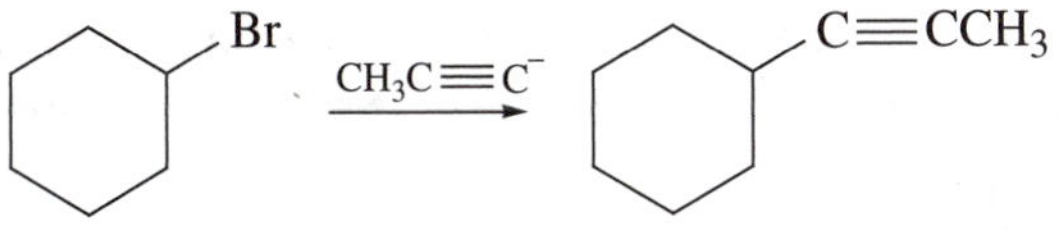

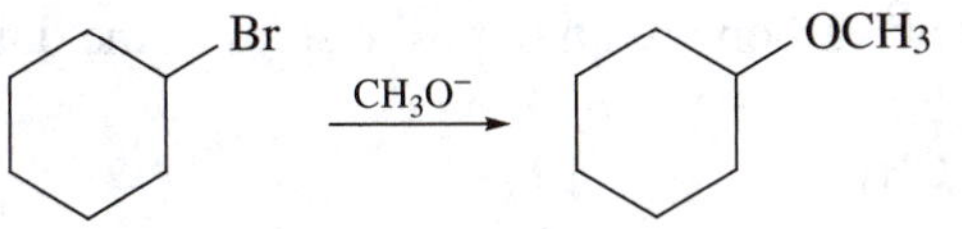

d.

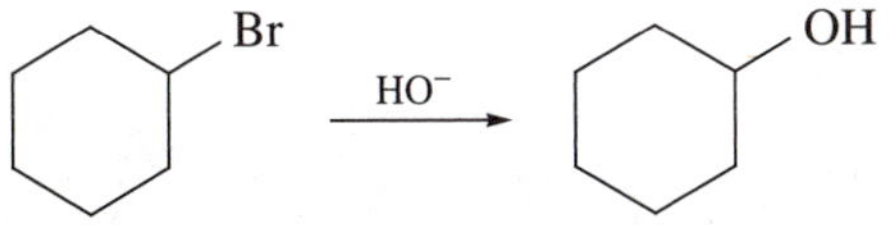

b.

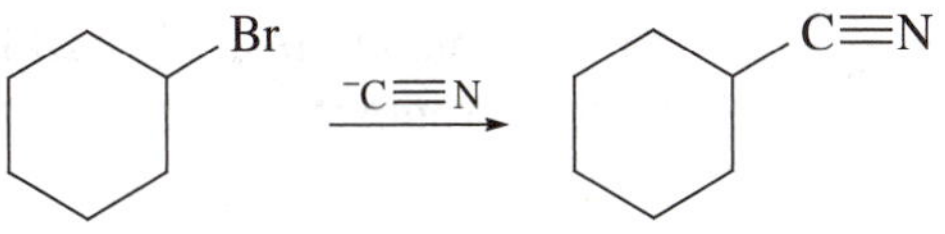

e.

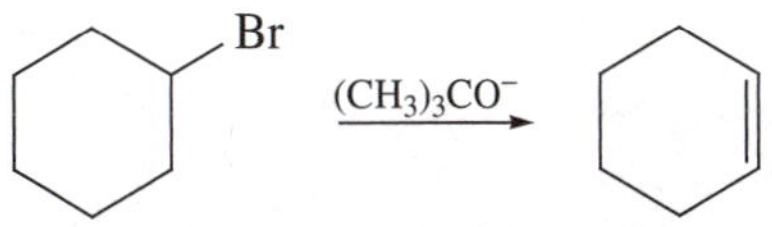

A bulky base is used
to minimize the amount
of substitution product and,
therefore, maximize the amount
of elimination product.

c.

8.   The stereoisomer obtained in greatest yield is the one with the larger of the two substituents on one sp^2 carbon trans to the larger of the two substituents on the other sp^2 carbon.

a.

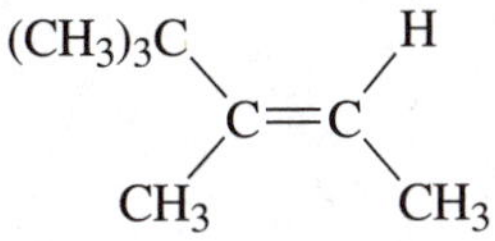

b.

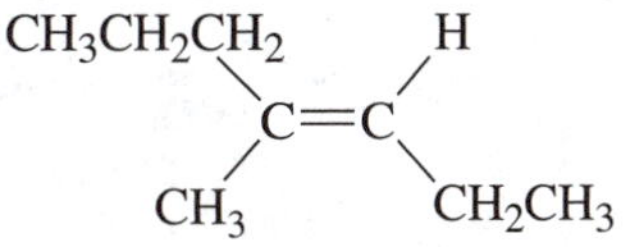

9.   **a.** CH_3O^-

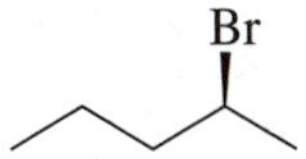

c.

b.

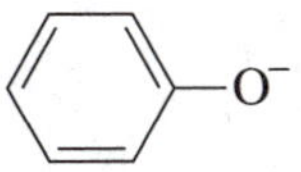

d.

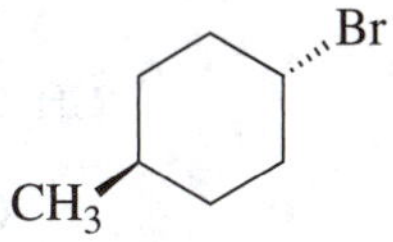

0.   These are all E2 reactions, because a high concentration of a good nucleophile is employed.

a.   The stereoisomer formed in greater yield is the one in which the larger group attached to one sp^2 carbon and the larger group attached to the other sp^2 carbon are on opposite sides of the double bond.

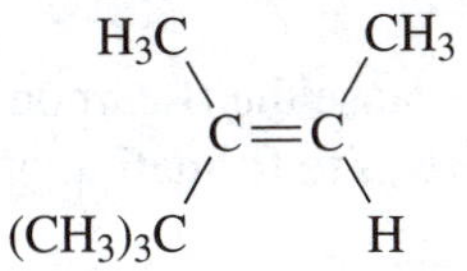

b.   No stereoisomers are possible for this compound, because one of the sp^2 carbons is bonded to two hydrogens.

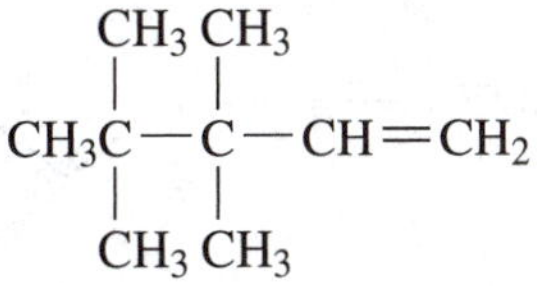

c. No stereoisomers are possible for this compound, because one of the sp^2 carbons is bonded to tw○ methyl groups.

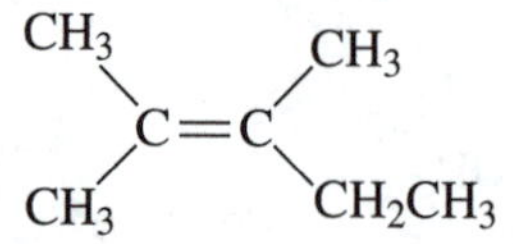

d. The stereoisomer formed in greater yield is the one in which the larger group attached to one sp^2 carbon and the larger group attached to the other sp^2 carbon are on opposite sides of the double bond.

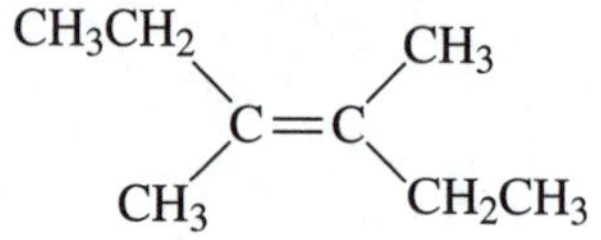

51.

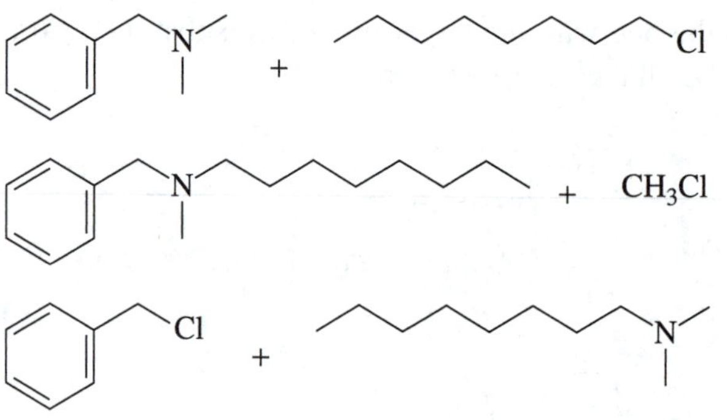

52.

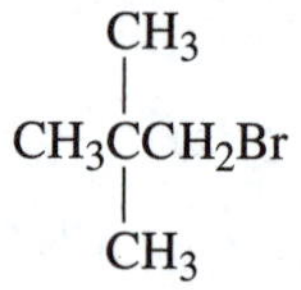

1-bromo-2,2-dimethylpropane

a. The bulky *tert*-butyl substituent blocks the backside of the carbon bonded to the bromine to nucleo philic attack, making an S_N2 reaction difficult. An S_N1 reaction is difficult because the carbocation formed when the bromide ion departs would be an unstable primary carbocation.

b. It cannot undergo an E2 reaction, because the β-carbon is not bonded to a hydrogen. It cannot undergo an E1 reaction, because that would require formation of a primary carbocation.

53. The second reaction will give a better yield of cyclopentyl methyl ether, because methyl bromide can form only the desired substitution product. In contrast, cyclopentyl bromide, the alkyl halide in the first reaction forms an elimination product in addition to the desired substitution product.

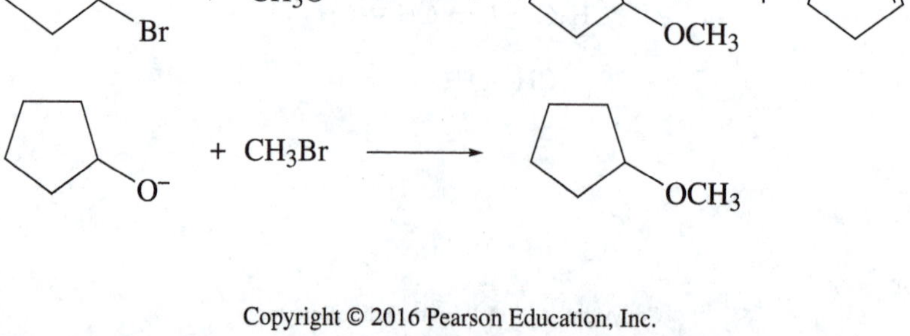

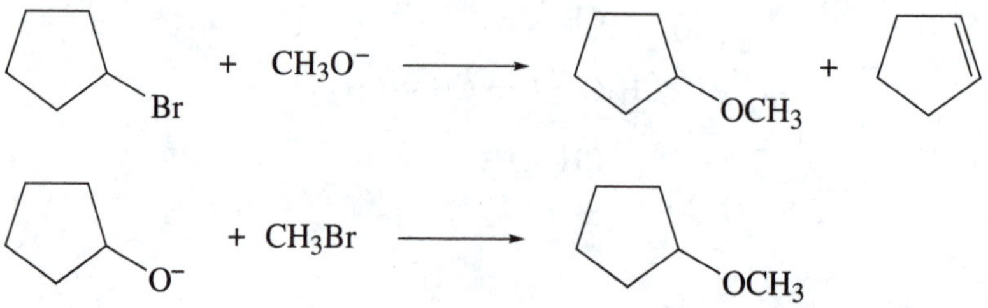

4. 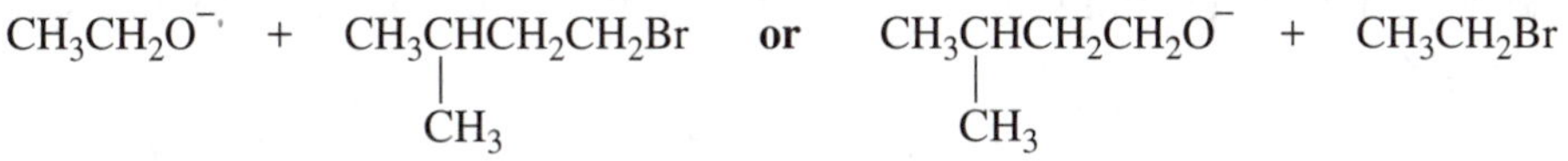 $CH_3CH_2O^-$ + $CH_3CHCH_2CH_2Br$ **or** $CH_3CHCH_2CH_2O^-$ + CH_3CH_2Br
　　　　　　　　　　　　|　　　　　　　　　　　　　　　　|
　　　　　　　　　　　CH_3　　　　　　　　　　　　　　CH_3

5. **a.** $CH_3CH_2CH_2CH_2Br$ $\xrightarrow{NH_3}$ $CH_3CH_2CH_2CH_2\overset{+}{N}H_3$

b.

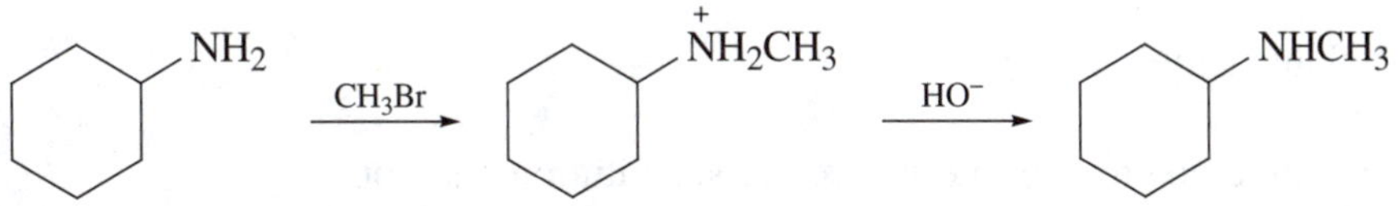

c.

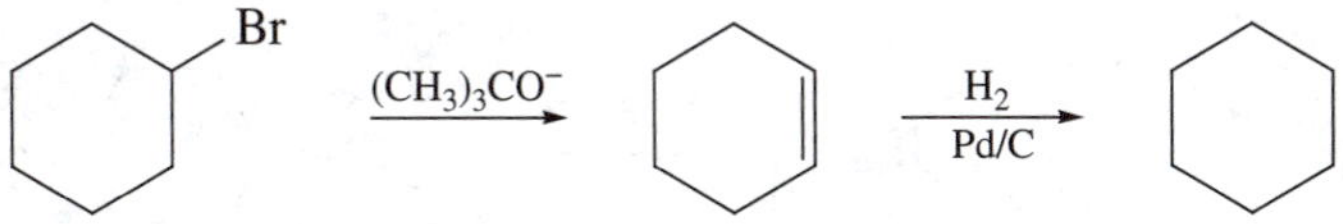

6. The reaction with ethoxide would have a higher substitution-product to elimination-product ratio, because the bulky *tert*-butoxide ion would have less access to the backside of the secondary carbocation, so it would form less substitution product.

7. The carbocation formed by 3-bromocyclohexene is stabilized by electron delocalization. Thus, it is more stable than the carbocation formed by bromocyclohexane. Therefore, 2-bromocyclohexene is more reactive in an E1 reaction, because the more stable the carbocation, the easier it can be formed.

8.

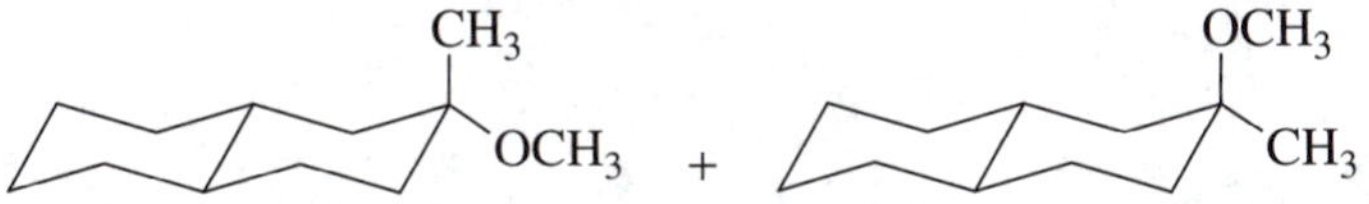

9. **a.** Both compounds form the same elimination products.

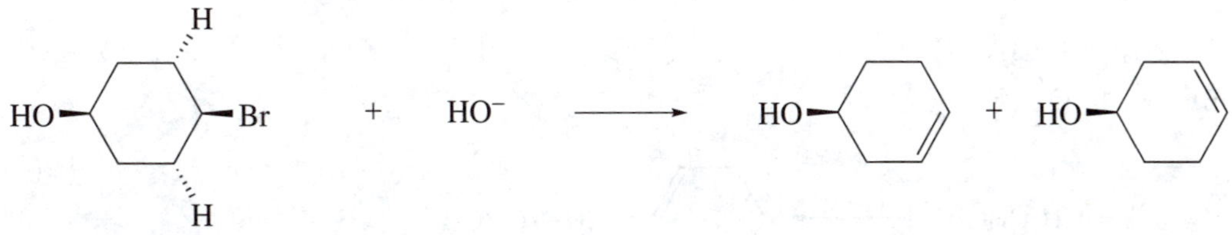

cis-4-bromocyclohexanol

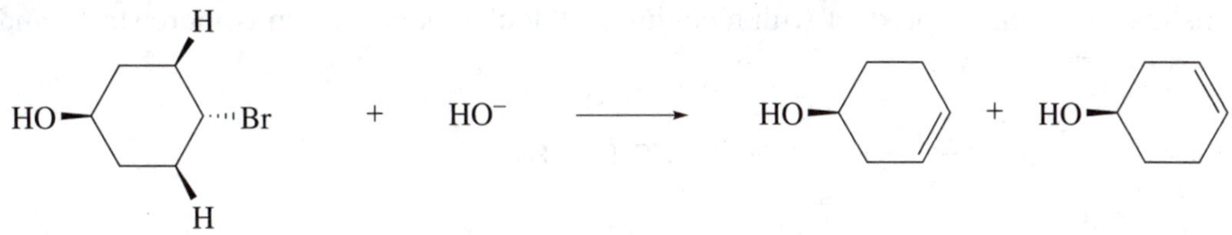

trans-4-bromocyclohexanol

b. Only the trans isomer can undergo an intramolecular substitution reaction, because the S_N2 reaction requires backside attack.

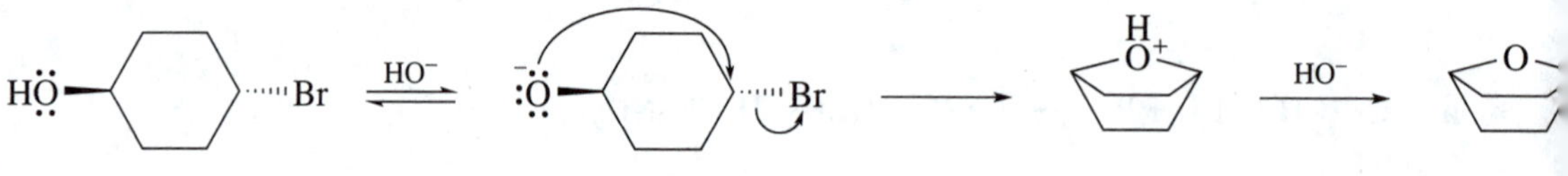

trans-4-bromocyclohexanol

The cis isomer can undergo only an intermolecular substitution reaction.

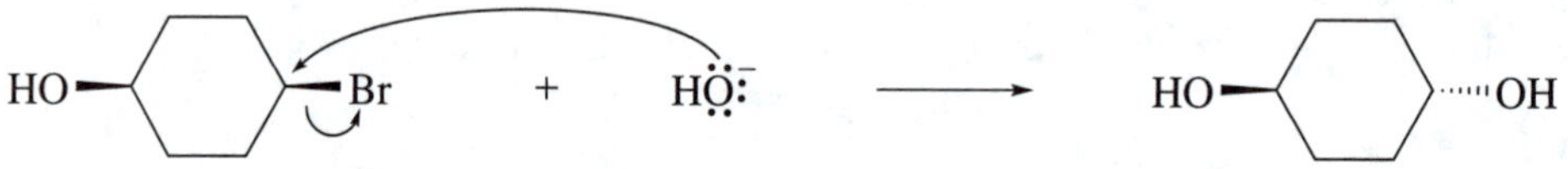

cis-4-bromocyclohexanol

c. The elimination reaction forms a pair of enantiomers, because the reaction creates an asymmetric center in the product. Both substitution reactions form a single stereoisomer, because the reaction does not create an asymmetric center in the product.

60.   **a.**

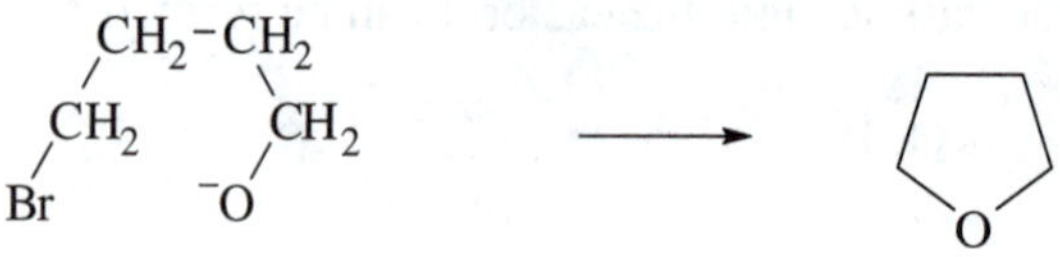

   b.

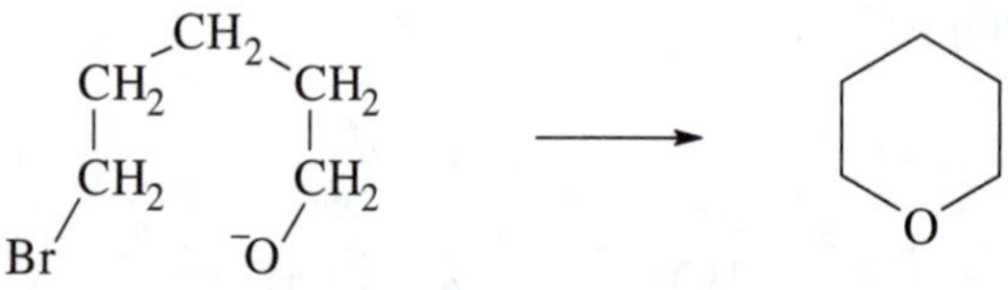

61.   **a.**

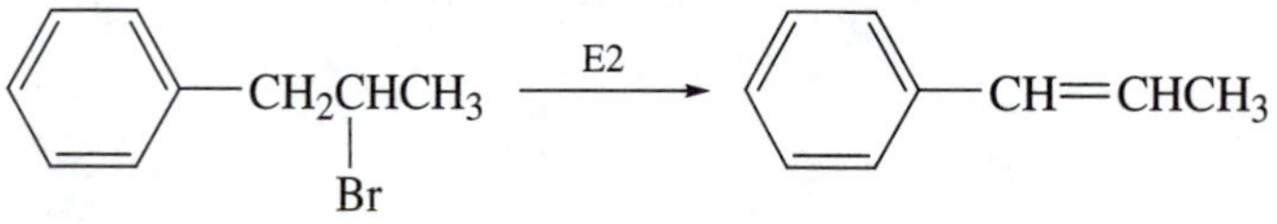

This reactant forms a product with a conjugated double bond, which is more stable and, therefore easier to form.

   b. $CH_2{=}CHCHCH_3$ $\xrightarrow{\text{E2}}$ $CH_2{=}CHCH{=}CH_2$

   with Br below the CHCH$_3$ carbon

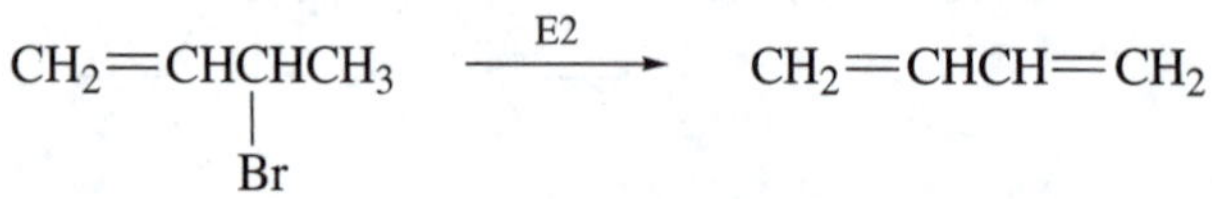

This reactant forms a product with a conjugated double bond, which is more stable and, therefore easier to form.

2. **a.**

b.

major   minor   major   minor

c.

major   minor   major   minor

d.

major   minor   major   minor

3. **a.**

$$CH_3CCH_3$$ with CH_3 and Br

$$CH_3\overset{+}{C}CH_3 \quad \xrightarrow[H_2O]{CH_3CH_2OH} \quad CH_3CCH_3 \quad + \quad CH_3CCH_3$$

$Br^- \ Cl^-$ OCH_2CH_3 OH

b. The products are obtained as a result of the nucleophiles reacting with the carbocation. 2-Bromo-2-methylpropane and 2-chloro-2-methylpropane form the same carbocation, so both alkyl halides form the same products.

4. **a.** The reaction with quinuclidine was faster, because quinuclidine is less sterically hindered as a result of the substituents on the nitrogen being pulled back into a ring structure.

b. The reaction with quinuclidine had the larger rate constant for the same reason given in part **a**.

5. The equilibrium constant is given by the relative stabilities of the products and the reactants. Therefore, any factor that stabilizes the products will increase the equilibrium constant.

$$K_{eq} = \frac{[\text{products}]}{[\text{reactants}]}$$

Ethanol will stabilize the charged products more than will diethyl ether, because ethanol is more polar. Therefore, the equilibrium will lie farther to the right (toward products) in ethanol.

66. The products will be less stabilized in a less polar solvent, so the K_a would decrease. Therefore, the p. would increase (it would be a weaker acid). (See Problem 28.)

$$K_a = \frac{[CH_3COO^-][H^+]}{[CH_3COOH]}$$

67.   a.

b. Two products are obtained because methanol can add to the top or the bottom of the planar doub bond.

c. One bromine is eliminated with the help of one of oxygen's lone pairs, forming a carbocation whe the positive charge is shared by a carbon and an oxygen. The oxygen cannot help the other bromi dissociate.

Chapter 8 Practice Test

. Which of the following is more reactive in an S_N1 reaction?

a. (benzene ring)—CHCH$_2$CH$_3$ or   (benzene ring)—CH$_2$CHCH$_3$
     |Br |Br

b. CH$_3$CH$_2$CHCH$_3$ (with CH$_3$ substituent)   or   CH$_3$CH=CHCHCH$_3$ (with Br substituent)

. Which of the following is more reactive in an S_N2 reaction?

a. CH$_3$CH$_2$CHBr (with CH$_3$ substituent)   or   CH$_3$CH$_2$CHBr (with CH$_2$CH$_3$ substituent)

b. (benzene ring)—CH$_2$CH$_2$Br   or   (benzene ring)—CH$_2$Br

. Indicate whether each of the following statements is true or false:

a. Increasing the concentration of the nucleophile favors an S_N1 reaction over
   an S_N2 reaction.   T   F

b. Ethyl iodide is more reactive than ethyl chloride in an S_N2 reaction.   T   F

c. In an S_N1 reaction, the product with the retained configuration is obtained
   in greater yield.   T   F

d. The rate of a substitution reaction in which none of the reactants is charged
   will increase if the polarity of the solvent is increased.   T   F

e. An S_N2 reaction is a two-step reaction.   T   F

f. The pK_a of a carboxylic acid is greater in water than it is in a mixture of
   dioxane and water.   T   F

g. 4-Bromo-1-butanol will form a cyclic ether faster than will 3-bromo-1-propanol.   T   F

. For each of the following pairs of S_N2 reactions, indicate the one that occurs with the greater rate constant
(that is, occurs faster):

a. CH$_3$CH$_2$CH$_2$Cl + HO$^-$ or   CH$_3$CHCH$_3$ + HO$^-$
                                          |Cl

b. CH$_3$CH$_2$CH$_2$Cl + HO$^-$ or   CH$_3$CH$_2$CH$_2$I + HO$^-$

c. CH$_3$CH$_2$CH$_2$Br + HO$^-$ or   CH$_3$CH$_2$CH$_2$Br + H$_2$O

d. CH_3CHCH_3 $\xrightarrow[\text{H}_2\text{O/CH}_3\text{OH}]{\text{CH}_3\text{O}^-}$ or CH_3CHCH_3 $\xrightarrow[\text{CH}_3\text{OH}]{\text{CH}_3\text{O}^-}$
  | |
  Br   Br

e. $BrCH_2CH_2CH_2CH_2NHCH_3$ or $BrCH_2CH_2CH_2NHCH_3$

5. How would increasing the polarity of the solvent affect the following?

   a. the rate of the S_N2 reaction of methylamine with 2-bromobutane

   b. the rate of the S_N1 reaction of methylamine with 2-bromo-2-methylbutane

   c. the rate of the S_N2 reaction of methoxide ion with 2-bromobutane

   d. the pK_a of acetic acid

   e. the pK_a of phenol

6. Which of the following is more reactive in an E2 reaction?

   a. (phenyl)$-CH_2CHCH_3$ or   (phenyl)$-CH_2CH_2CH_2Br$
                    |
                    Br

   b. $CH_3CH_2CHCH_3$ or $CH_2{=}CHCH_2CHCH_3$
              | |
              Br   Br

7. Which of the following would give the greater amount of substitution product under conditions that favo[r]
    $S_N2/E2$ reactions?

               CH_3 CH_3
                | |
            CH_3CBr or CH_3CHBr
                |
               CH_3

8. What products are obtained when (R)-2-bromobutane reacts with CH_3O^- / CH_3OH under conditions tha[t]
    favor $S_N2/E2$ reactions? Include the configuration of the products.

9. What alkoxide ion and what alkyl bromide should be used to synthesize the following ethers?

               CH_3
                |
   a. $CH_3CH_2COCH_2CH_2CH_3$
                |
               CH_3

   b. (cyclohexyl)$-O-CH_3$

0.   For each of the following pairs of E2 reactions, indicate the one that occurs with the greater rate constant:

a. $CH_3CH_2CH_2Cl + HO^-$ or $CH_3\underset{\underset{\displaystyle Cl}{|}}{C}HCH_3$ $+ HO^-$

b. $CH_3CH_2CH_2Cl + HO^-$ or $CH_3CH_2CH_2I$ $+ HO^-$

c. $CH_3CH_2CH_2Br + HO^-$ or $CH_3CH_2CH_2Br + H_2O$

d. $CH_3\underset{\underset{\displaystyle Br}{|}}{C}HCH_3$ $\xrightarrow[\text{H}_2\text{O/CH}_3\text{OH}]{\text{CH}_3\text{O}^-}$ or $CH_3\underset{\underset{\displaystyle Br}{|}}{C}HCH_3$ $\xrightarrow[\text{CH}_3\text{OH}]{\text{CH}_3\text{O}^-}$

e. $CH_3\underset{\underset{\displaystyle Br}{|}}{C}HCH_3 + HO^-$ or $CH_3\underset{\underset{\displaystyle Br}{|}}{\overset{\overset{\displaystyle CH_3}{|}}{C}}CH_3 + HO^-$

1.   What is the major product obtained from the E2 reaction of each of the following compounds with hydroxide ion?

a.

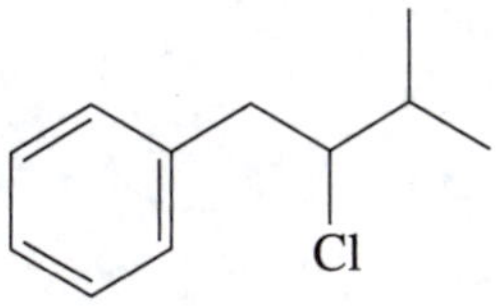

b.

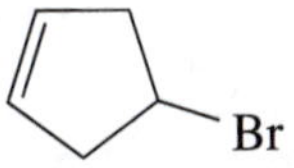

c.

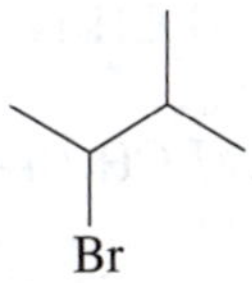

Answers to Chapter 8 Practice Test

1. **a.** —CHCH$_2$CH$_3$ with Br substituent

b. CH$_3$CH=CHCHCH$_3$ with Br on the CH

2. **a.** CH$_3$CH$_2$CHBr with CH$_3$ substituent

b. —CH$_2$Br

3. **a.** Increasing the concentration of the nucleophile favors an S$_N$1 reaction over an S$_N$2 reaction.   F

b. Ethyl iodide is more reactive than ethyl chloride in an S$_N$2 reaction.   T

c. In an S$_N$1 reaction, the product with the retained configuration is obtained in greater yield.   F

d. The rate of a substitution reaction in which none of the reactants is charged will increase if the polarity of the solvent is increased.   T

e. An S$_N$2 reaction is a two-step reaction.   F

f. The pK_a of a carboxylic acid is greater in water than it is in a mixture of dioxane and water.   F

g. 4-Bromo-1-butanol will form a cyclic ether faster than will 3-bromo-1-propanol.   T

4. **a.** CH$_3$CH$_2$CH$_2$Cl + HO$^-$

b. CH$_3$CH$_2$CH$_2$I + HO$^-$

c. CH$_3$CH$_2$CH$_2$Br + HO$^-$

d. CH$_3$CHCH$_3$ with Br substituent, $\xrightarrow[\text{CH}_3\text{OH}]{\text{CH}_3\text{O}^-}$

e. BrCH$_2$CH$_2$CH$_2$CH$_2$NHCH$_3$

5. **a.** The rate of the reaction would increase.

b. The rate of the reaction would increase.

c. The rate of the reaction would decrease.

d. The pK_a would decrease.

e. The pK_a would decrease.

6. **a.** —CH$_2$CHCH$_3$ with Br substituent

b. CH$_2$=CHCH$_2$CHCH$_3$ with Br substituent

7. CH$_3$CHBr with CH$_3$ substituent

8. 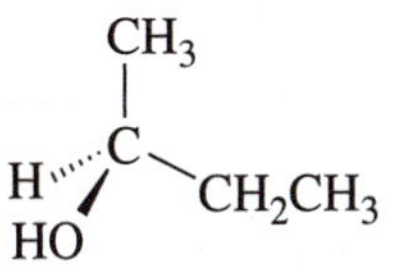 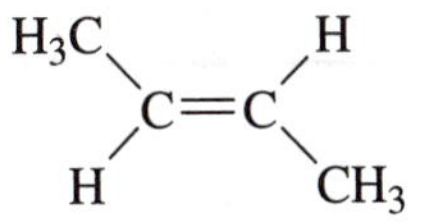 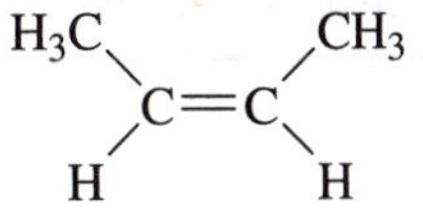

major minor

9. **a.** $CH_3CH_2\overset{CH_3}{\overset{|}{\underset{\underset{CH_3}{|}}{C}}}O^- + CH_3CH_2CH_2Br$

b. 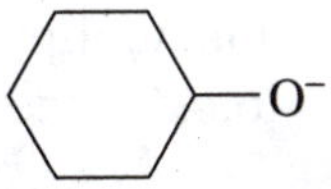 $+ CH_3Br$

10. **a.** $CH_3\underset{\underset{Cl}{|}}{C}HCH_3 + HO^-$

b. $CH_3CH_2CH_2I + HO^-$

c. $CH_3CH_2CH_2Br + HO^-$

d. $CH_3\underset{\underset{Br}{|}}{C}HCH_3 \xrightarrow[CH_3OH]{CH_3O^-}$

e. $CH_3\overset{CH_3}{\overset{|}{\underset{\underset{Br}{|}}{C}}}CH_3 + HO^-$

11. **a.** 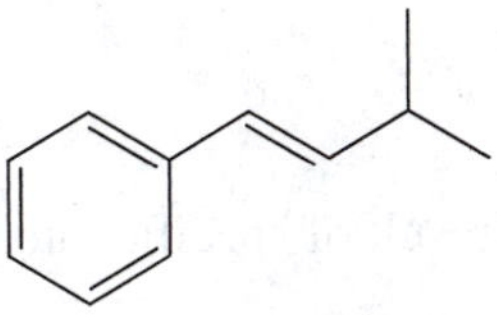

b. 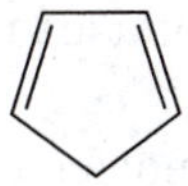

c. 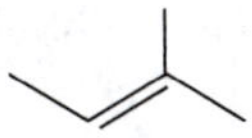

CHAPTER 9
Reactions of Alcohols, Ethers, Epoxides, Amines, and Thiols

Important Terms

alcohol	an organic compound with an OH functional group (ROH).
alkaloid	a natural product with a heterocyclic ring (where the heteroatom is a nitrogen found in the leaves, bark, roots, or seeds of plants.
dehydration	loss of water.
epoxide	an ether in which the oxygen is incorporated into a three-membered ring.
ether	a compound containing an oxygen bonded to two carbons (ROR).
lead compound	a prototype in a search for other physiologically active compounds.
mercapto group	an SH group.
molecular modification	changing the structure of a lead compound.
molecular recognition	the recognition of one molecule by another as a result of specific interactions.
sulfide (thioether)	the sulfur analogue of an ether (RSR).
sulfonium salt	$R_3S^+X^-$
thioether (sulfide)	the sulfur analogue of an ether (RSR).
thiol (mercaptan)	the sulfur analogue of an alcohol (RSH).

Solutions to Problems

1. CH_3OH
common = methyl alcohol
systematic = methanol

CH_3CH_2OH
common = ethyl alcohol
systematic = ethanol

$CH_3CH_2CH_2OH$
common = propyl alcohol or *n*-propyl alcohol
systematic = 1-propanol

$CH_3CH_2CH_2CH_2OH$
common = butyl alcohol or *n*-butyl alcohol
systematic = 1-butanol

$CH_3CH_2CH_2CH_2CH_2OH$
common = pentyl alcohol or *n*-pentyl alcohol
systematic = 1-pentanol

$CH_3CH_2CH_2CH_2CH_2CH_2OH$
common = hexyl alcohol or *n*-hexyl alcohol
systematic = 1-hexanol

2.   **a.** 1-pentanol
       primary

      b. 4-methylcyclohexanol
       secondary

      c. 5-methyl-3-hexanol
       secondary

      d. 2-ethyl-1-pentanol
       primary

      e. 5-chloro-2-methyl-2-pentanol
       tertiary

      f. 2,6-dimethyl-4-octanol
       secondary

3.

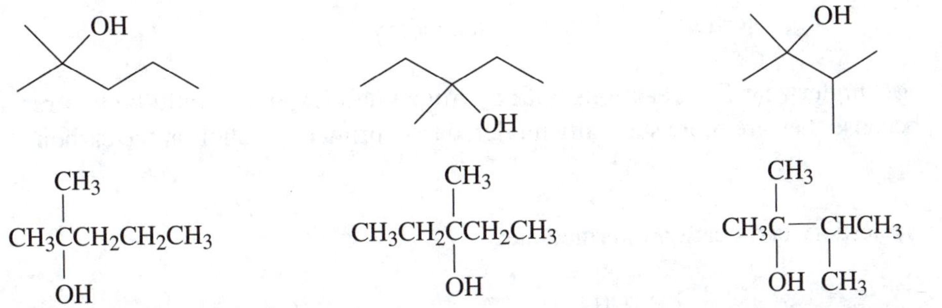

2-methyl-2-pentanol   3-methyl-3-pentanol   2,3-dimethyl-2-butanol

4.   They no longer have a lone pair of electrons.

5.   All four alcohols undergo an S_N1 reaction, because they are either secondary or tertiary alcohols. The arrows are shown for the first protonation step in part **a**, but are not shown in parts **b**, **c**, and **d**.

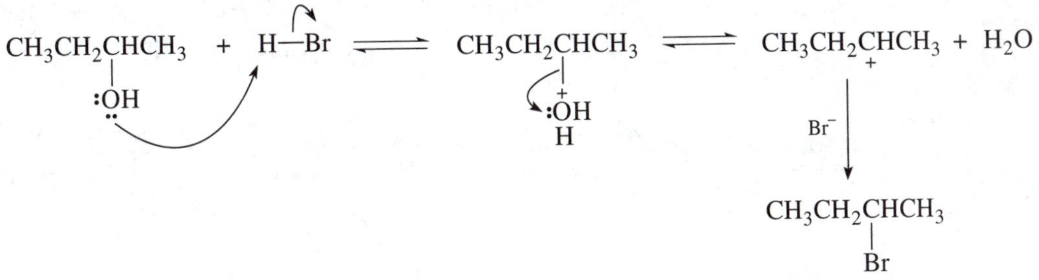

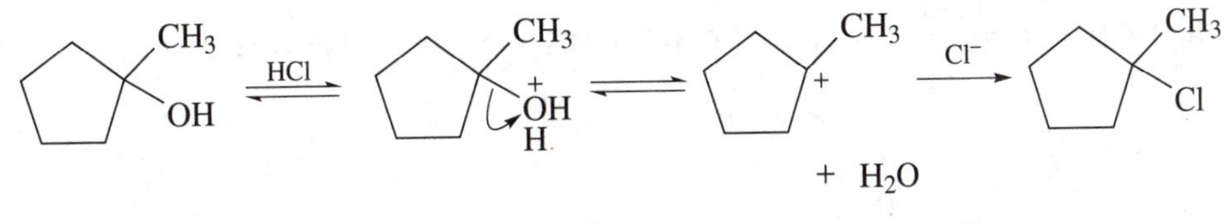

6.   **a.**   Solved in the text.

b.   The leaving group of $CH_3\overset{+}{O}H_2$ is H_2O; the conjugate acid of the leaving group is H_3O^+ with a p$K_a = -1.7$
The leaving group of CH_3OH is HO^-; the conjugate acid of the leaving group is H_2O with a p$K_a = 15.7$
Because H_3O^+ is a much stronger acid than H_2O, H_2O is a much weaker base than HO^- and, therefore
is the better leaving group. Therefore, $CH_3\overset{+}{O}H_2$ is more reactive than CH_3OH.

7.   **a.**   Solved in the text.

b.

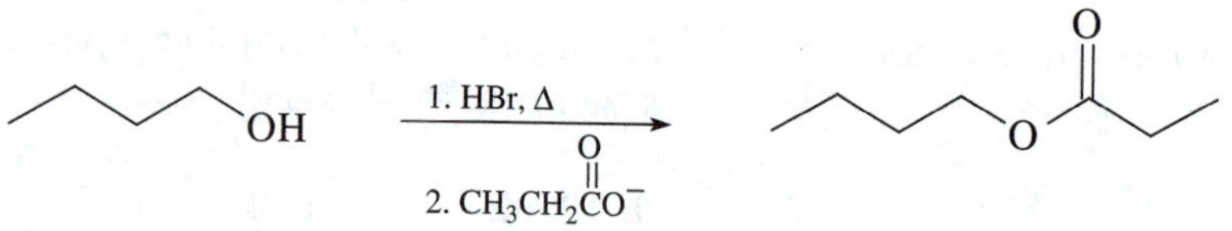

c.

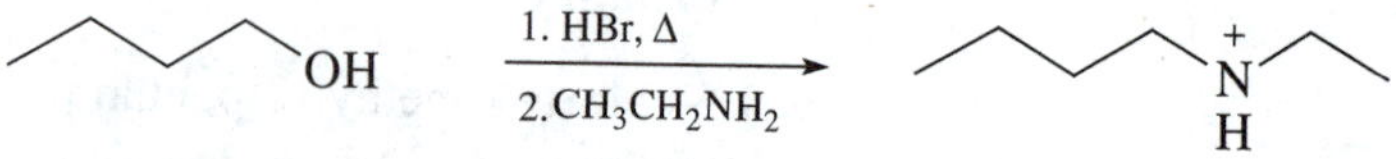

d. $CH_3CH_2CH_2CH_2OH$ $\xrightarrow[\text{2. }^-C\equiv N]{\text{1. HBr, }\Delta}$ $CH_3CH_2CH_2CH_2C\equiv N$

8.   The relative reactivity would be as follows:

$$\text{tertiary} > \text{primary} > \text{secondary}$$

If secondary alcohols underwent S_N2 reactions with hydrogen halides, they would be less reactive than
primary alcohols, because they are more sterically hindered than primary alcohols at the carbon attached to
the OH group.

9.   **B**, because it forms a tertiary carbocation intermediate.

10.   **a.**

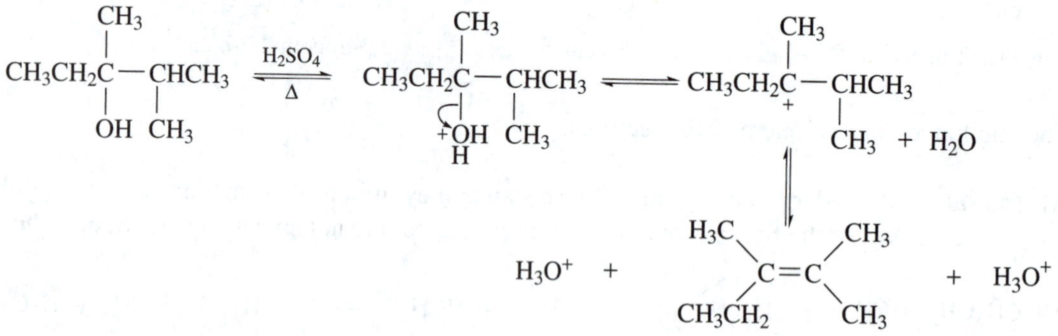

b.

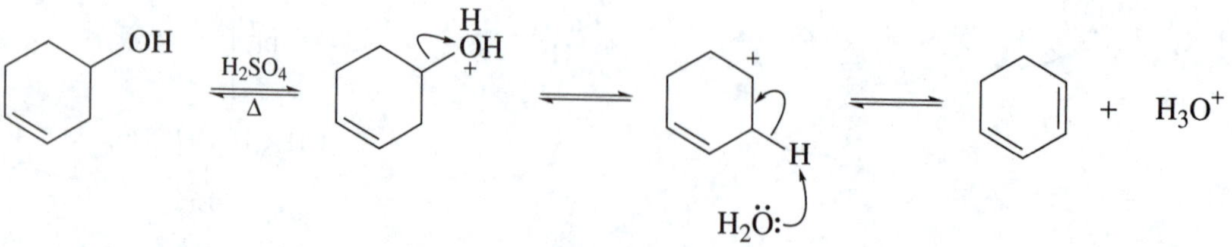

c. $CH_2{=}CHCH_2CH_2OH$ $\xrightleftharpoons[\Delta]{H_2SO_4}$ $CH_2{=}CHCH_2CH_2\overset{+}{O}H \atop H$ $\rightleftharpoons$ $CH_2{=}CHCH{=}CH_2$ $+$ H_3O^+

d. 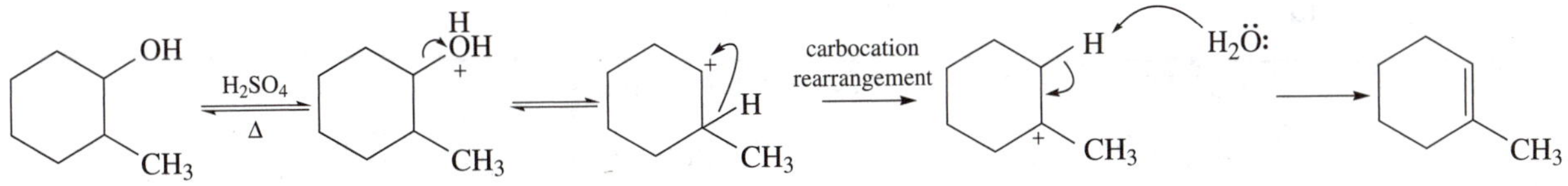

11. **a.** In order to synthesize an unsymmetrical ether (ROR′) by this method, two different alcohols (ROH and R′OH) would have to be heated with sulfuric acid. Therefore, three different ethers would be obtained as products. Consequently, the desired ether would account for considerably less than half of the total amount of ether that is synthesized.

$$ROH + R'OH \xrightarrow[\Delta]{H_2SO_4} ROR + \underset{\text{target molecule}}{ROR'} + R'OR'$$

b. It could be synthesized by a Williamson ether synthesis.

$$CH_3CH_2CH_2OH \xrightarrow{NaH} CH_3CH_2CH_2O^- \xrightarrow{CH_3CH_2Br} CH_3CH_2CH_2OCH_2CH_3 + Br^-$$

12. 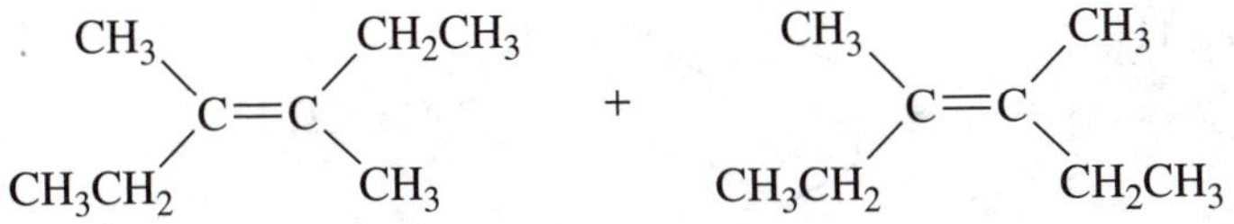

major minor

13. **a.** $CH_3CH_2\underset{CH_3}{\underset{|}{C}}{=}\underset{CH_2CH_3}{\underset{|}{C}}\overset{CH_3}{\overset{|}{C}}HCH_3$ **b.** 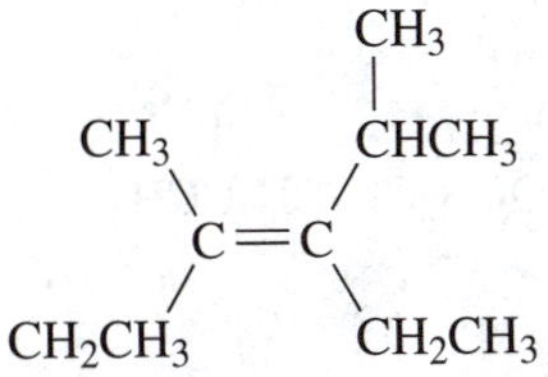

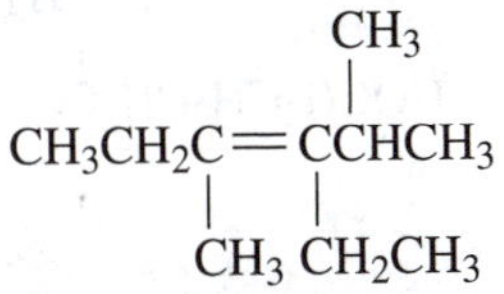

14. **a.** $CH_3CH_2\overset{O}{\overset{\|}{C}}CH_2CH_3$

b. A tertiary alcohol is not oxidized to a carbonyl compound.

c. $CH_3CH_2CH_2CH_2\overset{O}{\overset{\|}{C}}H$

15. **a.** $CH_3CH_2\underset{OH}{\underset{|}{C}}HCH_3$ **b.** 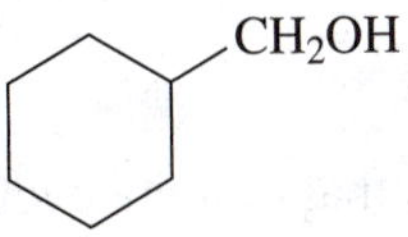 **c.** $CH_3CH_2CH_2CH_2OH$

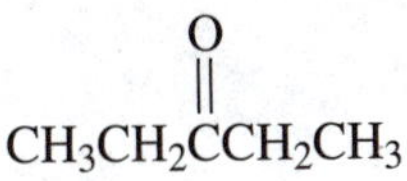

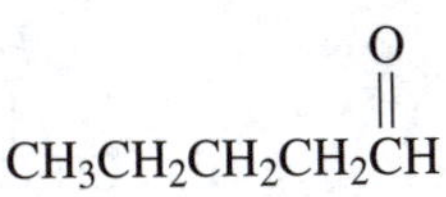

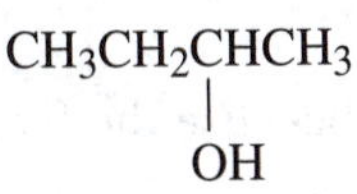

16. **a.** **1.** methoxyethane **3.** 4-methoxyoctane

2. ethoxyethane **4.** 1-propoxybutane

  b. no

  c. **1.** ethyl methyl ether   **3.** no common name

  2. diethyl ether   **4.** butyl propyl ether

17. Solved in the text.

18. **a.** Solved in the text.

  b. Cleavage occurs by an S_N2 pathway, because a primary carbocation is too unstable to be formed.

  c. Cleavage occurs by an S_N2 pathway, because a primary carbocation or a methyl cation is too unstable to be formed.

  d. Cleavage occurs by an S_N1 pathway, because the tertiary carbocation that is formed is relatively stable; I^- will react with the tertiary carbocation.

19. **a.**   **b.**

20. **a.**   **b.**   **c.**   **d.**

21. The reactivity of tetrahydrofuran is more similar to that of a noncyclic ether, because the five-membered ring does not have the strain that makes the epoxide reactive.

22. Solved in the text.

23. Each arene oxide will open in the direction that forms the most stable carbocation. Thus, the methoxy-substituted arene oxide opens so that the positive charge can be stabilized by electron donation from the methoxy group.

The nitro-substituted arene oxide opens in the direction that forms the most stable carbocation intermediate, the one where the positive charge is farther away from the electron-withdrawing NO_2 group.

24. The compound without the double bond in the second ring is more apt to be carcinogenic. It opens to form a less stable carbocation than the other compound, because it can be stabilized by electron delocalization only if the aromaticity of the benzene ring is destroyed. Because the carbocation is less stable, it is formed more slowly, giving the carcinogenic pathway a better chance to compete with ring opening.

25. The first compound is too insoluble.

The second compound is used clinically; it has fewer carbons than the first compound, so it is more soluble in water.

The third compound is less reactive than the second compound, because the lone pair on the nitrogen ca
be delocalized into the benzene ring, so the lone pair is less apt to displace a chloride ion and form th
three-membered ring that is needed for the compound to be an alkylating agent.

26.

a. $CH_3CH_2CH-\overset{\overset{\displaystyle CH_3}{|}}{\underset{\underset{\displaystyle OCH_3}{|}}{\underset{\displaystyle |}{C}}}CH_3$ with OH on the CH

d. $CH_3CH_2CH-\overset{\overset{\displaystyle CH_3}{|}}{\underset{\underset{\displaystyle OH}{|}}{\underset{\displaystyle |}{C}}}CH_3$ with CH_3O on the CH

b. $CH_3\overset{}{\underset{\underset{\displaystyle CH_3}{|}}{CH}}CH_2OH \;+\; CH_3I$

e. $\begin{array}{c}CH_3\\ CH_3\end{array}\!C\!=\!C\!\begin{array}{c}CH_3\\ CH_3\end{array}$

c. $CH_2\overset{\displaystyle O}{\overset{\|}{C}}OH$

f. $\overset{\displaystyle O}{\overset{\|}{C}}CH_3$

27.

a. isopropyl propyl ether
1-isopropoxypropane or 2-propoxypropane

c. *sec*-butyl methyl ether
2-methoxybutane

b. butyl ethyl ether
1-ethoxybutane

d. diisopropyl ether
2-isopropoxypropane

28.

a. CH_2CH_2OH

The other alcohol cannot undergo dehydration to form an alkene,
because its β-carbon is not bonded to a hydrogen.

b. CH_3 , OH

A tertiary carbocation is more stable than a secondary carbocation.

c.

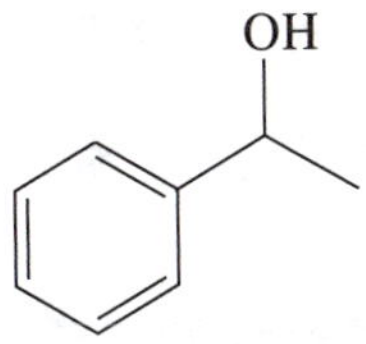

A secondary benzylic cation is more stable than a secondary carbocation.

d.

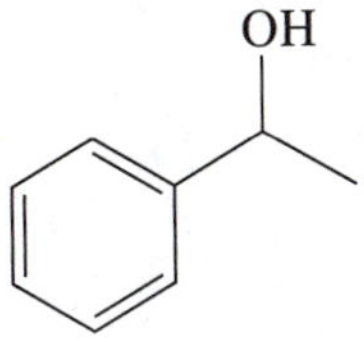

A secondary benzylic cation is more stable than a primary carbocation.

29. **a.** 3-ethoxyheptane

b. methoxycyclohexane

c. 4-methyl-1-pentanol

d. 1-isopropoxy-3-methylbutane

e. 3-ethylcyclohexanol

f. 2-isopropoxypentane

30. **a.** $CH_3CHOCHCH_3$
        $\qquad\ \ \ |\ \ \ |$
        $\qquad\ \ \ CH_3\ CH_3$

b. $CH_2{=}CHCH_2OCH{=}CH_2$

c. $CH_3CH_2CHOCH_2CHCH_3$
     $\qquad\ \ \ \ |\qquad\ \ |$
     $\qquad\ \ \ \ CH_3\qquad CH_3$

d.

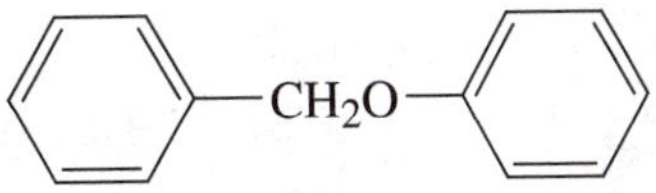

31. **a.** $\quad$ CH_3
         $\qquad\quad |$
         $CH_3CBr\ +\ CH_3CH_2OH$
         $\qquad\quad |$
         $\qquad\quad CH_3$

b. $CH_3CHCH_2OH\ +\ CH_3I$
     $\qquad\ |$
     $\qquad\ CH_3$

c.

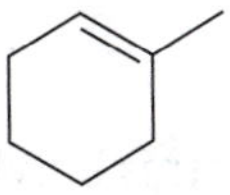

d.

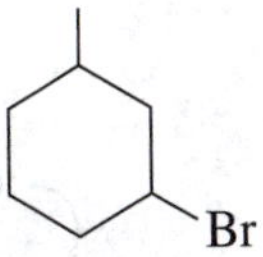

e.

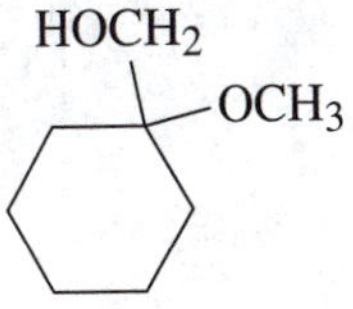

f.

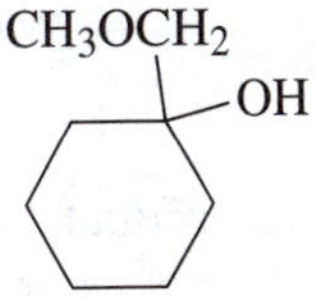

32.

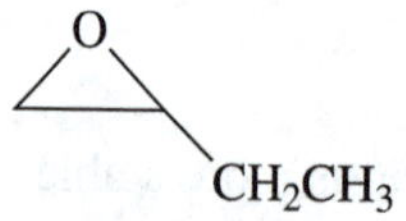

2-ethyloxirane

a. $CH_3CH_2CHCH_2OH$
          |
          OH

0.1 M HCl is a dilute solution
of HCl in water.

c. $CH_3CH_2CHCH_2OH$
          |
          OH

b. $CH_3CH_2CHCH_2OH$
          |
          OCH_3

d. $CH_3CH_2CHCH_2OCH_3$
          |
          OH

33. **a.** CH_3 ⎯ ⬡ ···· OH

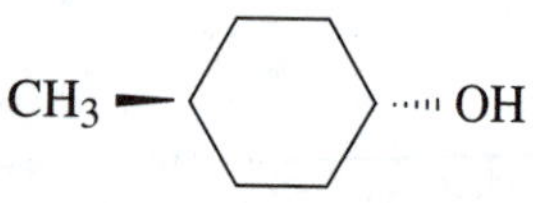

b. $CH_3CH_2OCH_2CH_2CH_2OH$

34. **a.** $CH_3\overset{O}{\overset{||}{C}}CHCH_2CH_3$
                  |
                  CH_3

b. $CH_3CH_2CH_2\overset{O}{\overset{||}{C}}OH$

c.
          CH_3

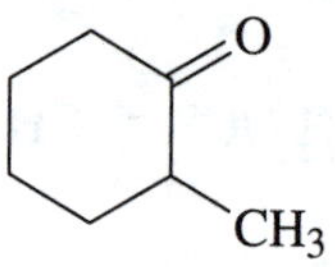

35.

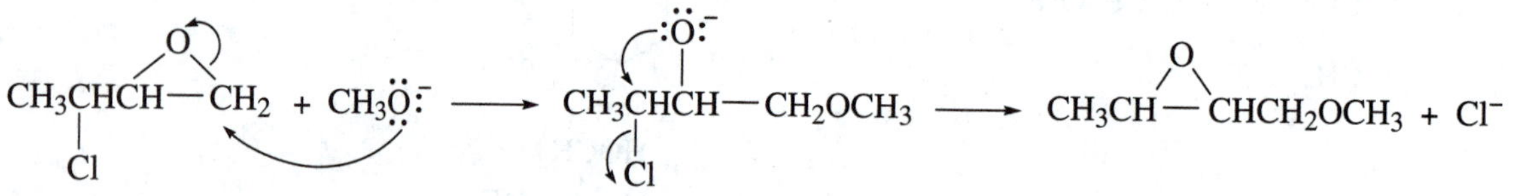

36. **a.** 1-propanol   **b.** 4-propyl-1-nonanol   **c.** 1-methoxy-5-methyl-3-propylheptane

37.   Ethyl alcohol is not obtained as a product, because it reacts with the excess HI and forms ethyl iodide.

$$CH_3CH_2OCH_2CH_3 \xrightarrow[\Delta]{HI} CH_3CH_2I + CH_3CH_2OH \xrightarrow[\Delta]{HI} CH_3CH_2I + H_2O$$

38. Cyclopropane does not react with HO^-, because cyclopropane does not contain a leaving group; a carbanion is far too basic to serve as a leaving group.

Ethylene oxide reacts with HO^-, because ethylene oxide contains an RO^- leaving group.

39. **a.**

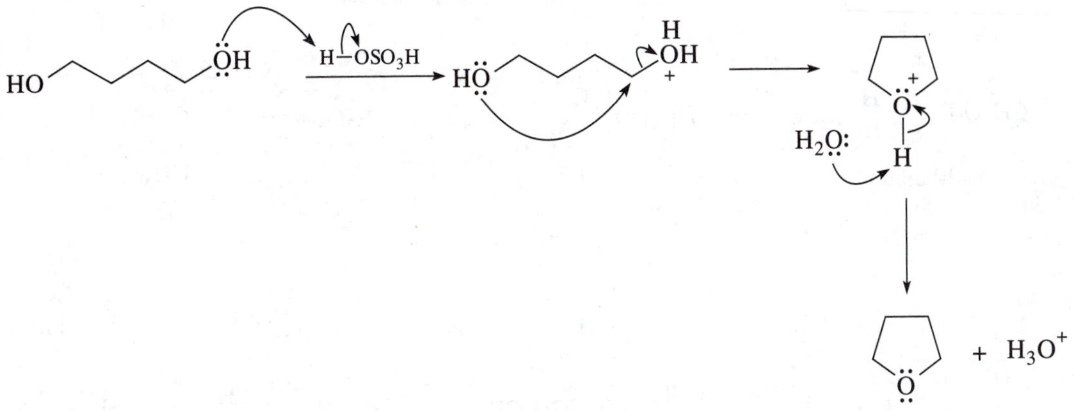

b.

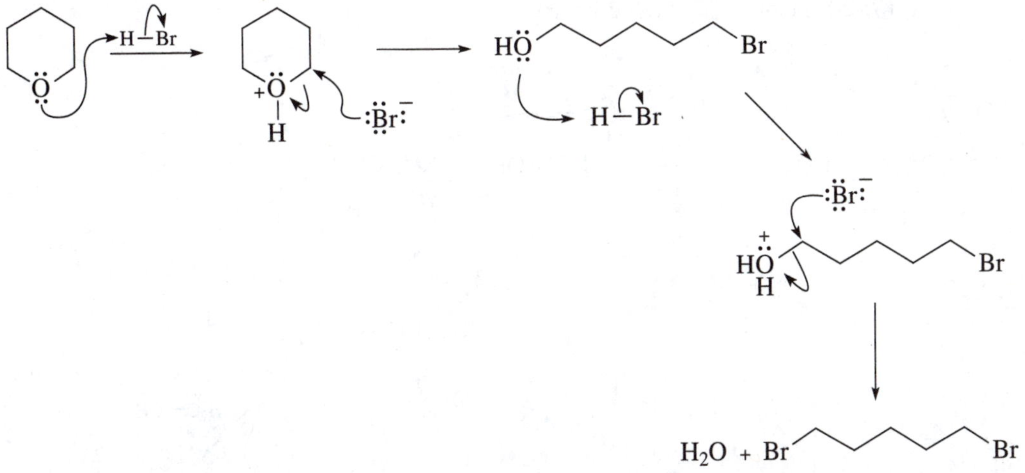

40. **Diethyl ether** is the ether that would be obtained in greatest yield, because it is a symmetrical ether. Since it is symmetrical, only one alcohol is used in its synthesis. Therefore, it is the only ether that would be formed. In contrast, the synthesis of an unsymmetrical ether requires two different alcohols. Therefore, the unsymmetrical ether is one of three different ethers that would be formed.

41. **a.**

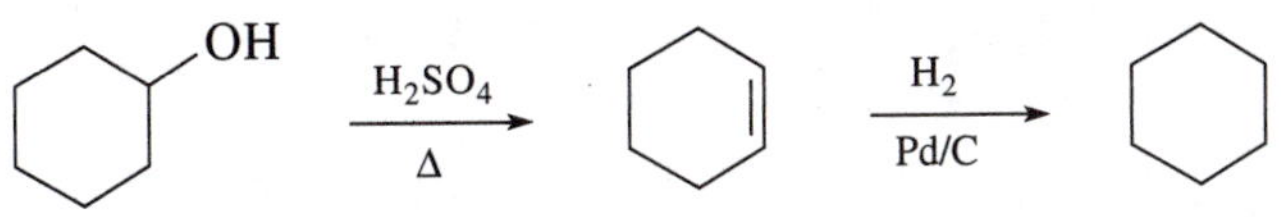

b. $CH_3CH_2CH_2CH_2Br \xrightarrow{HO^-} CH_3CH_2CH_2CH_2OH \xrightarrow{H_2CrO_4} CH_3CH_2CH_2\overset{\displaystyle O}{\overset{\displaystyle \|}{C}}OH$

42. When (*S*)-2-butanol loses water as a result of being heated in sulfuric acid, the asymmetric center in the reactant becomes a planar *sp*² carbon. Therefore, the chirality is lost. When water attacks the carbocation, it can attack from either side of the planar carbocation, forming (*S*)-2-butanol and (*R*)-2-butanol with equal ease.

43.

44.   The two different carbocations formed by naphthalene oxide differ in stability.

One carbocation is more stable than the other because it can be stabilized by electron delocalization without disrupting the aromaticity of the adjacent ring. The more stable carbocation leads to the major product.

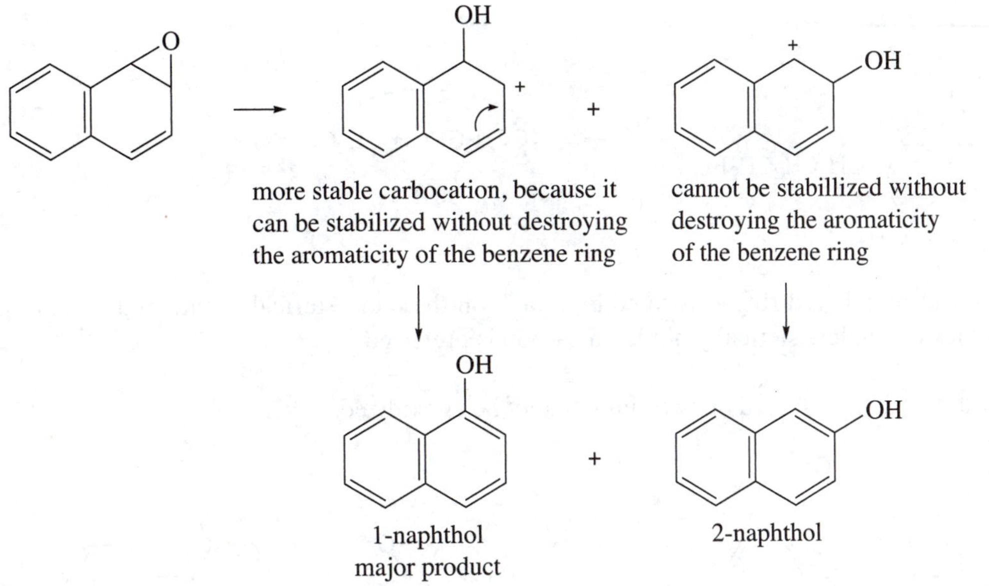

more stable carbocation, because it
can be stabilized without destroying
the aromaticity of the benzene ring

cannot be stabillized without
destroying the aromaticity
of the benzene ring

1-naphthol
major product

2-naphthol

45.   **a.**

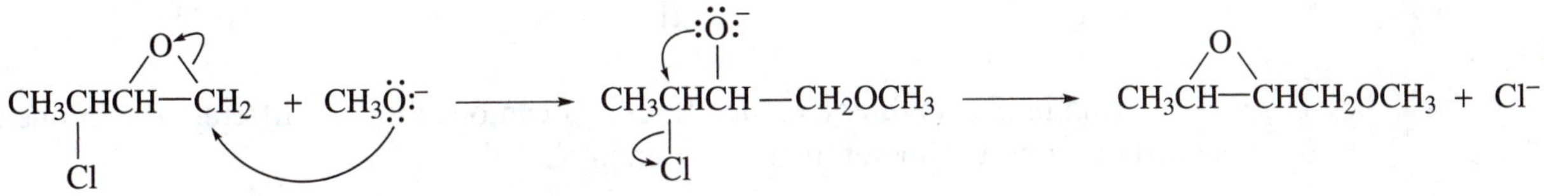

b.

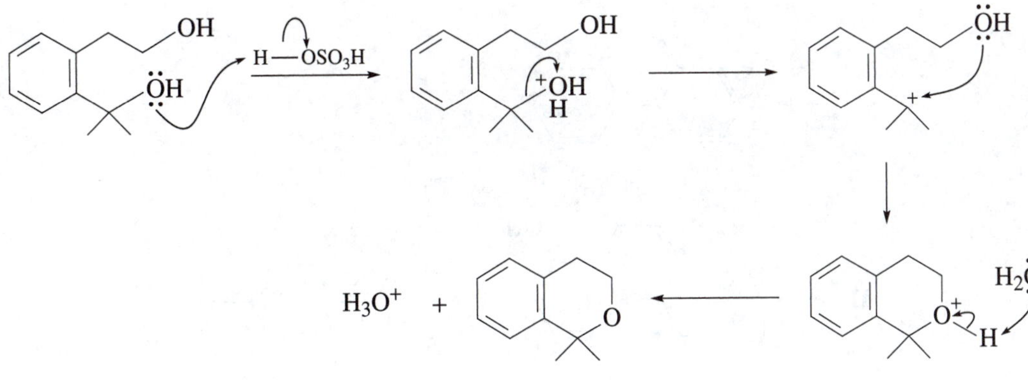

46. **a.**

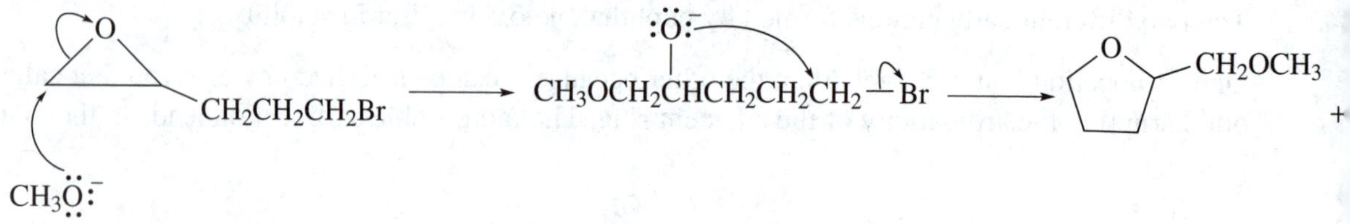

b.

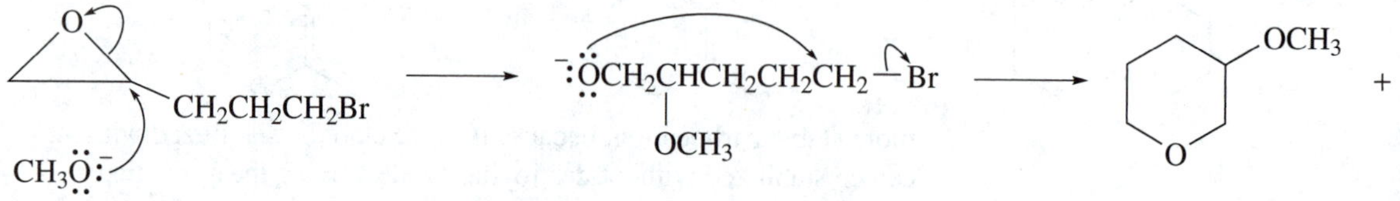

c. The six-membered ring is formed by attack on the more sterically hindered carbon of the epoxide. Attack on the less sterically hindered carbon is preferred.

47. **a.** Note that a bond shared by two rings cannot be epoxidized.

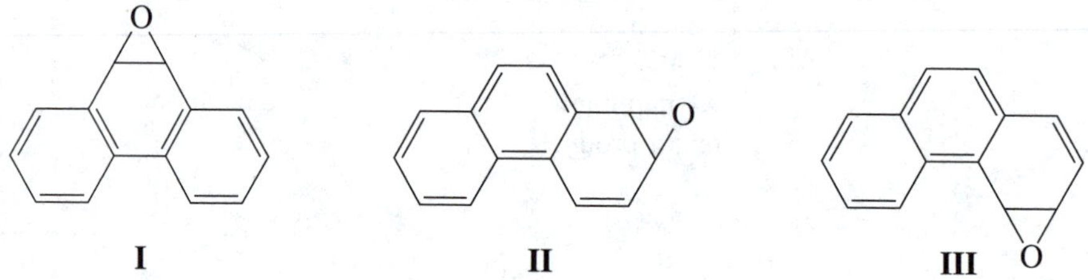

I   **II**   **III**

b. The epoxide ring in phenanthrene oxides II and III can open in two different directions to give two different carbocations and, therefore, two different phenols.

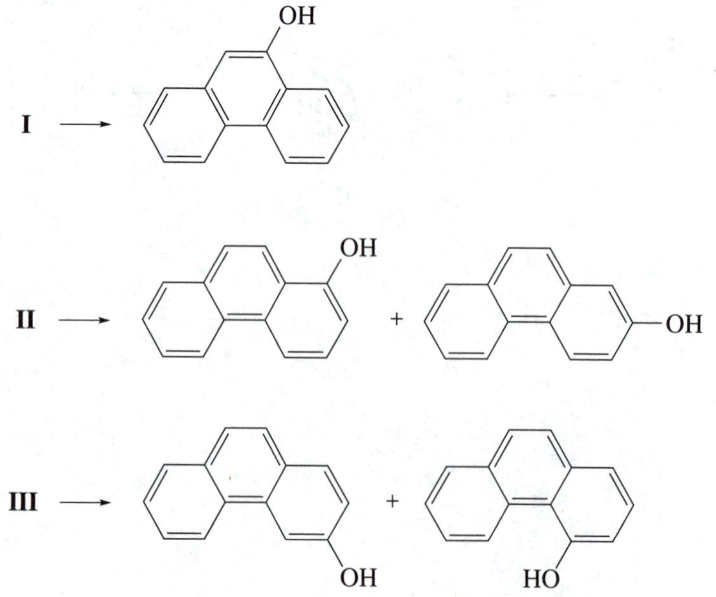

I →

II →   +

III →   +

c. The two different carbocations formed by phenanthrene oxides II and III differ in stability. One carbocation is more stable than the other because it can be stabilized by electron delocalization without disrupting the aromaticity of the adjacent ring. The more stable carbocation leads to the major product.

major product   minor product

major product   minor product

d. Phenanthrene oxide I is the most carcinogenic, because it is the only one that opens to form a carbocation that cannot be stabilized without disrupting the aromaticity of the other ring(s).

48.   **B** is the fastest reaction; **A** is the slowest reaction.

In order to form the epoxide, the alkoxide ion must attack the back side of the carbon that is bonded to Br. This means that the OH and Br substituents must both be in axial positions. To be 1,2-diaxial, they must be trans to each other.

A does not form an epoxide, because the OH and Br substituents are cis to each other.
B and **C** can form epoxides because the OH and Br substituents are trans to each other.

When the OH and Br substituents are in the required diaxial position, the large *tert*-butyl substituent is in the equatorial position in **B** and in the axial position in **C.**

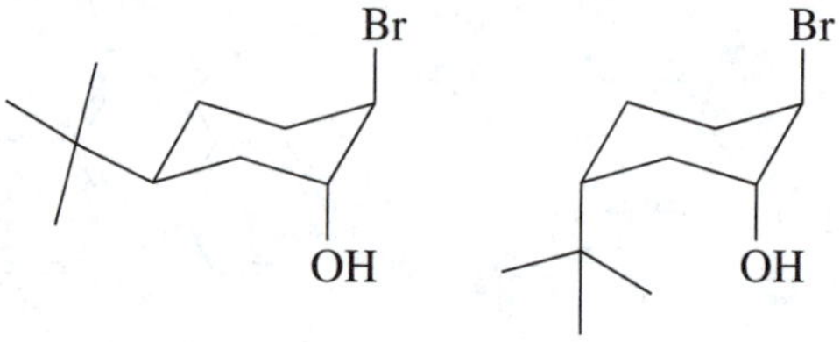
Br
OH
Br
OH

B **C**

Because the more stable conformer is the one with the large *tert*-butyl group in the equatorial position, the OH and Br substituents are in the required diaxial position in the more stable conformer of **B,** whereas the OH and Br substituents are in the required diaxial position in the less stable conformer of **C.** Therefore, **B** reacts faster than **C** because, at any one time, **B** will have a large concentration of the reactive conformer, and **C** will have a low concentration of the reactive conformer.

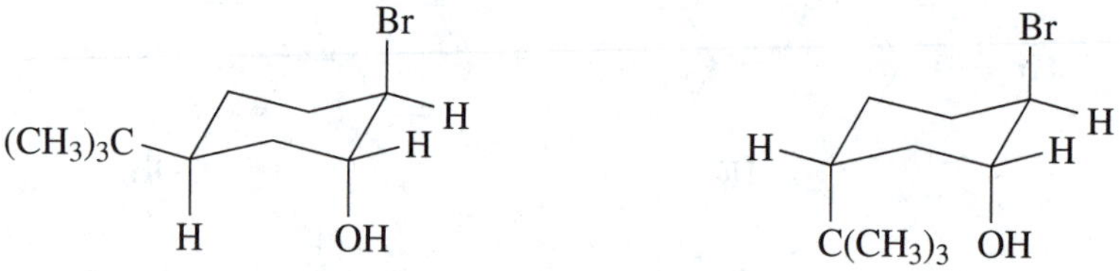
Br
(CH3)3C
H
H
H
OH
Br
H
H
H
C(CH3)3
OH

B **C**

49. **a.** The reaction of 2-chlorobutane with HO^- is an intermolecular reaction, so the two compounds have to find one another in the solution.

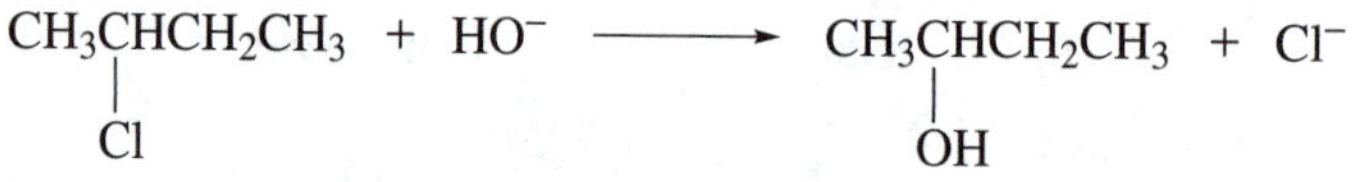
CH3CHCH2CH3 + HO−
Cl
CH3CHCH2CH3 + Cl−
OH

The following reaction takes place in two steps. The first is an intramolecular S_N2 reaction; the reaction is much faster than the above reaction because the two reactants are in the same molecule and can find one another relatively easily.

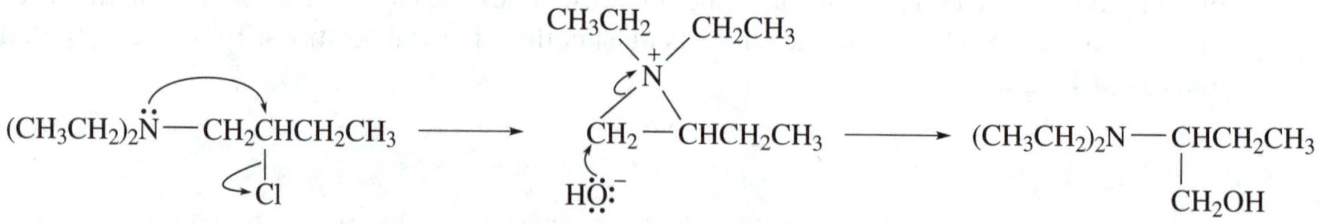
(CH3CH2)2N—CH2CHCH2CH3
Cl
CH3CH2
CH2CH3
N
CH2—CHCH2CH3
HO:−
(CH3CH2)2N—CHCH2CH3
CH2OH

The second reaction is also an S_N2 reaction and is fast because the strain of the three-membered ring and the positive charge on the nitrogen make the amine a very good leaving group.

b. The HO group is bonded to a different carbon, because HO^- attacks the least sterically hindered carbon of the three-membered ring.

0. **a.**

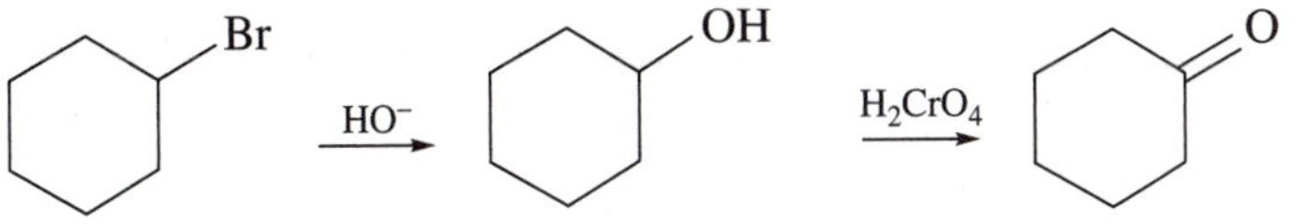

b.

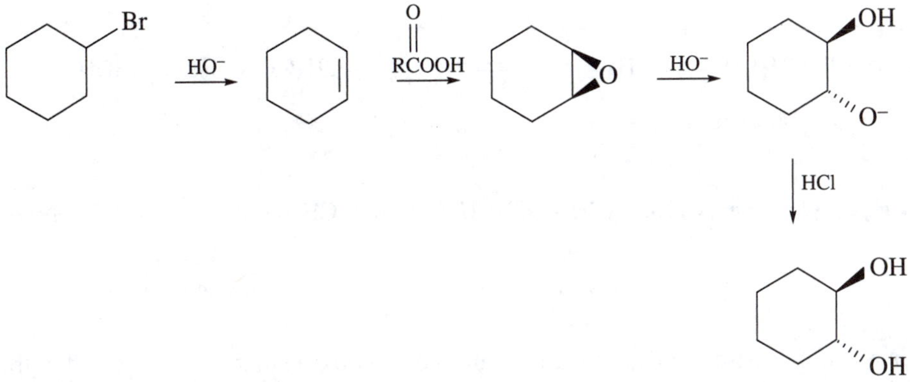

1.

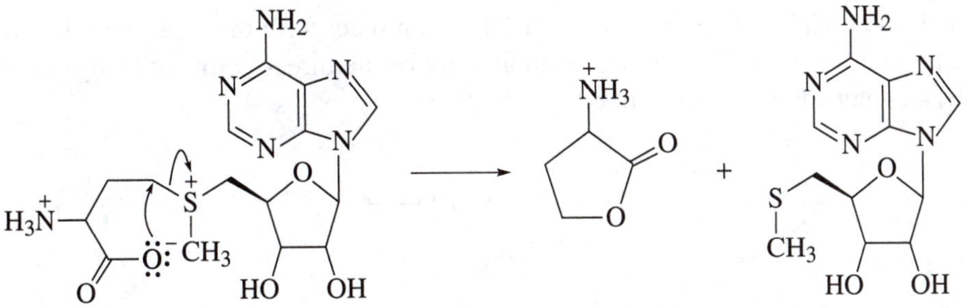

52. :B is any base in the solution (HSO_4^-, H_2O, ROH).

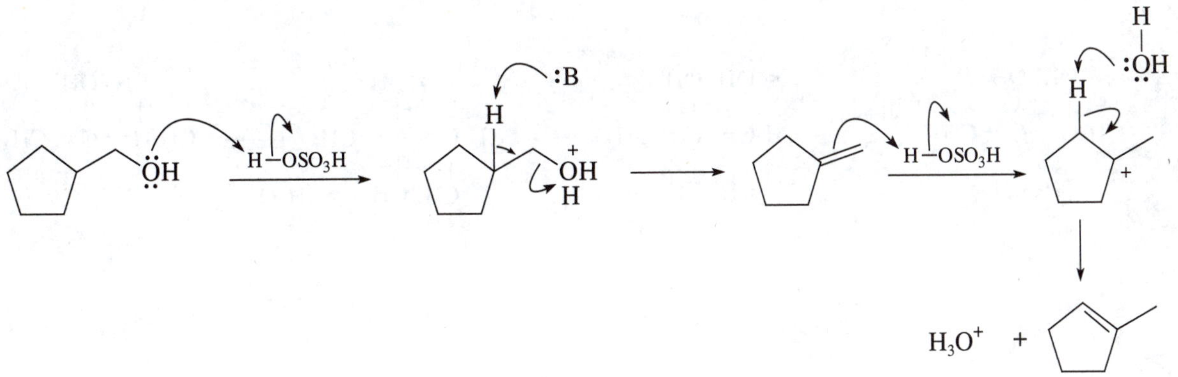

53. 1-Butanol dehydrates in an E2 reaction to form 1-butene. In the acidic solution, 1-butene is protonated to form a secondary carbocation, which loses a proton from the β-carbon bonded to the fewest hydrogens, resulting in the formation of 2-butene.

$$CH_3CH_2CH_2CH_2OH \underset{\Delta}{\overset{H_2SO_4}{\rightleftharpoons}} \underset{\text{1-butene}}{CH_3CH_2CH{=}CH_2} + H_2O \overset{H^+}{\rightleftharpoons} CH_3CH_2\overset{+}{C}HCH_3 \rightleftharpoons \underset{\text{2-butene}}{CH_3CH{=}CHCH_3} + H_3O^+$$

54.   2-hexene and 3-hexene

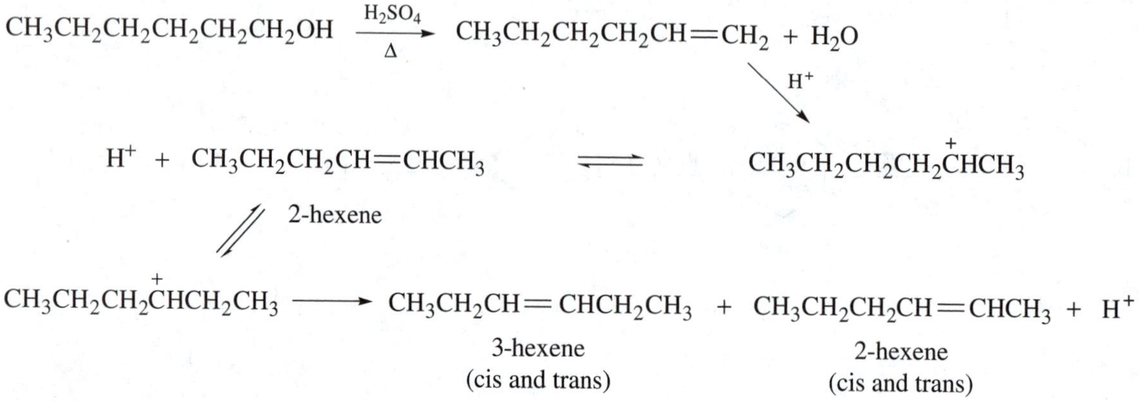

$$CH_3CH_2CH_2CH_2CH_2CH_2OH \xrightarrow[\Delta]{H_2SO_4} CH_3CH_2CH_2CH_2CH=CH_2 + H_2O$$

H^+

$$H^+ + CH_3CH_2CH_2CH=CHCH_3 \rightleftharpoons CH_3CH_2CH_2CH_2\overset{+}{C}HCH_3$$

2-hexene

$$CH_3CH_2CH_2\overset{+}{C}HCH_2CH_3 \longrightarrow CH_3CH_2CH=CHCH_2CH_3 + CH_3CH_2CH_2CH=CHCH_3 + H^+$$

3-hexene   2-hexene

(cis and trans)   (cis and trans)

55.

a.   A nitrogen is a stronger base than an oxygen, so unlike an epoxide that can be opened without the oxygen being protonated, the three-membered nitrogen-containing ring has to be protonated to improve the leaving propensity of the group.

b.   A nucleophile such as an NH_2 group on a chain of DNA can react with the three-membered ring. If a nucleophile on another chain of DNA reacts with another of the three-membered rings in this compound, the two DNA chains will be crosslinked.

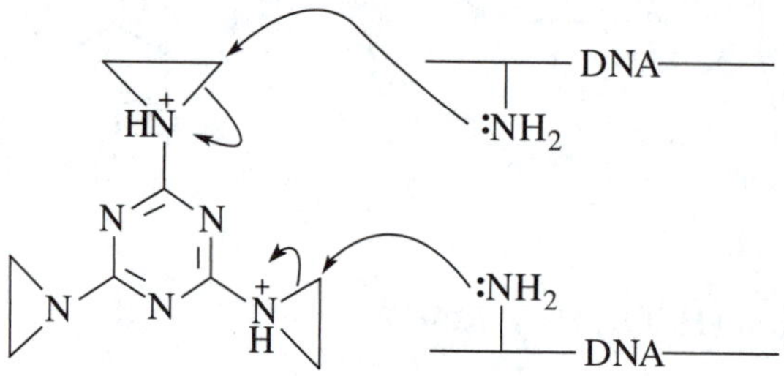

56.

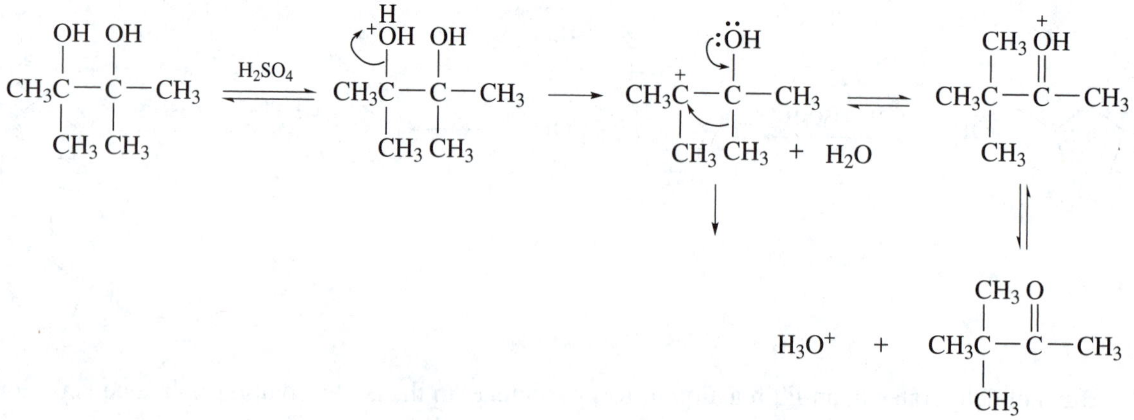

7. **a.**

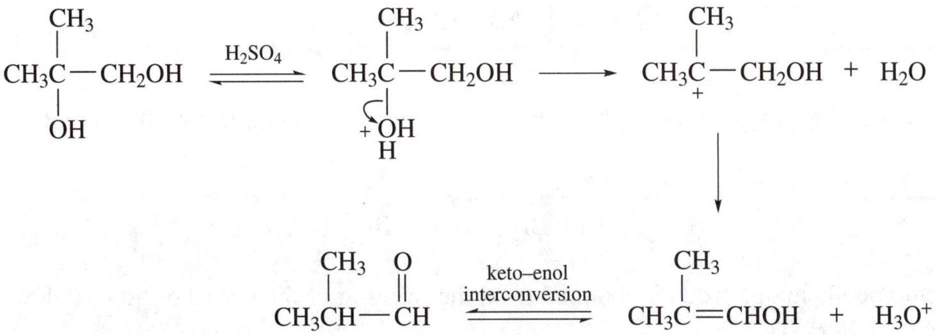

$$CH_3C(CH_3)(OH)—CH_2OH \xrightarrow{H_2SO_4} CH_3C(CH_3)(\overset{+}{O}H H)—CH_2OH \longrightarrow CH_3\overset{+}{C}(CH_3)—CH_2OH + H_2O$$

$$CH_3CH(CH_3)—CHO \underset{\text{interconversion}}{\overset{\text{keto–enol}}{\rightleftharpoons}} CH_3C(CH_3)=CHOH + H_3O^+$$

b. Dehydration of the primary alcohol group cannot occur, because it cannot lose water via an E1 pathway, since a primary carbocation cannot be formed. It cannot lose water via an E2 pathway because the β-carbon is not bonded to a hydrogen. However, dehydration of the tertiary alcohol group can occur. The product is an enol that tautomerizes to an aldehyde.

8.

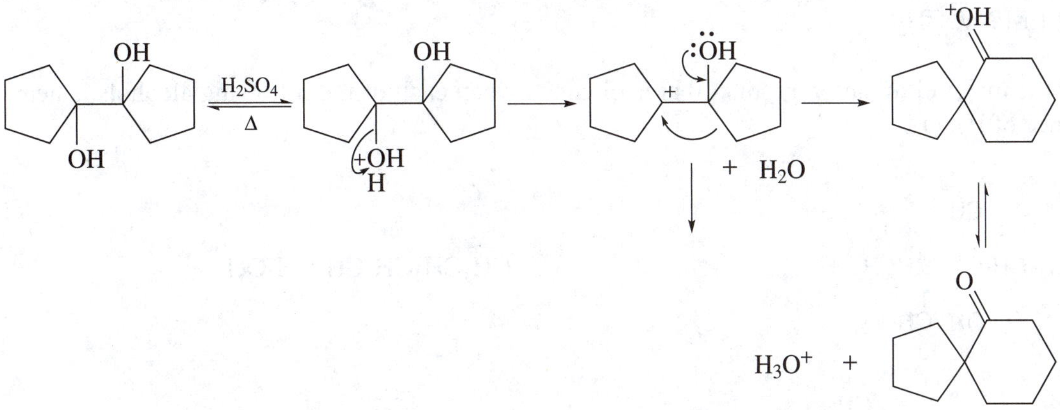

Chapter 9 Practice Test

1. Which of the following reagents is the best one to use in order to convert methyl alcohol into methyl bromide?

$$Br^- \quad HBr \quad Br_2 \quad NaBr \quad Br^+$$

2. **a.** What would be the major product obtained from the reaction of the epoxide shown below in methanol containing 0.1 M HCl?

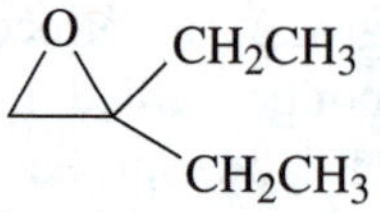

   b. What would be the major product obtained from the reaction of the epoxide in methanol containing 0.1 M NaOCH$_3$?

3. Draw the major elimination product that is obtained when each of the following alcohols is heated in the presence of H$_2$SO$_4$:

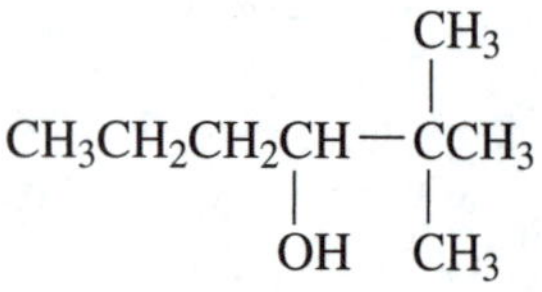

   a.

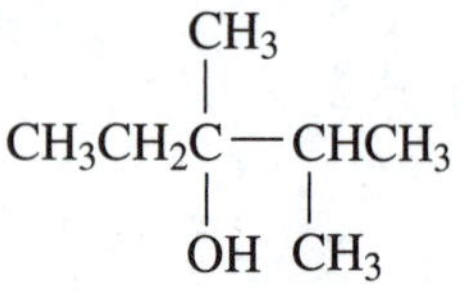

   c. CH$_3$CH$_2$CH$_2$CH$_2$CH$_2$OH

   b.

   d.

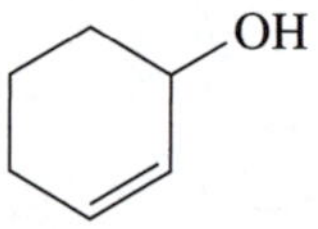

4. Indicate whether each of the following statements is true or false:

   a. Tertiary alcohols are easier to dehydrate than secondary alcohols.   T   F

   b. Alcohols are more acidic than thiols.   T   F

   c. Alcohols have higher boiling points than thiols.   T   F

   d. The acid-catalyzed dehydration of a primary alcohol is an S$_N$1 reaction.   T   F

   e. 1-Methylcyclohexanol reacts more rapidly than 2-methylcyclohexanol with HBr.   T   F

   f. 1-Butanol forms a ketone when it is oxidized by chromic acid.   T   F

What products would be obtained from heating the following ethers with one equivalent of HI?

a. 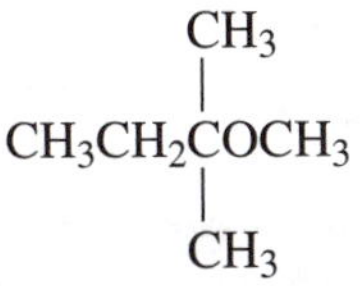 $\underset{\displaystyle \underset{CH_3}{|}}{\overset{\displaystyle \overset{CH_3}{|}}{CH_3CH_2COCH_3}}$

b. 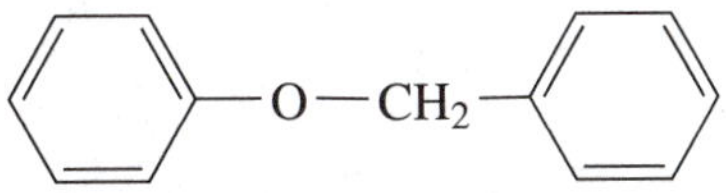   [benzene ring]—O—CH$_2$—[benzene ring]

Draw the major product of each of the following reactions:

a. 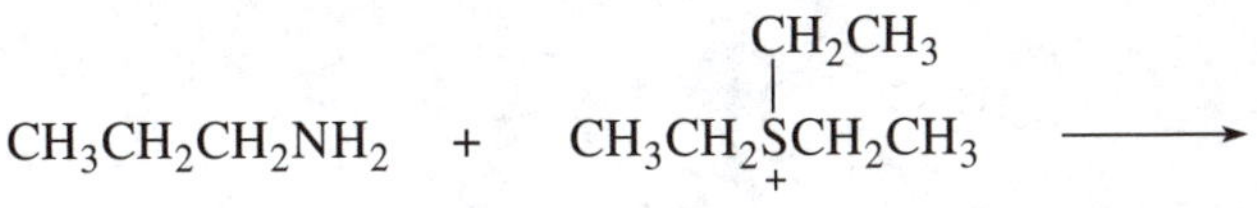 $CH_3CH_2CH_2NH_2$ + $\underset{+}{CH_3CH_2\overset{\displaystyle \overset{CH_2CH_3}{|}}{S}CH_2CH_3}$ $\longrightarrow$

b. $CH_3CH_2CH_2CH_2CH_2CH_2OH$ $\xrightarrow[\Delta]{H_2SO_4}$

c. $CH_3CH_2CH_2CH_2OH$ $\xrightarrow[\underset{0\,°C}{CH_3COOH}]{NaOCl}$

d. $CH_3CH_2CH_2CH_2OH$ $\xrightarrow{H_2CrO_4}$

Name the following compounds:

a. 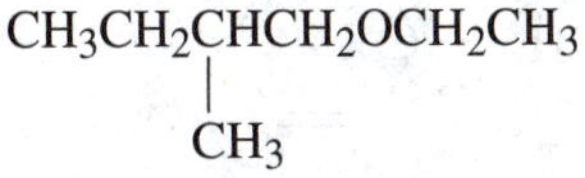 $\underset{\displaystyle \underset{CH_3}{|}}{CH_3CH_2CHCH_2OCH_2CH_3}$

b. 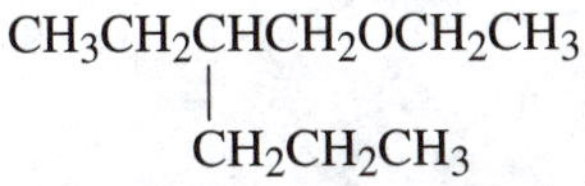 $\underset{\displaystyle \underset{CH_2CH_2CH_3}{|}}{CH_3CH_2CHCH_2OCH_2CH_3}$

c. 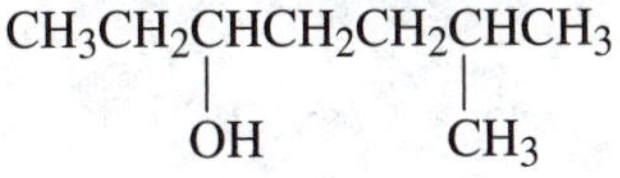 $\underset{\displaystyle \underset{OH}{|}}{CH_3CH_2CHCH_2CH_2}\underset{\displaystyle \underset{CH_3}{|}}{CHCH_3}$

d. 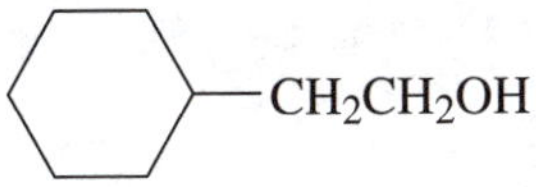   [cyclohexane ring]—CH_2CH_2OH

Answers to Chapter 9 Practice Test

1.   HBr

2.   a. $HOCH_2CCH_2CH_3$ with CH_2CH_3 above and OCH_3 below
      b. $CH_3OCH_2CCH_2CH_3$ with CH_2CH_3 above and OH below

3.   a. $CH_3CH_2C{=}CCH_3$ with CH_3 above and CH_3 below
      b. $CH_3CH_2CH_2C{=}CCH_3$ with CH_3 above and CH_3 below
      c. $CH_3CH_2CH{=}CHCH_3$
      d. (cyclohexadiene ring structure)

c. will form both trans and cis, but more trans.

4.   a. Tertiary alcohols are easier to dehydrate than secondary alcohols.   T
      b. Alcohols are more acidic than thiols.   F
      c. Alcohols have higher boiling points than thiols.   T
      d. The acid-catalyzed dehydration of a primary alcohol is an S_N1 reaction.   F
      e. 1-Methylcyclohexanol reacts more rapidly than 2-methylcyclohexanol with HBr.   T
      f. 1-Butanol forms a ketone when it is oxidized by chromic acid   F

5.   a. $CH_3CH_2C{-}I$ with CH_3 above and CH_3 below $+$ CH_3OH
      b. (phenyl)${-}OH$ $+$ $ICH_2{-}$(phenyl)

6.   a. $CH_3CH_2CH_2NHCH_2CH_3$
      b. $CH_3CH_2CH_2CH{=}CHCH_3$ $+$ $CH_3CH_2CH{=}CHCH_2CH_3$

      c. $CH_3CH_2CH_2$ $\overset{\displaystyle O}{\overset{\displaystyle \|}{C}}$ H

      d. $CH_3CH_2CH_2$ $\overset{\displaystyle O}{\overset{\displaystyle \|}{C}}$ OH

7.   a. 1-ethoxy-2-methylbutane
      b. 1-ethoxy-2-ethylpentane
      c. 6-methyl-3-heptanol
      d. 2-cyclohexyl-1-ethanol

CHAPTER 10
Determining the Structure of Organic Compounds

Important Terms

absorption band	a peak in a spectrum that occurs as a result of absorption of energy.
applied magnetic field	the externally applied magnetic field.
base peak	the peak in a mass spectrum with the greatest intensity.
bending vibration	a vibration that does not occur along the line of the bond.
chemically equivalent protons	protons with the same connectivity relationship to the rest of the molecule.
chemical shift	location of a signal occurring in an nuclear magnetic resonance (NMR) spectrum. It is measured downfield from a reference compound (most often TMS).
^{13}C NMR	nuclear magnetic resonance from carbon (^{13}C) nuclei.
coupled protons	protons that split each other. Coupled protons have the same coupling constant.
doublet	an NMR signal that is split into two peaks.
doublet of doublets	an NMR signal that is split into four peaks of approximately equal height. A doublet of doublets is caused by splitting a signal into a doublet by one hydrogen and into another doublet by another (nonequivalent) hydrogen.
effective magnetic field	the magnetic field that a nucleus "senses" through the surrounding cloud of electrons.
electromagnetic radiation	radiant energy that displays wave properties.
fragment ion peak	a positively charged fragment of a molecular ion.
frequency	the velocity of a wave divided by its wavelength.
^{1}H NMR	nuclear magnetic resonance from hydrogen nuclei.
infrared (IR) radiation	electromagnetic radiation familiar to us as heat.
IR spectroscopy	spectroscopy that uses infrared energy to provide a knowledge of the functional groups in a compound.
IR spectrum	a plot of relative absorption versus wavenumber (or wavelength) of absorbed infrared radiation.
λ_{max}	the wavelength at which there is maximum UV/Vis absorbance.

253

magnetic resonance imaging (MRI)	NMR used in medicine. The difference in the way water is bound in different tissues produces the signal variation between organs as well as between healthy and diseased states.
mass spectrometry	an instrumental technique that provides a knowledge of the molecular weight and certain structural features of a compound.
mass spectrum	a plot of the relative abundance of the positively charged fragments produced in a mass spectrometer versus their m/z values.
methine hydrogen	a tertiary hydrogen.
molecular ion (M)	the radical cation formed by removing one electron from a molecule.
MRI scanner	an NMR spectrometer used in medicine for whole-body NMR.
multiplet	an NMR signal split by two (or more) nonequivalent sets of protons.
multiplicity	the number of peaks in an NMR signal.
$N+1$ rule	a rule that states that an ^1H NMR signal for a hydrogen with N equivalent hydrogens bonded to an adjacent carbon is split into $N+1$ peaks; a proton-coupled ^{13}C NMR signal for a carbon bonded to N hydrogens is split into $N+1$ peaks.
NMR spectroscopy	the absorption of rf radiation by nuclei in an applied magnetic field to determine the structural features of an organic compound. In the case of ^1H NMR spectroscopy, it reveals the carbon–hydrogen framework.
nominal molecular mass	mass to the nearest whole number.
proton-coupled ^{13}C NMR spectrum	a ^{13}C NMR spectrum in which each signal for a carbon is split by the hydrogens bonded to that carbon.
quartet	an NMR signal that is split into four equally spaced peaks with an integral ratio of 1:3:3:1.
radical cation	a species with a positive charge and an unpaired electron.
rule of 13	a rule that allows possible molecular formulas to be determined from the m/z value of the molecular ion.
shielding	the electrons around a proton shield the proton from the full effect of the applied magnetic field. The more a proton is shielded, the farther to the right its signal appears in an NMR spectrum.
singlet	an unsplit NMR signal.

spectroscopy	study of the interaction of matter and electromagnetic radiation.
α-spin state	nuclei in this spin state have their magnetic moments oriented in the same direction as the applied magnetic field.
β-spin state	nuclei in this spin state have their magnetic moments oriented opposite to the direction of the applied magnetic field.
stretching vibration	a vibration occurring along the line of the bond.
triplet	an NMR signal that is split into three equally spaced peaks with an integral ratio of 1:2:1.
ultraviolet light	electromagnetic radiation with wavelengths ranging from 180 to 400 nm.
UV/Vis spectroscopy	the absorption of electromagnetic radiation that is useful in determining information about conjugated systems.
visible light	electromagnetic radiation with wavelengths ranging from 400 to 780 nm.
wavelength	distance from any point on one wave to the corresponding point on the next wave.
wavenumber	the number of waves in 1 cm.

Solutions to Problems

1. Only positively charged fragments are accelerated through the analyzer tube.

$$CH_3CH_2\overset{+}{C}H_2 \qquad [CH_3CH_2CH_3]^{+\bullet} \qquad \overset{+}{C}H_2CH{=}CH_2$$

2. The peak at $m/z = 57$ will be more intense for 2,2-dimethylpropane than for 2-methylbutane or pentane. The peak at $m/z = 57$ is due to loss of a methyl radical: loss of a methyl radical from 2,2-dimethylpropane forms a tertiary carbocation, whereas loss of a methyl radical from 2-methylbutane and pentane forms a less stable secondary carbocation and primary carbocation, respectively.

$$\begin{bmatrix} \overset{\displaystyle CH_3}{\underset{\displaystyle CH_3}{\mid}} \\ CH_3\overset{\mid}{\underset{\mid}{C}}CH_3 \end{bmatrix}^{+\bullet} \longrightarrow \underset{\substack{m/z=57 \\ \text{a tertiary carbocation}}}{CH_3\overset{CH_3}{\underset{+}{C}}CH_3} \quad + \quad \bullet CH_3$$

$$\begin{bmatrix} \overset{\displaystyle CH_3}{\underset{\displaystyle |}{\mid}} \\ CH_3\overset{\mid}{C}HCH_2CH_3 \end{bmatrix}^{+} \longrightarrow \underset{\substack{m/z=57 \\ \text{a secondary carbocation}}}{CH_3\overset{+}{C}HCH_2CH_3} \quad + \quad \bullet CH_3$$

$$\left[CH_3CH_2CH_2CH_2CH_3\right]^{+\bullet} \longrightarrow \underset{\substack{m/z=57 \\ \text{a primary carbocation}}}{CH_3CH_2CH_2\overset{+}{C}H_2} \quad + \quad \bullet CH_3$$

Notice that the mass spectrum of 2-methylbutane can be distinguished from those of the other isomers by the peak at $m/z = 43$. The peak at $m/z = 43$ will be most intense for 2-methylbutane, because such a peak is due to loss of an ethyl radical, which forms a secondary carbocation. Pentane gives a less intense peak at $m/z = 43$, because loss of an ethyl radical from pentane forms a primary carbocation. 2,2-Dimethylpropane will not show a peak at $m/z = 43$, because it does not have an ethyl group.

$$\begin{bmatrix} \overset{\displaystyle CH_3}{\underset{\displaystyle |}{\mid}} \\ CH_3\overset{\mid}{C}HCH_2CH_3 \end{bmatrix}^{+\bullet} \longrightarrow \underset{\substack{m/z=43 \\ \text{a secondary carbocation}}}{CH_3\overset{CH_3}{\underset{+}{C}}H} \; + \; CH_3\overset{\bullet}{C}H_2$$

$$\left[CH_3CH_2CH_2CH_2CH_3\right]^{+\bullet} \longrightarrow \underset{\substack{m/z=43}}{CH_3CH_2\overset{+}{C}H_2} \; + \; CH_3\overset{\bullet}{C}H_2$$

3. Intense peaks should occur at $m/z = 57$ for loss of an ethyl radical $(86 - 29)$, and at $m/z = 71$ for loss of a methyl radical $(86 - 15)$.

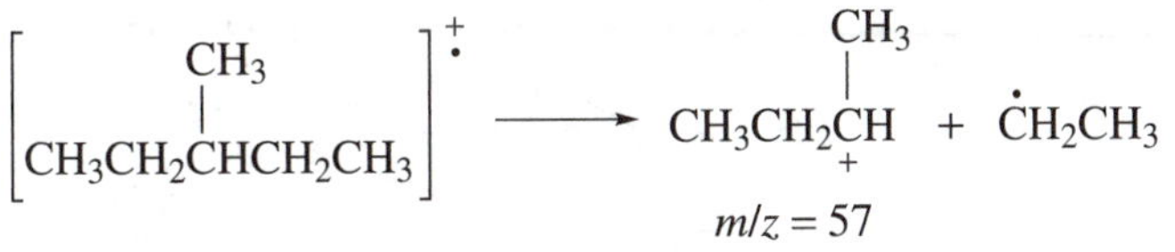

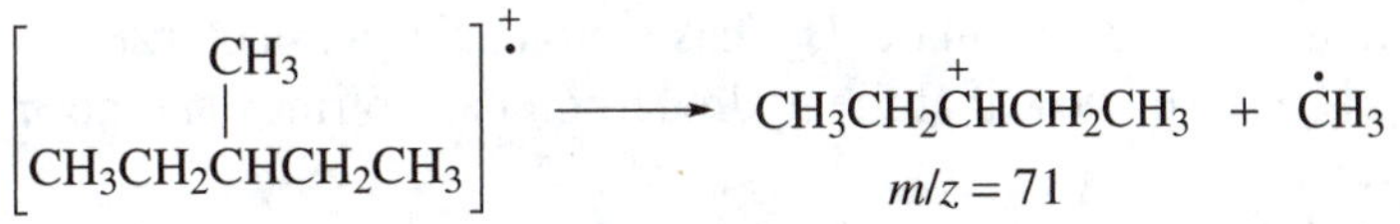

$$\left[\begin{array}{c} CH_3 \\ | \\ CH_3CH_2CHCH_2CH_3 \end{array}\right]^{+\cdot} \longrightarrow \underset{\underset{m/z\,=\,57}{+}}{CH_3CH_2CH} + \dot{C}H_2CH_3$$

where the product is $CH_3CH_2\overset{+}{C}H$ with a CH_3 substituent.

$$\left[\begin{array}{c} CH_3 \\ | \\ CH_3CH_2CHCH_2CH_3 \end{array}\right]^{+\cdot} \longrightarrow \underset{m/z\,=\,71}{CH_3CH_2\overset{+}{C}HCH_2CH_3} + \dot{C}H_3$$

A secondary carbocation is formed in both cases. Because an ethyl radical is more stable than a methyl radical, the base peak will most likely be at $m/z = 57$.

4. Solved in the text.

5.

a. Dividing 72 by 13 gives 5 with 7 left over. Thus, the base value is C_5H_{12}. Because the compound contains only carbons and hydrogens, we know that the base value is also the molecular formula of the compound.

b. Dividing 100 by 13 gives 7 with 9 left over. Thus, the base value is C_7H_{16}. Because the compound contains one oxygen, an O must be added to the base value and one C and four Hs must be subtracted. Therefore, the molecular formula is $C_6H_{12}O$.

c. Dividing 102 by 13 gives 7 with 11 left over. Thus, the base value is C_7H_{18}. Because the compound contains two oxygens, two Os must be added to the base value and two Cs and eight Hs must be subtracted. Therefore, the molecular formula is $C_5H_{10}O_2$.

d. Dividing 115 by 13 gives 8 with 11 left over. Thus, the base value is C_8H_{19}. Because the compound contains one oxygen, an O must be added to the base value and one C and four Hs must be subtracted. Because it contains an N, an N must be added to the base value and one C and three Hs must be subtracted. Therefore, the molecular formula is $C_6H_{12}NO$.

6. Dividing 86 by 13 gives 6 with 8 left over. Thus, the base value is C_6H_{14}.

If the compound contains only carbons and hydrogens, the base value is also the molecular formula of the compound. Some possible structures would be the following:

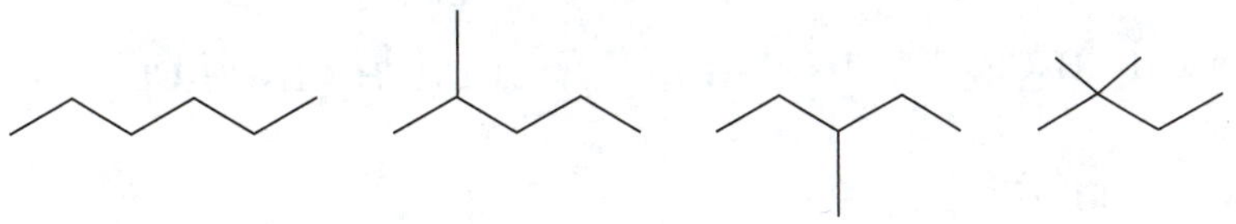

If the compound contains one oxygen, the molecular formula would be $C_5H_{10}O$. Some possible structures would be the following:

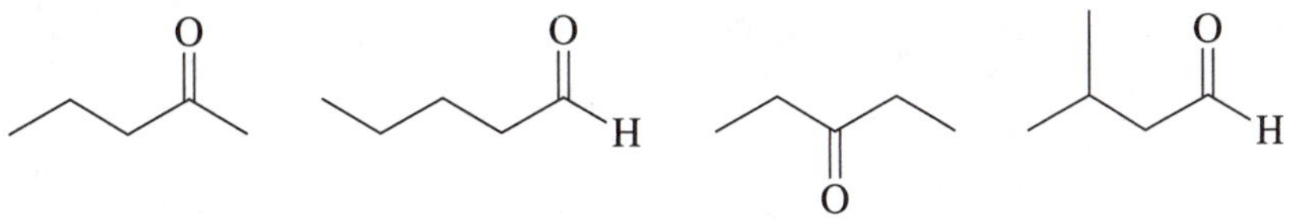

If the compound contains two oxygens, the molecular formula would be $C_4H_6O_2$. Some possible structure would be the following:

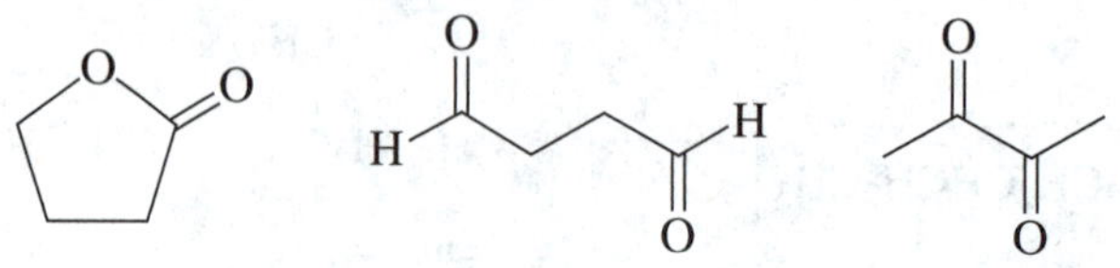

7.   A hydrocarbon with molecular formula C_9H_{20} has a molecular mass of 128.
Since $C_9H_{20} = C_nH_{2n+2}$, we know that the hydrocarbon has no rings and no π bonds. The hydrocarbon is **2,6-dimethylheptane.**

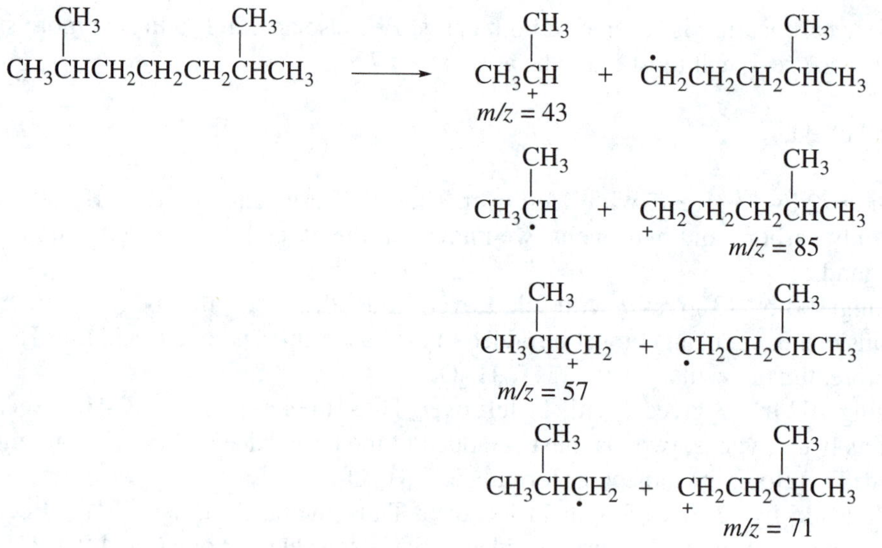

2-Methyloctane would also be expected to give a base peak of $m/z = 43$ because it, too, will form a secondary (isopropyl carbocation) together with a primary radical. All other cleavages that form primary radicals form primary carbocations. However, we would expect fragments with $m/z = 29$ and 99 to be present to the same extent as those with $m/z = 57$, 85, and 71. Because fragments with $m/z = 29$ and 99 are not mentioned, we can conclude that the hydrocarbon shown below is less likely than the one shown above.

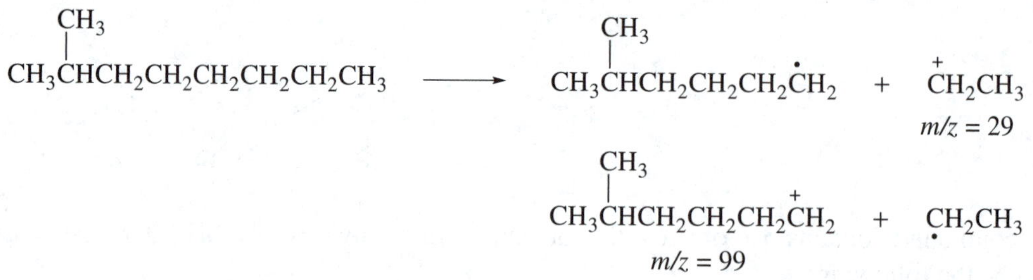

8.   1-bromopropane

9. Because the compound contains chlorine, the M+2 peak is one-third the size of the M peak. Breaking the weak C—Cl bond heterolytically and therefore losing a chlorine atom from either the M+2 peak $(80 - 37)$ or the M peak $(78 - 35)$ gives the base peak with $m/z = 43$ ($CH_3CH_2CH_2^+$). It is more difficult to predict the other peaks.

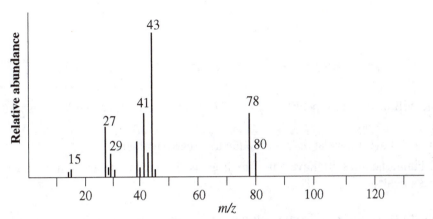

Mass spectrum of 1-chloropropane

10. The calculated exact masses show that only C_6H_{14} has an exact mass of 86.10955.

C_6H_{14} $6(12.00000) = 72.00000$

$14(1.007825) = \underline{14.10955}$

86.10955

$C_4H_6O_2$ $4(12.00000) = 48.00000$

$6(1.007825) = 6.04695$

$2(15.9949) = \underline{31.9898}$

86.03675

$C_4H_{10}N_2$ $4(12.00000) = 48.00000$

$10(1.007825) = 10.07825$

$2(14.0031) = \underline{28.0064}$

86.08465

11. **a.** A low-resolution spectrometer cannot distinguish between them, because they both have the same molecular mass (29).

b. A high-resolution spectrometer can distinguish between these ions, because they have different exact molecular masses; one has an exact molecular mass of 29.039125, and the other has an exact molecular mass of 29.002725.

12. The molecular ion with $m/z = 86$ indicates that the ketone has the molecular formula $C_5H_{10}O$. The spectrum shows a base peak at $m/z = 43$ for loss of a propyl (or isopropyl) radical $(86 - 43)$, indicating that it is the mass spectrum of either 2-pentanone or 3-methyl-2-butanone, since each of these has a propyl or an isopropyl group. The fact that the spectrum has a peak at $m/z = 58$, indicating loss of ethene $(86 - 28)$, indicates that the compound has a γ-hydrogen that enables it to undergo a rearrangement. Therefore, the ketone must be 2-pentanone, since 3-methyl-2-butanone does not have a γ-hydrogen.

13.   All three ketones will have a molecular ion with $m/z = 86$.

3-pentanone   2-pentanone   3-methyl-2-butanone

$86 - 29 = 57$ $86 - 43 = 43$ $86 - 43 = 43$

3-Pentanone will have a base peak at $m/z = 57$, whereas the other two ketones will have base peaks at $m/z = 43$.

2-Pentanone will have a peak at $m/z = 58$ due to a rearrangement.

3-Methyl-2-butanone does not have any γ-hydrogens. Therefore, it cannot undergo a rearrangement, so it will not have a peak at $m/z = 58$.

14.   The wavelength is the distance from the top of one wave to the top of the next wave. We see that **A** has a longer wavelength than **B**.

Infrared radiation has longer wavelengths than does visible light, because infrared radiation is lower in energy. Therefore, **A** depicts infrared radiation and **B** depicts visible light.

15.   **a.** 2000 cm^{-1} (The larger the wavenumber, the higher the energy.)
     b. $8 \ \mu\text{m}$ (The shorter the wavelength, the higher the energy.)

16.   **a.** $C\equiv C$ stretch   A triple bond is stronger than a double bond, so it takes more energy to stretch a triple bond.
     b. $C-H$ stretch   It requires more energy to stretch a given bond than to bend it.
     c. $C=N$ stretch   A double bond is stronger than a single bond, so it takes more energy to stretch a double bond.
     d. $C=O$ stretch   A double bond is stronger than a single bond, so it takes more energy to stretch a double bond.

17.   **a.**   The carbon–oxygen stretch of phenol, because it has partial double-bond character as a result of electron delocalization.

  b.   The carbon–oxygen double-bond stretch of a ketone, because it has more double-bond character. The double-bond character of the carbonyl group of an amide is reduced by electron delocalization.

c. The C—N stretch of aniline, because it has partial double-bond character.

8. A carbonyl group bonded to an sp^3 carbon will exhibit an absorption band at a larger wavenumber, because a carbonyl group bonded to an sp^2 carbon of an alkene will have greater single-bond character as a result of electron delocalization.

19. The C—O bond of the alcohol is a pure single bond. In contrast, the C—O bond of the carboxylic acid has double-bond character, so it is a stronger bond and therefore it takes more energy to stretch it.

20. Ethanol dissolved in carbon disulfide will show the oxygen–hydrogen stretch at a larger wavenumber. There is extensive hydrogen bonding in the undiluted alcohol, and an oxygen–hydrogen bond is easier to stretch if it is hydrogen bonded, so the O—H stretch will be at a smaller wavenumber.

21. The absence of an absorption band at 3400 cm^{-1} indicates that the compound does not have an N—H bond.

The absence of a carbonyl absorption band between 1700 cm^{-1} and 1600 cm^{-1} indicates that the compound is not an amide. The compound, therefore, must be a **tertiary amine.**

22. **a.** An aldehyde would show absorption bands at 2820 and 2720 cm^{-1}. A ketone would not have these absorption bands.

   b. Cyclohexene would show an sp^3 C—H stretch slightly to the right of 3000 cm^{-1}. Benzene would not show an absorption band in this region.

   c. Cyclohexene would show a carbon–carbon double-bond stretching vibration at 1680—1600 cm^{-1} and an sp^2 carbon–hydrogen stretching vibration at 3100–3020 cm^{-1}. Cyclohexane would not show these absorption bands.

   d. A primary amine would show a nitrogen–hydrogen stretch at 3500–3300 cm^{-1} and a tertiary amine would not have this absorption band.

23. **a.** The alcohol would have an intense absorption band at ~3400 cm^{-1} and the amine would have a much less intense absorption at that wavelength.

   b. An absorption band at 3300–2500 cm^{-1} due to an O—H stretching vibration would be present for the carboxylic acid and absent for the ester.

    c. Only the terminal alkyne would show an absorption band at 3300 cm^{-1} due to an sp C—H stretchin
    vibration.

    d. An absorption band at 1780–1650 cm^{-1} due to a C=O stretching vibration would be present for th
    carboxylic acid and absent for the alcohol.

24.   The absorption band at ~1700 cm^{-1} indicates that the compound has a carbonyl group, and the absorp
tion band at ~1600 cm^{-1} indicates that the compound has a carbon–carbon double bond. The absorptio
bands in the vicinity of 3000 cm^{-1} indicate that the compound has hydrogens attached to both sp^2 and sp
carbons. The absorption band at ~1380 cm^{-1} indicates that the compound has a methyl group. Because th
compound has only four carbons and one oxygen, it must be **methyl vinyl ketone.** Notice that the carbony
stretch is at a lower frequency (1700 cm^{-1}) than expected for a ketone (1720 cm^{-1}), because the carbony
group has partial single-bond character due to electron delocalization.

25. **a.**

      b.

26. **a.** Blue results from absorption of light that has a longer wavelength than the light that produces purpl
      when absorbed. The compound on the right has two N(CH$_3$)$_2$ auxochromes that will cause it to ab
      sorb at a longer wavelength than the compound on the left, which has only one N(CH$_3$)$_2$ auxochrome
      Therefore, the compound on the right is the blue compound.
    b. They will be the same color at pH = 3, because the N(CH$_3$)$_2$ groups will be protonated and, therefore
      will not possess the lone pair that causes the compound to absorb light of a longer wavelength.

27.   The two absorption bands that produce an observed green color correspond to the absorption bands tha
produce yellow and blue. So mixing yellow and blue produces green.

28.   NADH is formed as a product; it absorbs light at 340 nm. Therefore, the rate of the oxidation reaction ca
be determined by monitoring the increase in absorbance at 340 nm as a function of time.

29.   When the pH of the solution equals the pK_a of the compound, the concentration of the acidic form of the
compound is the same as the concentration of the basic form of the compound.

    From the data given, we see that the absorbance of the acid is 0. We also see that the absorbance ceases
to change with increasing pH after the absorbance reaches 1.60. That means that all of the compound is
in the basic form when the absorbance is 1.60. Therefore, when the absorbance is half of 1.60 (or 0.80),

half of the compound is in the basic form and half is in the acidic form; in other words, the concentration in the acid form is the same as the concentration in the basic form. We see that the absorbance is 0.80 at pH = 5.0. Therefore, the pK_a of the compound is 5.0.

30.

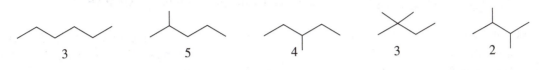

31.   **a.** 2   **c.** 4   **e.** 3
          b. 1   **d.** 3   **f.** 3

32.   **A** would give two signals, **B** would give one signal, and **C** would give three signals.

33.   Magnesium is less electronegative than silicon. Therefore, the peak for $(CH_3)_2Mg$ would be upfield from the TMS peak.

34.   **a.** and **b.**

1. $CH_3CH_2CH_2Cl$

          most   least
          shielded   shielded

2. $CH_3CH_2\overset{\overset{O}{\|}}{C}OCH_3$

          most   least
          shielded   shielded

3. $CH_3CHCHBr$
                $|$ $|$
                Br Br

          most   least
          shielded   shielded

35.   Recall that the highest-frequency signal is the one that is farthest to the left on the spectrum. The proton or protons that are underlined in the answer give the higher-frequency signal.

a. $CH_3CHCHBr$
              $|$ $|$
              Br Br

b. CH_3CHOCH_3
              $|$
              CH_3

c. $CH_3CH_2CHCH_3$
                  $|$
                  Cl

36.   **a.** $CH_3CH_2CH_2Cl$

b. $CH_3CH_2CHCH_3$
                  $|$
                  Cl

37.   **a.** $\overset{a}{C}H_3\overset{b}{C}H_2\overset{d}{C}H_2\overset{\overset{O}{\|}}{\underset{}{C}}\overset{c}{C}H_3$

b. $\overset{a}{C}H_3\overset{b}{C}H_2\overset{d}{C}H\overset{b}{C}H_2\overset{a}{C}H_3$
                                      $|$
                                $\overset{}{O}CH_3$
                                      c

c. $\overset{a}{C}H_3\overset{b}{C}H_2\overset{c}{C}H_2\overset{\overset{O}{\|}}{\underset{}{C}}\overset{d}{\underset{}{O}}CH_3$

d. $\overset{a}{C}H_3\overset{c}{C}H_2\overset{d}{C}H_2\overset{e}{O}\overset{b}{C}HCH_3$
                                              $|$
                                        CH_3
                                          b

e. $\overset{a}{C}H_3\overset{c}{C}H\overset{d}{|}\overset{b}{C}HCH_3$
                  $|$
              CH_3
                a

(with Cl above: $\overset{Cl}{|}$)

f. $\overset{a}{C}H_3\overset{b}{C}H\overset{d}{C}H_2\overset{c}{O}CH_3$
                  $|$
              CH_3
                a

38. Each of the compounds would show two signals, but the ratio of the integrals for the two signals would be different for each of the compounds. The ratio of the integrals for the signals given by the first compound would be 2:9 (or 1:4.5), the ratio of the integrals for the signals given by the second compound would be 1:3, and the ratio of the integrals for the signals given by the third compound would be 1:2.

39. The heights of the integrals for the signals in the spectrum shown in figure are about 3.5 and 5.2. The ratio of the integrals, therefore, is 5.2/3.5 = 1.5. This matches the ratio of the integrals calculated for **B**. (Later we will see that a signal at ~7 ppm is characteristic of a benzene ring.)

$$\frac{4}{2} = 2$$

$$\frac{6}{4} = 1.5$$

$$\frac{4}{4} = 1.0$$

$$\frac{4}{2} = 2$$

40. The signal farthest downfield in both spectra is the signal for the hydrogens bonded to the carbon that is also bonded to the halogen. Because chlorine is more electronegative than iodine, the farthest downfield signal should be farther downfield in the ^{1}H NMR spectrum for 1-chloropropane than in the ^{1}H NMR spectrum for 1-iodopropane. Therefore, the **first spectrum** is the ^{1}H NMR spectrum for **1-iodopropane** and the **second spectrum** is the ^{1}H NMR spectrum for **1-chloropropane.**

41. **C** is easiest to distinguish, because it will have **two** signals, whereas **A** and **B** will each have **three** signals.

A and **B** can be distinguished by looking at the splitting of their signals.

The signals in the ^{1}H NMR spectrum of **A** will be (left to right across the spectrum) **triplet, triplet, multiplet.**

The signals in the ^{1}H NMR spectrum of **B** will be (left to right across the spectrum) **doublet, multiplet, doublet.**

42. The signal at 12.2 ppm indicates that the compound is a carboxylic acid. From the molecular formula and the splitting patterns of the signals, the spectra can be identified as the ^{1}H NMR spectrum of the following:

43.   **a.** three signals   **b.** three signals   **c.** two signals   **d.** three signals

44.

Br
a | a   **b a b**
CH₃CCH₃   BrCH₂CH₂CH₂Br
|
Br

Br
|
CH₃CH₂CHBr
a   b   c

Br
|
CH₃CHCH₂Br
a   c   b

one signal   two signals

a singlet   a quintet (a) and
           a triplet (b)

three signals

two triplets (a and c)
and a multiplet (b)

three signals

two doublets (a and b)
and a multiplet (c)

45.

      t q
      ↓ ↓
a. CH₃CH₂CH₂CH₃

b. BrCH₂CH₂Br
        s
        ↓

c.
    t m t   O
    ↓ ↓ ↓   ‖   s
    CH₃CH₂CH₂   C   ↓
                  CH₃

      quin m t
      ↓   ↓ ↓
d. CH₃CH₂CHCH₂CH₃
           |
           Cl

      d m t
      ↓ ↓ ↓
e. CH₃CHCH₂CHCH₃
         | |
        CH₃ CH₃

f.
          d of d   d
          ↓   ↓
    t → ⬡—NO₂

46. Each spectrum is described going from left to right.

a. BrCH₂CH₂CH₂CH₂Br

    triplet   triplet (This triplet will not be split by the adjacent equivalent protons.)

b. two triplets (close to each other) singlet multiplet
(The table "¹H NMR Chemical Shifts" in the Study Section of Mastering indicates that a methylene group adjacent to an RO and a methylene group adjacent to a Br appear at about the same place.)

c. quartet triplet

d. singlet quartet triplet

e. doublet multiplet doublet

f. triplet singlet quintet

g. singlet (Equivalent Hs do not split each other's signals.)

h. quartet singlet triplet

47. The two singlets in the ¹H NMR spectrum that each integrate to three protons suggest two methyl groups, one of which is adjacent to an oxygen. That the benzene ring protons (6.7–7.1 ppm) consist of two doublets (each integrating to two hydrogens) indicates a 1,4-disubstituted benzene ring. The IR spectrum also indicates a benzene ring (1600 and 1500 cm⁻¹) with hydrogens bonded to sp^2 carbons and no carbonyl group. The absorption bands in the 1250–1000 cm⁻¹ region suggest there are two C—O single bonds, one with no double-bond character and one with some double-bond character.

CH₃O—⬡—CH₃

48. **a.** **1.** 3   **3.** 4   **5.** 3
            2. 2   **4.** 3   **6.** 3

b. An arrow is drawn to the carbon that gives the signal at the lowest frequency.

1.   CH$_3$CH$_2$CH$_2$Br

3.   CH$_3$CH$_2$—C(=O)—OCH$_3$

5.   (H$_3$C)(H$_3$C)(CH$_3$)C—O—CH$_3$

2.   CH$_3$CHCH$_3$ | Br

4.   CH$_3$CH(CH$_3$)—C(=O)—H

6.   CH$_3$—C(=O)—CH$_2$CH$_2$—C(=O)—CH$_3$

49.   Each spectrum is described going from left to right:

1.   triplet triplet quartet
2.   doublet quartet
4.   doublet doublet quartet

50.   The signal at 210 ppm is for the carbonyl carbon of a ketone. There are 10 other carbons in the compound but only five other signals. That suggests the compound is a symmetrical ketone with identical five-carbon alkyl groups.

CH$_3$CH$_2$CH$_2$CH$_2$CH$_2$—C(=O)—CH$_2$CH$_2$CH$_2$CH$_2$CH$_3$

51.   The molecular ion peak for these compounds is $m/z = 86$; the peak at $m/z = 57$ is due to loss of an ethyl radical (86 − 29), and the peak at $m/z = 71$ is due to loss of a methyl radical (86 − 15).

a.   3-Methylpentane will be more apt to lose an ethyl radical (forming a secondary carbocation and a primary radical) than a methyl radical (forming a secondary carbocation and a methyl radical). In addition, 3-methylpentane has two pathways to lose an ethyl radical. Therefore, the peak at $m/z = 57$ would be more intense than the peak at $m/z = 71$.

CH$_3$CH$_2$CHCH$_2$CH$_3$
|
CH$_3$

3-methylpentane

b.   2-Methylpentane has two pathways to lose a methyl radical (forming a secondary carbocation and a methyl radical in each pathway), and it cannot form a secondary carbocation by losing an ethyl radical. (Loss of an ethyl radical would form a primary carbocation and a primary radical.) Therefore, it will be more apt to lose a methyl radical than an ethyl radical, so the peak at $m/z = 71$ would be more intense than the peak at $m/z = 57$.

CH$_3$CHCH$_2$CH$_2$CH$_3$
|
CH$_3$

2-methylpentane

52.
   a. An absorption band at 3650–3200 cm^{-1} due to an O—H stretching vibration would be present for the alcohol and absent for the ether.

   b. An absorption band at 3500–3300 cm^{-1} due to an N—H stretching vibration would be present for the amide and absent for the ester.

   c. The alkene would have absorption bands at 1680–1600 cm^{-1} due to a C=C stretching vibration and at 3100–3020 cm^{-1} due to an sp^2 C—H stretching vibration that the alkyne would not have. The alkyne would have an absorption band at 2260–2100 cm^{-1} that the alkene would not have.

   d. The C=O absorption band will be at a larger wavenumber for the β,γ-unsaturated ketone (1720 cm^{-1}) than for the α,β-unsaturated ketone (1680 cm^{-1}), since the double bonds in the latter are conjugated.

53.   Dividing 128 by 13 gives 9 with 11 left over. Thus, the base value is C$_9$H$_{20}$. Because the compound is a saturated hydrocarbon, we know that the base value is also the molecular formula of the compound. Some possible structures are the following:

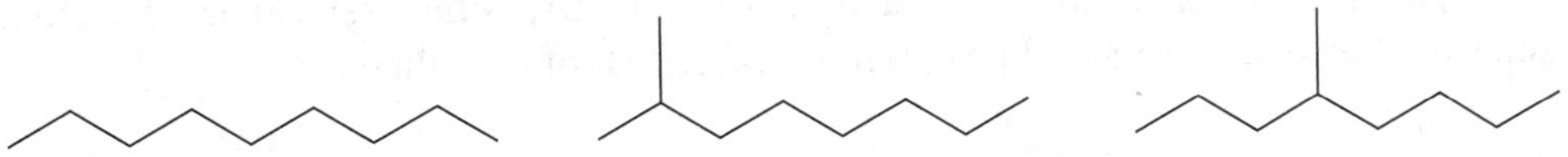

54.   The fact that the abundance of the M+2 peak is 30% of the abundance of the M peak indicates that the compound has one chlorine atom. The peak at $m/z = 77$ is due to loss of the chlorine atom ($112 - 35 = 77$). The fact that the peak at $m/z = 77$ does not fragment indicates it is a phenyl cation. Therefore, the compound is chlorobenzene.

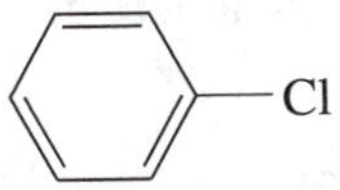

55.   Dividing 112 by 13 gives 8 with 8 left over. Thus, the base value is C$_8$H$_{16}$, and because the compound is a hydrocarbon, this is also its molecular formula. The molecular formula indicates that it has one degree of unsaturation, which is accounted for by the fact that we know it has a six-membered ring. Possible structures are shown here. Possible stereoisomers are not shown: the second and third structures have three stereoisomers and the fourth structure has two stereoisomers.

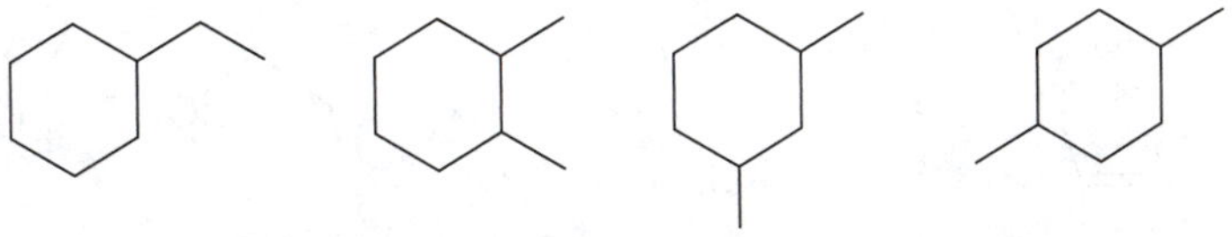

56.   **a.**

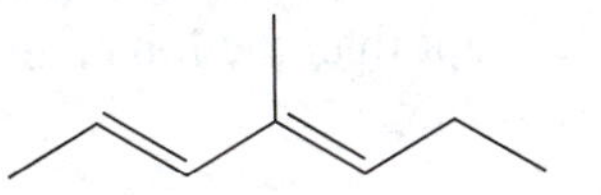

and

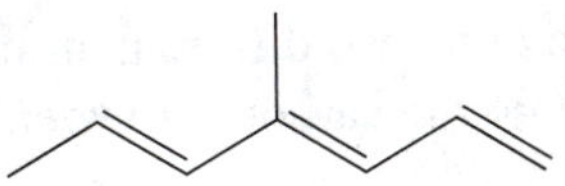

The λ_{max} will be at a longer wavelength, because there are three conjugated double bonds.

   b.

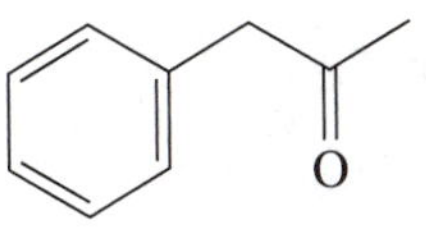

and

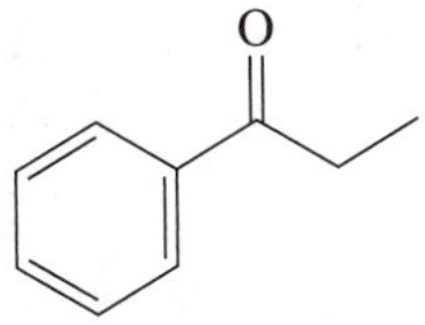

The λ_{max} will be at a longer wavelength because the carbonyl group is conjugated with the benzene ring.

57. Enovid would have its carbonyl stretch at a higher frequency. The carbonyl group in Norlutin has some single-bond character because of electron delocalization as a result of having conjugated double bonds. The single-bond character causes the carbon–oxygen bond to be easier to stretch than the carbon–oxygen bond in Enovid, which has isolated double bonds and therefore no electron delocalization.

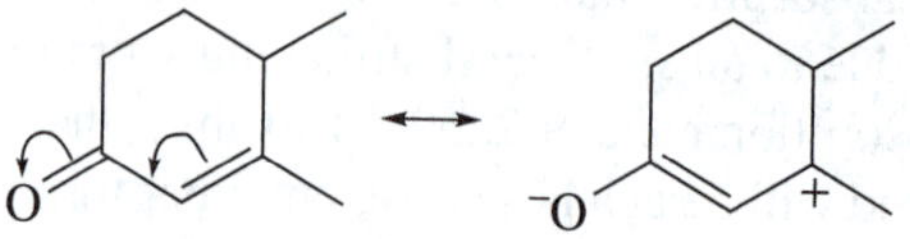

Norlutin

58. The ratio is 1:2:1.

To get the M peak, both Br atoms must be ^{79}Br. To get the M+4 peak, they both must be ^{81}Br. To get the M + 2 peak, the first Br atom can be ^{79}Br and the second ^{81}Br, or the first can be ^{81}Br and the second ^{81}Br. So the relative intensity of the M+2 peak will be twice that of the others.

M	M+2	M+4
^{79}Br ^{79}Br	^{79}Br ^{81}Br ^{81}Br ^{79}Br	^{81}Br ^{81}Br

59. The absorption bands at $\sim 2700 \text{ cm}^{-1}$ for the aldehyde hydrogen and at $\sim 1380 \text{ cm}^{-1}$ for the methyl group would distinguish the compounds.

A would have the band at $\sim 2700 \text{ cm}^{-1}$ but not the one at $\sim 1380 \text{ cm}^{-1}$.
B would have the band at $\sim 1380 \text{ cm}^{-1}$ but not the one at $\sim 2700 \text{ cm}^{-1}$.
C would have both the band at $\sim 2700 \text{ cm}^{-1}$ and the one at $\sim 1380 \text{ cm}^{-1}$.

60. The $C\!=\!O$ absorption band of the three compounds decreases in the following order:

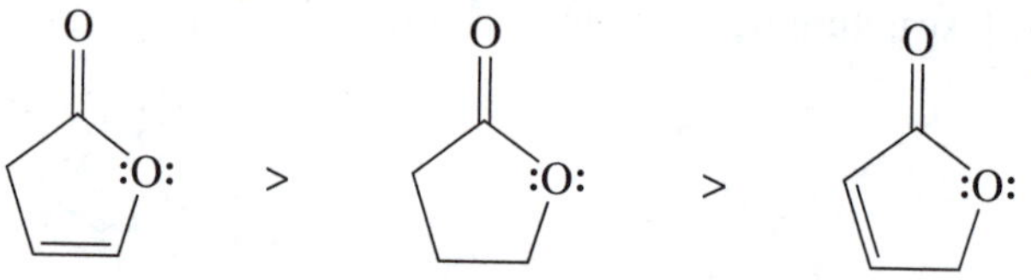

The first compound has the $C\!=\!O$ absorption band at the largest wavenumber, because a lone pair on the ring oxygen can be delocalized onto two different atoms; thus, it is less apt than the lone pair in the other compounds to be delocalized onto the carbonyl oxygen atom.

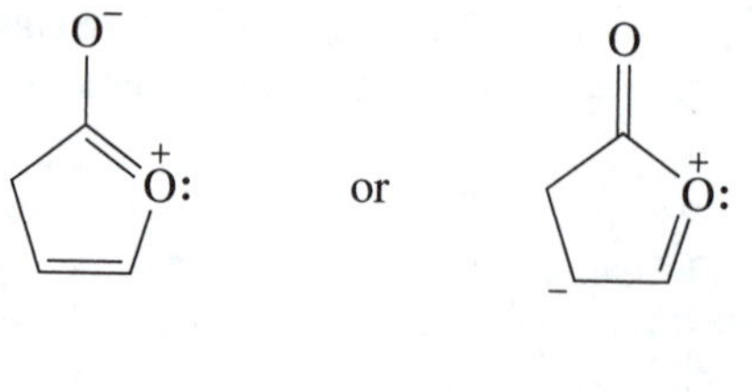

The third compound has the C=O absorption band at the smallest wavenumber because its carbonyl group has more single-bond character due to contributions from two other resonance contributors.

61. If addition of HBr to propene follows the rule that says that the electrophile adds to the sp^2 carbon that is bonded to the most hydrogens, the product of the reaction will give an NMR spectrum with two signals (a doublet and a septet). If addition of HBr does not follow the rule, the product will give an NMR spectrum with three signals (two triplets and a multiplet).

$$CH_3CH{=}CH_2 \ + \ HBr \ \longrightarrow$$

Br	
|	
CH_3CHCH_3	$CH_3CH_2CH_2Br$
follows the rule	does not follow the rule
two signals	**three signals**

62.

A	**B**	**C**	**D**
three signals	three signals	four signals	three signals
singlet, quartet, triplet	singlet, quartet, triplet	singlet, triplet multiplet,	singlet, septet
(singlet farthest downfield)	(quartet farthest downfield)	triplet	doublet

C can be distinguished from **A, B,** and **D,** because **C** has four signals and the others have three signals.

D can be distinguished from **A** and **B,** because the three signals of **D** are a singlet, a septet, and a doublet, whereas the three signals of **A** and **B** are a singlet, a quartet, and a triplet.

A and **B** can be distinguished, because the highest-frequency signal in **A** is a singlet, whereas the highest-frequency signal in **B** is a quartet.

63.

$CH_3CH_2CH{=}CH_2$	cis-2-butene	2-methylpropene
1-butene		

It would be better to use ^{13}C NMR, because you would have to look only at the number of signals in each spectrum: 1-butene will show four signals, cis-2-butene will show two signals, and 2-methylpropene will show three signals. (In the ^1H NMR spectrum, 1-butene will show five signals, and cis-2-butene and 2-methylpropene will both show two signals.)

64.

$$CH_3$$
$$|$$
$$CH_3CCH_3$$
$$|$$
$$Cl$$

tert-butyl chloride
Compound A

$$CH_3CHCH_2CH_3$$
$$|$$
$$Cl$$

sec-butyl chloride
Compound B

65.   **a.**   The absorption band at ~ 2100 cm^{-1} indicates a carbon–carbon triple bond, and the absorption band at ~ 3300 cm^{-1} indicates a hydrogen bonded to an *sp* carbon.

$$CH_3CH_2CH_2CH_2C\equiv CH$$

b.   The absence of an absorption band at ~ 2700 cm^{-1} indicates that the compound is not an aldehyde, and the absence of a broad absorption band in the vicinity of 3000 cm^{-1} indicates that the compound is not a carboxylic acid. The ester and the ketone can be distinguished by the absorption band at ~ 1200 cm^{-1} that indicates the carbon–oxygen single bond of an ester.

66.   **a.** **1.** 5   **3.** 4   **b.** **1.** 7   **3.** 5
             2. 5   **4.** 2   **2.** 7   **4.** 2

67.   **a.**   The broad absorption band at ~ 3300 cm^{-1} is characteristic of the oxygen–hydrogen stretch of an alcohol, and the absence of absorption bands at ~ 1600 cm^{-1} and ~ 3100 cm^{-1} indicates that it is not the alcohol with a carbon–carbon double bond.

b.   The absorption band at ~ 1685 cm^{-1} indicates a carbon–oxygen double bond. The absence of a strong and broad absorption band at ~ 3000 cm^{-1} rules out the carboxylic acid, and the absence of an absorption band at ~ 2700 cm^{-1} rules out the aldehyde. Thus, it must be one of the ketones. The ketone with the conjugated carbonyl group would be expected to show a C=O stretch at ~ 1685 cm^{-1}, whereas the ketone with the isolated carbonyl group would show a C=O stretch at ~ 1720 cm^{-1}. Thus, the compound is the ketone with the conjugated carbonyl group.

c.   The absorption band at ~ 1700 cm^{-1} indicates a carbon–oxygen double bond. The absence of an absorption band at ~ 1600 cm^{-1} rules out the ketones with the benzene or cyclohexene rings. The absence of absorption bands at ~ 2100 cm^{-1} and ~ 3300 cm^{-1} rules out the ketone with the carbon–carbon triple bond. Thus, it must be 4-ethylcyclohexanone.

68. The broad absorption band at ~3300 cm^{-1} indicates that the compound has an OH group. The absorption bands at ~2900 cm^{-1} indicate that the compound has hydrogens attached to an sp^3 carbon. The compound, therefore, is **benzyl alcohol.**

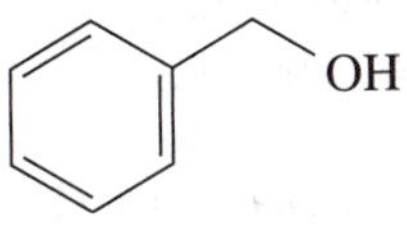

benzyl alcohol

69.   **a.** 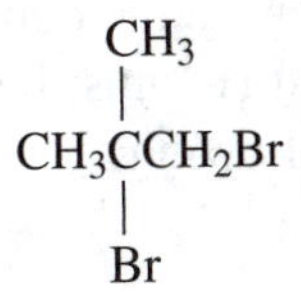   CH$_3$CCH$_2$Br (with CH$_3$ above and Br below the central C) 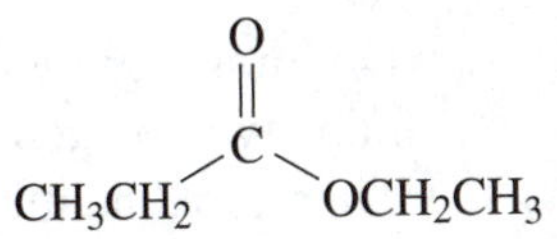   **b.** (benzene ring)–CHCH$_3$ with Br above   **c.**   CH$_3$CH$_2$ C(=O) OCH$_2$CH$_3$

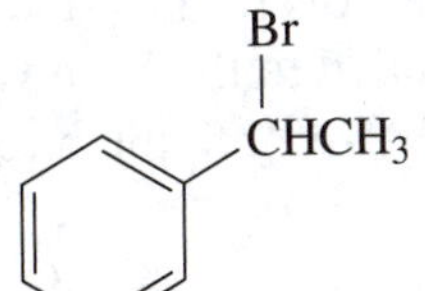

70. In acidic solutions, the three benzene rings are isolated from one another, so phenolphthalein is colorless. In basic solutions, loss of the proton from one of the OH groups causes the five-membered ring to open. As a result, the number of conjugated double bonds increases, which causes the solution to become colored.

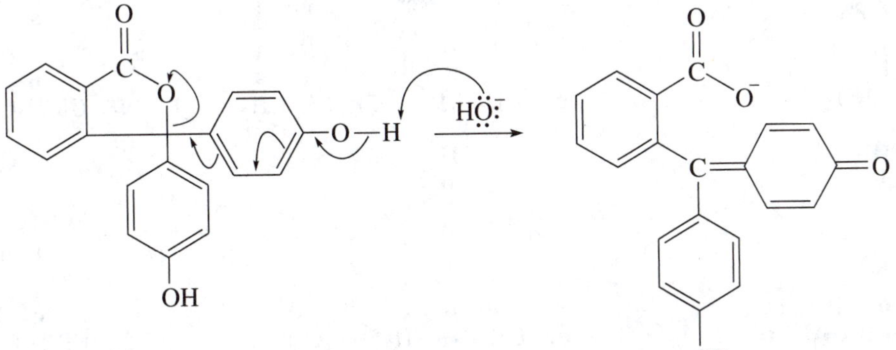

71. The broad absorption band at ~3300 cm^{-1} indicates that the compound has an OH group. The absence of absorbance at ~1700 cm^{-1} shows that the compound does not have a carbonyl group. The absence of absorption at ~2950 cm^{-1} indicates that the compound does not have any hydrogens bonded to sp^3 carbons. Therefore, the compound is **phenol.**

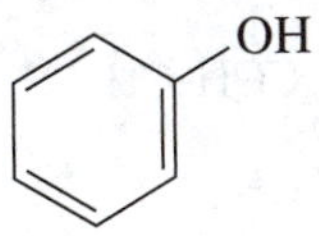

phenol

72. 1-Hexyne will show absorption bands at ~3300 cm^{-1} for a hydrogen bonded to an sp carbon and at ~2100 cm^{-1} for the triple bond.

$$CH_3CH_2CH_2CH_2C\equiv CH$$

1-hexyne

2-Hexyne will show the absorption band at $\sim 2100 \text{ cm}^{-1}$ but not the one at $\sim 3300 \text{ cm}^{-1}$.

$$C \equiv C$$

3-Hexyne will show neither the absorption band at $\sim 3300 \text{ cm}^{-1}$ nor the one at $\sim 2100 \text{ cm}^{-1}$ (there is n change in dipole moment when the $C \equiv C$ stretches).

$$CH_2CH_2C \equiv CCH_2CH_3$$
3-hexyne

73.   Dividing 116 by 13 gives 8 with 12 left over. Thus, the base value is C_8H_{20}. Because the compound con tains two oxygens, two Os must be added to the base value and two Cs and eight Hs must be subtracted Therefore, the molecular formula is $C_6H_{12}O_2$. Some possible structures are the following:

74.   **a.** CH_3CHNO_2 (with d septet, a b labels, CH_3 = a)

   c. CH_3CH—$CH_2CH_2CH_3$ (d septet, t m t; b e d c a; CH_3 = b)

   e. $ClCH_2CCHCl_2$ (s, s, a; CH_3 = b; c; CH_3 = a, s)

   b. $CH_3CH_2CH_2OCH_3$ (t m t s; a b d c)

   d. $CH_3CH_2CH_2$—CH_2Cl (t m t s; a b c d)

   f. $ClCH_2CH_2CH_2CH_2CH_2Cl$ (t m quintet; c b a b c)

75.   **a.**   and

   four signals   two signals

   b. CH_3CH—$CHCH_3$ (with CH_3 CH_3)   and $CH_3CCH_2CH_3$ (with CH_3, CH_3)

   two signals   three signals

   c. CH_3O—⟨benzene ring⟩—CH_3 and CH_3—⟨benzene ring⟩—CH_3

   this would have the
   highest-frequency signal

d. Br⌒⌒Br   and   Br⌒⌒NO$_2$

two signals   three signals

e.

CH$_3$ O
  | ||
CH$_3$C—COCH$_3$ and
  |
CH$_3$

OCH$_3$
   |
CH$_3$CCH$_3$
   |
OCH$_3$

two signals with an   two signals with an
integral ratio of 3 : 1   integral ratio of 1 : 1

f.   and

two signals   three signals

6. **a.** The spectrum must be that of **2-bromopropane,** because the NMR spectrum has two signals and the lowest-frequency (farthest upfield) signal is a doublet and the other signal is given by a single hydrogen.

    b. The spectrum must be that of **1-nitropropane,** because the NMR spectrum has three signals and both the lowest-frequency (farthest upfield) and highest-frequency (farthest downfield) signals are triplets.

    c. The spectrum must be that of **ethyl methyl ketone,** because the NMR spectrum has three signals and the signals are a triplet, a singlet, and a quartet.

77. By dividing the value of the integration steps by the smallest one, the ratios of the hydrogens are found to be 3:2:1:9.

$$\frac{40.5}{13} = 3.1 \quad \frac{27}{13} = 2.1 \quad \frac{13}{13} = 1 \quad \frac{118}{13} = 9.1$$

Because the ratios are given in the highest-frequency to lowest-frequency direction, a possible compound could be the following ester:

CH$_3$ O
   | ||
CH$_3$C—CHCOCH$_3$
   | |
CH$_3$ CH$_2$Cl

78. **a.** CH$_3$CH$_2$CHBr
               |
               CH$_3$

b. CH$_3$CH$_2$CH$_2$CH$_2$Br

c. CH$_3$CHCH$_2$Br
       |
       CH$_3$

79. **a.**
        O
        ||
        C   CH$_3$
  CH$_3$ CH$_2$CCH$_3$
              |
              CH$_3$

b.
        O
        ||
        C
  CH$_3$CH   CH$_2$CH$_2$CH$_3$
      |
      CH$_3$

c.
        O
        ||
        C
  CH$_3$CH   CHCH$_3$
      | |
      CH$_3$ CH$_3$

80.   **a.** $CH_3OCH_2CH_2OCH_3$ **b.** There are three possibilities: **c.**

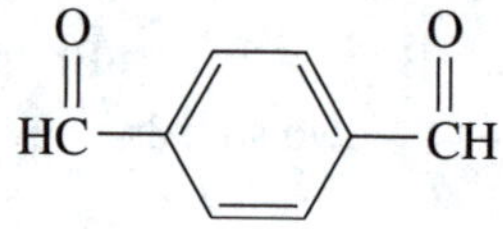

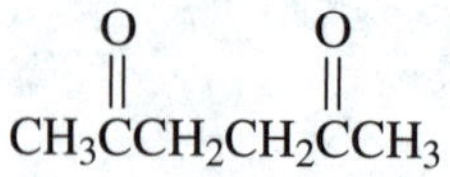

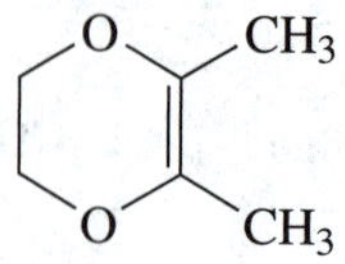

$$CH_3\overset{\overset{\displaystyle O}{\|}}{C}CH_2CH_2\overset{\overset{\displaystyle O}{\|}}{C}CH_3$$

$CH_3OCH_2C\equiv CCH_2OCH_3$

81.   It is the ^1H NMR spectrum of *tert*-butyl methyl ether.

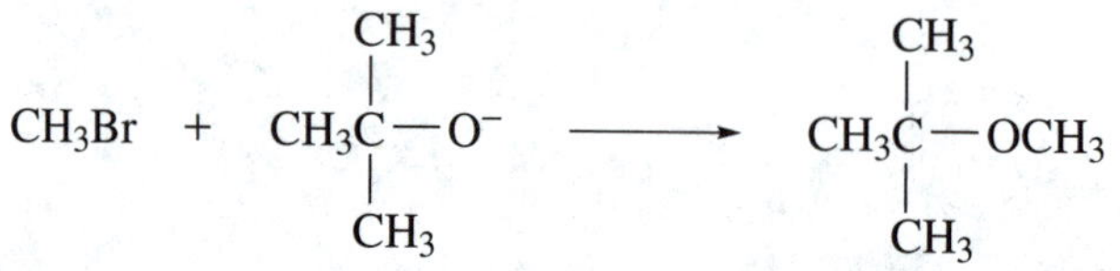

tert-butyl methyl ether

82.

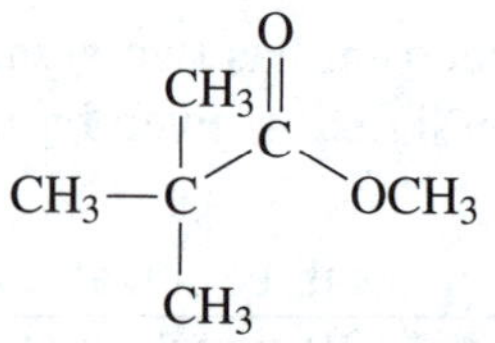

Chapter 10 Practice Test

Give one IR absorption band that could be used to distinguish each of the following pairs of compounds. Indicate the compound for which the band would be present.

a. 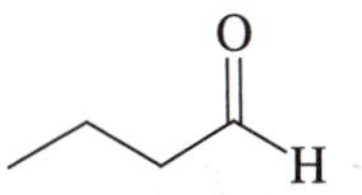 and

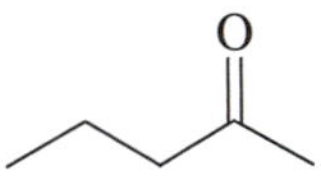

b. 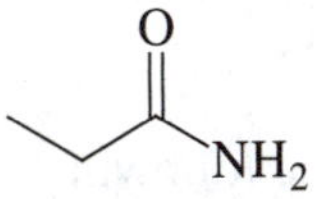 and

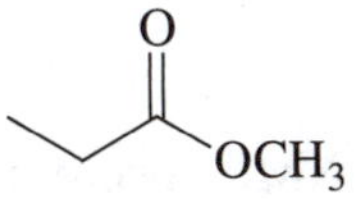

c. $CH_3CH_2CH_2CH_2OH$ and $CH_3CH_2CH_2OCH_3$

d. 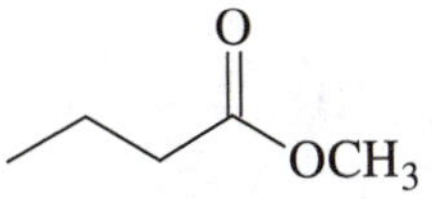 and

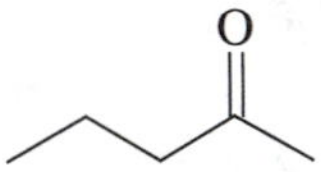

e. $CH_3CH_2CH=CHCH_3$ and $CH_3CH_2C\equiv CCH_3$

f. $CH_3CH_2C\equiv CH$ and $CH_3CH_2C\equiv CCH_3$

Indicate whether each of the following is true or false:

a. The O—H stretch of a concentrated solution of an alcohol occurs at a higher frequency than the O—H stretch of a dilute solution.   T   F

b. Light of 2 μm is of higher energy than light of 3 μm.   T   F

c. It takes more energy for a bending vibration than for a stretching vibration.   T   F

d. The M+2 peak of an alkyl chloride is half the height of the M peak.   T   F

How could you distinguish between the IR spectra of the following compounds?

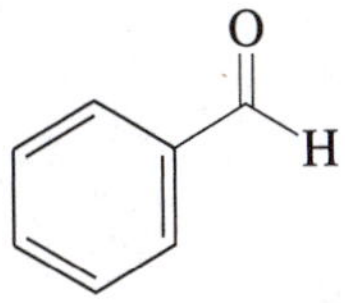

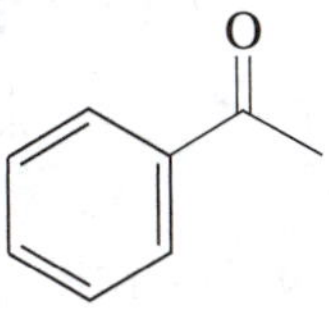

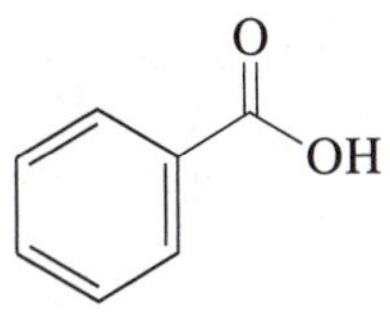

a. 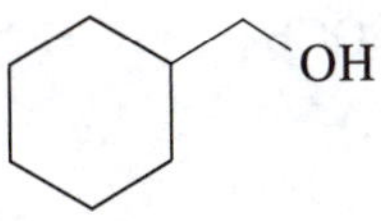 and

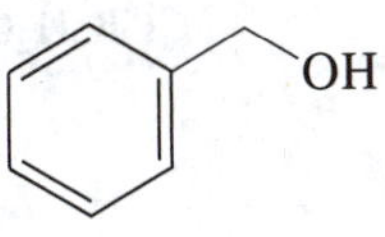

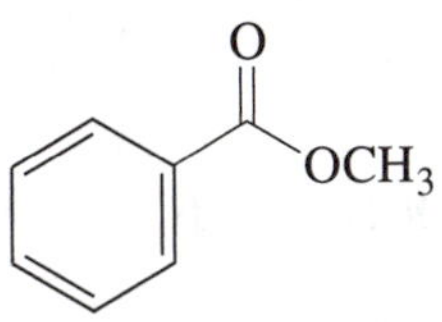

d. 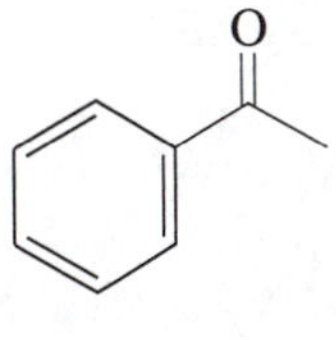 and

b. and

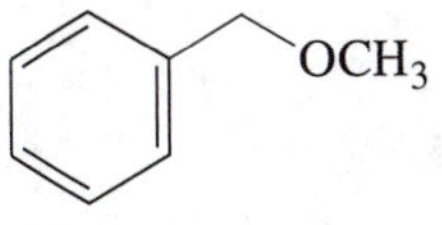

e. 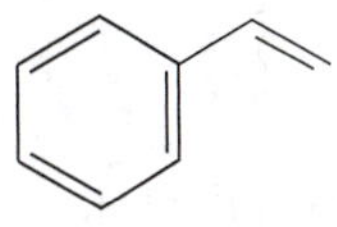 and

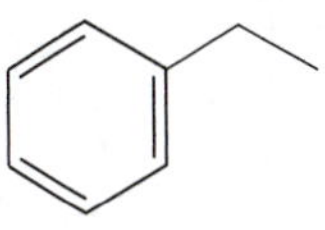

c. and

4.   Which compound has the greater λ_{max}? (300 nm is a greater λ_{max} than 250 nm.)

a.   and

c. NH_2 and $\overset{+}{N}H_3$

b. with $\overset{H}{\underset{}{N}}$ and NH_2

5.   How many signals would you expect to see in the ^1H NMR spectrum of each of the following compounds

$$CH_3CH_2CH_2\overset{\overset{\displaystyle O}{\|}}{C}CH_3$$

$$CH_3CH_2\overset{\underset{\displaystyle Cl}{|}}{C}HCH_2CH_3$$

$$CH_2{=}CH\overset{\overset{\displaystyle O}{\|}}{C}H$$

$-NO_2$

(dichlorobenzene with Cl and Cl)

$$CH_3\overset{\underset{\displaystyle CH_3}{|}}{C}HCH_2\overset{\underset{\displaystyle CH_3}{|}}{C}HCH_3$$

6.   Indicate the multiplicity of each of the indicated sets of protons. (That is, indicate whether it is a singlet, doublet, triplet, quartet, quintet, multiplet, or doublet of doublets.)

$$CH_3CH_2\overset{\overset{\displaystyle O}{\|}}{C}CH_3$$
↑

H$-$$-NO_2$
↑

$$CH_3CH_2\overset{\overset{\displaystyle O}{\|}}{C}OCH_2CH_3$$
↑

$$CH_3\overset{\underset{\displaystyle CH_3}{|}}{C}HCH_2Cl$$
↑

$ClCH_2CH_2CH_2OCH_3$
↑

$CH_3OCH_2CH_2CH_2OCH_3$
↑

$BrCH_2CH_2Br$
↑

7.   How could you distinguish the following compounds using ^1H NMR spectroscopy?

$$CH_3\overset{\overset{\displaystyle O}{\|}}{C}OCH_2CH_3$$

$$CH_3CH_2\overset{\overset{\displaystyle O}{\|}}{C}OCH_3$$

$$H\overset{\overset{\displaystyle O}{\|}}{C}OCH_2CH_2CH_3$$

Indicate whether each of the following statements is true or false:

a. The signals on the right of an NMR spectrum are deshielded compared to the signals on the left.   T   F

b. Dimethyl ketone has the same number of signals in its ^{1}H NMR spectrum as in its ^{13}C NMR spectrum.   T   F

c. The greater the frequency of the signal, the greater its chemical shift in ppm.   T   F

For each compound,

a. indicate the number of signals you would expect to see in its ^{1}H NMR spectrum.
b. indicate the hydrogen or set of hydrogens that would give the highest-frequency (farthest downfield) signal.
c. indicate the multiplicity of that signal.

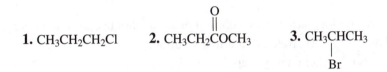

1. $CH_3CH_2CH_2Cl$ **2.** $CH_3CH_2COCH_3$ (with O double-bonded to the C)   **3.** CH_3CHCH_3 (with Br below the central C)

0. For each compound of Problem 9,

a. indicate the number of signals you would expect to see in its ^{13}C NMR spectrum.
b. indicate the carbon that would give the highest-frequency (farthest downfield) signal.
c. indicate the multiplicity of that signal in a proton-coupled ^{13}C NMRspectrum.

Answers to Chapter 10 Practice Test

1. **a.** 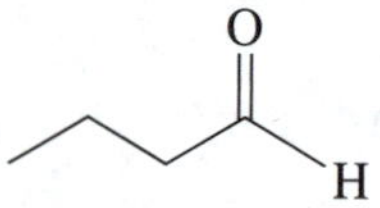

~2770 cm^{-1}

d. 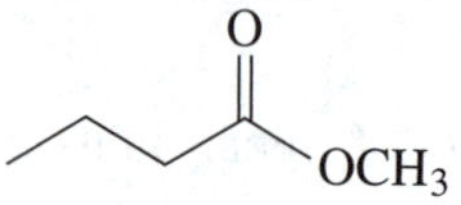

~1050 or ~1250 cm^{-1}

b. 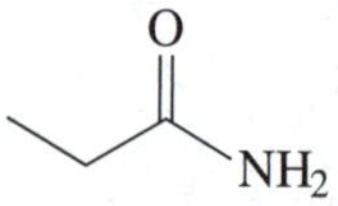

~3300 cm^{-1}

e. $CH_3CH_2CH{=}CHCH_3$ ~1600 cm^{-1}, ~3100 cm^{-1} $CH_3CH_2C{\equiv}CCH_3$ ~2100 cm^{-1}

c. $CH_3CH_2CH_2CH_2OH$ ~3600–3200 cm^{-1}

f. $CH_3CH_2C{\equiv}CH$ ~3300 cm^{-1}

2.
a. The O—H stretch of a concentrated solution of an alcohol occurs at a higher frequency than the O—H stretch of a dilute solution. F
b. Light of 2 μm is of higher energy than light of 3 μm. T
c. It takes more energy for a bending vibration than for a stretching vibration. F
d. The M+2 peak of an alkyl chloride is half the height of the M peak. F

3. **a.** 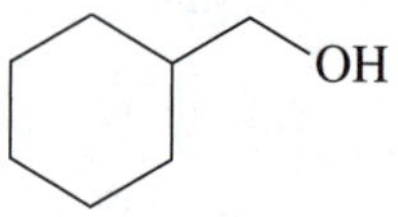

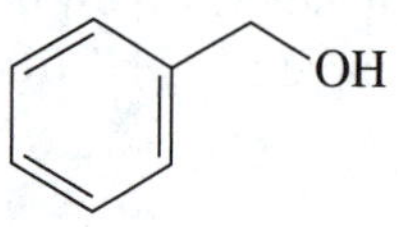

~3100 cm^{-1}
~1500 cm^{-1}

d. 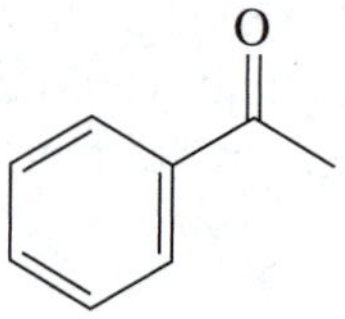

~1700 cm^{-1}

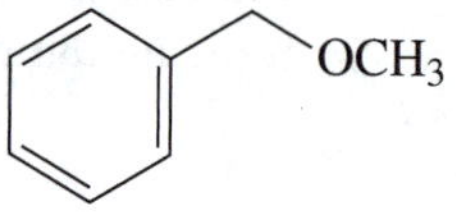

~1050 cm^{-1}

b. 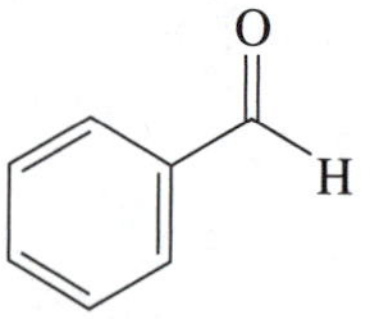

~2700 cm^{-1}

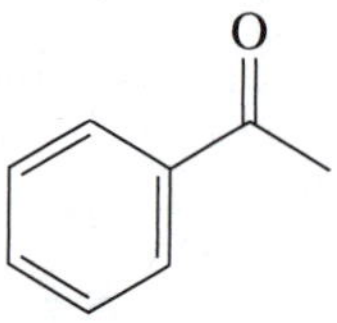

~2900 cm^{-1}
~1380 cm^{-1}

e. 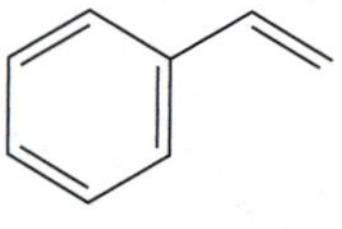

~3100 cm^{-1}

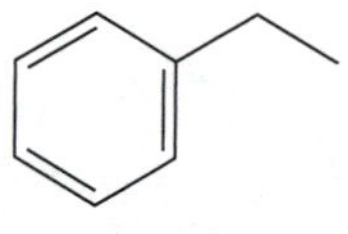

~2900 cm^{-1}
~1380 cm^{-1}

c. 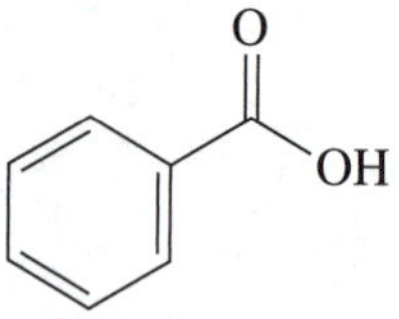

~3300–2500 cm^{-1}

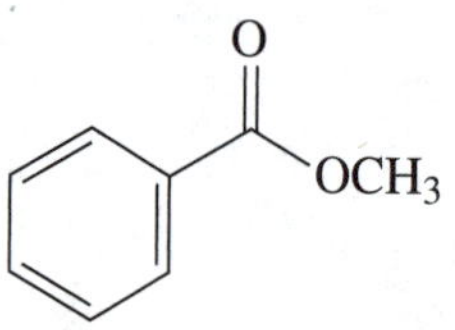

~2900 cm^{-1}
~1380 cm^{-1}
~1050 cm^{-1}

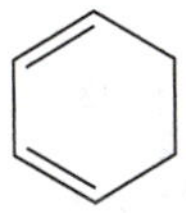

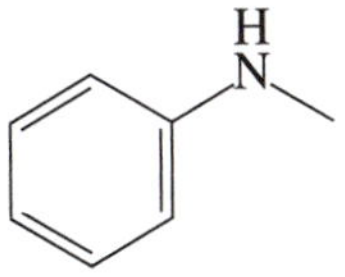

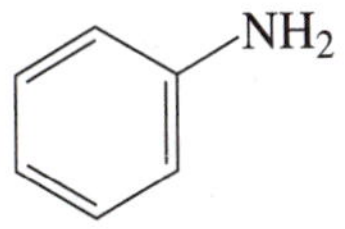

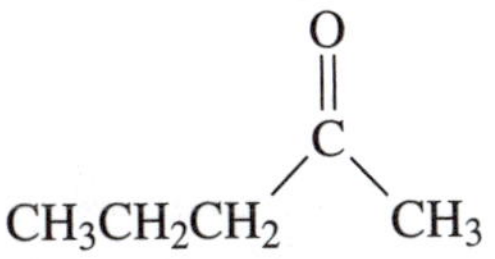

a.

b.

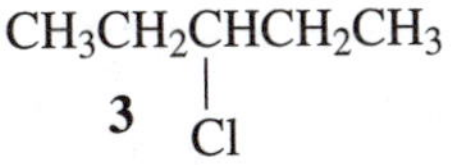

c.

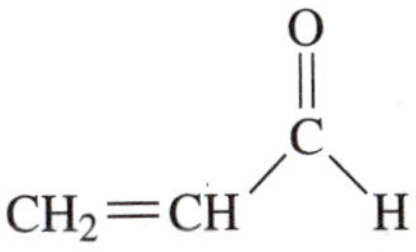

$CH_3CH_2CH_2$ CH_3

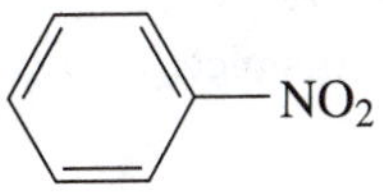

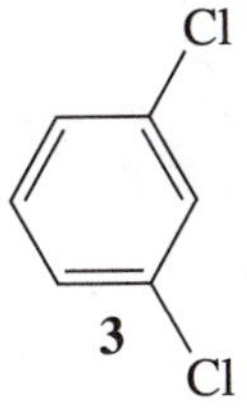

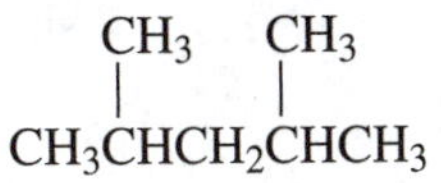

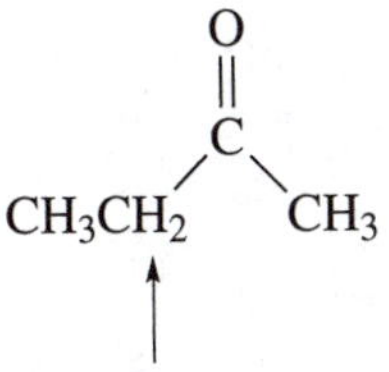

4

$CH_3CH_2CHCH_2CH_3$

3 Cl

$CH_2=CH$ H

4

—NO_2

3

3

CH_3 CH_3

$CH_3CHCH_2CHCH_3$

3

CH_3CH_2 CH_3

↑

quartet

H——NO_2

↑

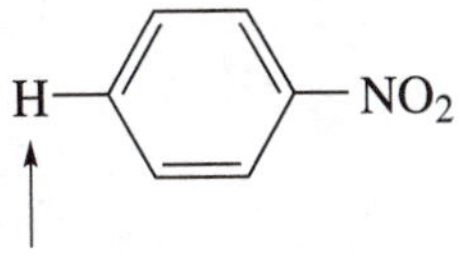

triplet

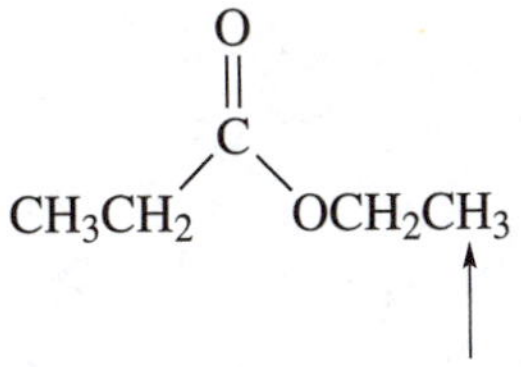

CH_3CH_2 OCH_2CH_3

↑

triplet

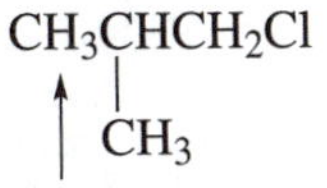

CH_3CHCH_2Cl

↑ CH_3

doublet

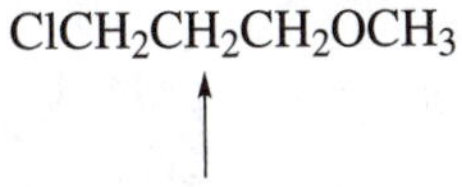

$ClCH_2CH_2CH_2OCH_3$

↑

multiplet

$CH_3OCH_2CH_2CH_2OCH_3$

↑

quintet

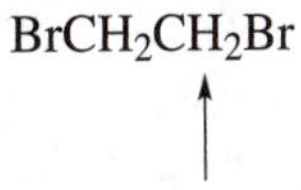

$BrCH_2CH_2Br$

↑

singlet

7.

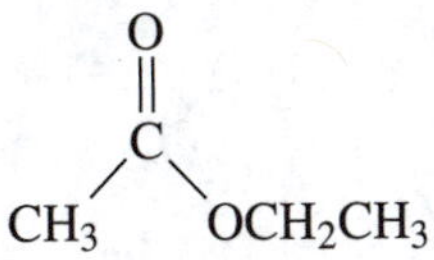

CH_3 OCH_2CH_3

3 signals

CH_3CH_2 OCH_3

3 signals

H $OCH_2CH_2CH_3$

4 signals

The signal at the highest frequency (farthest downfield) is a quartet.

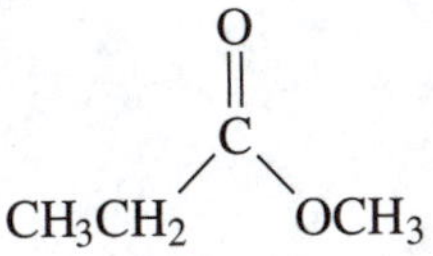

The signal at the highest frequency (farthest downfield) is a singlet.

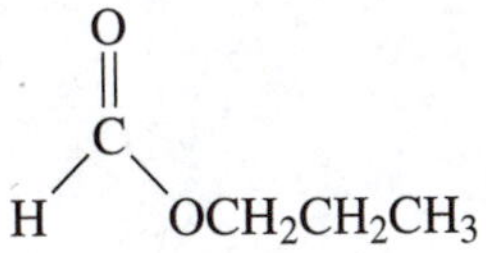

8.

a. The peaks on the right of an NMR spectrum are deshielded compared to the peaks on the left.   F

b. Dimethyl ketone has the same number of signals in its ^{1}H NMR spectrum as in its ^{13}C NMR spectrum.   F

c. The greater the frequency of the signal, the greater its chemical shift in ppm.   T

9.   **a.** CH$_3$CH$_2$CH$_2$Cl

triplet

3 signals

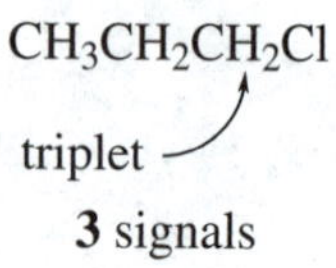

b. $\overset{\displaystyle O}{\overset{\displaystyle \|}{\text{CH}_3\text{CH}_2\text{COCH}_3}}$

singlet

3 signals

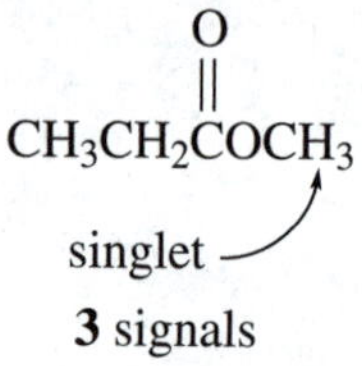

c. CH$_3$CHCH$_3$ — septet

|
Br

2 signals

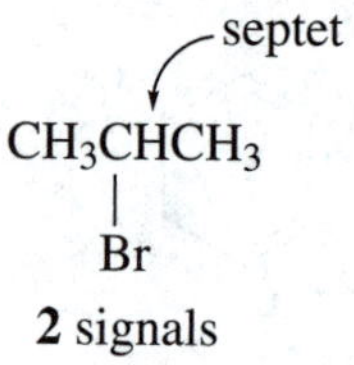

10.   **a.** CH$_3$CH$_2$CH$_2$Cl

3 signals

triplet

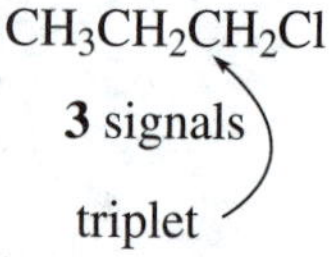

b. $\overset{\displaystyle O}{\overset{\displaystyle \|}{\text{CH}_3\text{CH}_2\text{COCH}_3}}$

4 signals

singlet

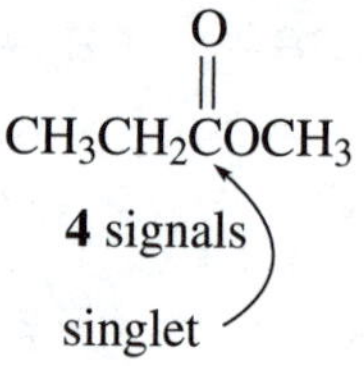

c. CH$_3$CHCH$_3$

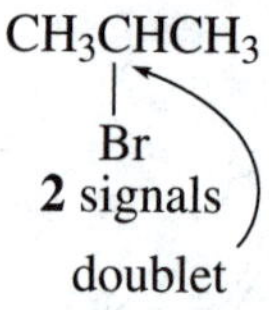

|
Br

2 signals

doublet

CHAPTER 11
Reactions of Carboxylic Acids and Carboxylic Acid Derivatives

Important Terms

acid anhydride

acyl adenylate a carboxylic acid derivative with AMP as the leaving group.

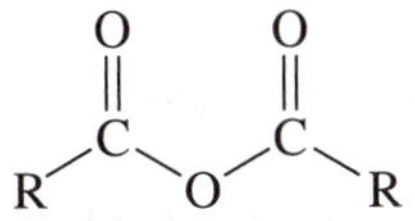

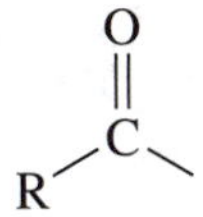

acyl group

acyl chloride

acyl phosphate a carboxylic acid derivative with a phosphate leaving group.

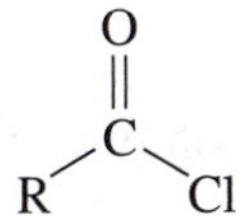

alcoholysis a reaction with an alcohol that converts one compound into two compounds.

amide

amino acid a protonated α-amino carboxylic acid.

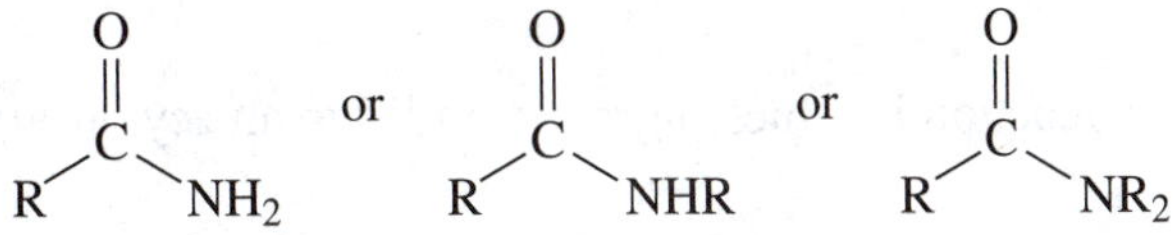

aminolysis a reaction with an amine that converts one compound into two compounds.

biosynthesis synthesis that occurs in a biological system.

carbonyl carbon the carbon of a C=O group.

carbonyl compound a compound that contains a C=O group.

carbonyl group a carbon doubly bonded to an oxygen (C=O).

carbonyl oxygen the oxygen of a C=O group.

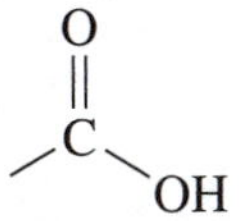

carboxyl group

carboxylic acid

$$\underset{R}{\overset{\overset{\displaystyle O}{\|}}{\underset{\qquad}{C}}}\ OH$$

carboxylic acid derivative   a compound that is hydrolyzed to a carboxylic acid.

carboxyl oxygen   the single-bonded oxygen of a carboxylic acid or ester.

catalyst   a species that increases the rate of a reaction without being consumed in the reaction.

ester

$$\underset{R}{\overset{\overset{\displaystyle O}{\|}}{\underset{\qquad}{C}}}\ OR'$$

hydrolysis   a reaction with water that converts one compound into two compounds.

mixed anhydride   an acid anhydride with two different R groups.

nitrile   a compound that contains a carbon–nitrogen triple bond.

$$R-C\equiv N$$

nucleophilic acyl substitution reaction   a reaction in which a group bonded to an acyl group is substituted by another group.

phosphoanhydride bond   the bond holding two phosphate groups together within a diphosphate or triphosphate.

symmetrical anhydride   an acid anhydride with identical R groups.

tetrahedral intermediate   the intermediate formed in a nucleophilic acyl substitution reaction (or a nucleophilic addition-elimination reaction)

thioester   the sulfur analog of an ester.

$$\underset{R}{\overset{\overset{\displaystyle O}{\|}}{\underset{\qquad}{C}}}\ SR'$$

transesterification reaction   the reaction of an ester with an alcohol to form a different ester.

olutions to Problems

a. benzyl acetate   b. isopentyl acetate   c. methyl butyrate

a. potassium butanoate   d. 5-methylhexanoic acid
    potassium butyrate   δ-methylcaproic acid

b. isobutyl butanoate   e. propanamide
    isobutyl butyrate   propionamide

c. pentanoyl chloride   f. *N,N*-dimethylhexanamide
    valeryl chloride

a.

d. $CH_3CH_2CH_2CH$ (with Cl substituent) — OCH_2CH_3, C=O

b. CH_3—C(=O)—$NHCH_2$—phenyl

e. CH_3CHCH_2 (with Br substituent) —C(=O)— NH_2

c. $CH_3CH_2CHCH_2CH_2$—C(=O)—OH, with CH_3 substituent

f. structure with Cl, C=O, OH

B is a correct statement. The delocalization energy (resonance energy) is greater for the amide than for the ester, because the second resonance contributor of the amide has a greater predicted stability and so contributes more to the overall structure of the amide; nitrogen is less electronegative than oxygen, so it is more stable with a positive charge.

The carbon–oxygen single bond in an alcohol is longer because, as a result of electron delocalization, the carbon–oxygen single bond in a carboxylic acid has some double-bond character.

$$RCH_2\text{—}OH \qquad \text{longer} \qquad \text{shorter}$$

6. **a.** Because HCl is a stronger acid than H_2O, Cl^- is a weaker base than HO^-.
Therefore, Cl^- will be eliminated from the tetrahedral intermediate, so the product of the reaction wil
be acetic acid. Because the solution is basic, acetic acid will be in its basic form as a result of losing a
proton.

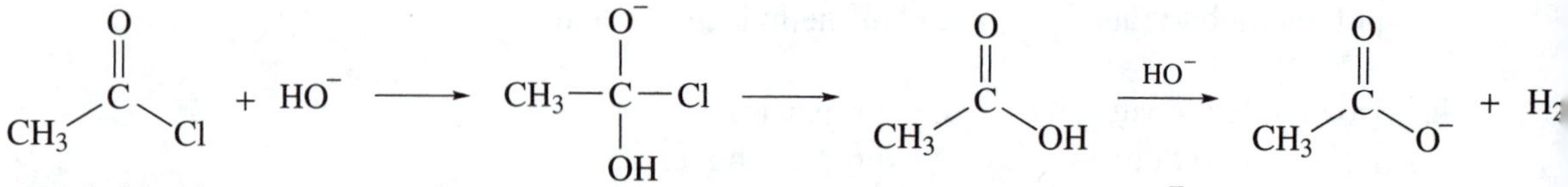

acetyl chloride   acetic acid + Cl^-

b. Because H_2O is a stronger acid than NH_3, HO^- is a weaker base than $^-NH_2$.
Therefore, HO^- will be eliminated from the tetrahedral intermediate, so the reactant will reform. In
other words, no reaction will take place.

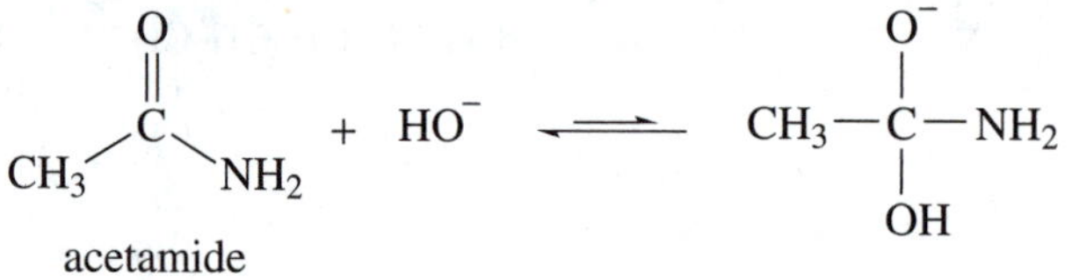

acetamide

7. **a.** a new carboxylic acid derivative

b. no reaction

c. a mixture of two carboxylic acid derivatives

8. **a.**

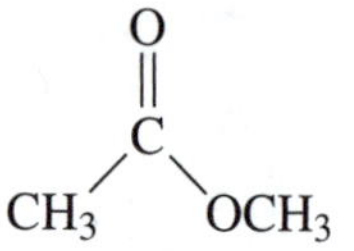

+ NaCl $\longrightarrow$ no reaction, because Cl^- is weaker base than CH_3O^-

b.

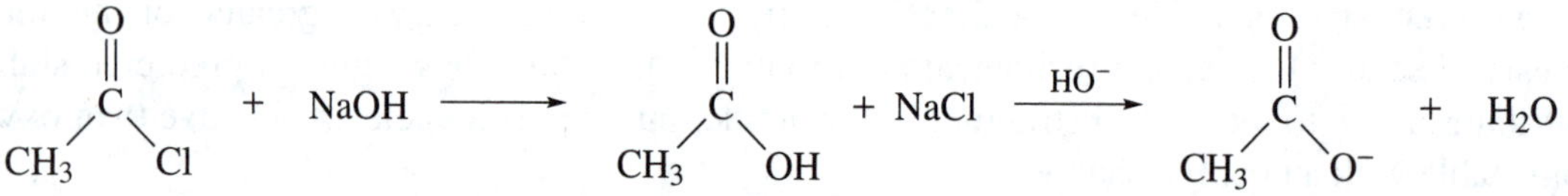

+ NaOH $\longrightarrow$ ... + NaCl $\xrightarrow{HO^-}$ ... + H_2O

c.

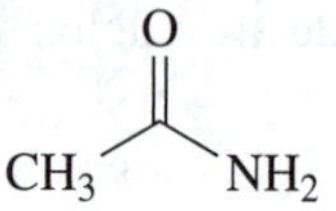

+ NaCl $\longrightarrow$ no reaction, because Cl^- is a weaker base than $^-NH_2$

d.

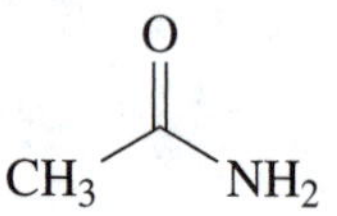

+ NaOH $\longrightarrow$ no reaction, because HO^- is a weaker base than $^-NH_2$

9. **a.** H_2O **c.** $CH_3CH_2CH_2OH$ **e.** HO—⬡

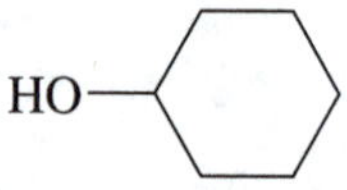

b. NH_3 **d.** $(CH_3)_2NH$ **f.** 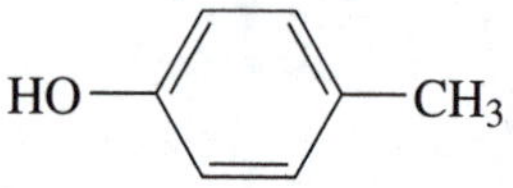 HO—⬡—CH_3

0.   **a.**

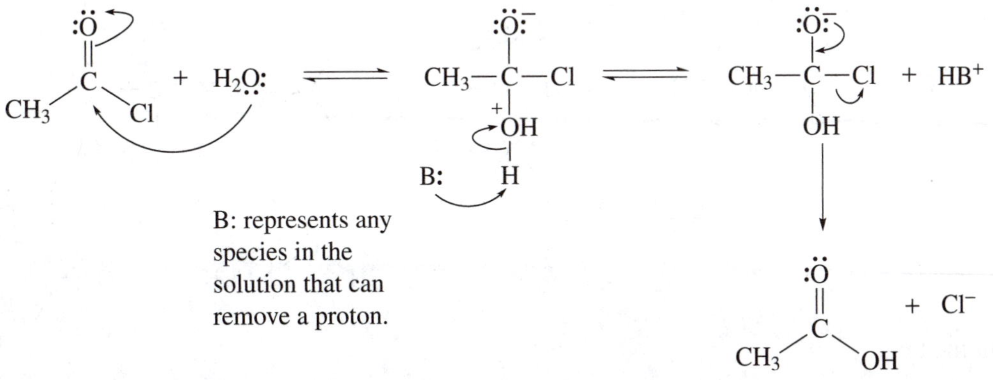

B: represents any
species in the
solution that can
remove a proton.

b.

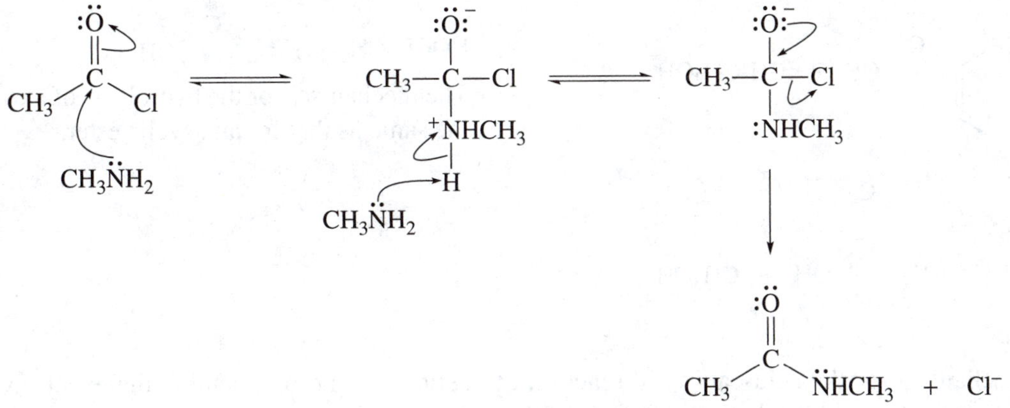

1.   **a.** 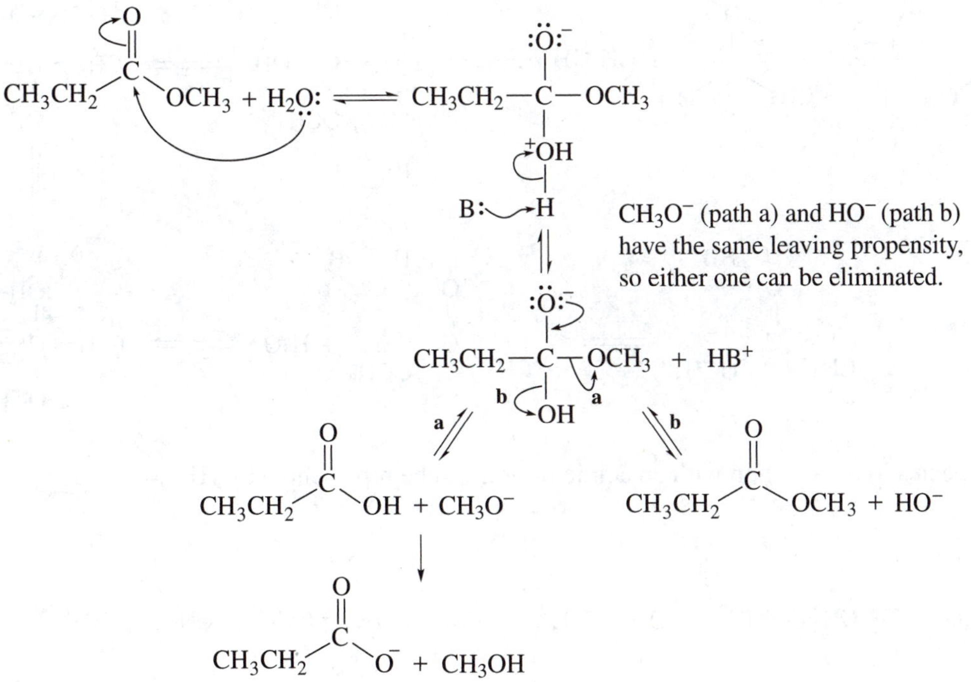 CH_3CH_2—C(=O)—$OCH_3 + H_2O$: ⇌ CH_3CH_2—C—OCH_3

CH₃O⁻ (path a) and HO⁻ (path b)
have the same leaving propensity,
so either one can be eliminated.

b.

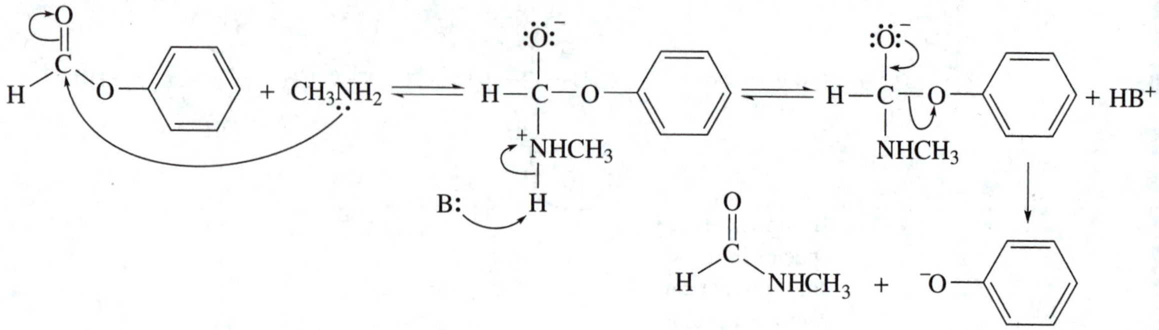

H—C(=O)—O—C₆H₅ + CH₃NH₂ ⇌ H—C(O⁻)(O—C₆H₅)(⁺NHCH₃) ⇌ H—C(O⁻)—O—C₆H₅ (NHCH₃) + HB⁺

B: H

H—C(=O)—NHCH₃ + ⁻O—C₆H₅

12. Solved in the text.

13. **a.** 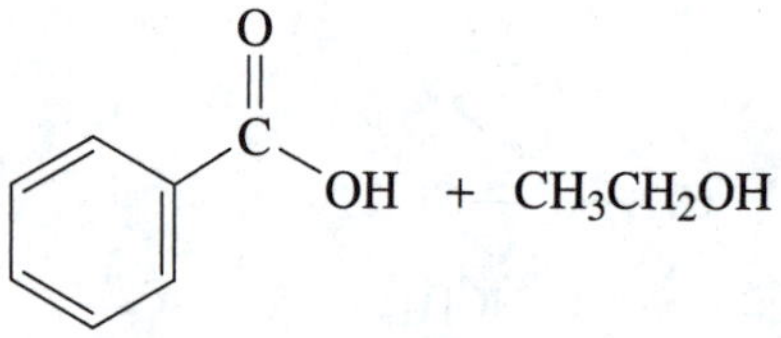 C₆H₅—C(=O)—OH + CH₃CH₂OH

c. HOCH₂CH₂CH₂CH₂—C(=O)—OH

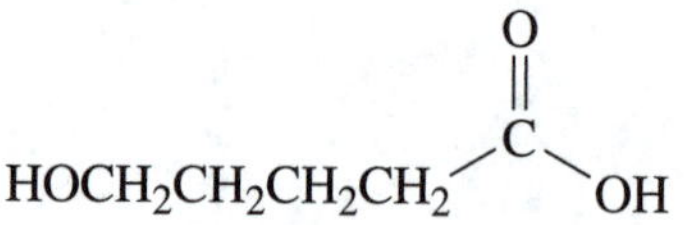

The mechanism for the hydrolysis of a cyclic ester is the same as that for an acyclic ester.

b. 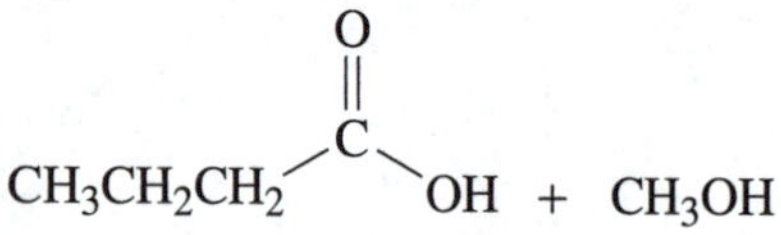 CH₃CH₂CH₂—C(=O)—OH + CH₃OH

14. The mechanism for the acid-catalyzed reaction of acetic acid and methanol is the exact reverse of the mechanism for the acid-catalyzed hydrolysis of methyl acetate.

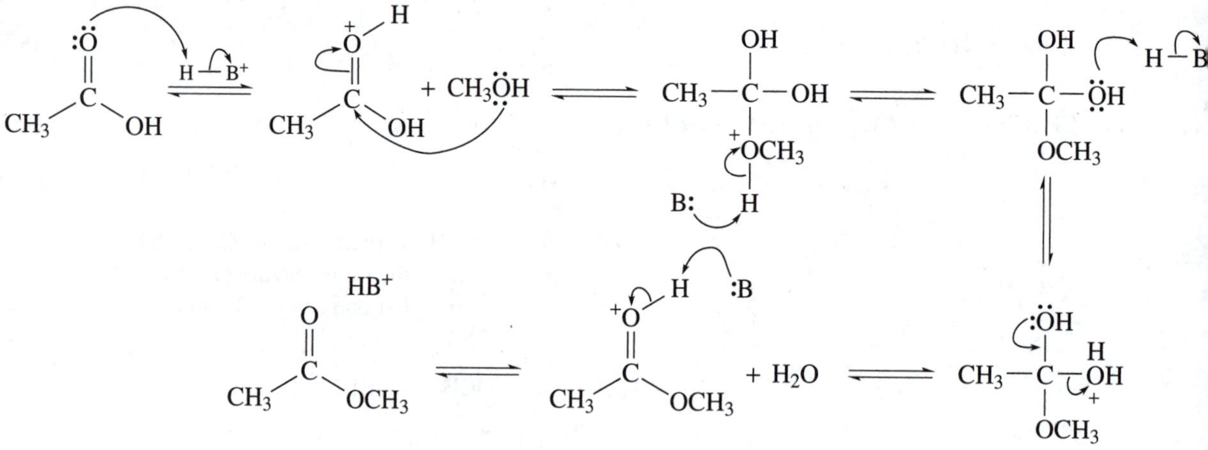

CH₃—C(=O)—OH + H—B⁺ ⇌ CH₃—C(=⁺OH)—OH + CH₃ÖH ⇌ CH₃—C(OH)(OH)(⁺OCH₃) ⇌ CH₃—C(OH)(ÖH)(OCH₃) + H—B

B: H

CH₃—C(=O)—OCH₃ ⇌ CH₃—C(=⁺O)—OCH₃ + H₂O ⇌ CH₃—C(ÖH)(OH)(⁺OCH₃)(H)

HB⁺   :B

15. **a.** Any species in the solution with an acidic proton can be represented by HB⁺.

H₃O⁺   CH₃ÖH₂   CH₃—C(=O)—OH   CH₃—C(=⁺OH)—OH   CH₃C(OH)(OCH₃)(⁺OH H)   CH₃C(OH)(⁺OCH₃ H)

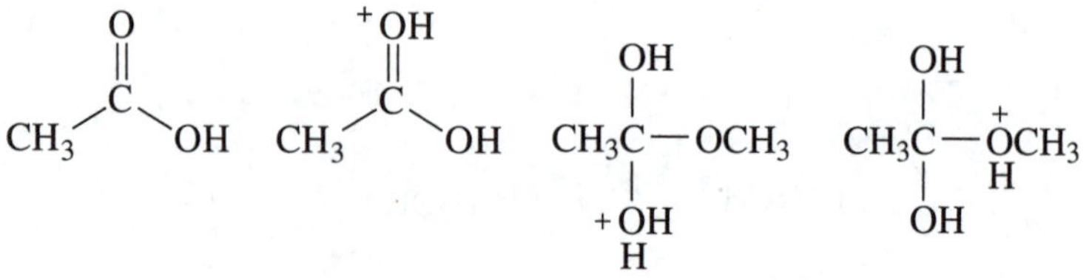

b. Any species in the solution with a lone pair can be represented by :B.

$$H_2O \quad CH_3OH \quad CH_3\overset{\displaystyle O}{\overset{\|}{C}}OH \quad CH_3\underset{\underset{OH}{|}}{\overset{\overset{OH}{|}}{C}}-OCH_3$$

c. H_3O^+

d. H_3O^+ if excess water is used, $CH_3\overset{+}{O}H_2$ if excess water is not used.

6. These are transesterification reactions.

a. $CH_3CH_2CH_2\overset{\displaystyle O}{\overset{\|}{C}}OCH_2CH_3 + CH_3\underset{\underset{OH}{|}}{C}HCH_3 \;\overset{HCl}{\rightleftharpoons}\; CH_3CH_2CH_2\overset{\displaystyle O}{\overset{\|}{C}}O\underset{\underset{CH_3}{|}}{C}HCH_3 + CH_3CH_2OH$

b. $CH_3\overset{\displaystyle O}{\overset{\|}{C}}O-\text{C}_6\text{H}_5 + CH_3CH_2OH \;\overset{HCl}{\rightleftharpoons}\; CH_3\overset{\displaystyle O}{\overset{\|}{C}}OCH_2CH_3 + \text{C}_6\text{H}_5\text{OH}$

17.

18.

a. The conjugate base ($CH_3CH_2CH_2O^-$) of the reactant alcohol ($CH_3CH_2CH_2OH$) can be used to increase the rate of the reaction, because it is the nucleophile that we want to become attached to the carbonyl carbon.

b. If H^+ is used as a catalyst, the amine will be protonated in the acidic solution and therefore will not be able to react as a nucleophile.

If HO^- is used as a catalyst, HO^- will be the best nucleophile in the solution, so it will add to the ester, and the product of the reaction will be a carboxylate ion rather than an amide.

If RO^- is used as a catalyst, RO^- will be the nucleophile, and the product of the reaction will be an ester rather than an amide.

19.   Solved in the text.

20.   a.   CH$_3$CH$_2$CH$_2$—C(=O)—O$^-$ $\xrightarrow{\text{CH}_3\text{Br}}$ CH$_3$CH$_2$CH$_2$—C(=O)—OCH$_3$

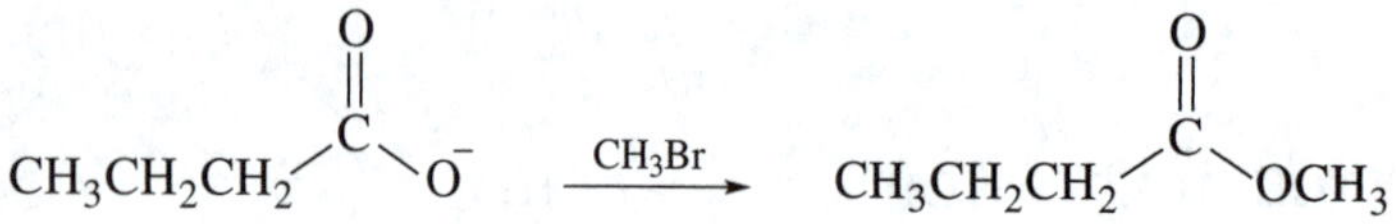

or

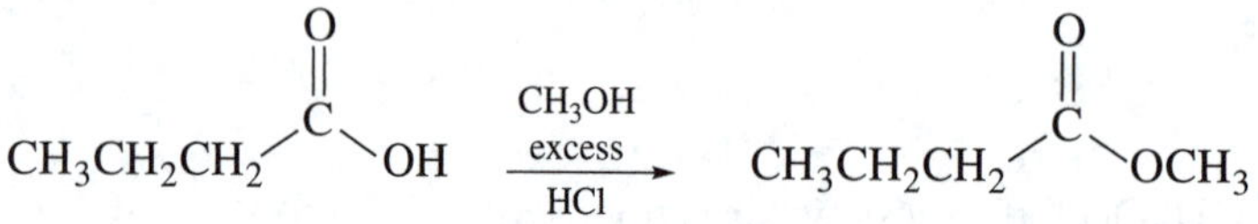

CH$_3$CH$_2$CH$_2$—C(=O)—OH $\xrightarrow[\text{HCl}]{\overset{\text{CH}_3\text{OH}}{\text{excess}}}$ CH$_3$CH$_2$CH$_2$—C(=O)—OCH$_3$

b.   CH$_3$—C(=O)—O$^-$ $\xrightarrow{\text{CH}_3(\text{CH}_2)_7\text{Br}}$ CH$_3$—C(=O)—O(CH$_2$)$_7$CH$_3$

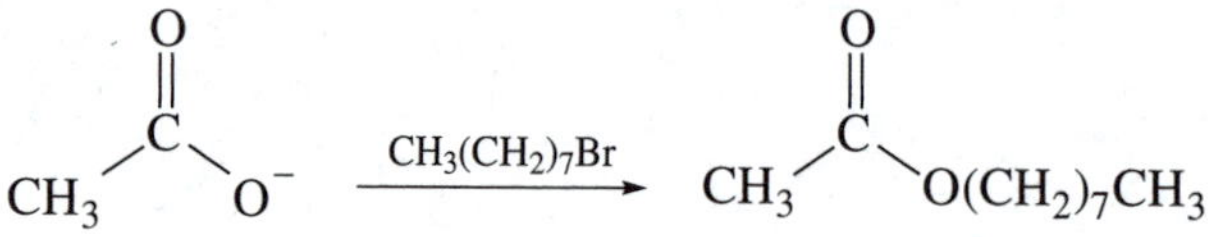

or

CH$_3$—C(=O)—OH $\xrightarrow[\text{HCl}]{\overset{\text{CH}_3(\text{CH}_2)_7\text{OH}}{\text{excess}}}$ CH$_3$—C(=O)—O(CH$_2$)$_7$CH$_3$

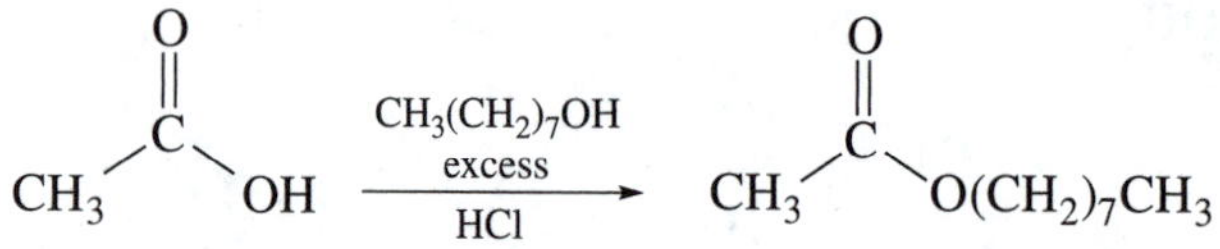

21.   Only B and E will form amides.

22.

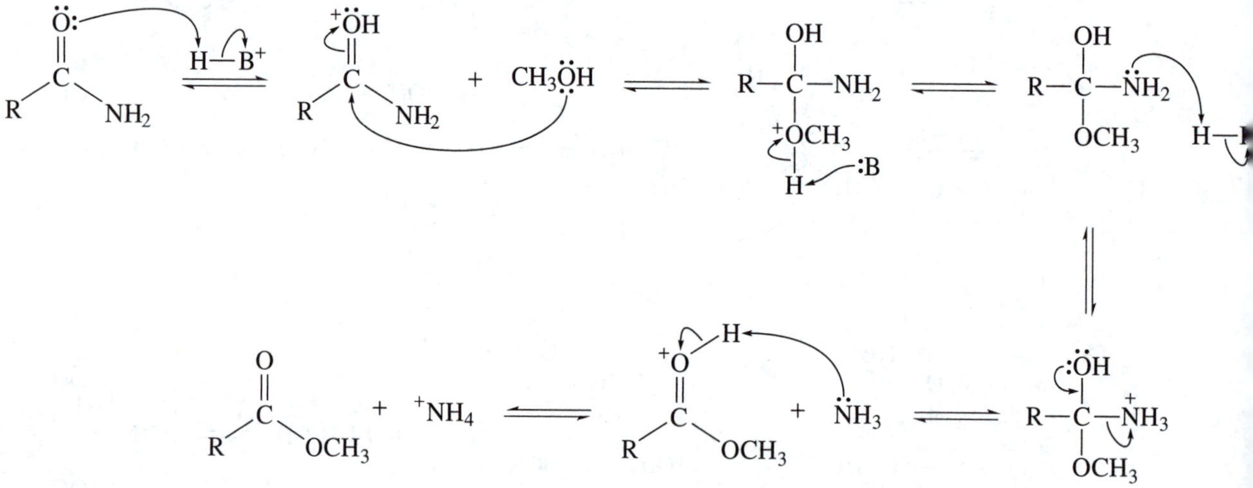

23.   The relative reactivities of the amides depend on the basicities of their leaving groups: the weaker the base, the more reactive the amide. (*para*-Nitroanilinium ion has a pK_a value of 0.98, and the anilinium ion has a pK_a value of 4.58.)

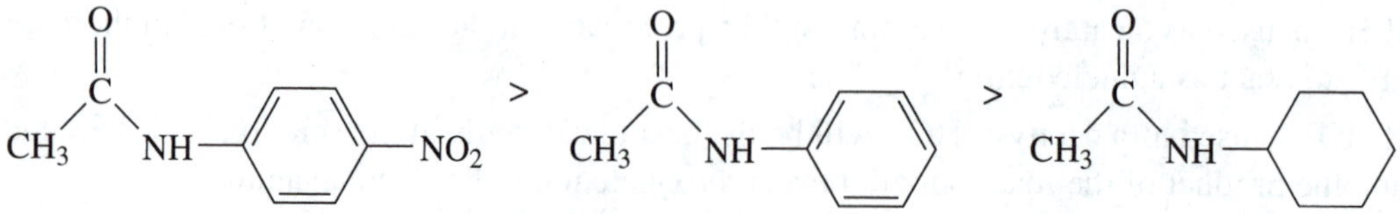

24. **a.** butanenitrile
           propyl cyanide

b. 4-methylpentanenitrile
        isopentyl cyanide

25. Notice that the alkyl halide has one fewer carbon than the target carboxylic acid, because the alkyl halide will obtain a carbon from the cyanide ion.

a. $CH_3CH_2CH_2Br$

b. 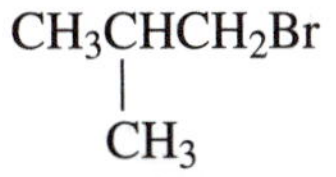 CH_3CHCH_2Br
              |
              CH_3

c. $CH_3CH_2CH_2CH_2CH_2Br$

26. **a.**

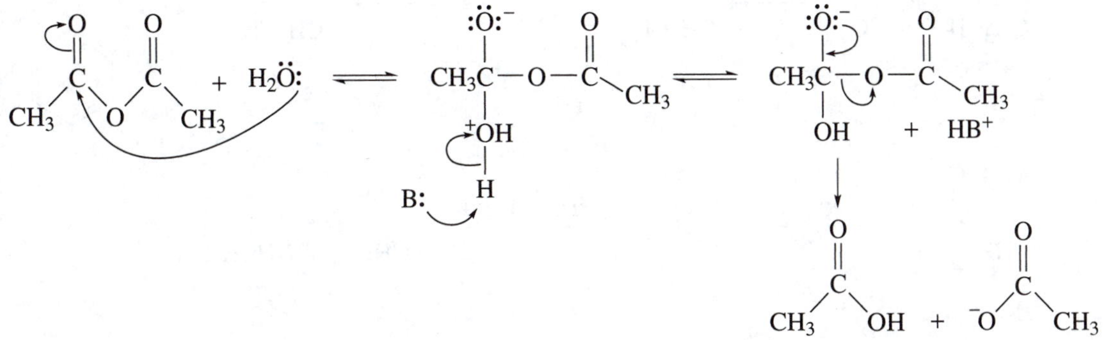

b. The mechanisms are exactly the same.

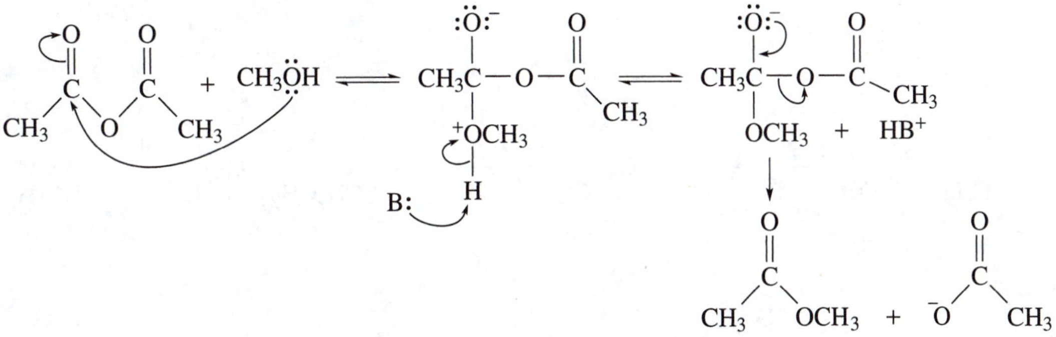

27.

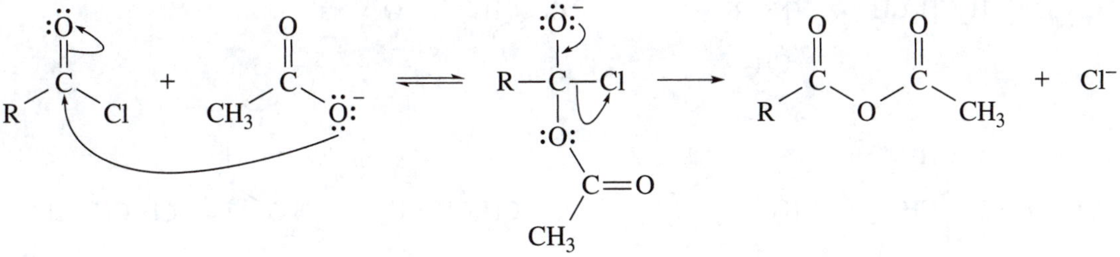

28. **a.**

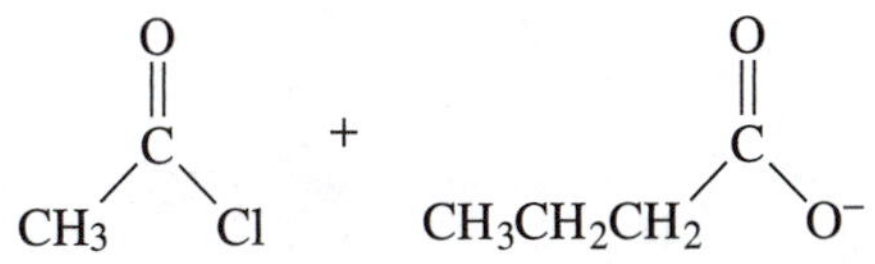

b.

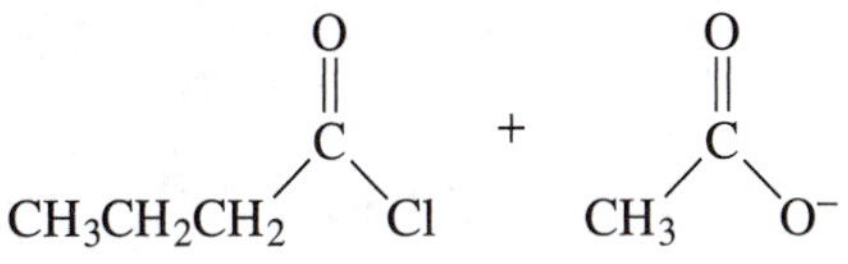

29.   **a.**

$$\underset{CH_3CH_2CH_2}{\overset{\displaystyle O}{\overset{\|}{C}}}\text{—Cl} \;+\; CH_3CH_2NH_2$$

b.

$$\underset{CH_3}{\overset{\displaystyle O}{\overset{\|}{C}}}\text{—Cl} \;+\; \underset{\overset{|}{CH_3}}{CH_3NH}$$

30.   **a.**

$$\underset{CH_3CH_2}{\overset{\displaystyle O}{\overset{\|}{C}}}\text{—O}^- \;\xrightarrow[\Delta]{PCl_3}\; \underset{CH_3CH_2}{\overset{\displaystyle O}{\overset{\|}{C}}}\text{—Cl} \;\xrightarrow{\text{C}_6\text{H}_5\text{—OH}}\; \underset{CH_3CH_2}{\overset{\displaystyle O}{\overset{\|}{C}}}\text{—O—C}_6\text{H}_5$$

b.

$$\underset{CH_3}{\overset{\displaystyle O}{\overset{\|}{C}}}\text{—O}^- \;\xrightarrow[\Delta]{PCl_3}\; \underset{CH_3}{\overset{\displaystyle O}{\overset{\|}{C}}}\text{—Cl} \;\xrightarrow{2\;CH_3CH_2NH_2}\; \underset{CH_3}{\overset{\displaystyle O}{\overset{\|}{C}}}\text{—NHCH}_2\text{CH}_3$$

c.

$$\underset{CH_3CH_2}{\overset{\displaystyle O}{\overset{\|}{C}}}\text{—O}^- \;\xrightarrow[\Delta]{PCl_3}\; \underset{CH_3CH_2}{\overset{\displaystyle O}{\overset{\|}{C}}}\text{—Cl} \;\xrightarrow{\underset{CH_3}{\overset{O}{\overset{\|}{C}}}\text{—O}^-}\; \underset{CH_3CH_2}{\overset{\displaystyle O}{\overset{\|}{C}}}\text{—O—}\underset{CH_3}{\overset{\displaystyle O}{\overset{\|}{C}}}$$

or

$$\underset{CH_3}{\overset{\displaystyle O}{\overset{\|}{C}}}\text{—O}^- \;\xrightarrow[\Delta]{PCl_3}\; \underset{CH_3}{\overset{\displaystyle O}{\overset{\|}{C}}}\text{—Cl} \;\xrightarrow{\underset{CH_3CH_2}{\overset{O}{\overset{\|}{C}}}\text{—O}^-}\; \underset{CH_3CH_2}{\overset{\displaystyle O}{\overset{\|}{C}}}\text{—O—}\underset{CH_2CH_3}{\overset{\displaystyle O}{\overset{\|}{C}}}$$

31.   **a.** $\underset{\overset{|}{CH_3}}{CH_3CH_2CH_2CH_2CH_2\overset{\displaystyle O}{\overset{\|}{C}}\text{—NCH}_3}$

d. $CH_3\overset{\displaystyle O}{\overset{\|}{C}}\text{—O}^-\;Na^+$

b. $\underset{\overset{|}{CH_3}}{CH_3CH_2CH_2\overset{\overset{\textstyle CH_3}{|}}{C}CH_2\overset{\displaystyle O}{\overset{\|}{C}}\text{—NH}_2}$

e. $CH_3CH_2CH_2\overset{\displaystyle O}{\overset{\|}{C}}\text{—O—}\overset{\displaystyle O}{\overset{\|}{C}}CH_2CH_2CH_3$

c. $CH_3CH_2\overset{\displaystyle O}{\overset{\|}{C}}\text{—NH}_2$

f.   (CH₃)₂CHCH₂C≡N

2. a. pentanoyl chloride
        valeryl chloride

    b. *N,N*-dimethylbutanamide
        N,N-dimethylbutyramide

    c. propyl propanoate
        propyl propionate

    d. propanoic anhydride
        propionic anhydride

    e. 5-ethylheptanoic acid

    f. (*S*)-3-methylpentanoic acid
        (*S*)-β-methylvaleric acid

3. The alcohol component of phenyl acetate is a stronger acid (pK_a = 10.0) than is the alcohol component of methyl acetate (pK_a = 15.5). Therefore, the phenoxide ion is a weaker base than the methoxide ion, causing phenyl acetate to be the more reactive ester.

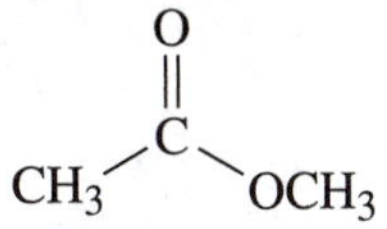

methyl acetate 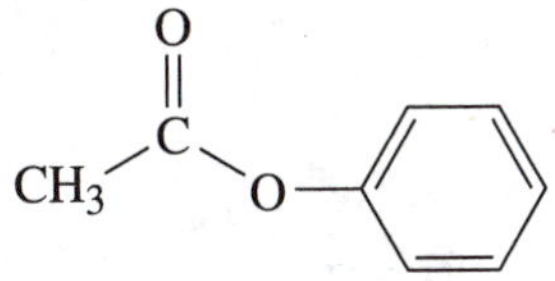   phenyl acetate

CH_3OH
pK_a = 15.5

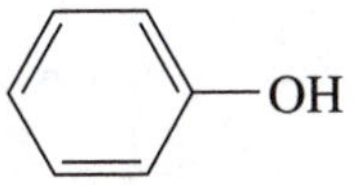

—OH

pK_a = 10.0

4. a.

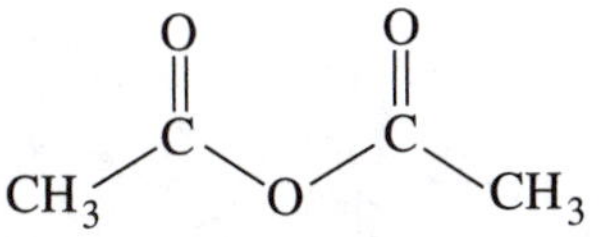

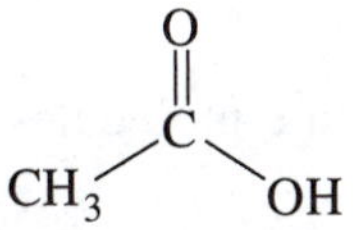

    b.

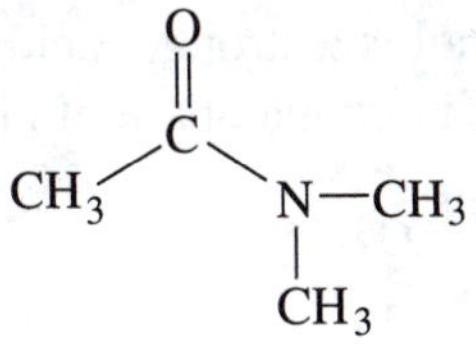

    c.

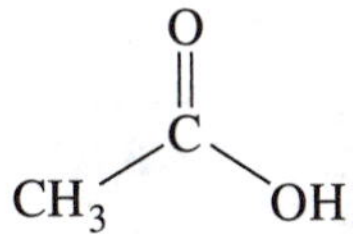

    d.

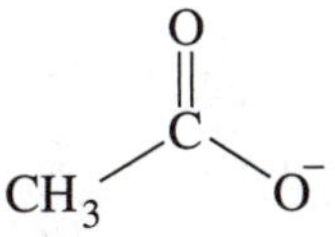

        A carboxylic acid will
        be in its basic form in
        a basic solution.

    e.

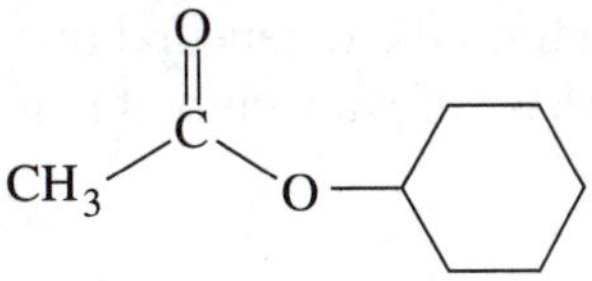

    f.

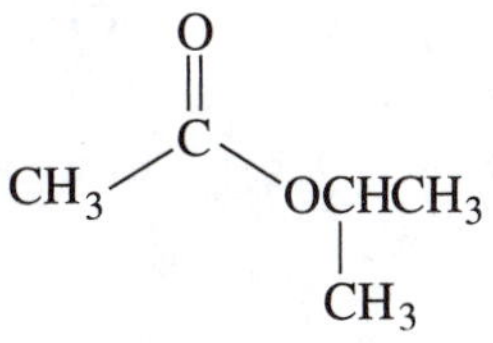

35. **a.** 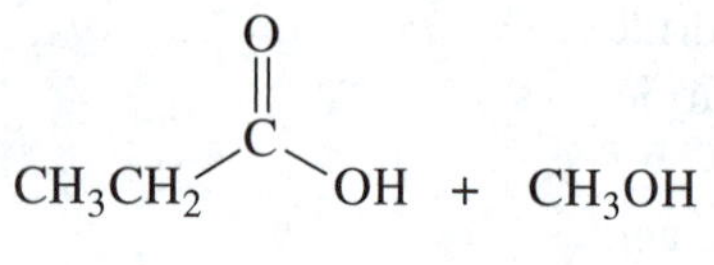 CH₃CH₂—C(=O)—OH + CH₃OH

c. —C(=O)—OH + CH₃CH₂NH₃⁺ Cl⁻

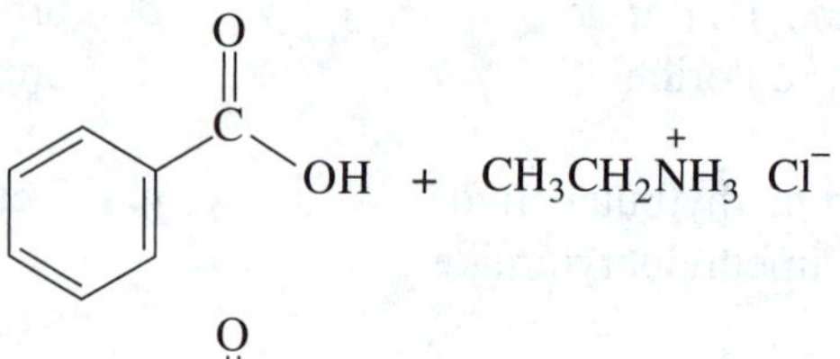

b. 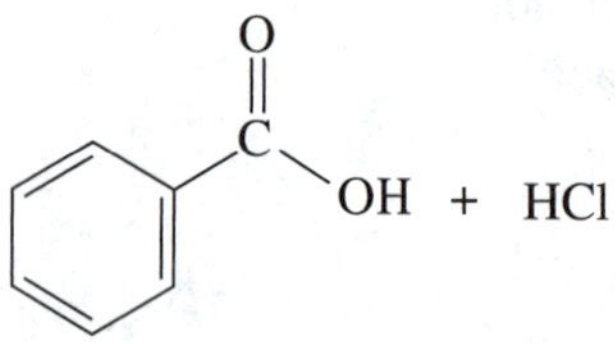 —C(=O)—OH + HCl

d. 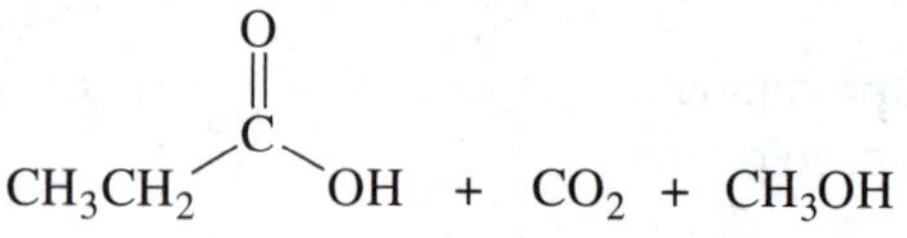 CH₃CH₂—C(=O)—OH + CO₂ + CH₃OH

36. The reaction of methylamine with propionyl chloride generates a proton that will protonate unreacte amine, thereby destroying the amine's nucleophilicity. If two equivalents of CH₃NH₂ are used, one equiv alent will remain unprotonated and be able to react as a nucleophile with propionyl chloride to for N-methylpropanamide.

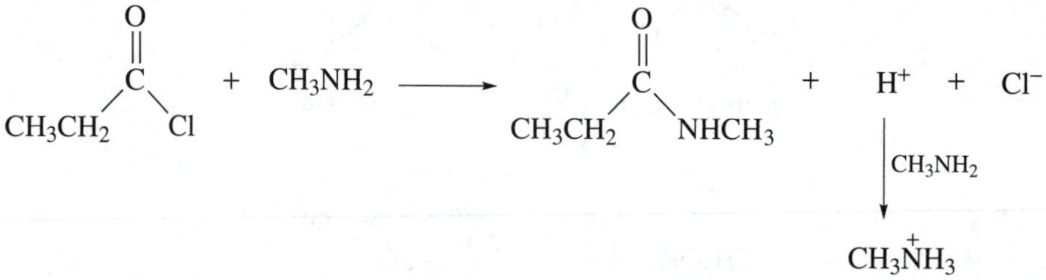

37. **a.** Methyl acetate has a resonance contributor that butanone does not have. This resonance contributo causes methyl acetate to be more polar than butanone. Because methyl acetate is more polar, it has th greater dipole moment.

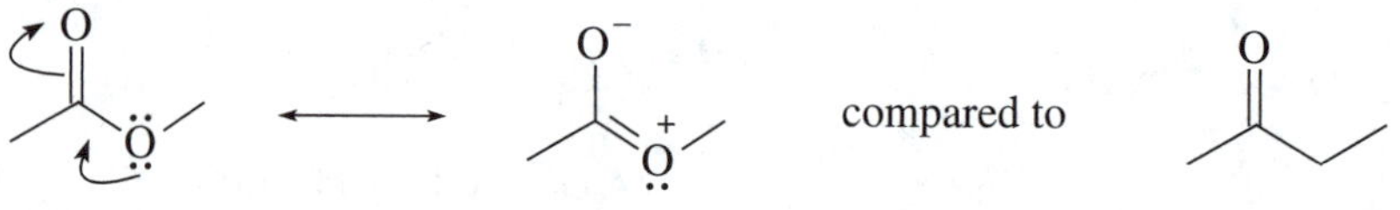

compared to

b. Because it is more polar, the intermolecular forces holding methyl acetate molecules together ar stronger, so we would expect methyl acetate to have a higher boiling point.

38. **a.** The weaker the base attached to the acyl group, the stronger its electron-withdrawing ability; therefore the easier it is to form the tetrahedral intermediate. *para*-Chlorophenol is a stronger acid than phenol so the conjugate base of *para*-chlorophenol is a weaker base than the conjugate base of phenol.

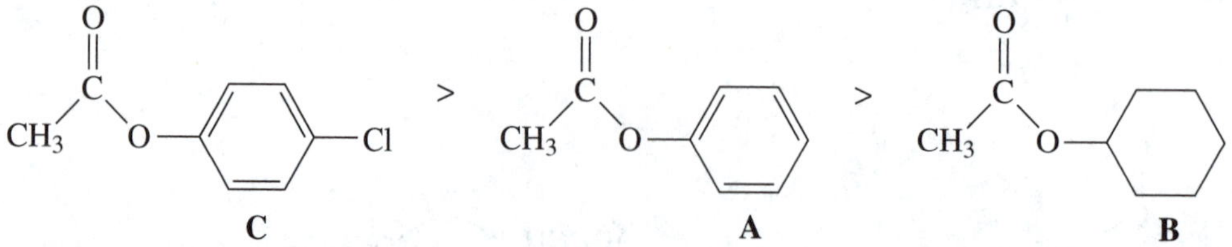

C   >   A   >   B

b. The tetrahedral intermediate collapses by eliminating the OR group. The weaker the OR group is as a base, the easier it is to eliminate it.

Thus, the rate of both formation of the tetrahedral intermediate and collapse of the tetrahedral intermediate is decreased by increasing the basicity of the OR group.

39.

a. The alcohol (CH_3CH_2OH) contained the ^{18}O label.

b. The carboxylic acid would have contained the ^{18}O label.

40.

a.

$CH_3CH_2 \overset{O}{\underset{}{\text{C}}} Cl + H_2O \longrightarrow CH_3CH_2 \overset{O}{\underset{}{\text{C}}} OH + HCl$

b.

$CH_3CH_2CH_2 \overset{O}{\underset{}{\text{C}}} OCH_2CH_3 + CH_3CH_2CH_2OH \xrightarrow{HCl} CH_3CH_2CH_2 \overset{O}{\underset{}{\text{C}}} OCH_2CH_2CH_3 + CH_3CH_2OH$

c.

$CH_3CH_2CH_2CH_2 \overset{O}{\underset{}{\text{C}}} OCH_2CH_3 + CH_3NH_2 \longrightarrow CH_3CH_2CH_2CH_2 \overset{O}{\underset{}{\text{C}}} NHCH_3 + CH_3CH_2OH$

41.

a.

$CH_3 \overset{O}{\underset{}{\text{C}}} O^- \xrightarrow[\Delta]{PCl_3} CH_3 \overset{O}{\underset{}{\text{C}}} Cl \xrightarrow{CH_3CH_2CH_2OH} CH_3 \overset{O}{\underset{}{\text{C}}} OCH_2CH_2CH_3$

or

$CH_3 \overset{O}{\underset{}{\text{C}}} O^- \xrightarrow{CH_3CH_2CH_2Br} CH_3 \overset{O}{\underset{}{\text{C}}} OCH_2CH_2CH_3$

b.

$CH_3 \overset{O}{\underset{}{\text{C}}} O^- \xrightarrow[\Delta]{PCl_3} CH_3 \overset{O}{\underset{}{\text{C}}} Cl \xrightarrow{(CH_3)_2CH(CH_2)_2OH} CH_3 \overset{O}{\underset{}{\text{C}}} O(CH_2)_2CH(CH_3)_2$

or

$CH_3 \overset{O}{\underset{}{\text{C}}} O^- \xrightarrow{(CH_3)_2CH(CH_2)_2Br} CH_3 \overset{O}{\underset{}{\text{C}}} O(CH_2)_2CH(CH_3)_2$

c.

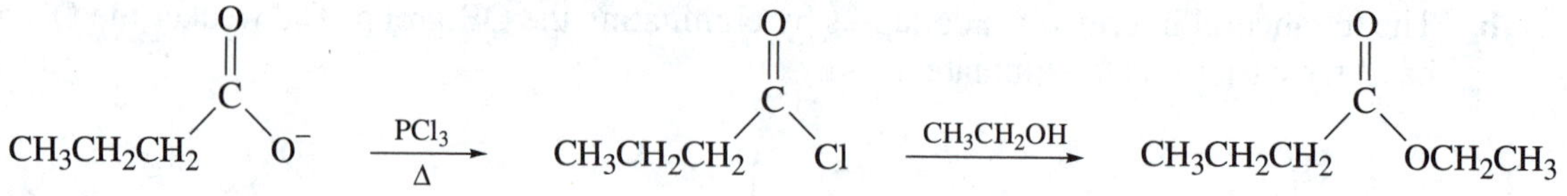

or

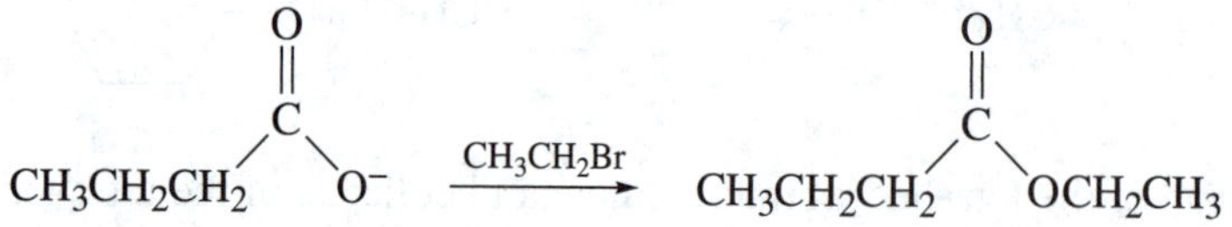

d.

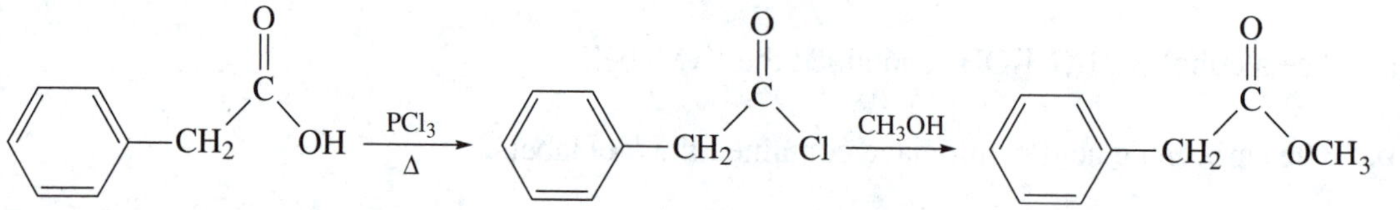

or

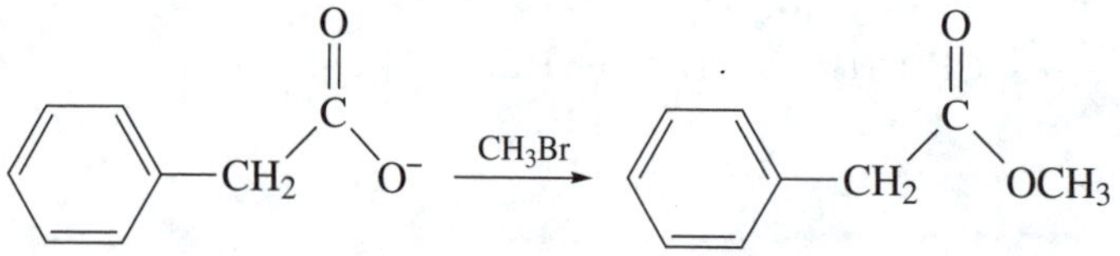

42.   **a.**   isopropyl alcohol and HCl   **c.**   ethylamine

       b.   aqueous sodium hydroxide   **d.**   water and HCl

43.   In the tetrahedral intermediate, the potential leaving groups will be a chloride ion and an alkoxide ion; the chloride ion is the one that will be eliminated. Therefore, it is easier for the intermediate to collapse back to starting materials, reforming the ester, than to eliminate the alkoxide ion and form an acyl chloride. This means that **a** represents the reaction coordinate diagram for the reaction.

44.   Aspartame has an amide group and an ester group that will be hydrolyzed in an aqueous solution of HCl. Because the hydrolysis is carried out in an acidic solution, the carboxylic acid groups and the amino groups in the products will be in their acidic forms.

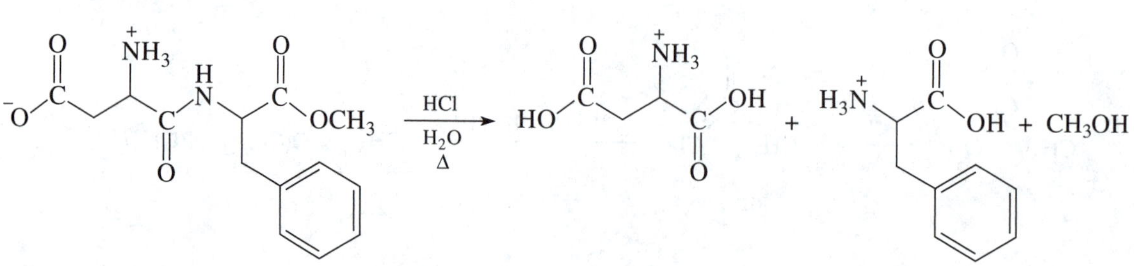

5.   If the amine is tertiary, the nitrogen in the amide product cannot get rid of its positive charge by losing a proton. An amide with a positively charged nitrogen is very reactive toward addition–elimination, because the positively charged nitrogen-containing group is an excellent leaving group. Thus, if water is added, it will immediately react with the amide and, because the $^+NR_3$ group is a better leaving group than the OH group, the $^+NR_3$ group will be eliminated. The product will be a carboxylic acid, which will lose its acidic proton to the amine.

46.   The acid-catalyzed hydrolysis of acetamide forms acetic acid and ammonium ion. It is an irreversible reaction, because the pK_a of an acetic acid is less than the pK_a of the ammonium ion. Therefore, it is impossible to have the carboxylic acid in its reactive acidic form and ammonia in its reactive basic form.

If the solution is sufficiently acidic to have the carboxylic acid in its acidic form, ammonia will also be in its acidic form, so it will not be a nucleophile.

no reaction
The ammonium ion is not a nucleophile.

If the pH of the solution is sufficiently basic to have ammonia in its nucleophilic basic form, the carboxylic acid will also be in its basic form; a negatively charged carboxylate ion cannot be attacked by nucleophiles.

no reaction
The carboxylate ion is not attacked by nucleophiles.

47.   The acyl chloride is formed in the first step, and then the second carboxylate group reacts with the acyl chloride to form the final product—an anhydride.

48. **a.** 1, 3, 4, 6, 7, and 9 will not form the indicated products under the given conditions.

b. 9 will form the product shown in the presence of an acid catalyst.

49. **a.**

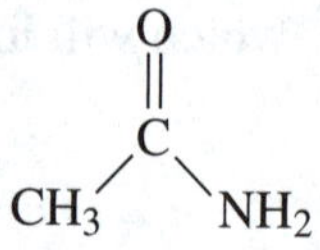

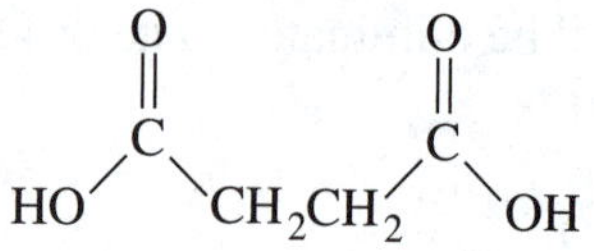

b.

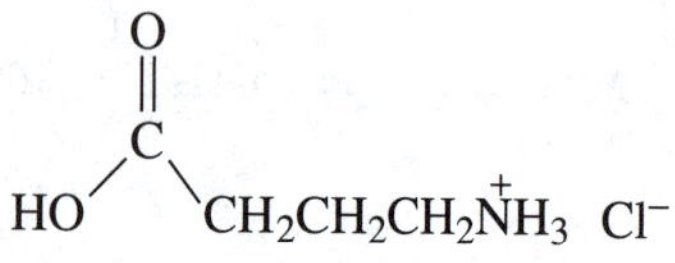

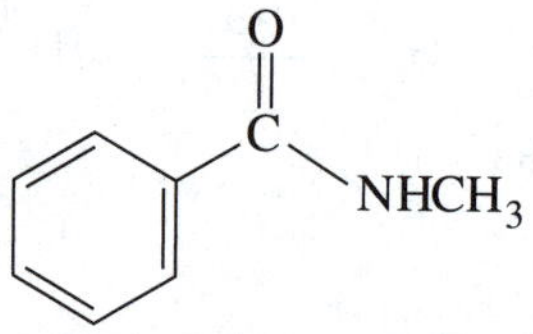

c.

d.

e.

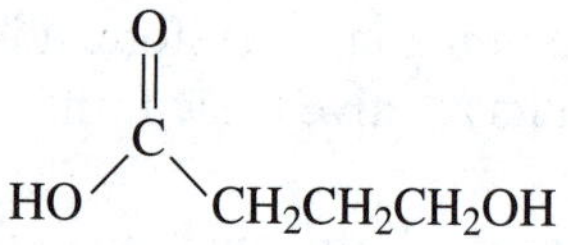

f.

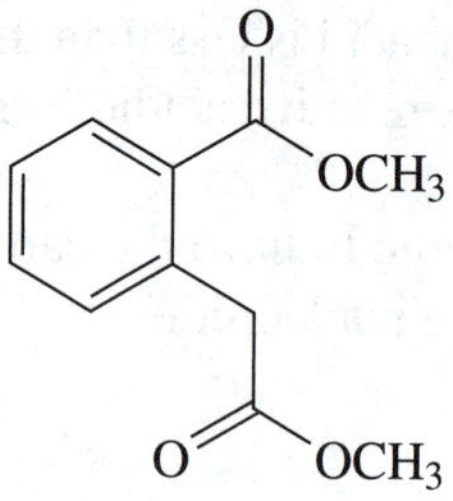

50. **a.**

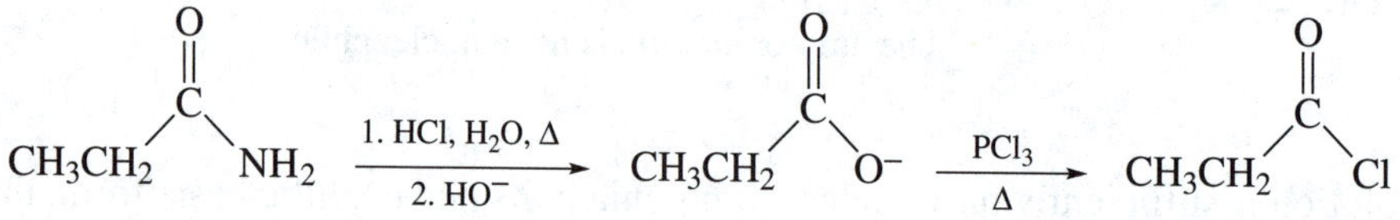

b.

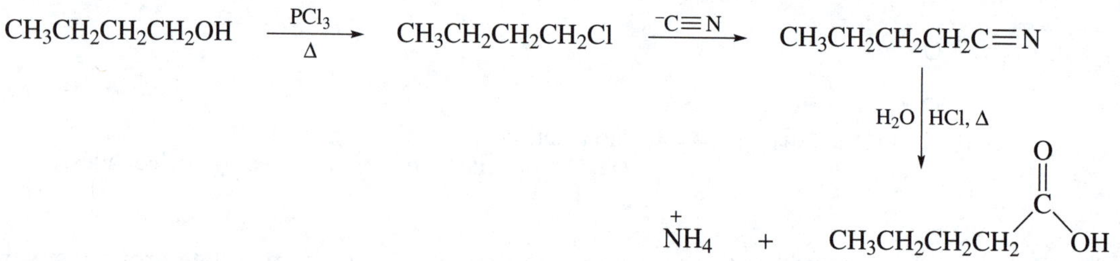

51. **a.**

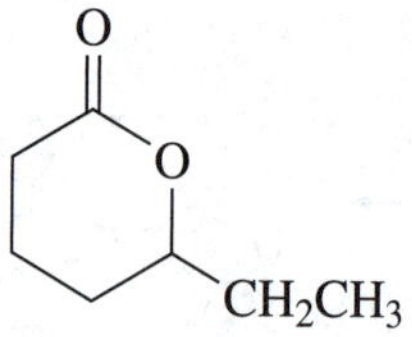

b.

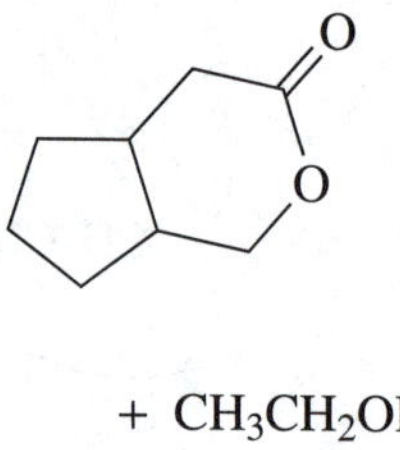

Notice that the reacting groups have to be cis in order to form the new ring.

$+ CH_3CH_2OH$

2.   **a.**

b.   The carboxyl oxygen will be labeled. Only one isotopically labeled oxygen can be incorporated into the ester, because the bond between the methyl group and the labeled oxygen does not break, so there is no way for the carbonyl oxygen to become labeled.

53.   This is the reaction of an acyl chloride with an alcohol.

Chapter 11 Practice Test

1. Circle the compound in each pair that is more reactive in a nucleophilic acyl substitution reaction.

a. 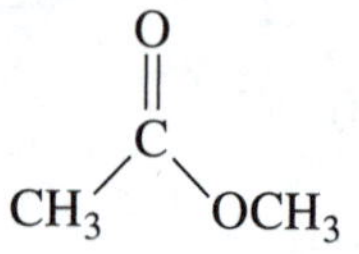   or

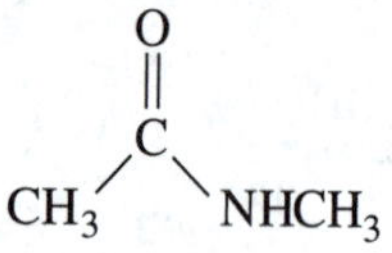

b. 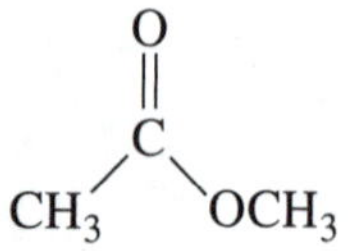   or

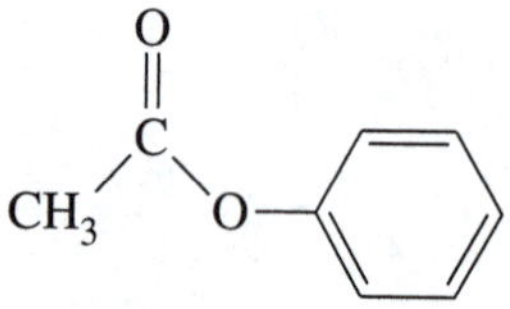

c. 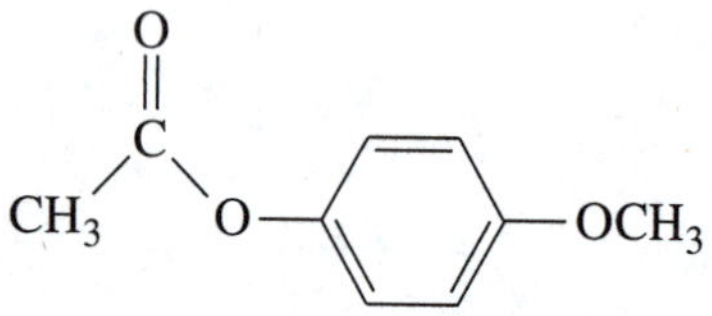   or

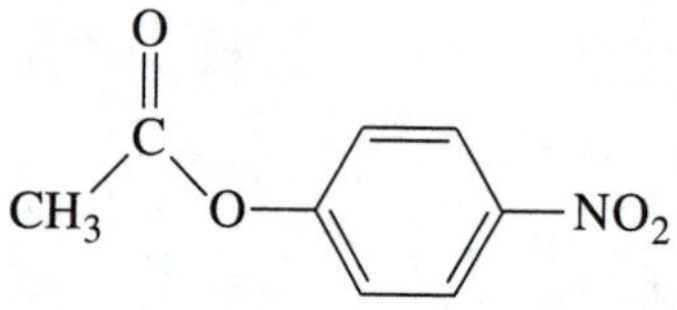

d. 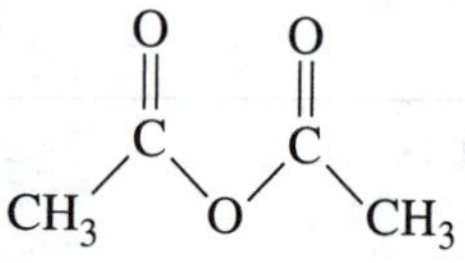   or

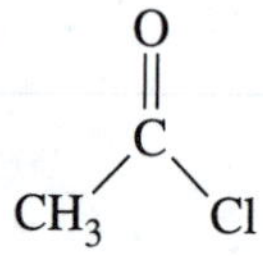

2. What is each compound's systematic name?

a.

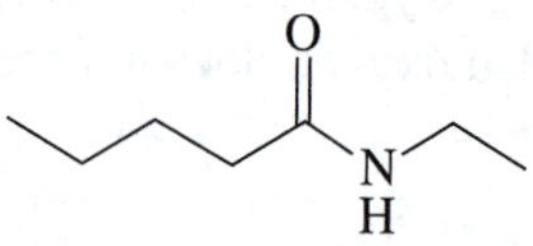

b.

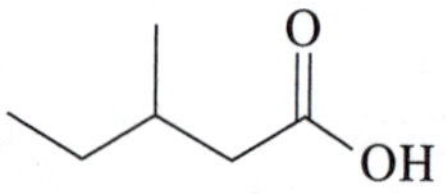

c. 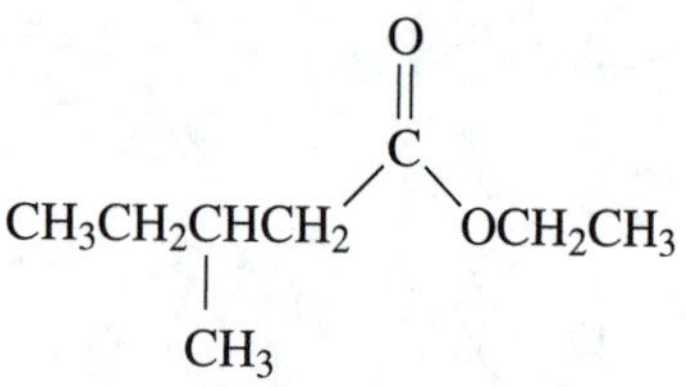 $CH_3CH_2CHCH_2$ $\overset{O}{\underset{}{C}}$ OCH_2CH_3
      |
      CH_3

d.

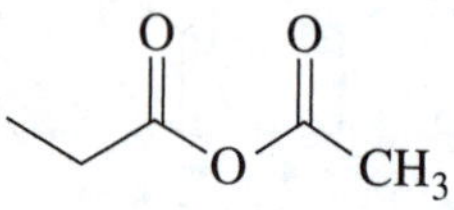

3. Give an example of each of the following:

   a. a symmetrical anhydride

   b. a hydrolysis reaction

   c. a transesterification reaction

   d. aminolysis of an ester

4. What carbonyl compound would be obtained from collapse of each of the following tetrahedral intermediates?

   a. 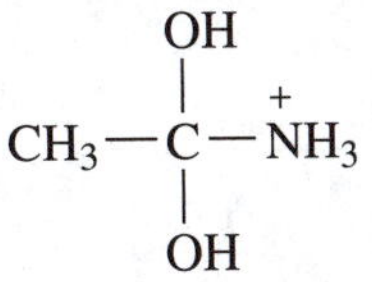
$$CH_3-\overset{\overset{\displaystyle OH}{|}}{\underset{\underset{\displaystyle OH}{|}}{C}}-\overset{+}{N}H_3$$

   c.
$$CH_3-\overset{\overset{\displaystyle \overset{-}{O}}{|}}{\underset{\underset{\displaystyle OH}{|}}{C}}-NH_2$$

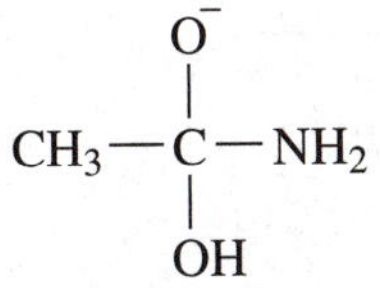

   b. 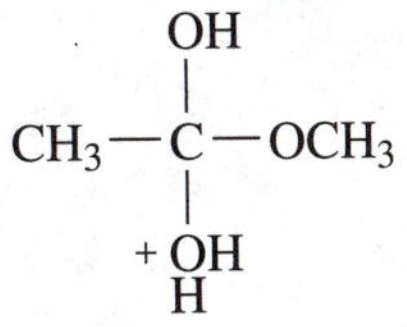
$$CH_3-\overset{\overset{\displaystyle OH}{|}}{\underset{\underset{\displaystyle \overset{+}{O}H}{\underset{\displaystyle H}{|}}}{C}}-OCH_3$$

   d. 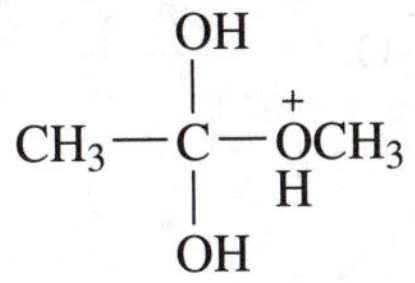
$$CH_3-\overset{\overset{\displaystyle OH}{|}}{\underset{\underset{\displaystyle OH}{|}}{C}}-\overset{+}{\underset{\underset{\displaystyle H}{}}{O}}CH_3$$

5. What is the product of each of the following reactions?

   a. $CH_3CH_2C\equiv N \; + \; H_2O \; \xrightarrow[\Delta]{HCl}$

   b. 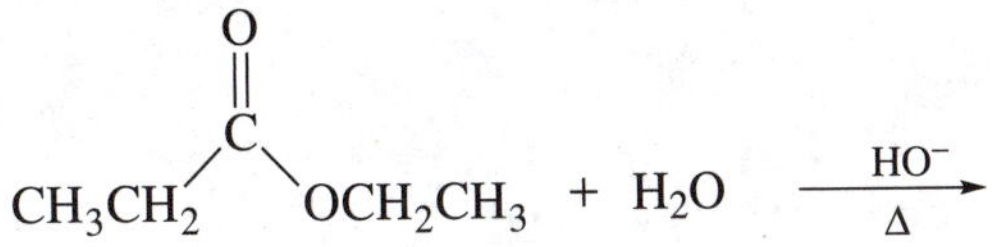 $CH_3CH_2\overset{\overset{\displaystyle O}{\|}}{C}\diagdown OCH_2CH_3 \; + \; H_2O \; \xrightarrow[\Delta]{HO^-}$

   c. $CH_3CH_2CH_2\overset{\overset{\displaystyle O}{\|}}{C}\diagdown O \diagup \overset{\overset{\displaystyle O}{\|}}{C}\diagdown CH_2CH_3 \; + \; H_2O \; \longrightarrow$

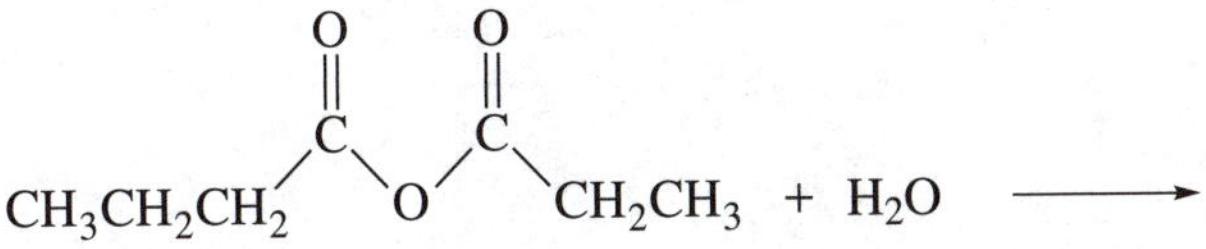

   d. 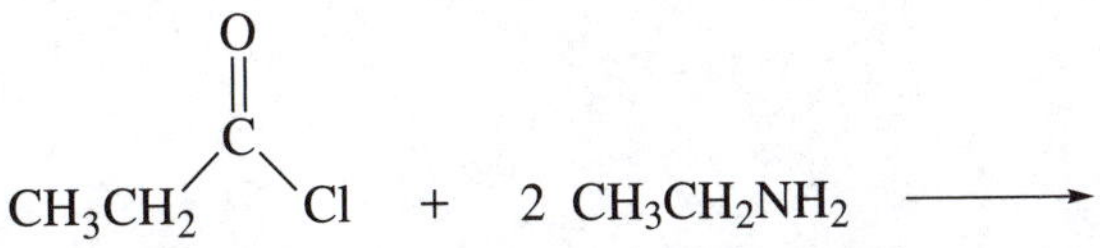 $CH_3CH_2\overset{\overset{\displaystyle O}{\|}}{C}\diagdown Cl \; + \; 2\;CH_3CH_2NH_2 \; \longrightarrow$

   e. 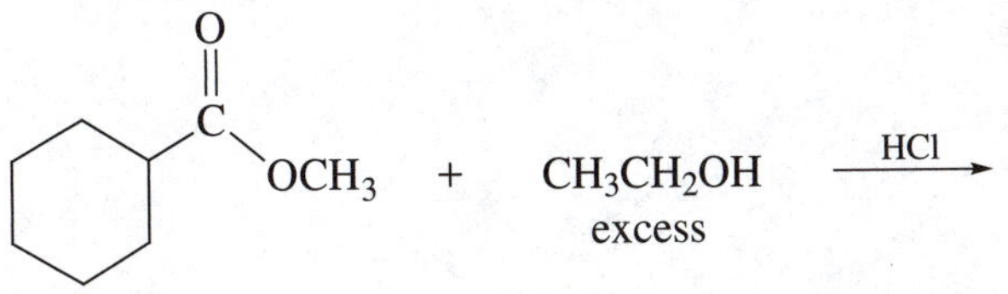 cyclohexyl$-\overset{\overset{\displaystyle O}{\|}}{C}\diagdown OCH_3 \; + \; CH_3CH_2OH \; \underset{\text{excess}}{\xrightarrow{HCl}}$

6. Show how the following compounds can be prepared from the given starting material.

   a. $CH_3CH_2CH_2Br \; \longrightarrow \; CH_3CH_2CH_2CH_2NH_2$

   b. 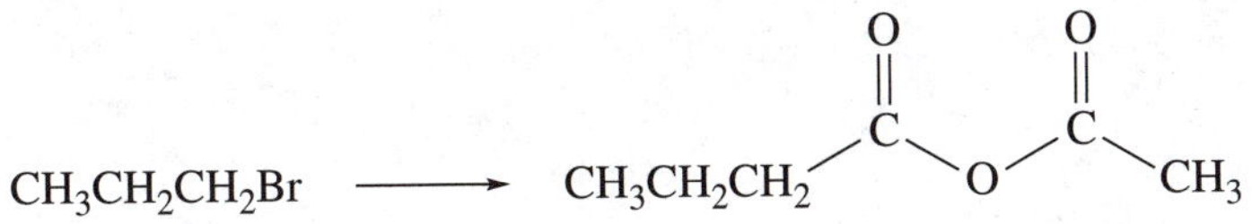 $CH_3CH_2CH_2Br \; \longrightarrow \; CH_3CH_2CH_2\overset{\overset{\displaystyle O}{\|}}{C}\diagdown O \diagup \overset{\overset{\displaystyle O}{\|}}{C}\diagdown CH_3$

7.   What is the product of each of the following reactions?

a. 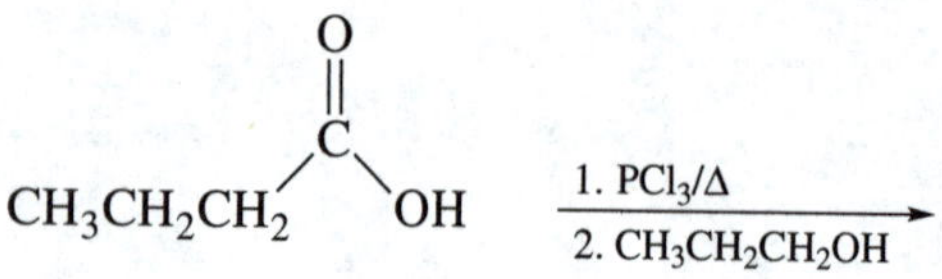 $CH_3CH_2CH_2$—C(=O)—OH $\xrightarrow[\text{2. CH}_3\text{CH}_2\text{CH}_2\text{OH}]{\text{1. PCl}_3/\Delta}$

b. CH_3CH_2—C(=O)—NHCH$_3$ + H_2O $\xrightarrow{\text{HCl}}{\Delta}$

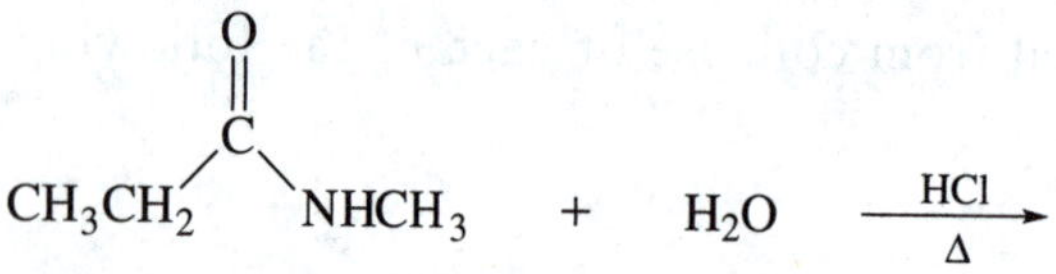

c. CH_3CH_2—C(=O)—O—C$_6$H$_5$ + H_2O excess $\xrightarrow{\text{HCl}}$

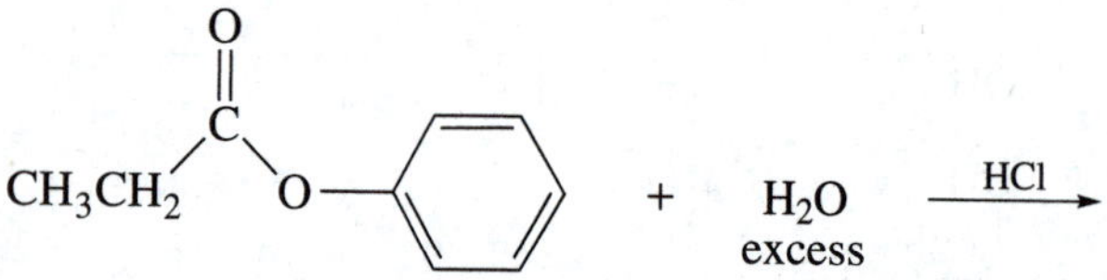

d. 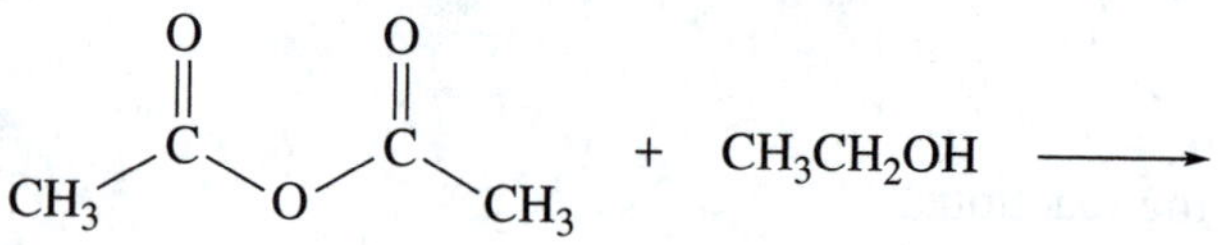 CH_3—C(=O)—O—C(=O)—CH$_3$ + CH_3CH_2OH $\longrightarrow$

Answers to Chapter 11 Practice Test

1.

a. $CH_3-\overset{\overset{\displaystyle O}{\|}}{C}-OCH_3$ **b.** $CH_3-\overset{\overset{\displaystyle O}{\|}}{C}-O-C_6H_5$ **c.** $CH_3-\overset{\overset{\displaystyle O}{\|}}{C}-O-C_6H_4-NO_2$ **d.** $CH_3-\overset{\overset{\displaystyle O}{\|}}{C}-Cl$

2.

a. *N*-ethylpentanamide

b. 3-methylpentanoic acid

c. ethyl 3-methylpentanoate

d. ethanoic propanoic anhydride

3.

a. $CH_3CH_2-\overset{\overset{\displaystyle O}{\|}}{C}-O-\overset{\overset{\displaystyle O}{\|}}{C}-CH_2CH_3$

b. $CH_3-\overset{\overset{\displaystyle O}{\|}}{C}-Cl$ + H_2O ⟶ $CH_3-\overset{\overset{\displaystyle O}{\|}}{C}-OH$ + HCl

Any reaction in which a reactant is cleaved as a result of reaction with water.

c. $CH_3-\overset{\overset{\displaystyle O}{\|}}{C}-OCH_3$ + CH_3CH_2OH $\overset{HCl}{\rightleftharpoons}$ $CH_3-\overset{\overset{\displaystyle O}{\|}}{C}-OCH_2CH_3$ + CH_3OH

d. $CH_3-\overset{\overset{\displaystyle O}{\|}}{C}-OCH_3$ + CH_3NH_2 $\overset{\Delta}{\longrightarrow}$ $CH_3-\overset{\overset{\displaystyle O}{\|}}{C}-NHCH_3$ + CH_3OH

4.

a. $CH_3-\overset{\overset{\displaystyle O}{\|}}{C}-OH$ **b.** $CH_3-\overset{\overset{\displaystyle O}{\|}}{C}-OCH_3$ **c.** $CH_3-\overset{\overset{\displaystyle O}{\|}}{C}-NH_2$ **d.** $CH_3-\overset{\overset{\displaystyle O}{\|}}{C}-OH$

5.

a. $CH_3CH_2-\overset{\overset{\displaystyle O}{\|}}{C}-OH$ + $\overset{+}{N}H_4$

d. $CH_3CH_2-\overset{\overset{\displaystyle O}{\|}}{C}-NHCH_2CH_3$ + Cl^- + $CH_3CH_2\overset{+}{N}H_3$

b. $CH_3CH_2-\overset{\overset{\displaystyle O}{\|}}{C}-O^-$ + CH_3CH_2OH

e. $C_6H_{11}-\overset{\overset{\displaystyle O}{\|}}{C}-OCH_2CH_3$ + CH_3OH

c. $CH_3CH_2CH_2-\overset{\overset{\displaystyle O}{\|}}{C}-OH$ + $CH_3CH_2-\overset{\overset{\displaystyle O}{\|}}{C}-OH$

6.   **a.** $CH_3CH_2CH_2Br \xrightarrow{^-C\equiv N} CH_3CH_2CH_2C\equiv N \xrightarrow[\text{Raney Ni}]{H_2} CH_3CH_2CH_2CH_2NH_2$

b. $CH_3CH_2CH_2Br \xrightarrow{^-C\equiv N} CH_3CH_2CH_2C\equiv N \xrightarrow[\Delta]{HCl, H_2O}$

$$CH_3CH_2CH_2\overset{\overset{\displaystyle O}{\|}}{C}{-}OH$$

$$\Big\downarrow \Delta \mid PCl_3$$

$$CH_3CH_2CH_2\overset{\overset{\displaystyle O}{\|}}{C}{-}O{-}\overset{\overset{\displaystyle O}{\|}}{C}{-}CH_3 \longleftarrow CH_3\overset{\overset{\displaystyle O}{\|}}{C}{-}O^- \qquad CH_3CH_2CH_2\overset{\overset{\displaystyle O}{\|}}{C}{-}Cl$$

7.   **a.** $CH_3CH_2CH_2\overset{\overset{\displaystyle O}{\|}}{C}{-}OCH_2CH_2CH_3$

b. $CH_3CH_2\overset{\overset{\displaystyle O}{\|}}{C}{-}OH \;+\; CH_3\overset{+}{N}H_3$

c. $CH_3CH_2\overset{\overset{\displaystyle O}{\|}}{C}{-}OH \;+\; HO{-}\langle\!\!\langle\bigcirc\rangle\!\!\rangle$

d. $CH_3\overset{\overset{\displaystyle O}{\|}}{C}{-}OCH_2CH_3 \;+\; CH_3\overset{\overset{\displaystyle O}{\|}}{C}{-}OH$

Important Terms

acetal	$\underset{\underset{\displaystyle OR}{|}}{\overset{\overset{\displaystyle OR}{|}}{R-C-H}}$ **or** $\underset{\underset{\displaystyle OR}{|}}{\overset{\overset{\displaystyle OR}{|}}{R-C-R}}$

aldehyde	$\underset{R}{\overset{\displaystyle O}{\overset{\|}{C}}}\diagdown_H$

conjugate addition (1,4-addition)	nucleophilic addition to the β-carbon of an α,β-unsaturated carbonyl compound.

cyanohydrin	$\underset{\underset{\displaystyle R'(H)}{|}}{\overset{\overset{\displaystyle OH}{|}}{R-C-C\equiv N}}$

direct addition (1,2-addition)	nucleophilic addition to the carbonyl carbon of an α,β-unsaturated carbonyl compound.

Grignard reagent	the compound that results when magnesium is inserted between the carbon and the halogen of an alkyl halide (RMgBr, RMgCl).

hemiacetal	$\underset{\underset{\displaystyle OR}{|}}{\overset{\overset{\displaystyle OH}{|}}{R-C-H}}$ **or** $\underset{\underset{\displaystyle OR}{|}}{\overset{\overset{\displaystyle OH}{|}}{R-C-R}}$

ketone	$\underset{R}{\overset{\displaystyle O}{\overset{\|}{C}}}\diagdown_R$

nucleophilic addition reaction	a reaction that involves the addition of a nucleophile to the carbonyl carbon of an aldehyde or a ketone.

nucleophilic addition–elimination reaction	a reaction involving nucleophilic addition to a carbonyl group to form a tetrahedral intermediate, which then undergoes elimination of a leaving group. Imine formation is an example: an amine adds to the carbonyl carbon, and water is eliminated.

organometallic compound	a compound with a carbon–metal bond.

303

reduction reaction

in the case of an organic molecule, a reaction in which the number of C—H bond is increased or the number of C—O, C—N, or C—X (X = halogen) bonds i decreased.

reductive amination

a reaction of an aldehyde or a ketone with ammonia or with a primary amine in th presence of a reducing agent.

olutions to Problems

a. 6-bromo-3-heptanone   **b.** 4-phenylbutanal   **c.** 5-methyl-2-heptanone

a. 4-heptanone, dipropyl ketone
b. 2-methyl-4-heptanone, isobutyl propyl ketone
c. 4-ethylhexanal, γ-ethylcaproaldehyde

If the carbonyl group were anywhere else in these compounds, they would not be ketones (they would be aldehydes) and, therefore, would not have the "one" suffix.

a. 2-Heptanone is more reactive, because it has less steric hindrance. There is little difference in the amount of steric hindrance provided by the propyl and the pentyl group at the carbonyl carbon (the site of nucleophilic addition), because they differ at a point somewhat removed from the site of nucleophilic addition. However, there is a significant difference in size between a methyl group and a propyl group at the site of nucleophilic addition.

2-heptanone   4-heptanone

b. Chloromethyl phenyl ketone is more reactive, because chlorine is more strongly electron withdrawing than bromine since chlorine is more electronegative. Withdrawing electrons inductively away from the carbonyl group makes the carbonyl carbon more electrophilic and, therefore, more reactive toward a nucleophile.

bromomethyl phenyl ketone   chloromethyl phenyl ketone

Notice that a Grignard reagent will react with an H bonded to an O or an N, forming an alkane.

a. $CH_3CH_2MgBr + H_2O \longrightarrow CH_3CH_3 + Mg^{2+} Br^- HO^-$

b. $CH_3CH_2MgBr + CH_3OH \longrightarrow CH_3CH_3 + Mg^{2+} Br^- CH_3O^-$

c. $CH_3CH_2MgBr + CH_3NH_2 \longrightarrow CH_3CH_3 + Mg^{2+} Br^- CH_3\bar{N}H$

d. $CH_3MgBr + HC\equiv CH \longrightarrow CH_4 + Mg^{2+}Br^-HC\equiv C^-$

7.   Only **C** can be used to form a Grignard reagent. Both **A** and **B** have an H bonded to an oxygen that would immediately react as an acid with the Grignard reagent, forming an alkane.

8.   **a.** $\underset{\overset{\displaystyle |}{OH}}{CH_3CH_2CHCH_3}$ **b.** $\underset{\overset{\displaystyle |}{CH_3}}{\overset{\overset{\displaystyle OH}{|}}{CH_3CH_2CH_2CCH_3}}$ **c.**

9. $CH_3 \quad CH_2CH_3 + CH_3CH_2CH_2MgBr$ or $CH_3CH_2 \quad CH_2CH_2CH_3 + CH_3MgBr$

10.   **a.** Two stereoisomers are obtained, because the reaction creates an asymmetric center in the product.

$\underset{\text{(S)-3-methyl-3-hexanol}}{CH_3\!-\!\!\underset{\overset{\displaystyle |}{CH_2CH_3}}{\overset{\overset{\displaystyle OH}{|}}{\rule{0pt}{1em}}}\!\!-\!CH_2CH_2CH_3}$ $\underset{\text{(R)-3-methyl-3-hexanol}}{CH_3\!-\!\!\underset{\overset{\displaystyle |}{OH}}{\overset{\overset{\displaystyle CH_2CH_3}{|}}{\rule{0pt}{1em}}}\!\!-\!CH_2CH_2CH_3}$

b. Only one compound is obtained, because the product does not have an asymmetric center.

11.   **a.** Solved in the text.

b.   **1.** Solved in the text.

   2. $CH_3 \quad OR + 2\,CH_3MgBr$

   4. $CH_3 \quad OR + 2\,CH_3CH_2MgBr$

   6. $CH_3 \quad OR + 2$ $MgBr$

12.   If a secondary alcohol is formed from the reaction of a formate ester with excess Grignard reagent, the two alkyl substituents of the alcohol will be identical, because they both come from the Grignard reagent. Therefore, only the following two alcohols (**B** and **D**) can be prepared that way.

$\underset{\overset{\displaystyle |}{OH}}{CH_3CHCH_3}$ $\underset{\overset{\displaystyle |}{OH}}{CH_3CH_2CHCH_2CH_3}$

3.

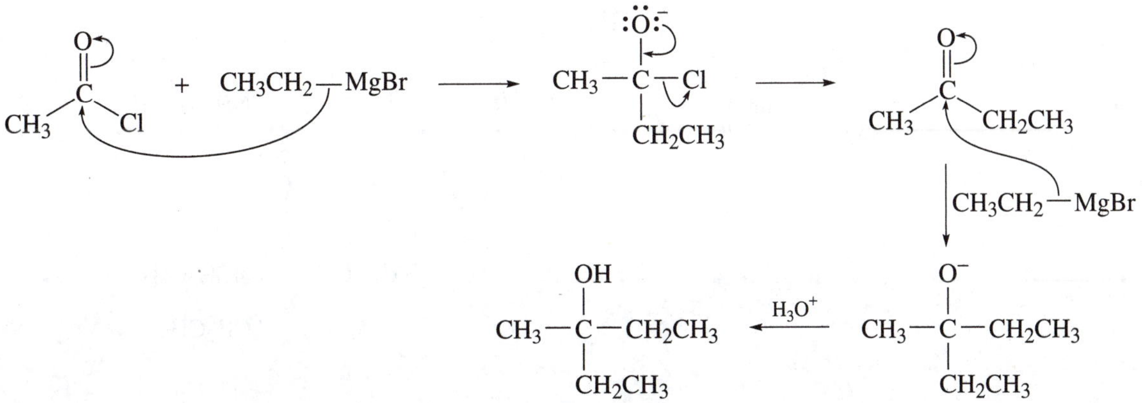

4. **A** and **C** will not undergo nucleophilic addition with a Grignard reagent. **A** has a H bonded to a nitrogen and **C** has a H bonded to an oxygen; these acidic hydrogens will react with the Grignard reagent, converting it into an alkane, before the Grignard reagent has a chance to react as a nucleophile.

B will undergo a nucleophilic addition–elimination reaction with the Grignard reagent to form a ketone. This will be followed by a nucleophilic addition reaction of the ketone with another equivalent of the Grignard reagent. The product will be a tertiary alcohol.

5. Solved in the text.

6. **1.** $CH_3CH_2Br \xrightarrow[Et_2O]{Mg} CH_3CH_2MgBr \xrightarrow[2.\ H_3O^+]{1.\ CO_2}$

2. $CH_3CH_2Br \xrightarrow{^-C\equiv N} CH_3CH_2C\equiv N \xrightarrow[\Delta]{HCl,\ H_2O}$

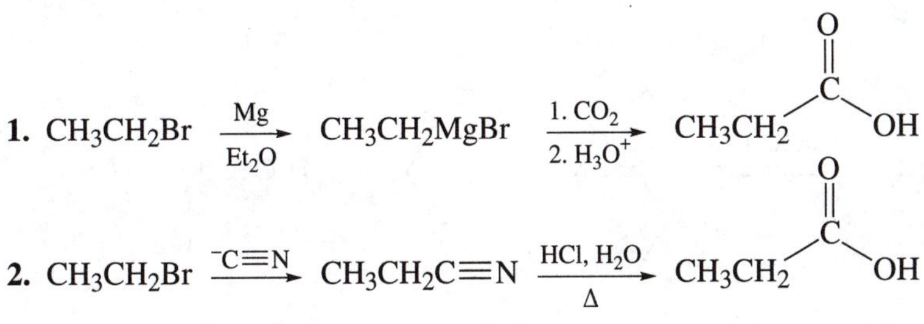

7. **a.** $CH_3\overset{\underset{|}{CH_3}}{C}HCH_2OH$ **b.**   **c.**   **d.**

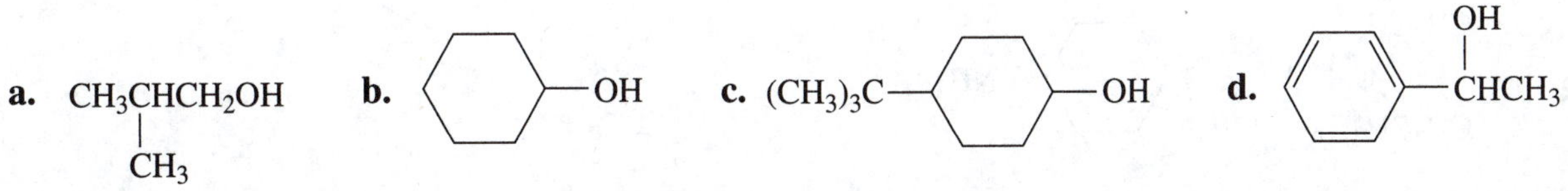

8. **a.** $CH_3CH_2CH_2CH_2OH$ + CH_3CH_2OH

b.   + CH_3OH

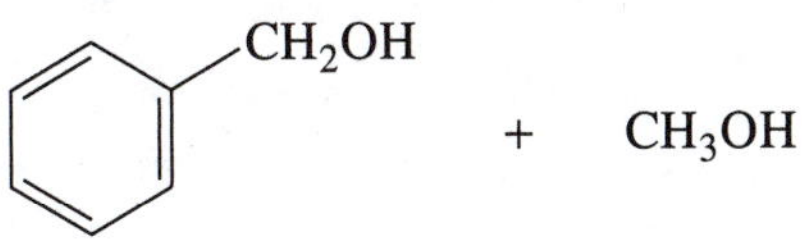

c. $CH_3CH_2CH_2CH_2CH_2OH$

19.   **a.** 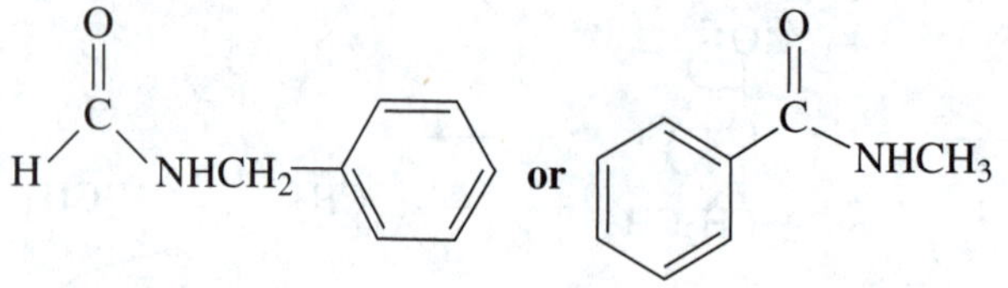   **or**   **c.**

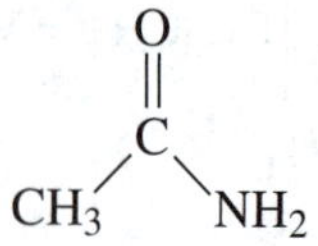

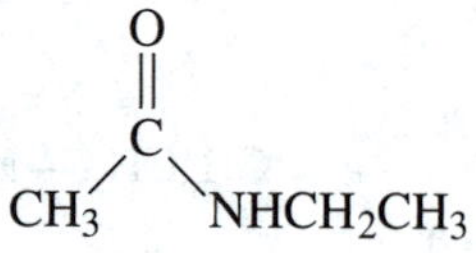

b.   **d.**

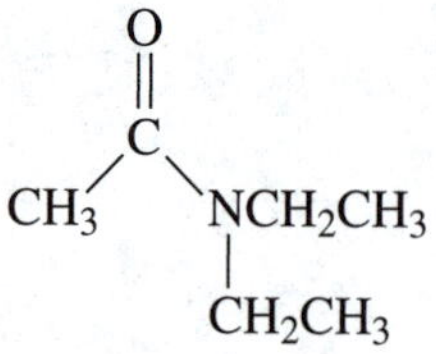

20.   **a.**

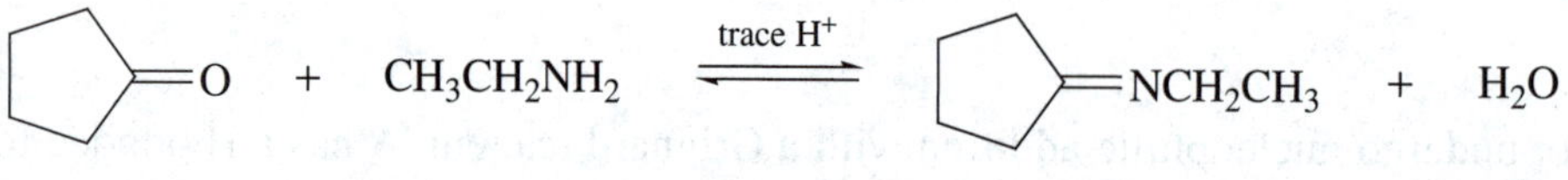

b.

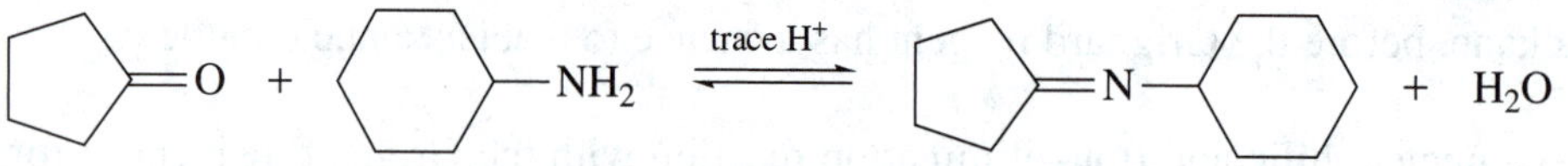

c.

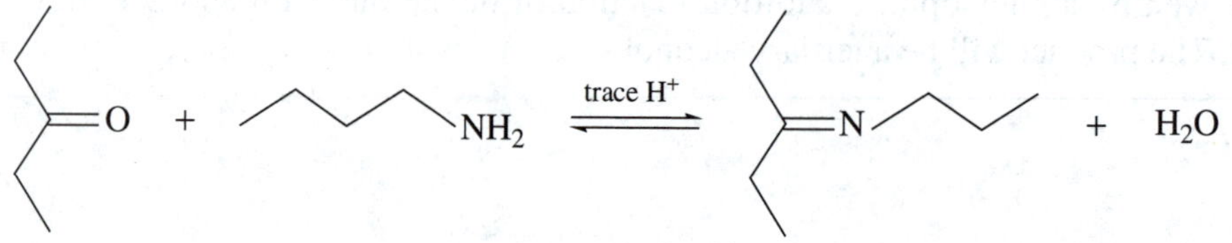

d.

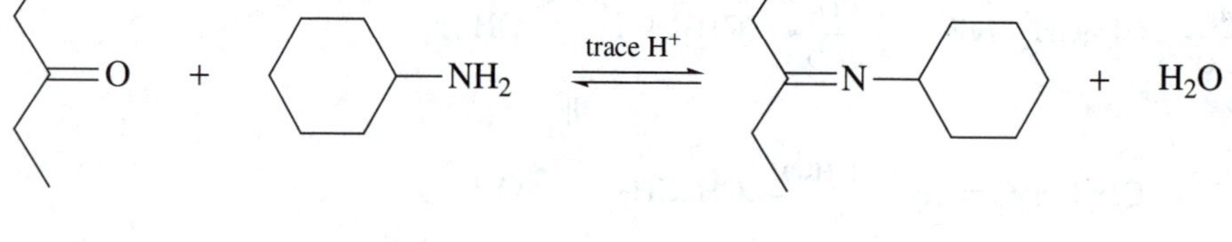

21.

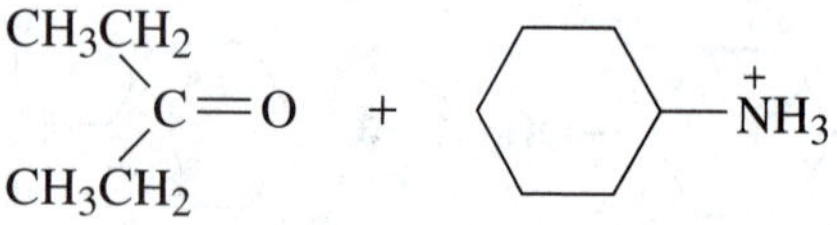

22.   **a.** 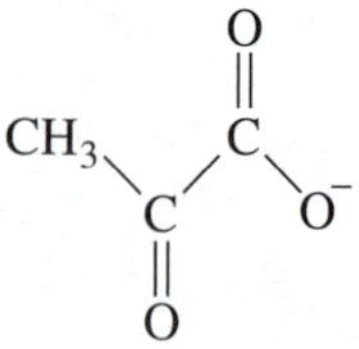   **b.**

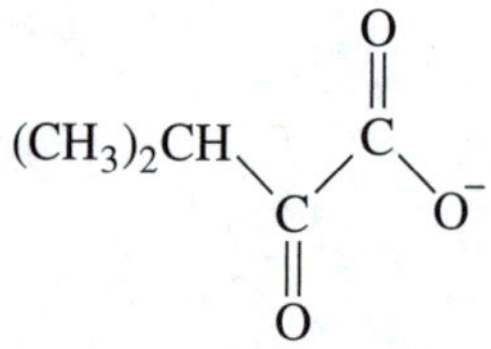

23.   **a.**   hemiacetals: 1, 5, 6

b.   acetals: 2, 3, 4

4. **a.** 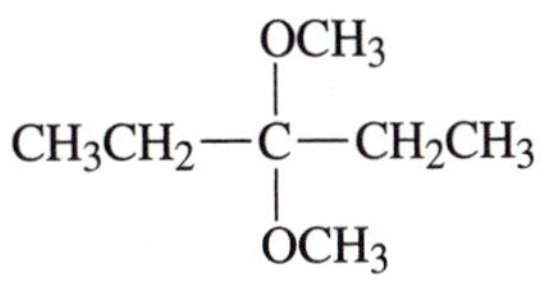   **b.**

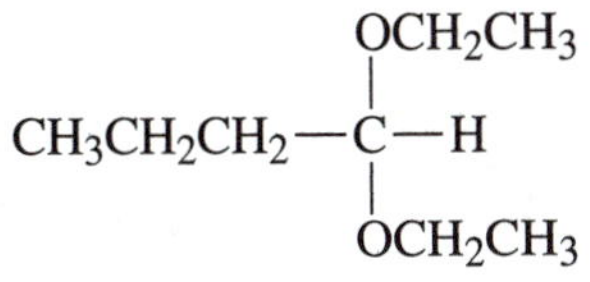

5. **a.**

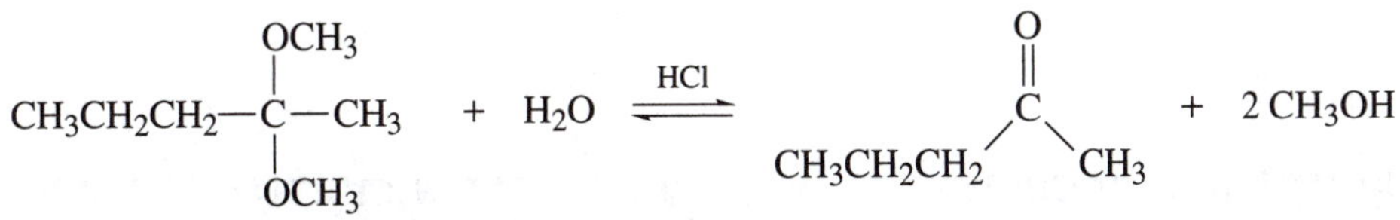

b.

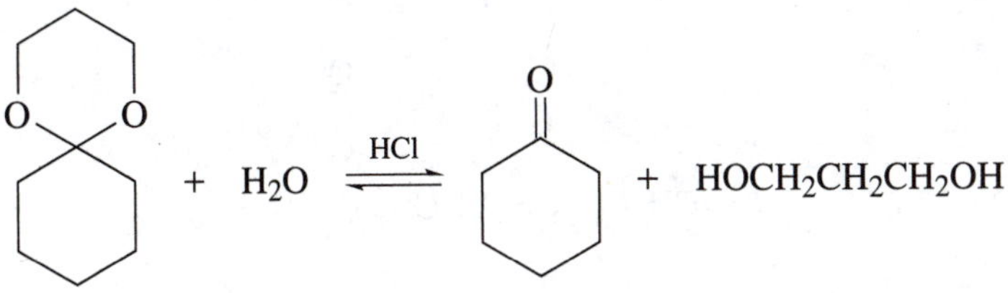

6.

a. 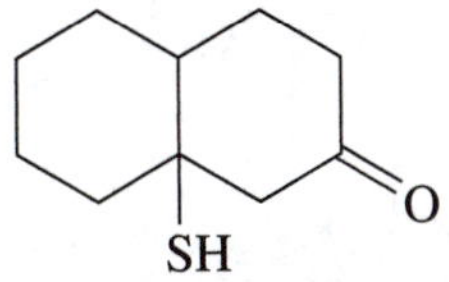
weak base favors
conjugate addition

b. 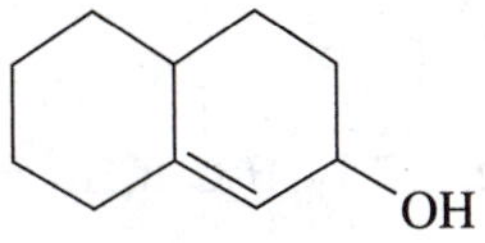
strong base
favors direct addition

c. 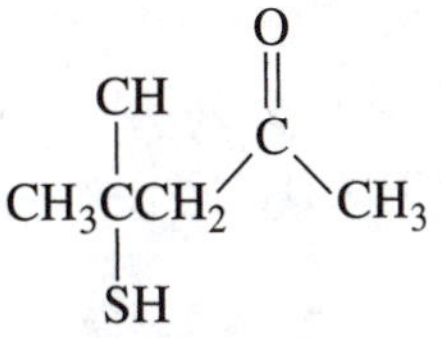
weak base favors
conjugate addition

d. $CH_3CH{=}CHCH_2OH$
strong base favors
direct addition

7. Conjugate addition occurs in **a** and **c,** because the nucleophile is a relatively weak base. Nucleophilic acyl substitution occurs in **b,** because the cabonyl group is very reactive. In part **d,** one equivalent of ammonia reacts in a nucleophilic acyl substitution reaction, one picks up the acid generated in the reaction, and one reacts in a conjugate addition reaction.

a. 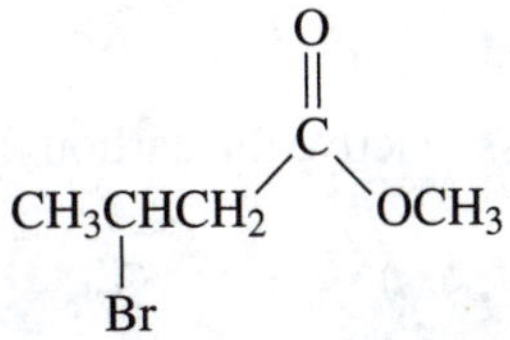   **c.**

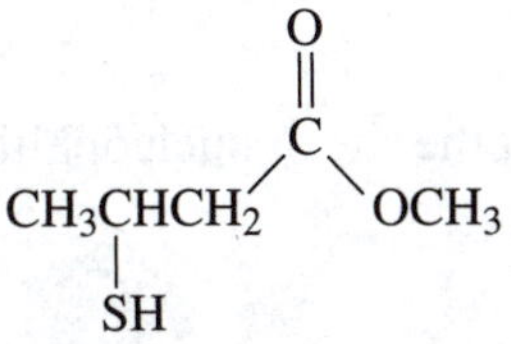

b. 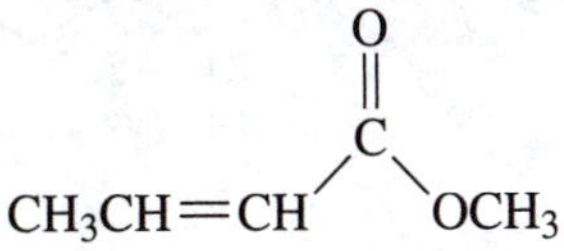   **d.**

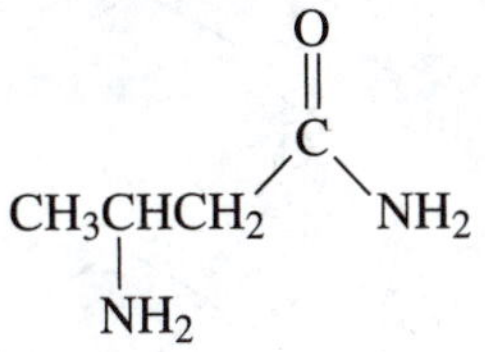

28.

a. 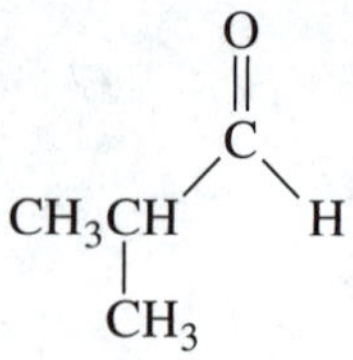 CH₃CH(CH₃)—C(=O)H

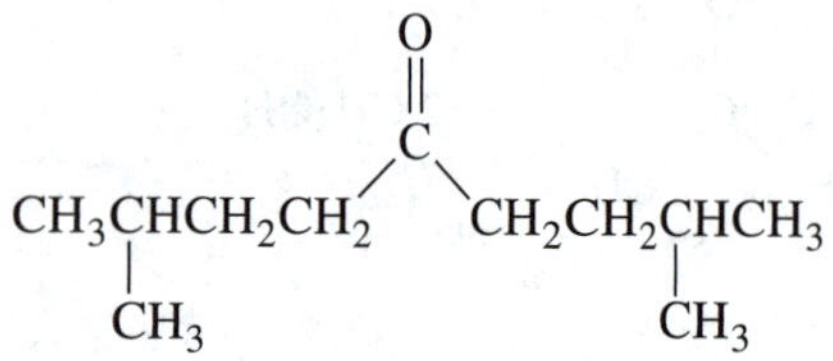

b. CH₃CHCH₂CH₂—C(=O)—CH₂CH₂CHCH₃
      | |
      CH₃   CH₃

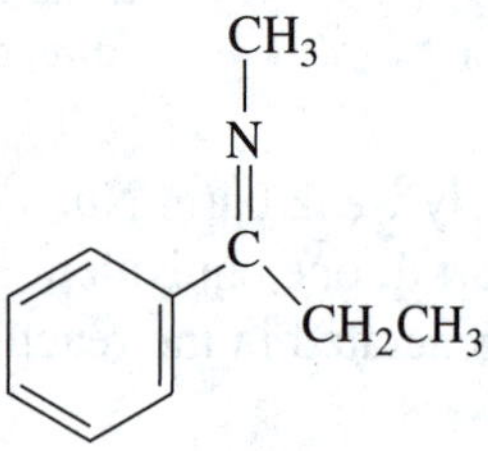

c. (cyclohexanone with CH₃ substituent)

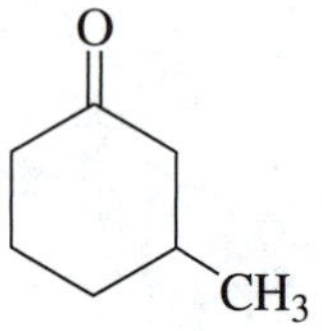

d. 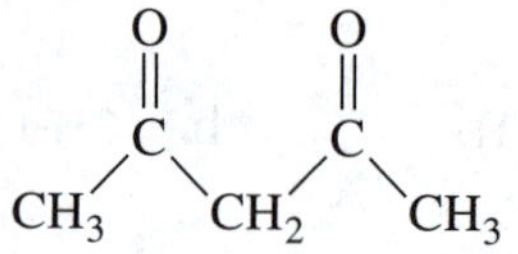 CH₃—C(=O)—CH₂—C(=O)—CH₃

e. 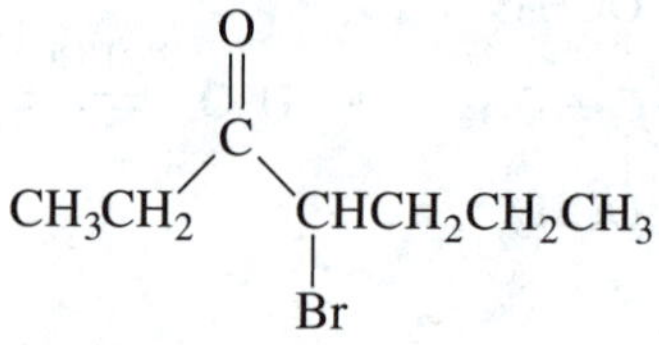 CH₃CH₂—C(=O)—CHCH₂CH₂CH₃
                       |
                      Br

f. 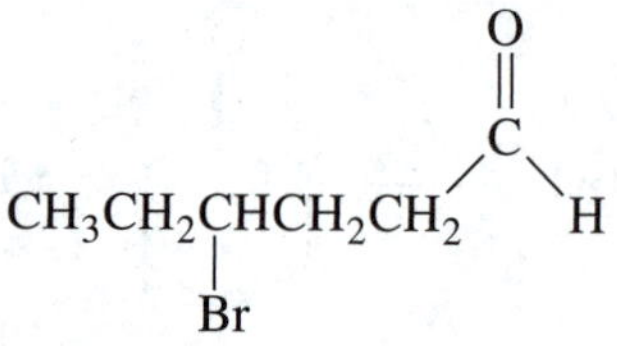 CH₃CH₂CHCH₂CH₂—C(=O)H
              |
             Br

29.

a. 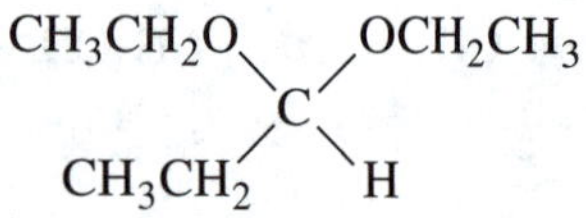
CH₃CH₂O   OCH₂CH₃
      \   /
       C 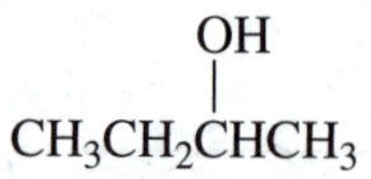
      /   \
CH₃CH₂   H

b. (phenyl)C(=N—CH₃)CH₂CH₃

c.
          OH
          |
CH₃CH₂CHCH₃

d. 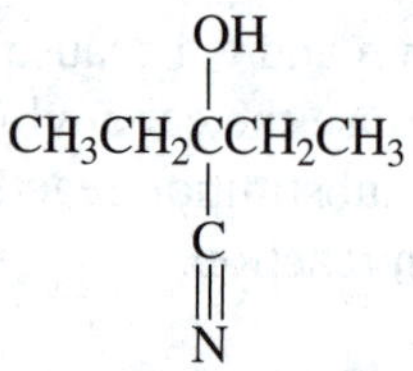
          OH
          |
CH₃CH₂CCH₂CH₃
          |
          C
          ‖
          N

30. The greater the steric hindrance at the site of nucleophilic addition, the less reactive the carbonyl compound

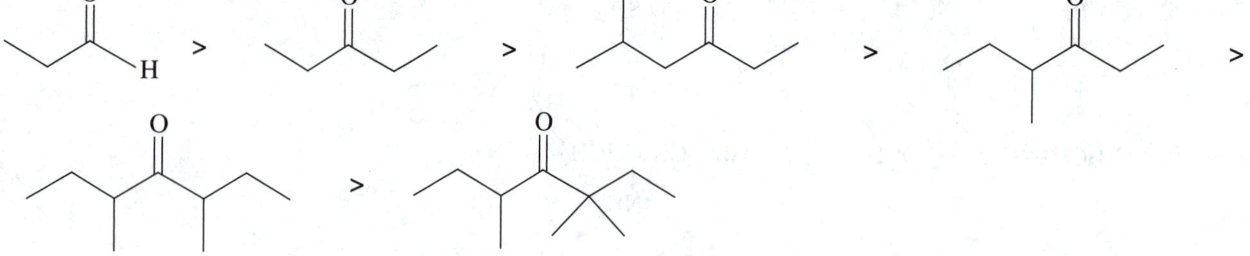

1.

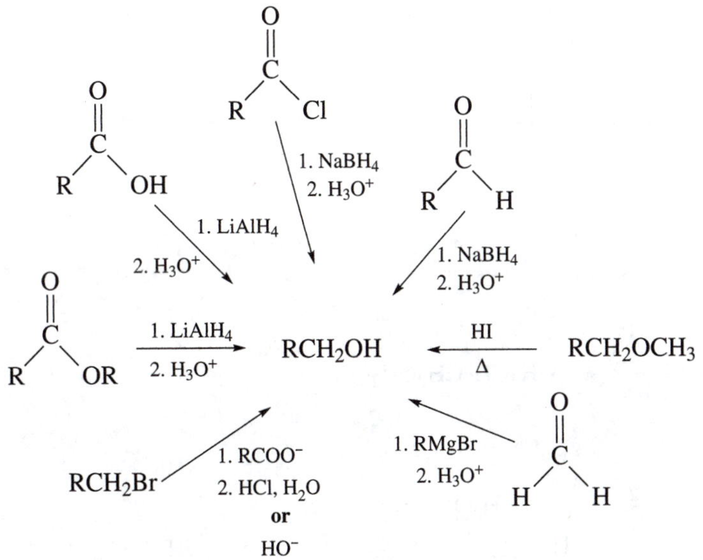

32.

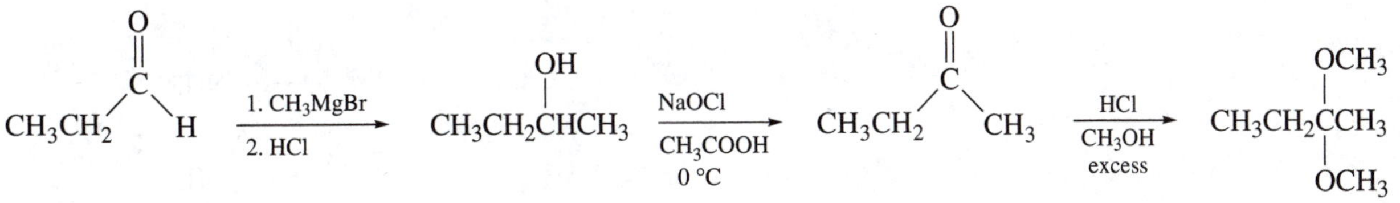

33. **a.**

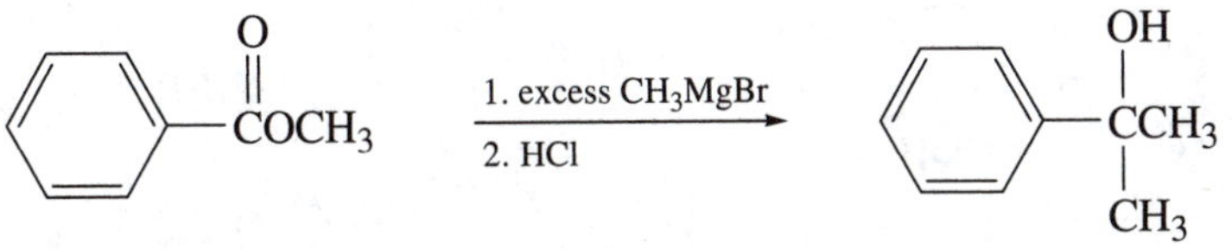

b.

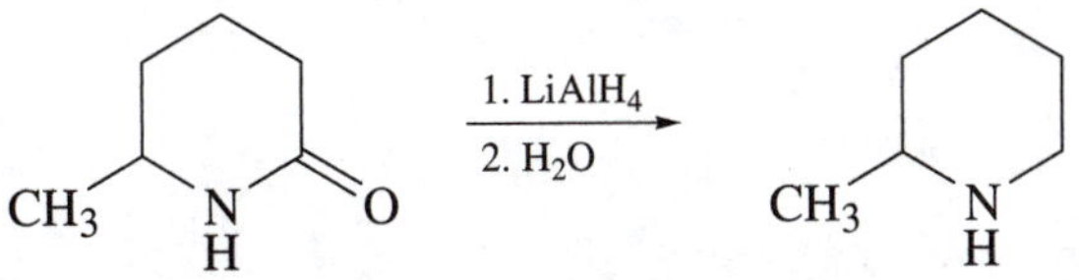

34. Methyl formate and excess Grignard reagent will form a secondary alcohol because, unlike other esters that have an alkyl or aryl group on the carbonyl carbon that cause them to form tertiary alcohols, methyl formate has a hydrogen.

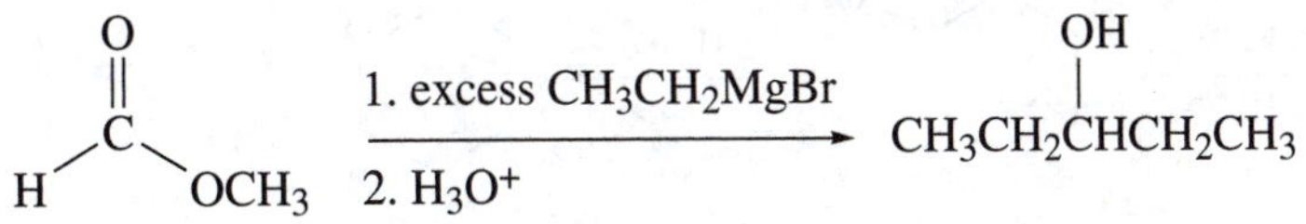

35.   **a.** 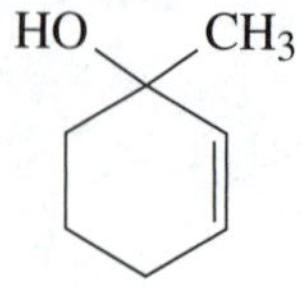   **c.**

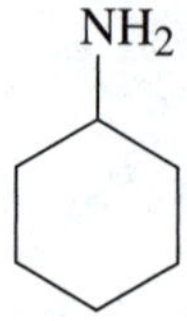

b. 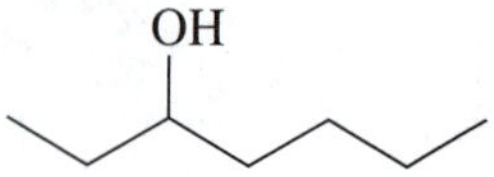   **d.** $CH_3CH_2CH_2CH_2OH$ + CH_3CH_2OH

36. 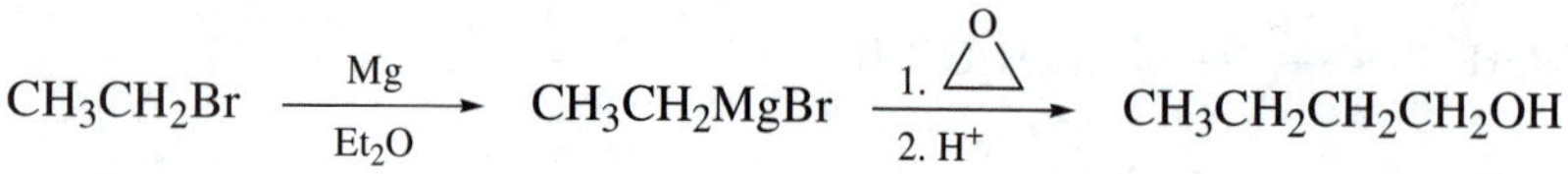 $CH_3CH_2Br \xrightarrow[Et_2O]{Mg} CH_3CH_2MgBr \xrightarrow[2.\ H^+]{1.\ \triangle} CH_3CH_2CH_2CH_2OH$

37.

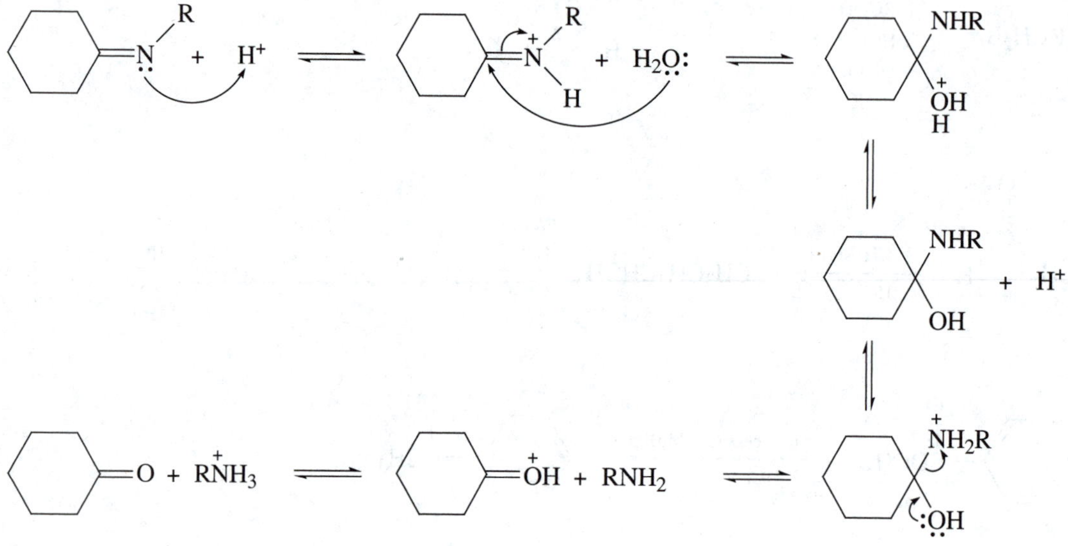

38.

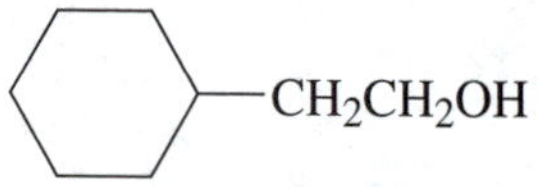

$\cancel{}$—CH_2CH_2OH

39.   **a.**

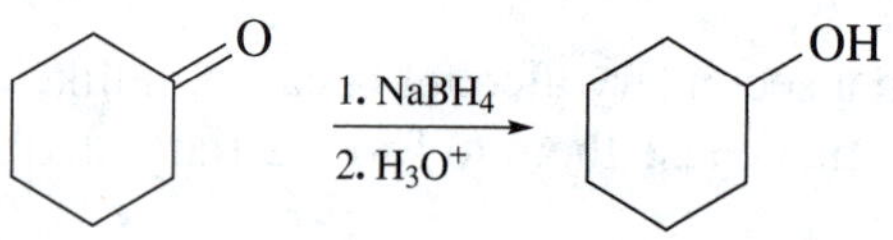

b.

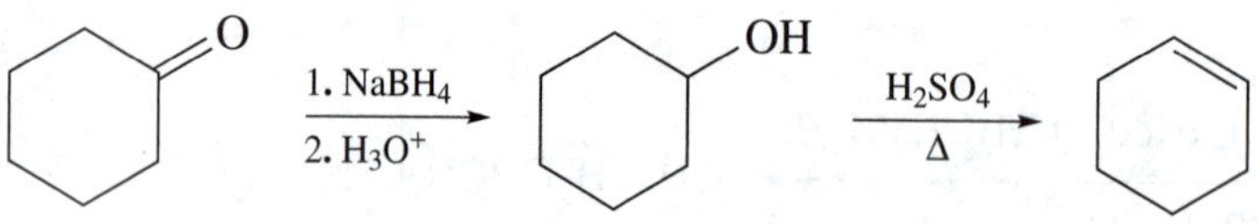

c.

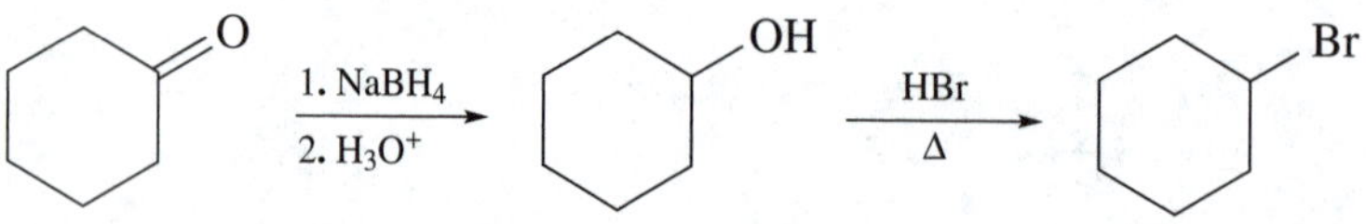

d.

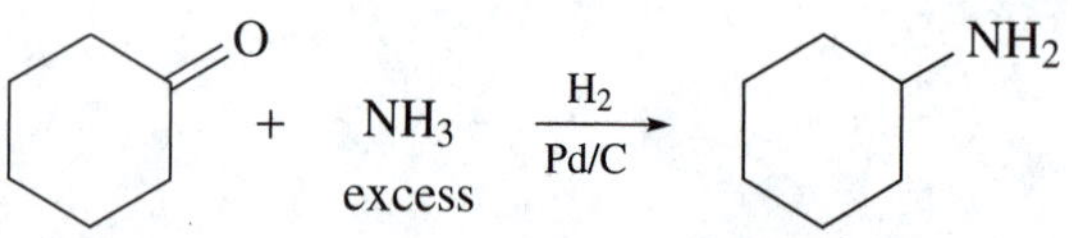

e.

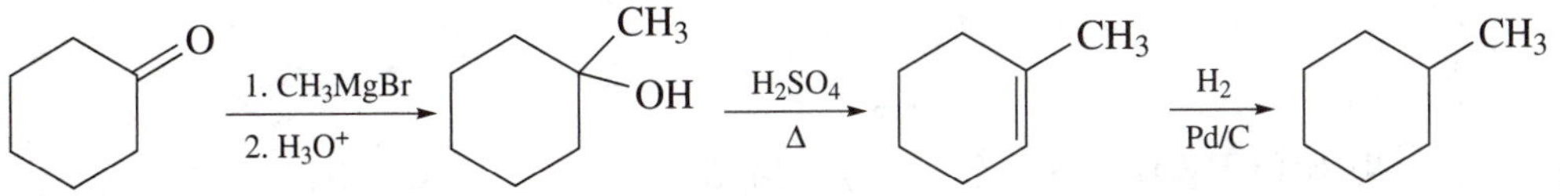

0.

a. $CH_3CH_2CH_2CH_2Br$ $\xrightarrow{^-C\equiv N}$ $CH_3CH_2CH_2CH_2C\equiv N$ $\xrightarrow[\Delta]{H^+,\ H_2O}$ $CH_3CH_2CH_2CH_2\overset{\displaystyle O}{\overset{\|}{C}}OH$

b. $CH_3CH_2CH_2CH_2Br$ $\xrightarrow[Et_2O]{Mg}$ $CH_3CH_2CH_2CH_2MgBr$ $\xrightarrow[2.\ H_3O^+]{1.\ CO_2}$ $CH_3CH_2CH_2CH_2CH_2\overset{\displaystyle O}{\overset{\|}{C}}OH$

---

1. CH_3OH $\xrightarrow[\Delta]{HBr}$ CH_3Br $\xrightarrow[Et_2O]{Mg}$ CH_3MgBr $\xrightarrow[2.\ HCl]{1.\ H_2C=O}$ CH_3CH_2OH

2.

a.

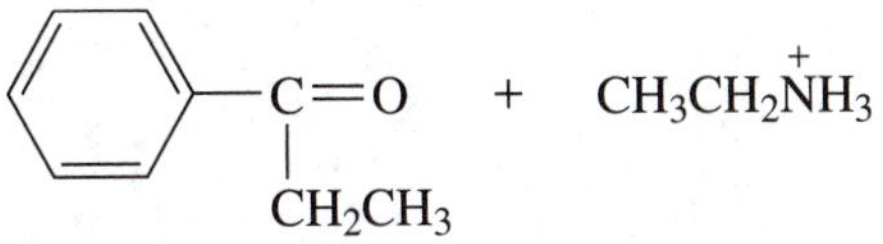

$+$ $CH_3CH_2\overset{+}{N}H_3$

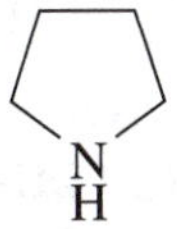

b. 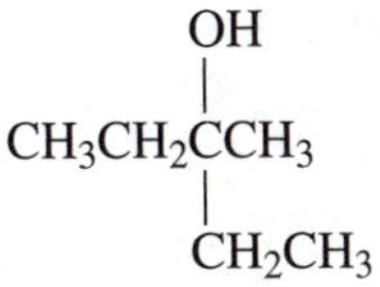 $CH_3CH_2\overset{\displaystyle OH}{\underset{\displaystyle CH_2CH_3}{\overset{|}{\underset{|}{C}}}}CH_3$

c.

d.

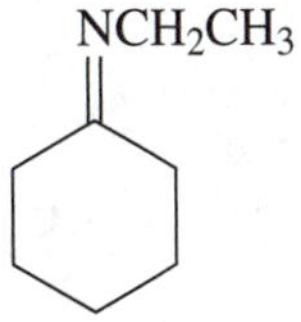

3.

a.

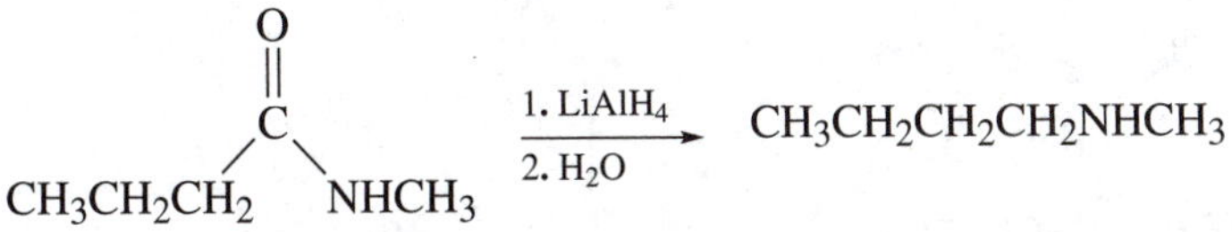

$\xrightarrow[2.\ H_2O]{1.\ LiAlH_4}$ $CH_3CH_2CH_2CH_2NHCH_3$

b.

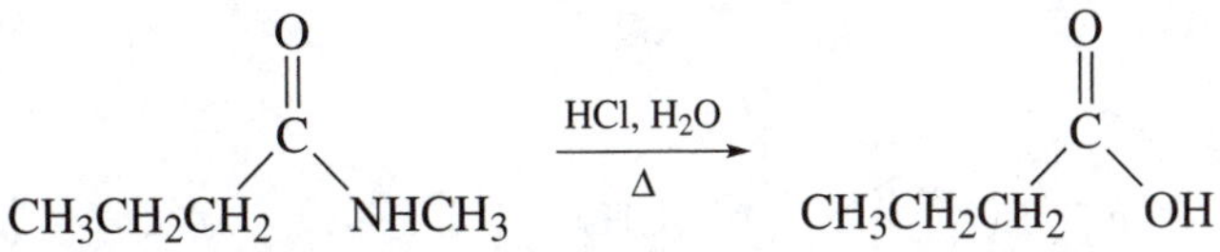

$\xrightarrow[\Delta]{HCl,\ H_2O}$

c.

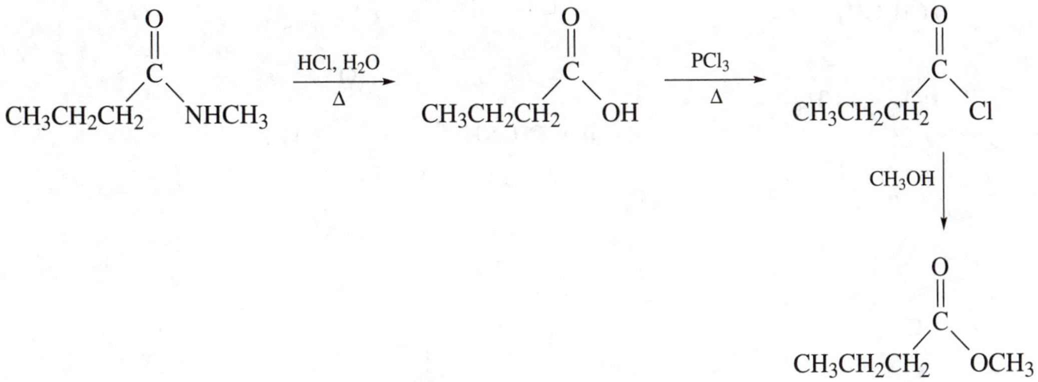

$\xrightarrow[\Delta]{HCl,\ H_2O}$

$\xrightarrow[\Delta]{PCl_3}$

$\downarrow CH_3OH$

d.

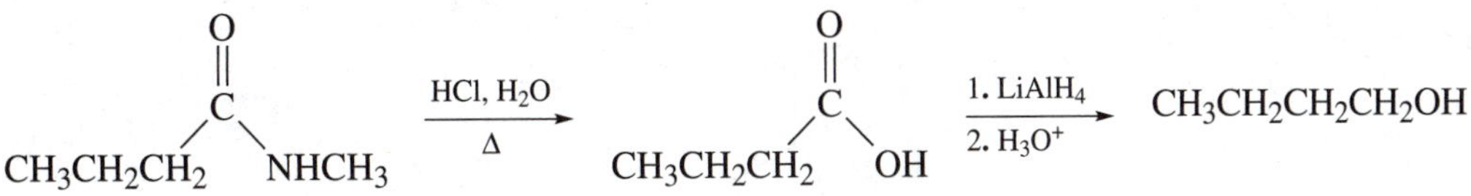

$\xrightarrow[\Delta]{HCl,\ H_2O}$

$\xrightarrow[2.\ H_3O^+]{1.\ LiAlH_4}$ $CH_3CH_2CH_2CH_2OH$

44. **a.** $CH_3CH_2\overset{\overset{\displaystyle O}{\|}}{C}CH_2CH_2CH_2CH_3$ + ⬡—MgBr

⬡—$\overset{\overset{\displaystyle O}{\|}}{C}CH_2CH_3$ + $CH_3CH_2CH_2CH_2MgBr$

⬡—$\overset{\overset{\displaystyle O}{\|}}{C}CH_2CH_2CH_2CH_3$ + CH_3CH_2MgBr

b. $CH_3CH_2\overset{\overset{\displaystyle O}{\|}}{C}CH_2CH_2CH_3$ + CH_3CH_2MgBr

$CH_3CH_2\overset{\overset{\displaystyle O}{\|}}{C}CH_2CH_3$ + $CH_3CH_2CH_2MgBr$

45. **a.** HO, CH₃ ring with CH₃ (3-methyl, 1-methyl cyclohexenol)

b. cyclohexanone with CH₃ and SCH₂CH₃

c. cyclohexanone with CH₃ and Br

46. **a.** $CH_3CH_2\overset{\overset{\displaystyle OH}{|}}{C}CH_2CH_3$ with $\overset{|}{CH_2CH_3}$

b. benzene ring—C(HO)(OCH₃)—CH₂CH₂CH₃

c. $Br\diagup\diagdown\overset{\overset{\displaystyle O}{\|}}{C}OCH_3$

d. $CH_3NH\diagup\diagdown\overset{\overset{\displaystyle O}{\|}}{C}OCH_3$

47. **a.** $H_3C,\,C,\,CH_3$ with $\overset{\|}{N}$ attached to tetrahydronaphthalene ring

b. CH₂=CH—C(=O)—NH attached to tetrahydronaphthalene ring

8.

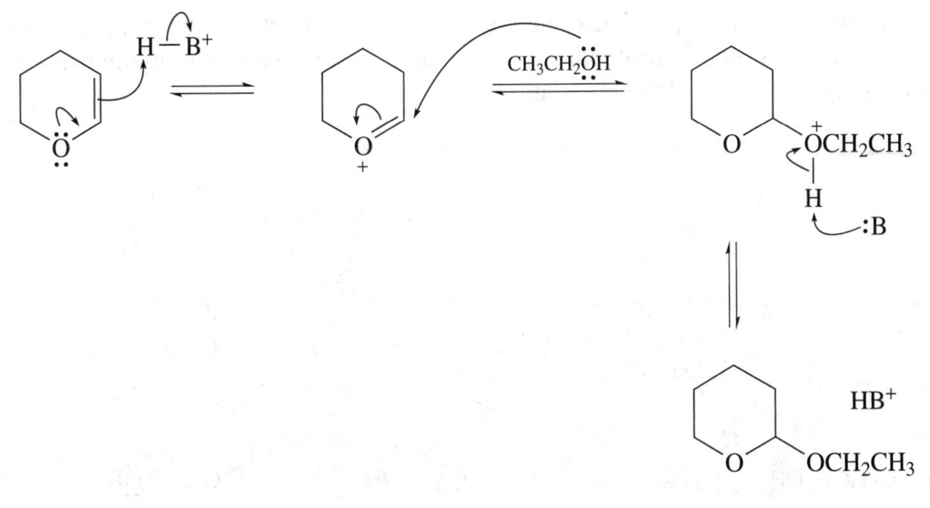

9.

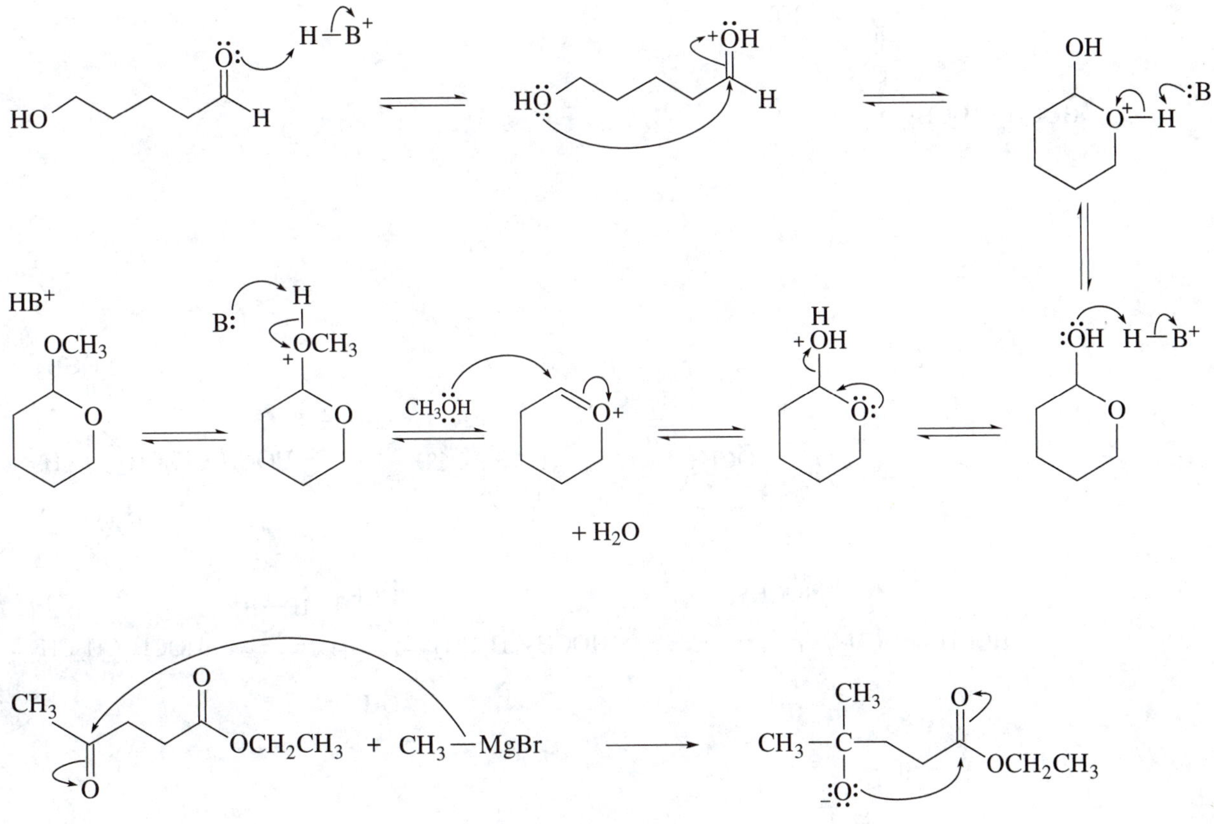

0.

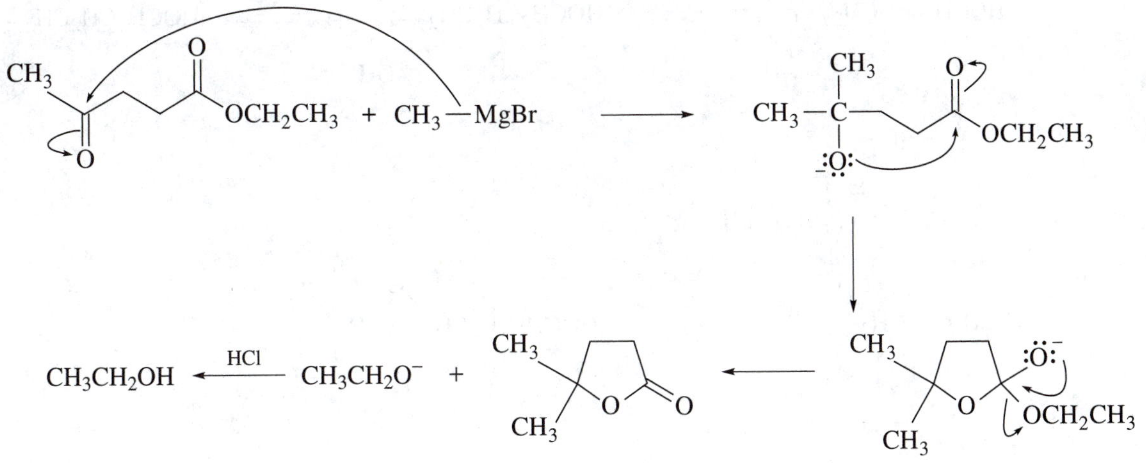

51.   Notice that in all the mechanisms in the problems, there is an equilibrium between a protonated intermedate, a neutral intermediate, and a second protonated intermediate, just as we saw in many previous acid catalyzed mechanisms in this chapter and the previous chapter.

a.

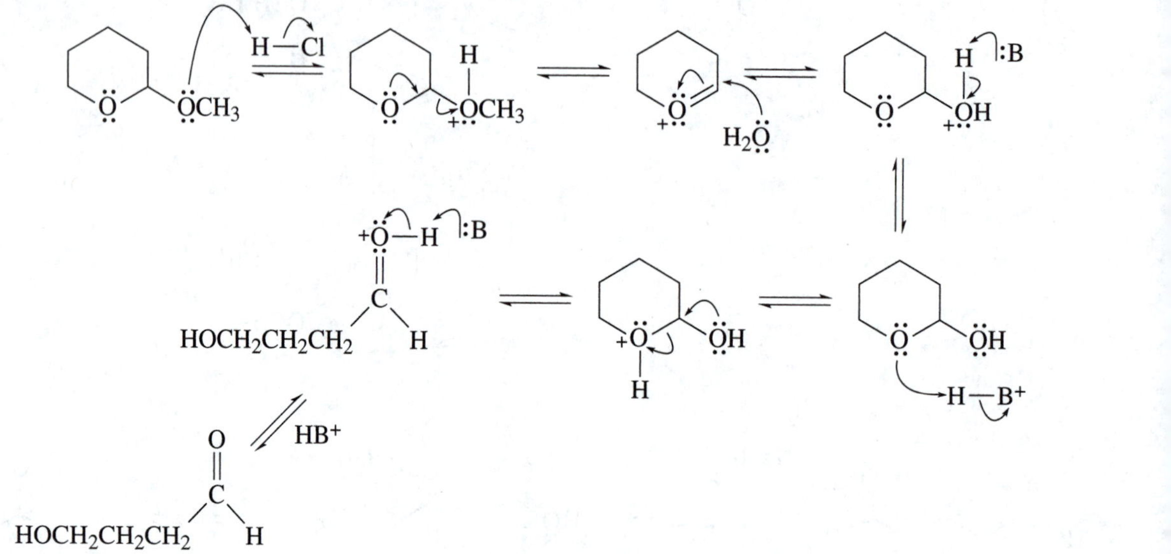

or

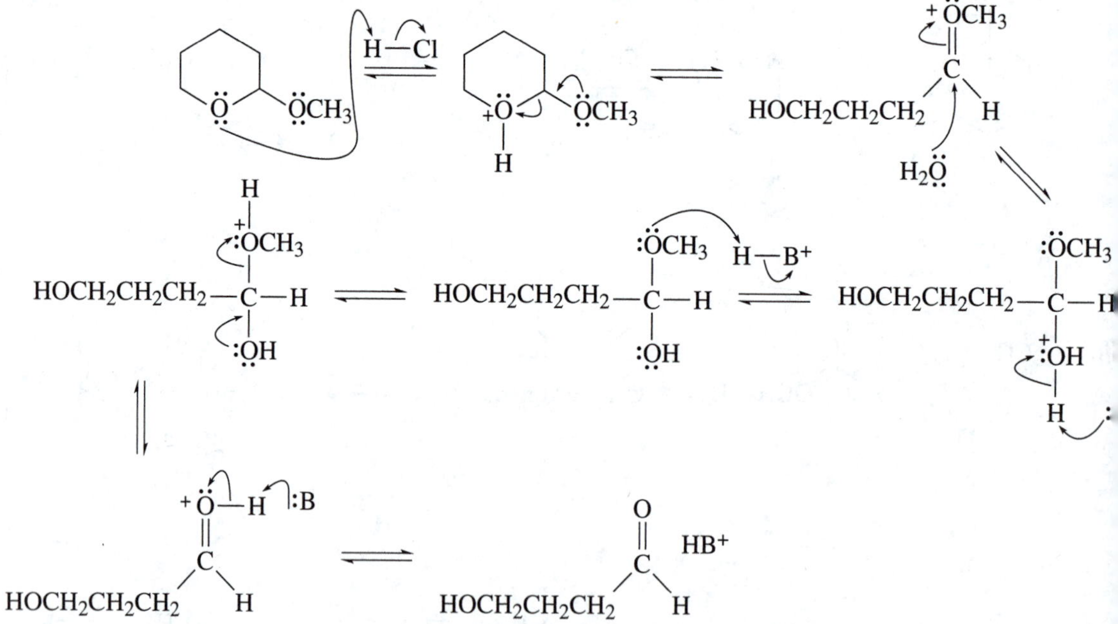

b.

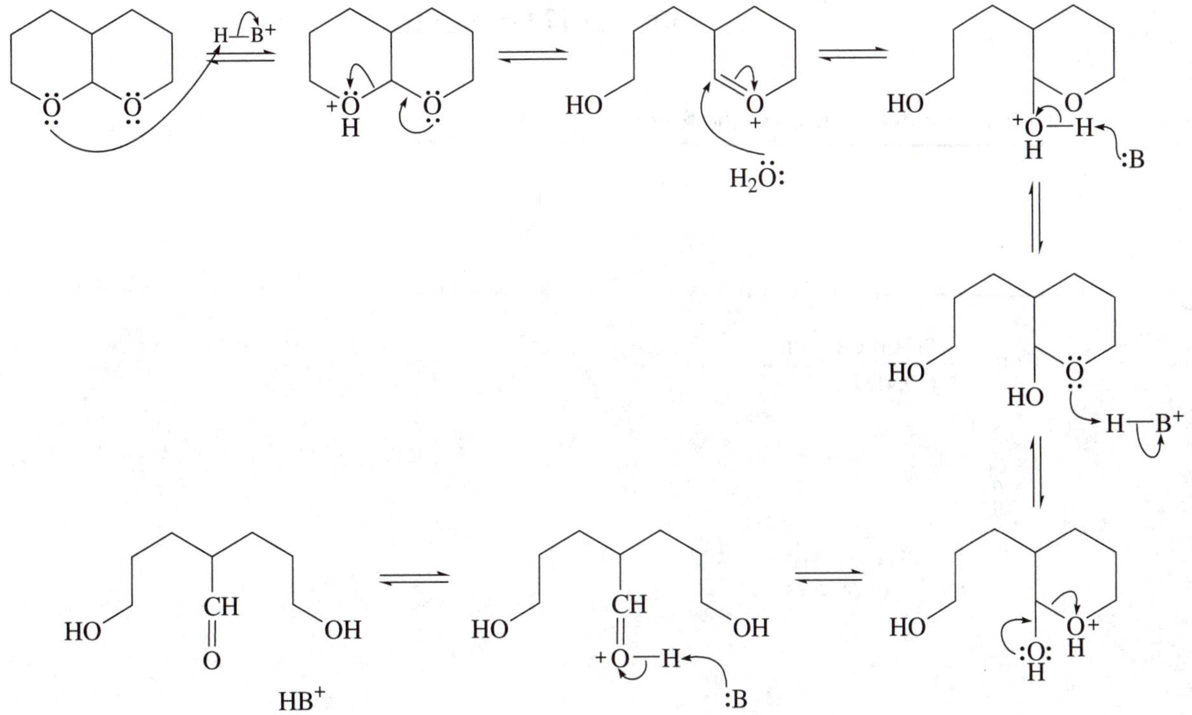

2.   a.

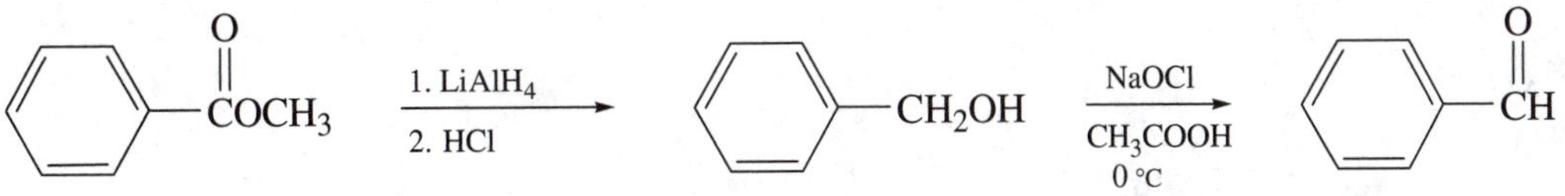

b.

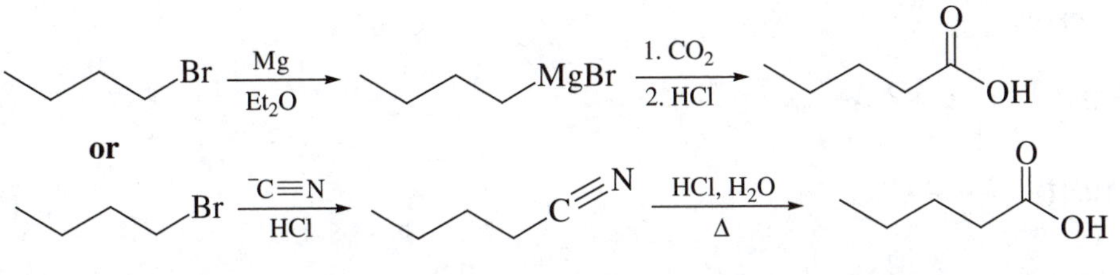

or

53.   Notice that in these problems, the two alcohol groups that form the acetal are in the same molecule.

a.

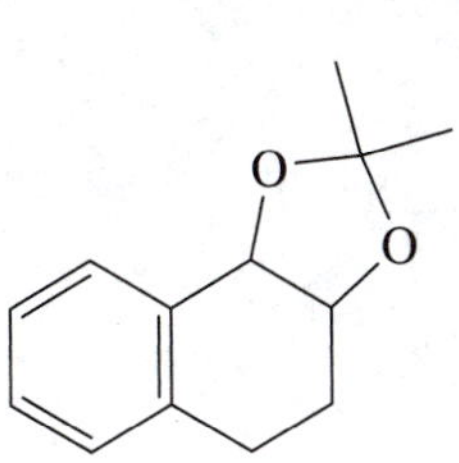

b.

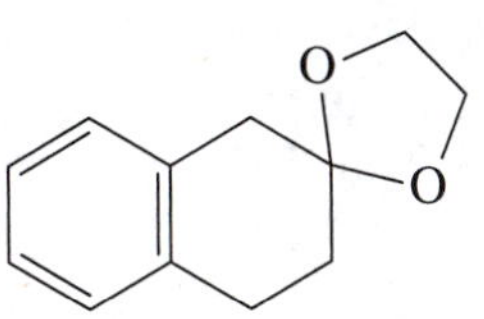

Chapter 12 Practice Test

1.   What is the product of each of the following reactions?

a. 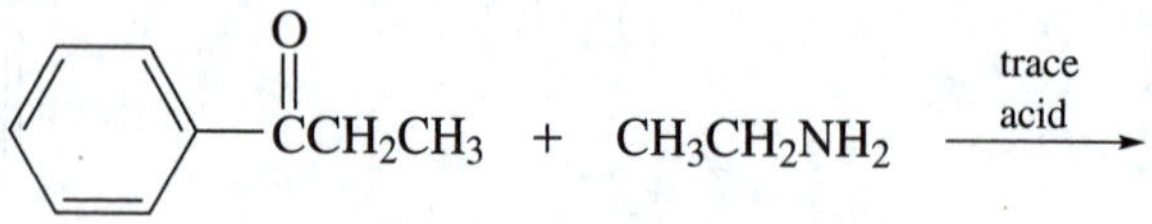   (phenyl)$-\overset{\overset{\text{O}}{\|}}{C}CH_2CH_3$ + $CH_3CH_2NH_2$ $\xrightarrow{\text{trace acid}}$

b. CO_2 $\xrightarrow[\text{2. HCl, H}_2\text{O}]{\text{1. CH}_3\text{CH}_2\text{CH}_2\text{MgBr}}$

c. 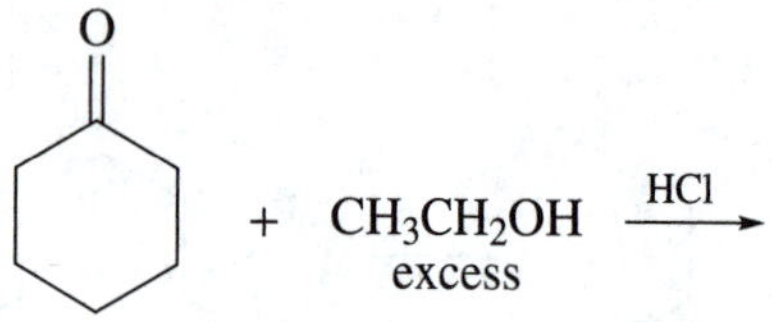   (cyclohexanone) + CH_3CH_2OH (excess) $\xrightarrow{\text{HCl}}$

d. $CH_3CH_2\overset{\overset{\text{O}}{\|}}{C}CH_2CH_3$ $\xrightarrow[\text{2. HCl}]{\text{1. CH}_3\text{MgBr}}$

e. 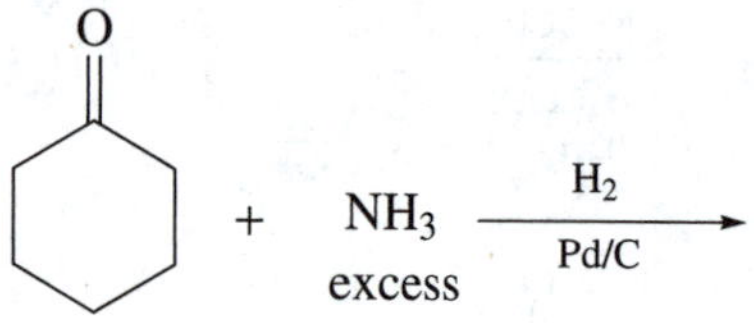   (cyclohexanone) + NH_3 (excess) $\xrightarrow[\text{Pd/C}]{\text{H}_2}$

f. $CH_3CH_2\overset{\overset{\text{O}}{\|}}{C}CH_2CH_3$ $\xrightarrow[\text{HCl}]{\text{}^-C\equiv N}$

g. $CH_3CH=CH\overset{\overset{\text{O}}{\|}}{C}CH_3$ + CH_3SH $\longrightarrow$

h. 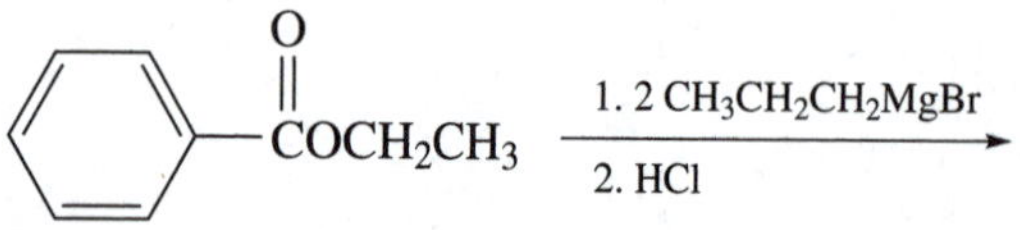   (phenyl)$-\overset{\overset{\text{O}}{\|}}{C}OCH_2CH_3$ $\xrightarrow[\text{2. HCl}]{\text{1. 2 CH}_3\text{CH}_2\text{CH}_2\text{MgBr}}$

2.   Which of the following alcohols cannot be prepared by the reaction of an ester with excess Grignard reagent?

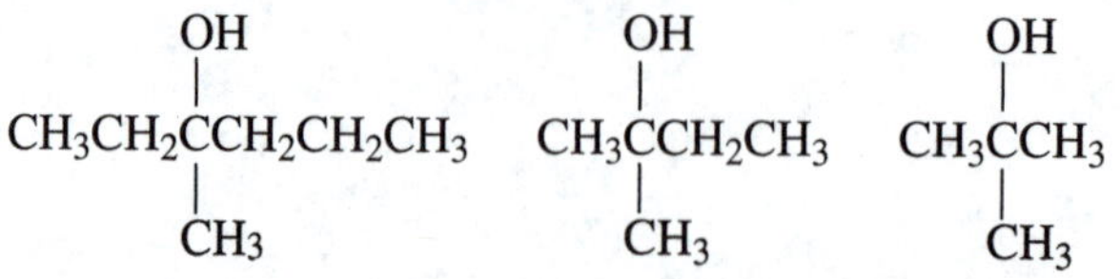

$$CH_3CH_2\underset{\underset{\text{CH}_3}{|}}{\overset{\overset{\text{OH}}{|}}{C}}CH_2CH_2CH_3 \qquad CH_3\underset{\underset{\text{CH}_3}{|}}{\overset{\overset{\text{OH}}{|}}{C}}CH_2CH_3 \qquad CH_3\underset{\underset{\text{CH}_3}{|}}{\overset{\overset{\text{OH}}{|}}{C}}CH_3$$

Name the following compounds:

a. 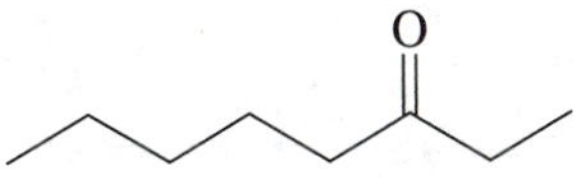   b.

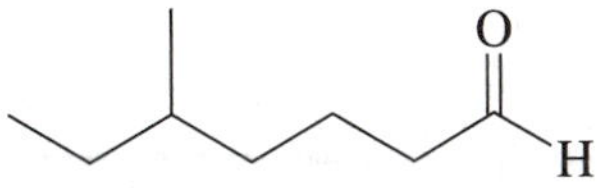

Give an example for each of the following:

a. an acetal

b. an imine

c. a hemiacetal

Which is more reactive toward nucleophilic addition?

a. butanal or methyl propyl ketone

b. 2-heptanone or 2-pentanone

Indicate how the following compounds could be prepared using the given starting material:

a. 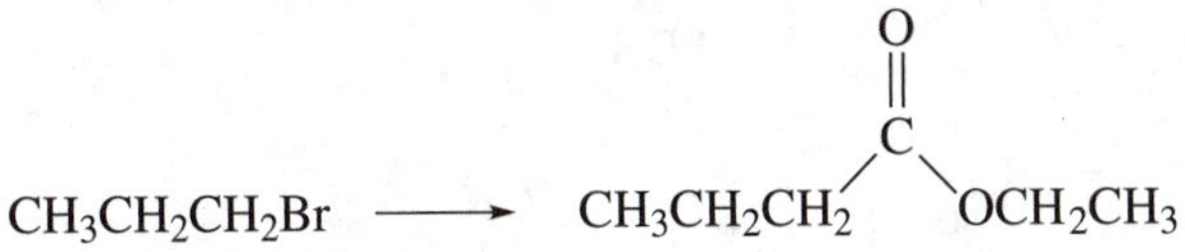 $CH_3CH_2CH_2Br \longrightarrow$

b. 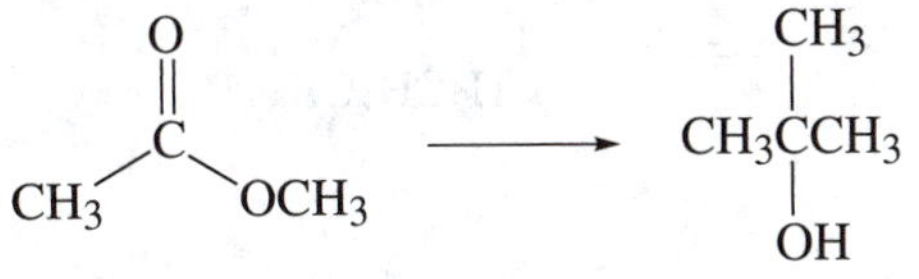 $\longrightarrow$

c. $CO_2 \longrightarrow CH_3CH_2OH$

d. $\longrightarrow$

Answers to Chapter 12 Practice Test

1. a. 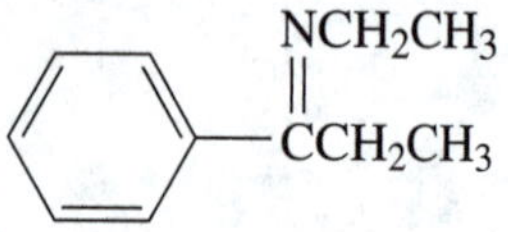 + H_2O

e. 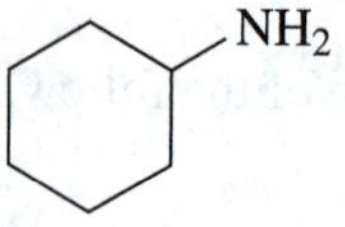

b. 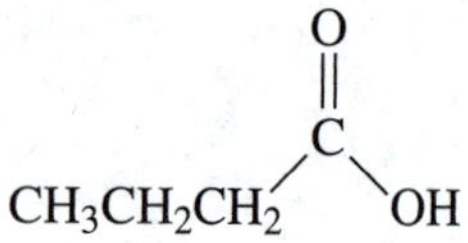

f. 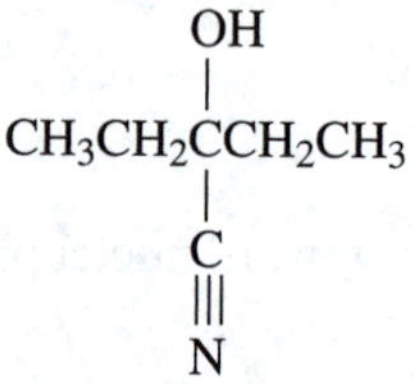

c. 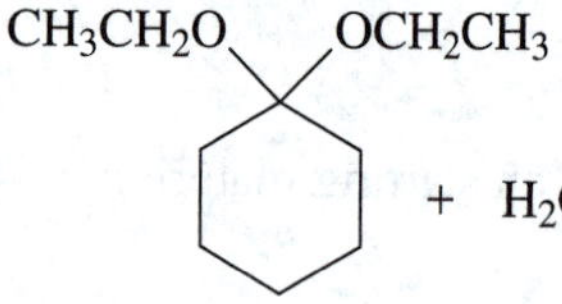 + H_2O

g. 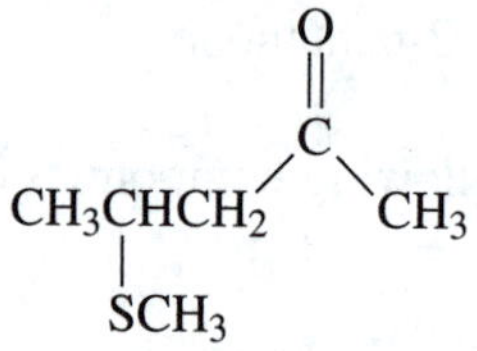

d. 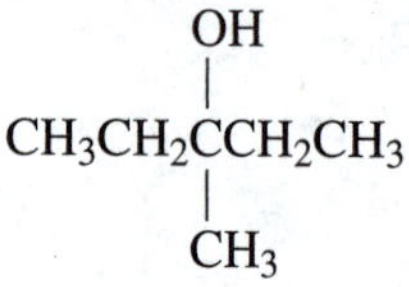

h. 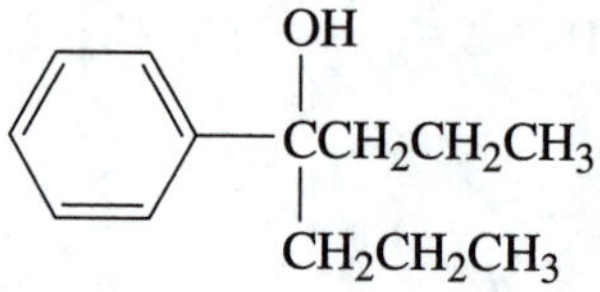

2. 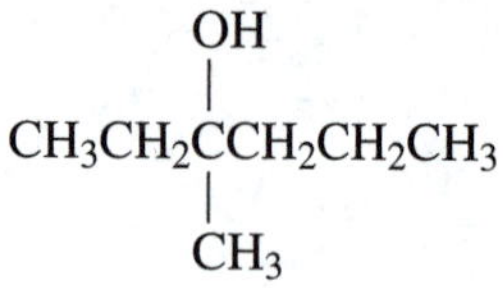

3. a. 3-octanone b. 5-methylheptanal

4. a. 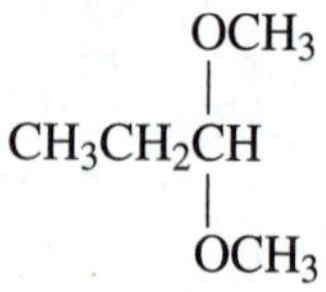 b. 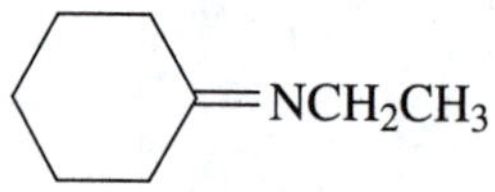 c. 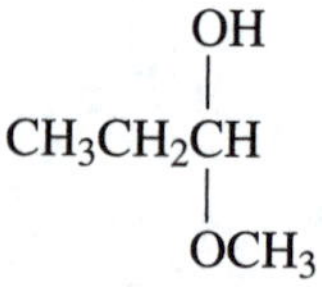

5. a. butanal b. 2-pentanone

a. $CH_3CH_2CH_2Br \xrightarrow{\text{Mg}} CH_3CH_2CH_2MgBr \xrightarrow{CO_2} CH_3CH_2CH_2\overset{\displaystyle O}{\underset{}{C}}O^-$

$\Delta \mid PCl_3$

$CH_3CH_2CH_2\overset{\displaystyle O}{\underset{}{C}}OCH_2CH_3 \xleftarrow{CH_3CH_2OH} CH_3CH_2CH_2\overset{\displaystyle O}{\underset{}{C}}Cl$

b. $CH_3\overset{\displaystyle O}{\underset{}{C}}OCH_3 \xrightarrow[\text{CH}_3\text{MgBr}]{\text{excess}} CH_3\underset{\underset{\displaystyle CH_3}{|}}{\overset{\overset{\displaystyle O^-}{|}}{C}}CH_3 \xrightarrow{H_3O^+} CH_3\underset{\underset{\displaystyle CH_3}{|}}{\overset{\overset{\displaystyle OH}{|}}{C}}CH_3$

c. $CO_2 \xrightarrow{CH_3MgBr} CH_3\overset{\displaystyle O}{\underset{}{C}}O^- \xrightarrow[H_3O^+]{LiAlH_4} CH_3CH_2OH$

d.

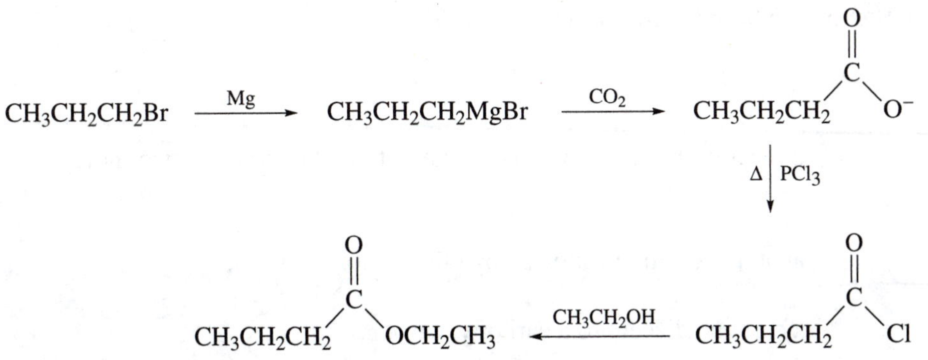

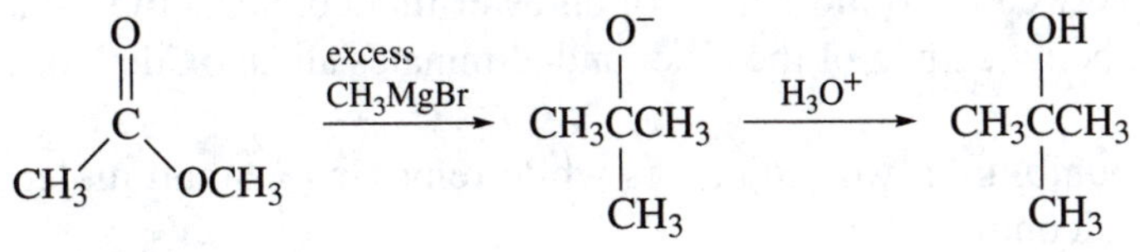

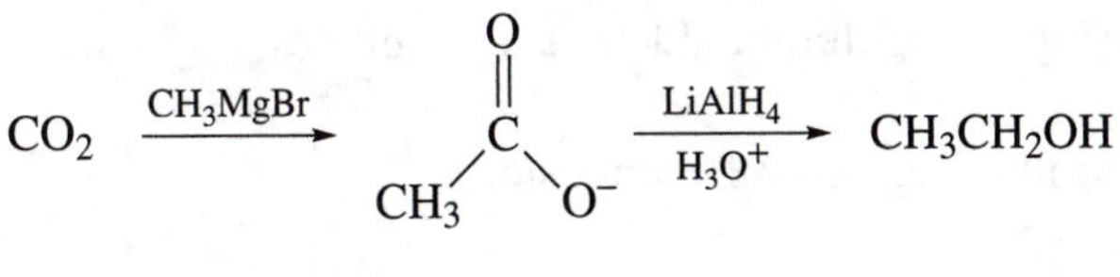

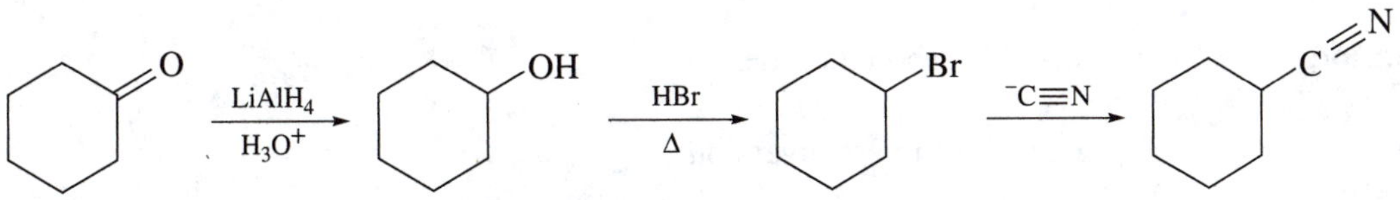

$\xrightarrow[H_3O^+]{LiAlH_4}$ (cyclohexanol, OH) $\xrightarrow[\Delta]{HBr}$ (cyclohexyl bromide, Br) $\xrightarrow{^-C\equiv N}$ (cyclohexane carbonitrile, $C\equiv N$)

CHAPTER 13
Reactions at the α-Carbon of Carbonyl Compounds

Important Terms

aldol addition	a reaction between two molecules of an aldehyde (or two molecules of a keton that connects the α-carbon of one with the carbonyl carbon of the other.
aldol condensation	an aldol addition followed by elimination of water.
α-carbon	a carbon adjacent to a carbonyl carbon.
Claisen condensation	a reaction between two molecules of an ester that connects the α-carbon of or with the carbonyl carbon of the other and eliminates an alkoxide ion.
condensation reaction	a reaction combining two molecules while removing a small molecule (usuall water or an alcohol).
crossed aldol addition	an aldol addition using two different aldehydes or ketones.
crossed Claisen condensation	a Claisen condensation using two different esters.
decarboxylation	loss of carbon dioxide.
enolization	keto–enol interconversion.
gluconeogenesis	the synthesis of D-glucose from pyruvate.
glycolysis	the breakdown of D-glucose into two molecules of pyruvate.
α-hydrogen	a hydrogen bonded to the carbon adjacent to a carbonyl carbon.
keto–enol tautomerism (keto–enol interconversion)	an interconversion of keto and enol tautomers.
β-keto ester	an ester with a carbonyl group at the β-position.
α-substitution reaction	a reaction that puts a substituent on an α-carbon in place of an α-hydrogen.
tautomers	constitutional isomers that are in rapid equilibrium—for example, keto and eno tautomers. The keto and enol tautomers differ only in the location of a double bon and a hydrogen.

olutions to Problems

a. The ketone is a stronger acid than the ester, because the electrons left behind when a proton is removed from the α-carbon of the ketone are more readily delocalized onto the carbonyl oxygen atom. When the ester loses a proton, the electrons have to compete with the lone pair on the alkoxy group oxygen for delocalization onto the carbonyl oxygen.

b. Because it is the weaker acid, the ester has the greater pK_a value.

2,4-Pentanedione is a stronger acid because the electrons left behind when a proton is removed can be readily delocalized onto two carbonyl oxygen atoms. The electrons left behind when a proton is removed from ethyl 3-oxobutyrate can be readily delocalized onto one carbonyl oxygen, but there is competition for delocalization onto the second carbonyl oxygen by the ethoxy group.

A proton cannot be removed from the α-carbon of N-methylethanamide or ethanamide, because these compounds have a hydrogen bonded to the nitrogen and this hydrogen is more acidic than the one attached to the α-carbon. Therefore, the hydrogen attached to the nitrogen will be the one removed by a base. In the case of N,N-dimethylethanamide, there is no N—H proton, so a proton can be removed from the α-carbon.

The following resonance contributors show why the hydrogen attached to the nitrogen is more acidic (the nitrogen has a partial positive charge) than the hydrogen attached to the α-carbon.

5.   Acyl chlorides are much more reactive than ketones or esters, so it is easier for hydroxide ion to attack th
reactive carbonyl group than to remove a hydrogen from an α-carbon.

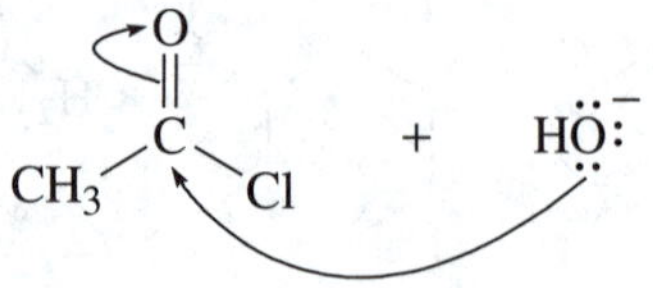

6.   **a.**

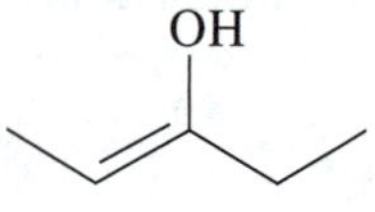

b.

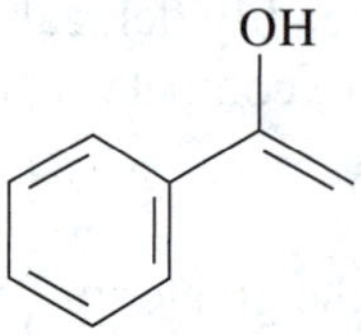

c.

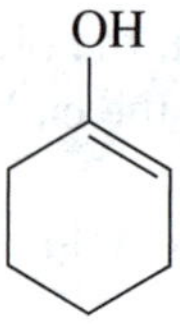

7.

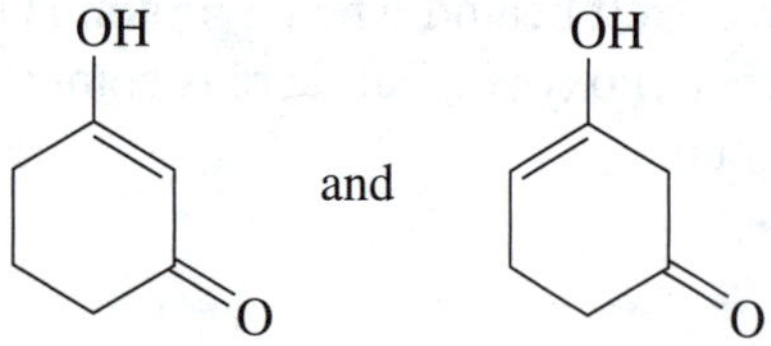

and

more stable,
because the double bonds
are conjugated

8.   **a.**

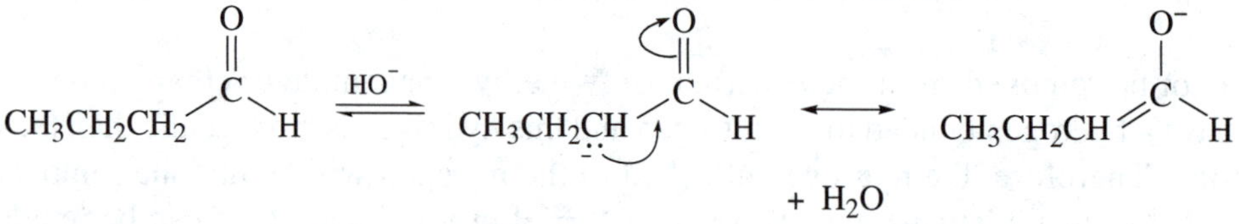

+ H$_2$O

b.

+ H$_2$O

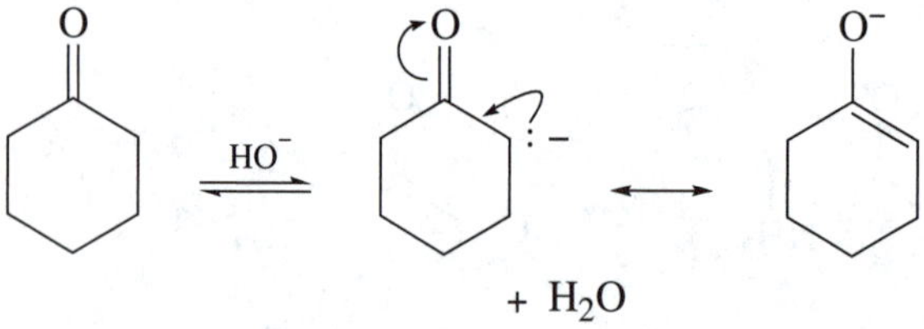

9.   The methyl hydrogens can be removed by a base ($^-$OD) and then are reprotonated by D$_2$O. The aldehyd
hydrogen cannot be removed by a base because the electrons left behind if it were to be removed cannot be
delocalized.

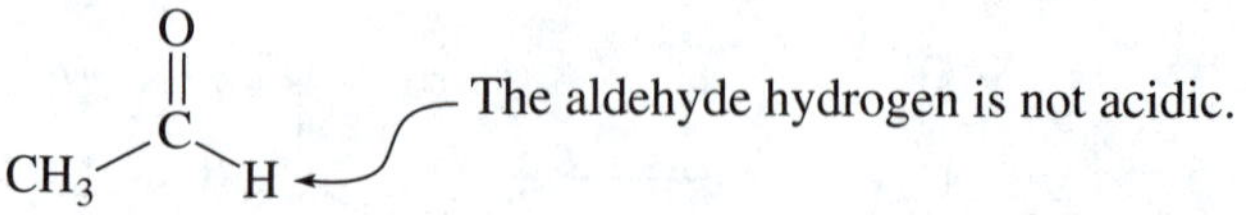

The aldehyde hydrogen is not acidic.

0.   **a.** 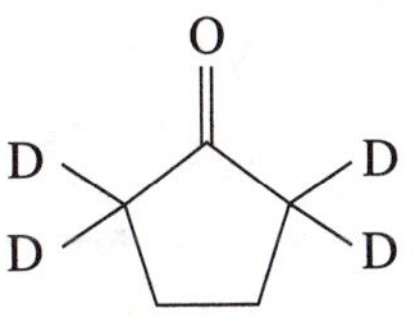   **b.** 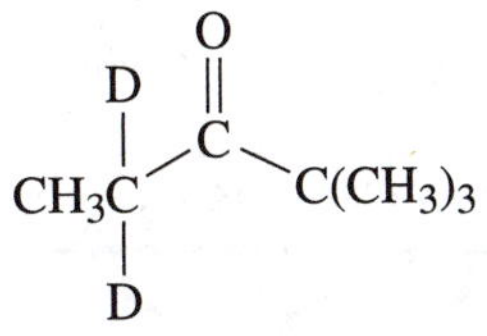   **c.**

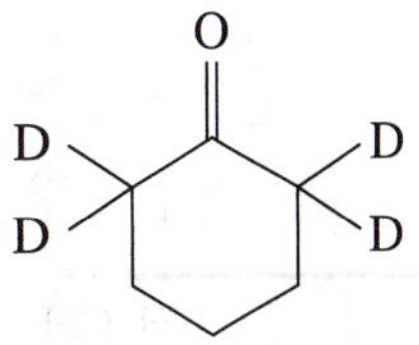

1.   Alkylation of an α-carbon is an S_N2 reaction. S_N2 reactions work best with primary alkyl halides because a primary alkyl halide has less steric hindrance than a secondary alkyl halide. S_N2 reactions do not work at all with tertiary alkyl halides because they are the most sterically hindered of the alkyl halides. Therefore, in the case of tertiary alkyl halides, the S_N2 reaction cannot compete with the E2 elimination reaction.

2.   **a.** 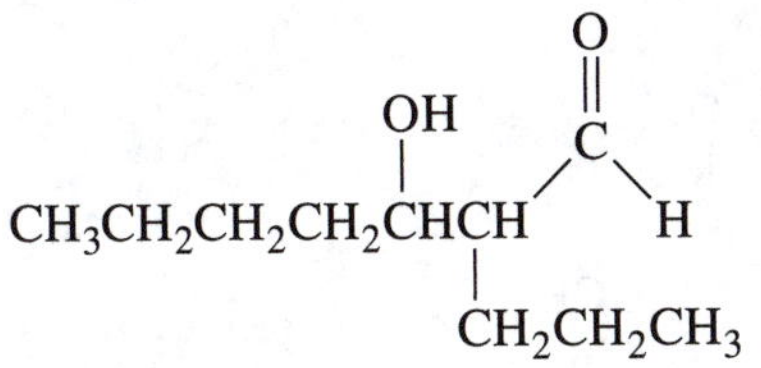   **b.** 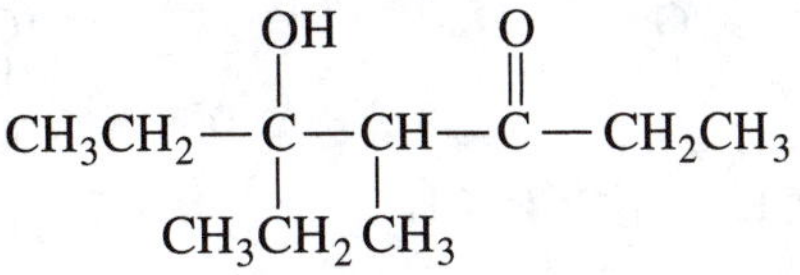   **c.**

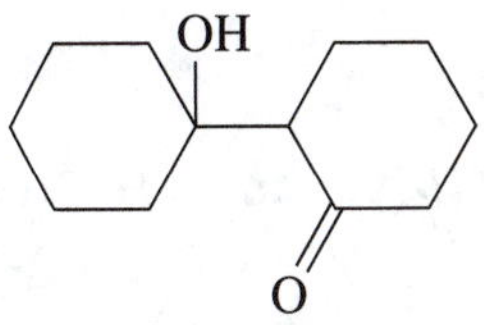

13.   **a.** 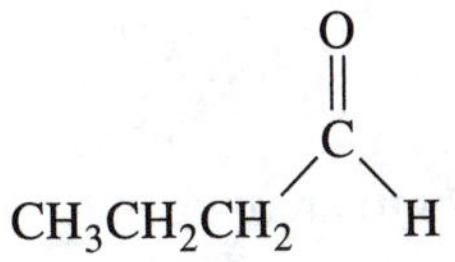   **b.**

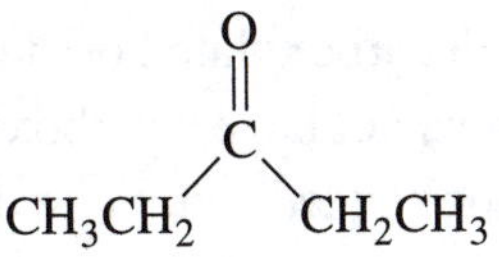

14.

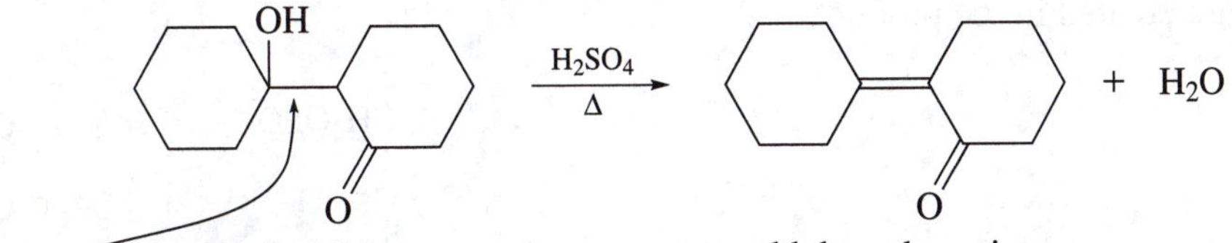

aldol addition   aldol condensation

A bond has formed between the α-carbon of one
molecule and the carbonyl carbon of another.

15.   **a.**   Solved in the text.

   b.

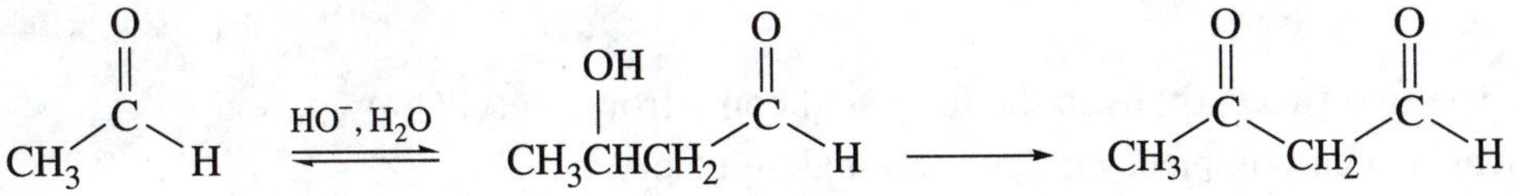

16.

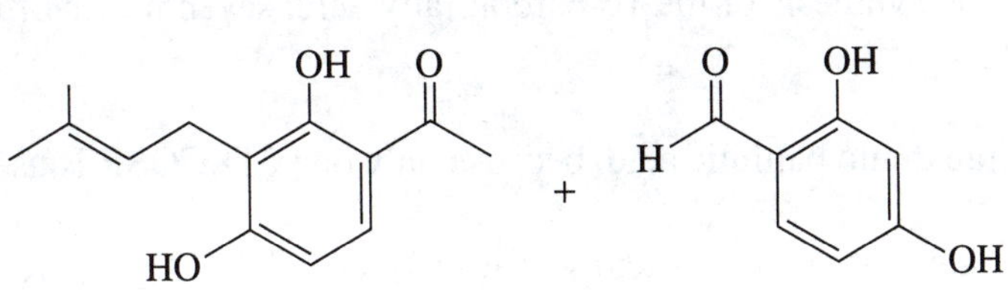

17. **a.**

$$CH_3CH_2CH_2-\underset{\underset{CH_2CH_3}{|}}{\overset{\overset{O}{\|}}{C}}-\underset{}{\overset{}{CH}}-\overset{\overset{O}{\|}}{C}-OCH_3$$

b.

$$CH_3CHCH_2-\overset{\overset{O}{\|}}{C}-\underset{\underset{\underset{CH_3}{|}}{CHCH_3}}{\overset{}{CH}}-\overset{\overset{O}{\|}}{C}-OCH_2CH_3 \quad (CH_3 \text{ on first CH})$$

18.   **A, B,** and **D** cannot undergo a Claisen condensation.

A cannot, because a proton cannot be removed from an sp^2 carbon.

B and **D** cannot, because they do not have an α-hydrogen.

19. **a.**

$$CH_3CH_2O-\overset{\overset{O}{\|}}{C}-\underset{\underset{CH_3}{|}}{\overset{}{CH}}-\overset{\overset{O}{\|}}{C}-OCH_2CH_3$$

b.

$$H-\overset{\overset{O}{\|}}{C}-\underset{\underset{CH_2CH_3}{|}}{\overset{}{CH}}-\overset{\overset{O}{\|}}{C}-OCH_3$$

20.   **A** and **D** are β-keto acids, so they can be decarboxylated on heating.

B cannot be decarboxylated, because it does not have a carboxyl group.

The electrons left behind if **C** were decarboxylated cannot be delocalized onto an oxygen.

21.   Because the catalyst is a hydroxide ion rather than an enzyme, four stereoisomers will be formed since two asymmetric centers are created in the product.

22.   **Seven** moles. The first two carbons in the fatty acid come from acetyl CoA.

Each subsequent two-carbon piece comes from malonyl CoA.

Because this amounts to 14 carbons for the synthesis of the 16-carbon fatty acid, seven moles of malonyl CoA are required.

23. **a.** **Three** deuteriums would be incorporated into palmitic acid, because only one CD_3COSR is used in the synthesis.

b. **Seven** deuteriums would be incorporated into palmitic acid, because seven $^-OOCCD_2COSR$ are used in the synthesis (for a total of 14 Ds), and each $^-OOCCD_2COSR$ loses one deuterium in the dehydration step $(14\,Ds - 7\,Ds = 7\,Ds)$.

24.

a.

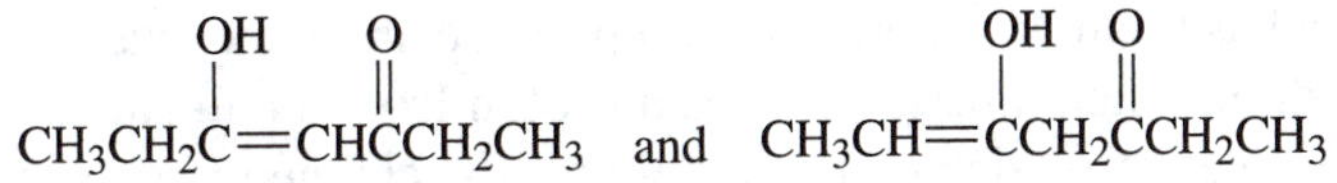

$$\underset{\text{more stable}}{CH_3CH_2\overset{\overset{\displaystyle OH}{|}}{C}=CH\overset{\overset{\displaystyle O}{\|}}{C}CH_2CH_3} \quad \text{and} \quad CH_3CH=\overset{\overset{\displaystyle OH}{|}}{C}CH_2\overset{\overset{\displaystyle O}{\|}}{C}CH_2CH_3$$

more stable

because the double bonds are conjugated

b.

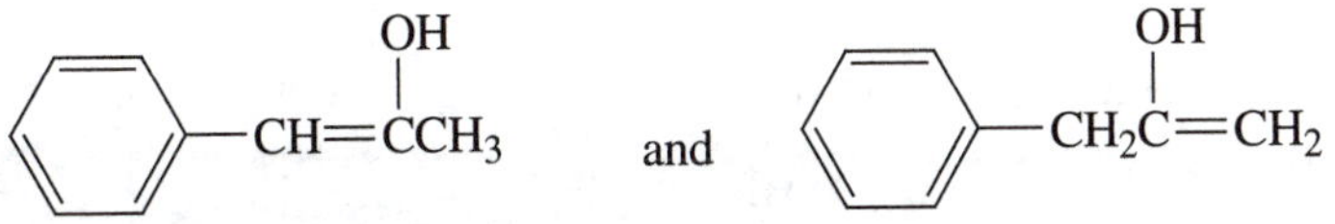

more stable
because the double bond is
conjugated with the benzene ring

c.

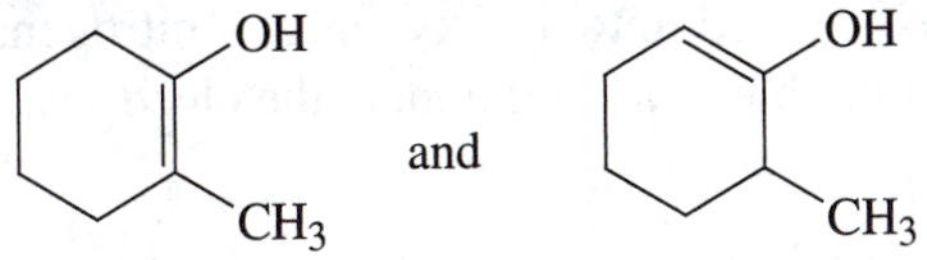

more stable

because the sp^2 carbons are
attached to a greater number
of alkyl substituents

25.

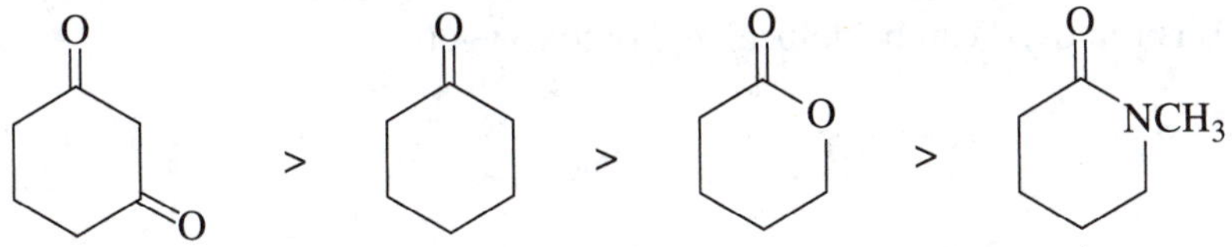

26. Electron delocalization of the lone pair on nitrogen or oxygen competes with electron delocalization of the electrons left behind on the α-carbon when it loses a proton. The lone pair on nitrogen is more delocalized than the lone pair on oxygen, because nitrogen is better able to accommodate a positive charge since it is less electronegative than oxygen. Therefore, the amide competes better with the carbanion for electron delocalization, so the α-carbon is less acidic.

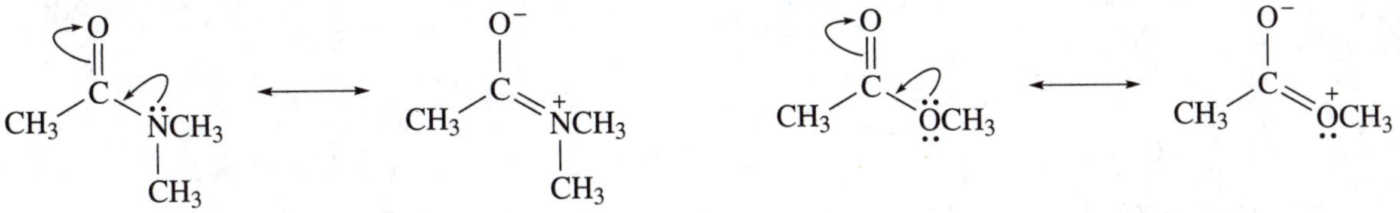

27.   The electrons left behind when a base removes a proton from propene are delocalized over three carbon. In contrast, the electrons left behind when a base removes a proton from an alkane are localized—the belong to a single carbon. Because electron delocalization allows the charge to be distributed over mor than one atom, it stabilizes the base, and the more stable the base, the stronger its conjugate acid. There fore, propene is a stronger acid than an alkane.

$$B\overset{\cdot\cdot}{}^- + CH_2{=}CHCH_3 \;\rightleftharpoons\; CH_2{=}CH\overset{\cdot\cdot}{C}H_2 \;\longleftrightarrow\; {}^-CH_2CH{=}CH_2$$
$$+\ HB$$

$$B\overset{\cdot\cdot}{}^- + CH_3CH_2CH_3 \;\rightleftharpoons\; {}^-\overset{\cdot\cdot}{C}H_2CH_2CH_3 + HB$$

Propene, however, is not as acidic as the carbon acids, because the electrons left behind when a base removes a proton from these carbon acids are delocalized onto an oxygen or a nitrogen, which are more electronegative than carbon and, therefore, are better able to accommodate the electrons.

$$B\overset{\cdot\cdot}{}^- + RCH_2{-}\overset{\displaystyle :\overset{\cdot\cdot}{O}}{\overset{\|}{C}}{-}R \;\longrightarrow\; RCH{-}\overset{\displaystyle :\overset{\cdot\cdot}{O}}{\overset{\|}{C}}{-}R \;\longleftrightarrow\; RCH{=}\overset{\displaystyle :\overset{\cdot\cdot}{O}{:}^-}{\overset{|}{C}}{-}R$$
$$+\ HB$$

28.   The compound on the far right loses CO_2 at the lowest temperature, because it is a β-keto acid. Thus, the electrons left behind when CO_2 is removed can be delocalized onto oxygen.

29.   a.   **b.**

30.

31. The initially formed addition product loses water immediately, because the new double bonds are particularly stable (and therefore easy to form) since each is conjugated with a benzene ring.

+ 2 H$_2$O

32. 18 carbons come from malonyl-CoA, so 9 moles of malonyl-CoA are required.

33.
a. Three deuteriums would be incorporated, all from the single molecule of CD$_3$COSR used at the beginning of the synthesis.
b. Nine deuteriums would be incorporated, one from each of the nine $^-$OOCCD$_2$COSR used in the synthesis. (Notice that one D is lost in the dehydration step.)

34. a. b. c.

35. Remember that there are no positively charged organic reactants, intermediates, or products in a basic solution, and no negatively charged organic reactants, intermediates, or products in an acidic solution.

a.

b.

36.

$$\underset{\underset{\underset{CH_2CH_3}{|}}{\overset{OH}{\overset{|}{C}}H}}{CH_3CH_2CH_2}\overset{O}{\overset{||}{C}H}$$

OH   O

CH₃CH₂CH₂CHCHCH
  |
  CH₂CH₃

OH   O

CH₃CH₂CH₂CHCHCH
  |
  CH₂CH₂CH₃

OH   O

CH₃CH₂CH₂CH₂CHCHCH
  |
  CH₂CH₃

OH   O

CH₃CH₂CH₂CH₂CHCHCH
  |
  CH₂CH₂CH₃

37.   In a basic solution, the ketone will be in equilibrium with its enol tautomer; when the enol tautomer forms the asymmetric center is lost. When the enol tautomer reforms the ketone, it can form the R and S enantio mer equally as easily, so a racemic mixture is obtained.

asymmetric center

O   CH₂CH₃   OH   O
|| | HO⁻ | H₂O || *
C—C   ⇌   C=CCH₂CH₃   ⇌   C—CHCH₂CH₃
CH₃CH₂ CH₃   H₂O   CH₃CH₂ CH₃   HO⁻   CH₃CH₂   CH₃
        H

(R)-2-methyl-1-
phenyl-1-butanone

38.   You need a ketone that has an α-carbon that is an asymmetric center. A racemic mixture will be formed when an α-hydrogen is removed from the asymmetric carbon.

O   — asymmetric center
||
CH₃CCHCH₂CH₃
    |
    CH₃

39.

O
||
CH₂CH₂CCH₃   HO⁻
|
CH₂CCH₃
  ||
  O

2,6-heptanedione

O
||
CH₂CH₂CCH₃
|
CH₂CCH₂
  ||
  O   + H₂O

                O⁻
                |
                CH₃
         ⇌
                O

H₂O   OH
⇌ |
                  CH₃

                  O   + HO⁻

O
||
CH₂CH₂CCH₃   HO⁻
|
CH₂
|
CH₂CH₂CCH₃
       ||
       O

2,8-nonanedione

O
||
CH₂CHCCH₃
|
CH₂
|
CH₂CH₂CCH₃
       ||
       O   + H₂O

                O
                ||
                CCH₃
         ⇌
                CH₃
                O⁻

H₂O   O
⇌ ||
                  CCH₃

                  CH₃
                  OH

                  + HO⁻

40.

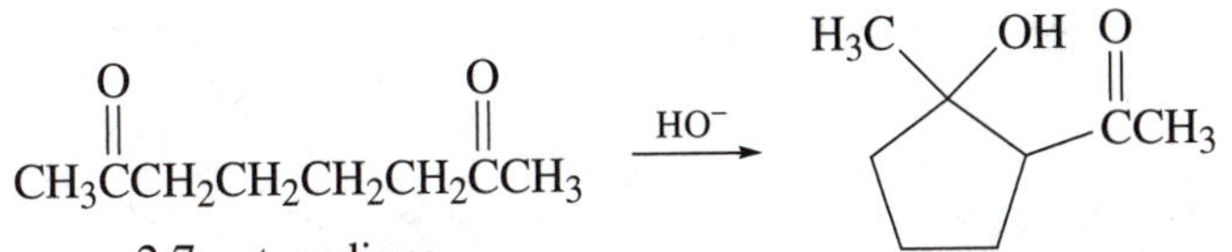

2,7-octanedione

41.

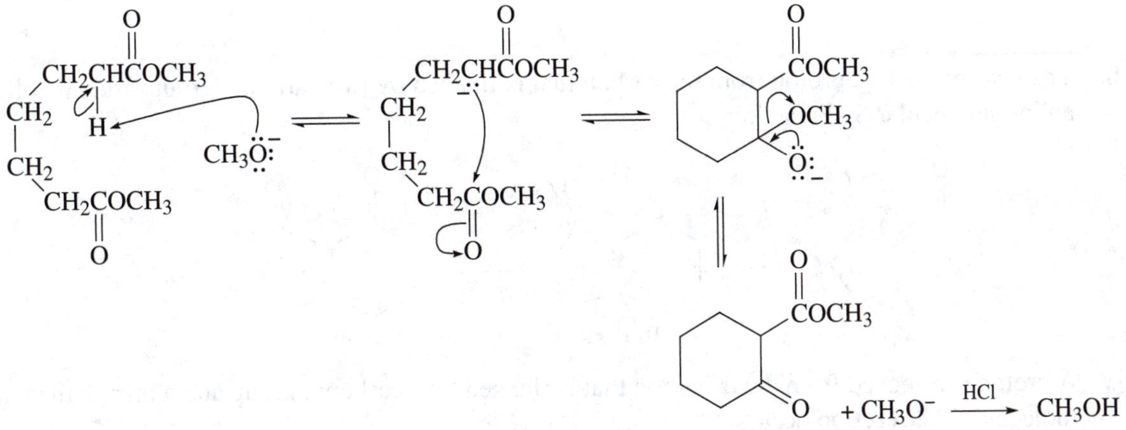

42.

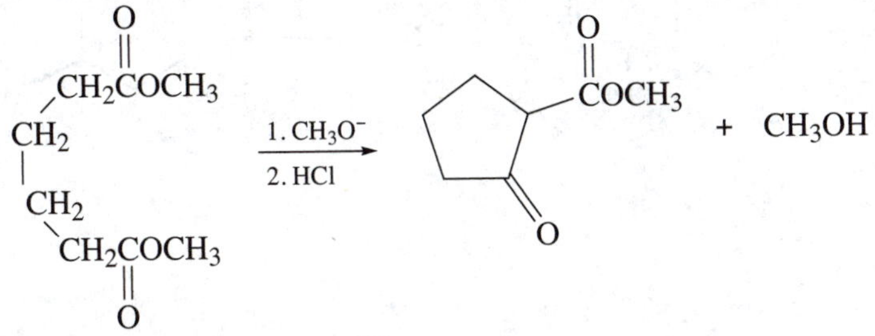

+ CH₃OH

43.

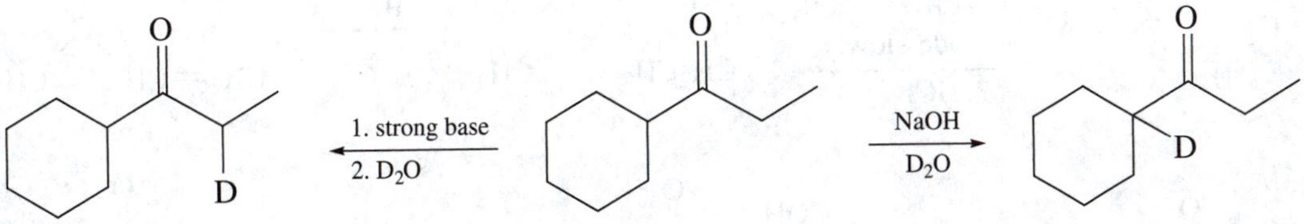

A strong base forms
the kinetic enolate,
which leads to substitution
at the least substituted α-carbon.

A relatively week base forms
the thermodynamic enolate,
which leads to substitution
at the most substituted α-carbon.

44.   **a.**   The reaction involves two successive aldol condensations.

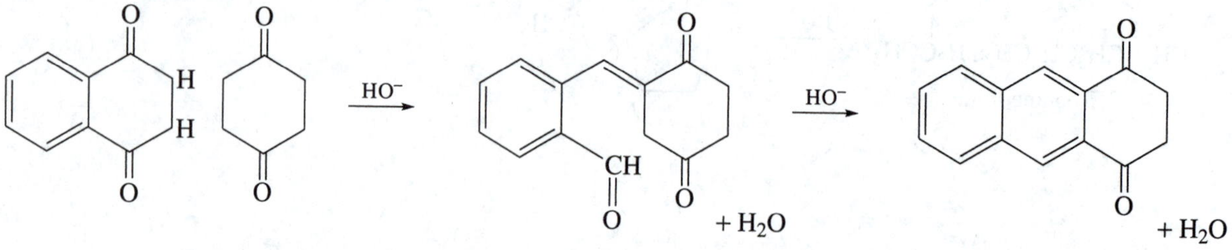

b.   The base removes a proton from the carbon that is flanked by two carbonyl groups; this is followed by an intramolecular S_N2 reaction.

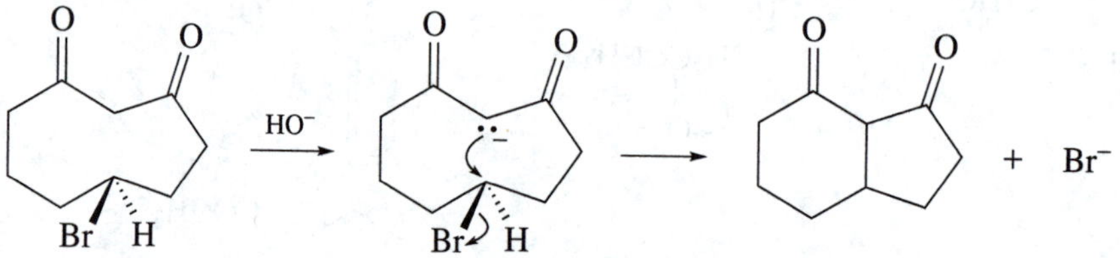

c.   A proton is removed from the α-carbon that is flanked by a carbonyl group and a nitrile; then, an intramolecular S_N2 reaction occurs.

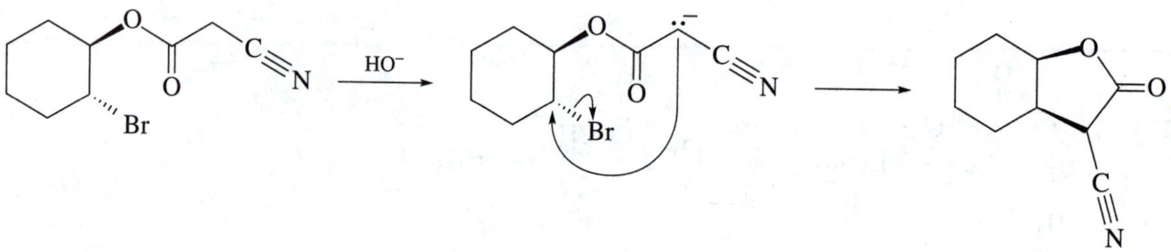

45.   **a.**

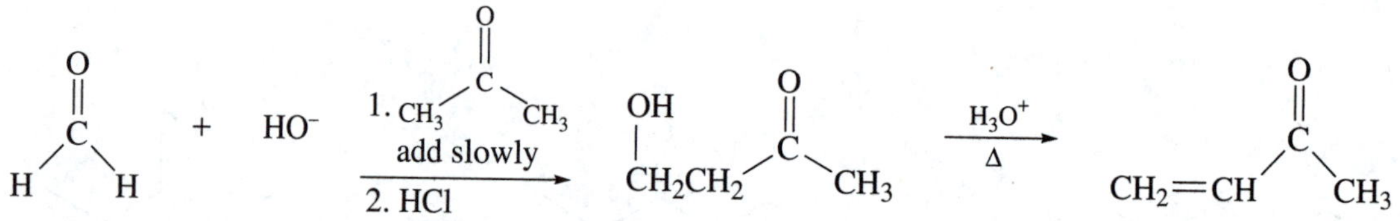

b.

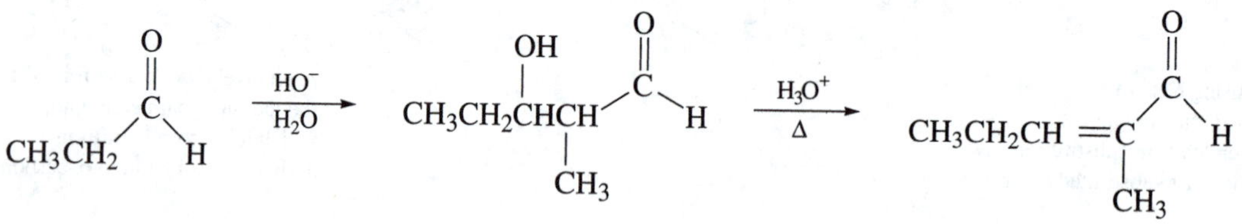

46.

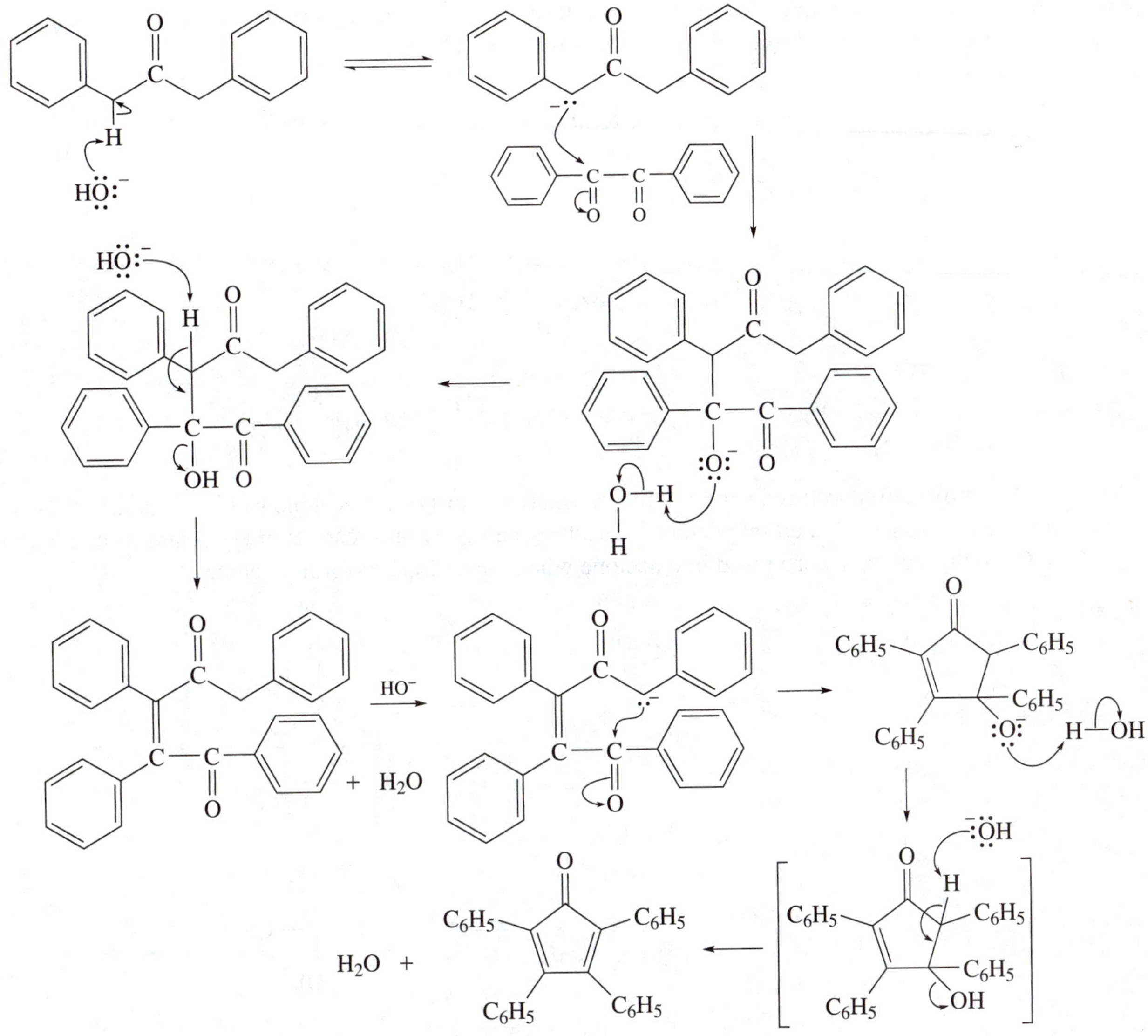

HO:⁻

HO:⁻

+ H₂O

HO⁻

C₆H₅ — C₆H₅

C₆H₅

C₆H₅   :O: H—OH

C₆H₅ — C₆H₅

C₆H₅

C₆H₅   :ÖH

H₂O +

C₆H₅ — C₆H₅

C₆H₅   C₆H₅

47. Decarboxylation of the β-dicarboxylic acid would require a higher temperature, because the electrons left behind when CO_2 is removed are not as readily delocalized onto the carbonyl oxygen since a lone pair on the second OH group can also be delocalized onto that oxygen.

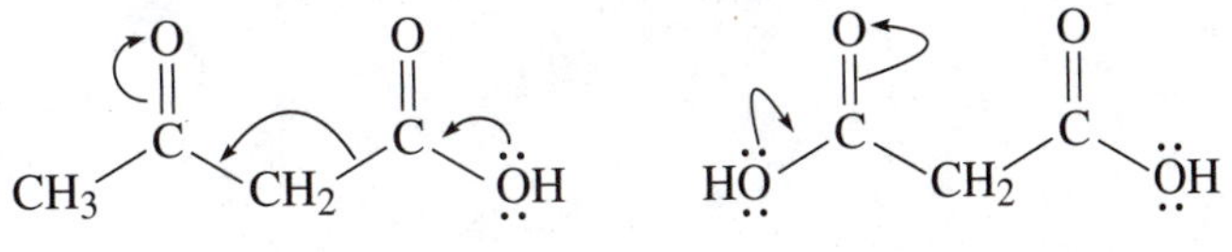

CH₃   CH₂   ÖH

β-keto acid

HÖ   CH₂   ÖH

β-dicarboxylic acid

48.   It tells you that an imine is formed as an intermediate. Because an imine is formed, the only source of oxygen for the carbonyl oxygen of acetone formation is $H_2^{18}O$.

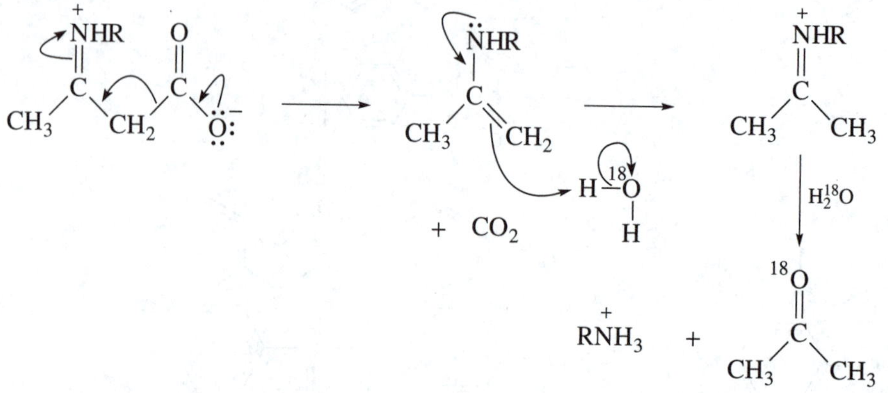

$+ \; CO_2$

$RNH_3^+ \quad + \quad$ (acetone with ^{18}O)

$H_2^{18}O$

If decarboxylation occurred without imine formation, most of the acetone would contain ^{16}O. There would be some ^{18}O incorporated into acetone, because acetone would react with $H_2^{18}O$ and form a hydrate. The ^{18}O would become incorporated into acetone when the hydrate re-forms acetone.

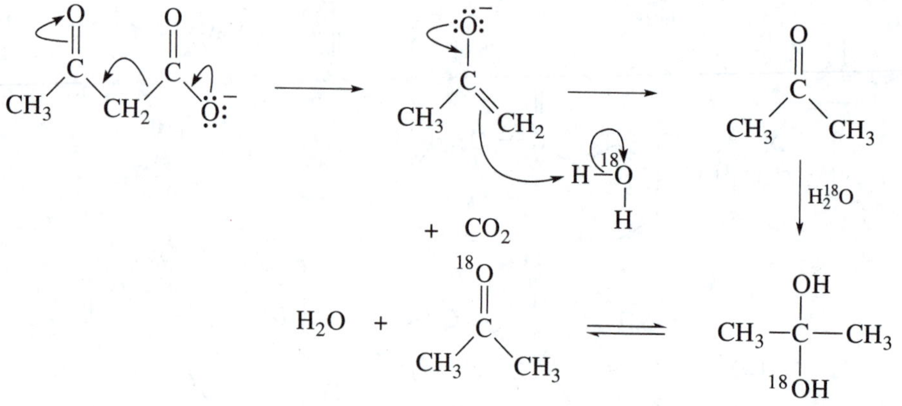

$+ \; CO_2$

$H_2O \quad + \quad$ (acetone with ^{18}O) $\rightleftharpoons$ $CH_3-\overset{OH}{\underset{^{18}OH}{C}}-CH_3$

$H_2^{18}O$

49.

$CH_3CH_2O^-$

$CH_3CH_2O^- \quad + \quad$

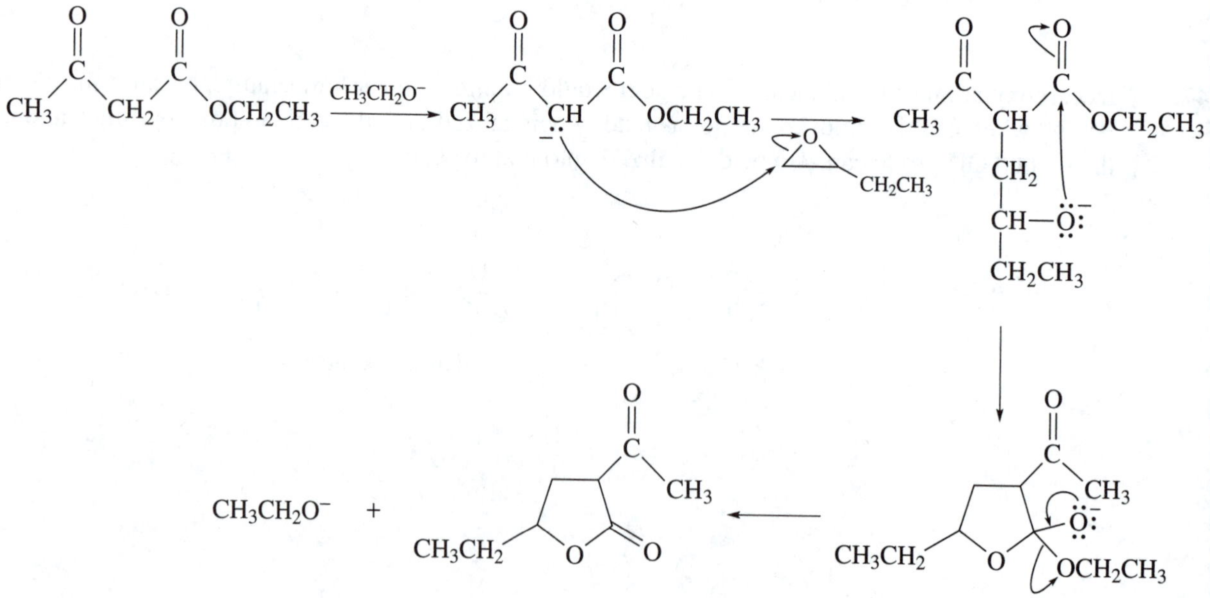

50. **a.** 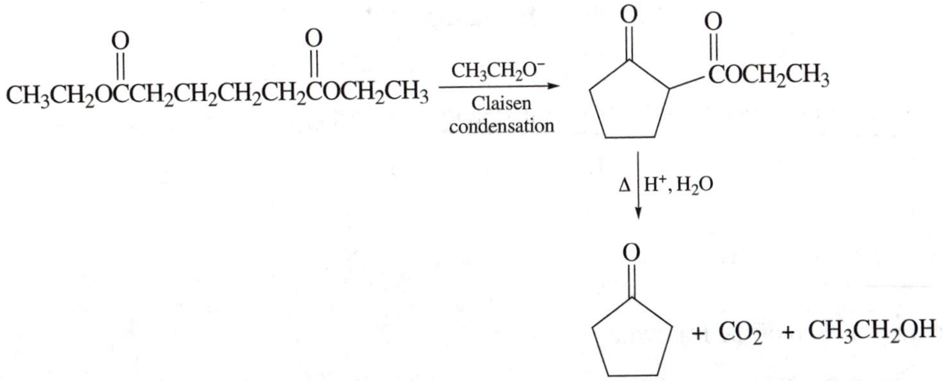 $CH_3CH_2OCCH_2CH_2CH_2CH_2COCH_2CH_3$ →[$CH_3CH_2O^-$ / Claisen condensation]

Δ | H^+, H_2O

$+ CO_2 + CH_3CH_2OH$

b. 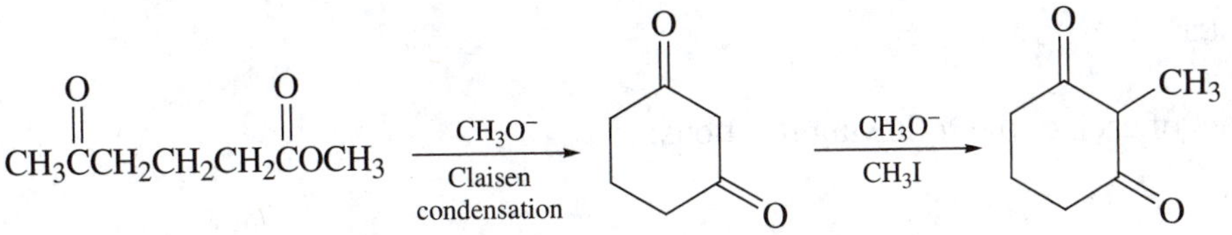 $CH_3CCH_2CH_2CH_2COCH_3$ →[Claisen condensation / CH_3O^-]

→[CH_3O^- / CH_3I]

51. **a.** 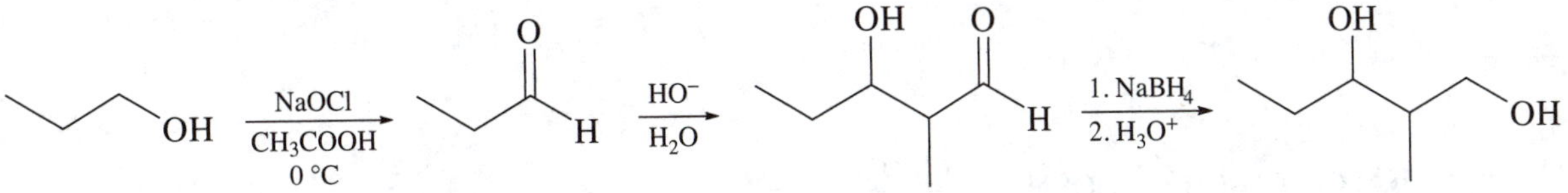 →[NaOCl / CH_3COOH / 0 °C] →[HO^- / H_2O] →[1. $NaBH_4$ / 2. H_3O^+]

b. 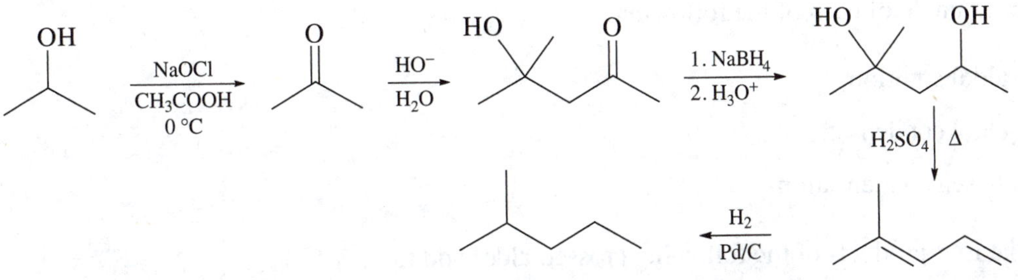 →[NaOCl / CH_3COOH / 0 °C] →[HO^- / H_2O] →[1. $NaBH_4$ / 2. H_3O^+]

H_2SO_4 | Δ

←[H_2 / Pd/C]

52. **a.**

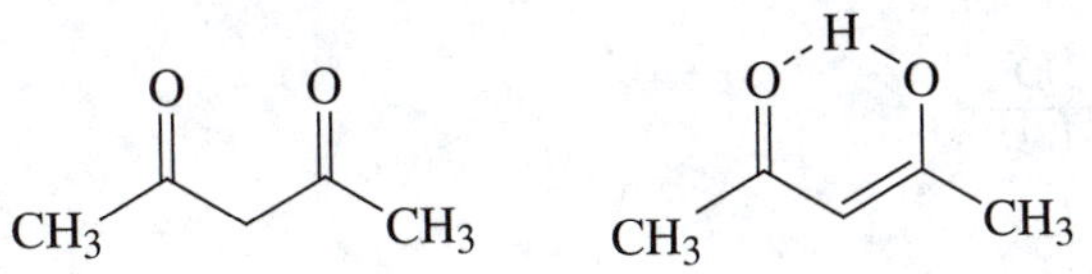

keto tautomer   enol tautomer

2,4-pentanedione

b. The enol tautomer of 2,4-pentanedione is more stable than most enol tautomers because it can form an intramolecular hydrogen bond that stabilizes it.

Chapter 13 Practice Test

1. Rank the following compounds in order from most acidic to least acidic:

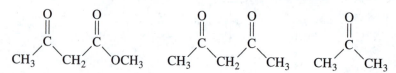

2. Draw a structure for each of the following:

   a. the most stable enol tautomer of 2,4-pentanedione

   b. a β-keto ester

3. Draw the product of each of the following reactions:

   a. $CH_3CH_2\overset{O}{\underset{}{\overset{||}{C}}}CH_2\overset{O}{\underset{}{\overset{||}{C}}}OH \xrightarrow{\Delta}$

   b. $CH_3CH_2CH_2\overset{O}{\underset{}{\overset{||}{C}}}OCH_3 \xrightarrow[\text{2. HCl}]{\text{1. } CH_3O^-}$

4. Give an example of each of the following:

   a. an aldol addition

   b. an aldol condensation

   c. a Claisen condensation

5. Draw the four products of the following crossed aldol addition:

   $CH_3CHCH_2CH_2\overset{O}{\underset{}{\overset{||}{C}}}H + CH_3CH_2CH_2\overset{O}{\underset{}{\overset{||}{C}}}H \xrightarrow[H_2O]{HO^-}$
   $\quad\quad|$
   $\quad\,CH_3$

6. What ester would be required to prepare each β-keto ester?

   a. $CH_3CH_2CH_2\overset{O}{\underset{}{\overset{||}{C}}}CH\overset{O}{\underset{}{\overset{||}{C}}}OCH_3$
   $\quad\quad\quad\quad\quad|$
   $\quad\quad\quad\quad CH_2CH_3$

   b.

Answers to Chapter 13 Practice Test

1. 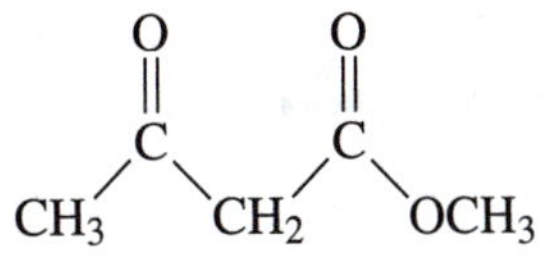
2

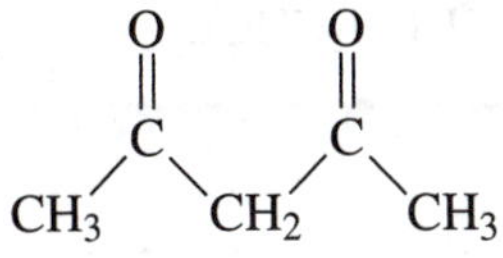
1

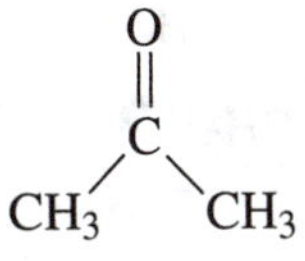
3

2. **a.** 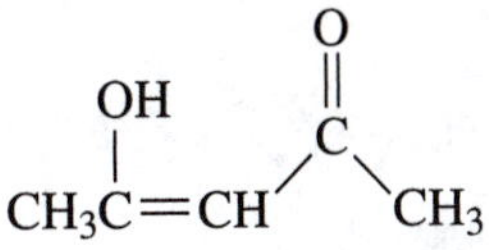 **b.** 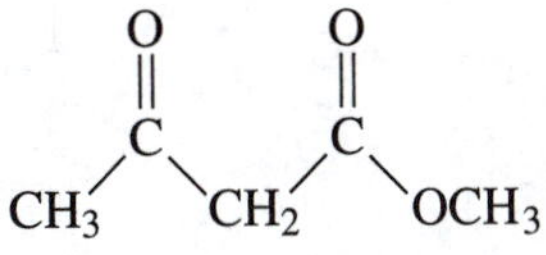

3. **a.** 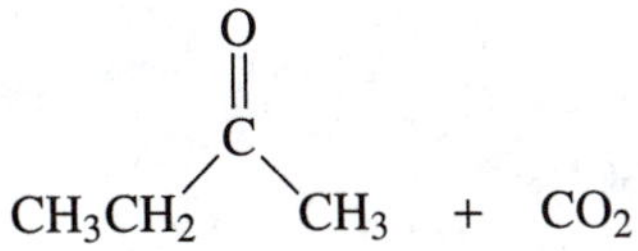 **b.** 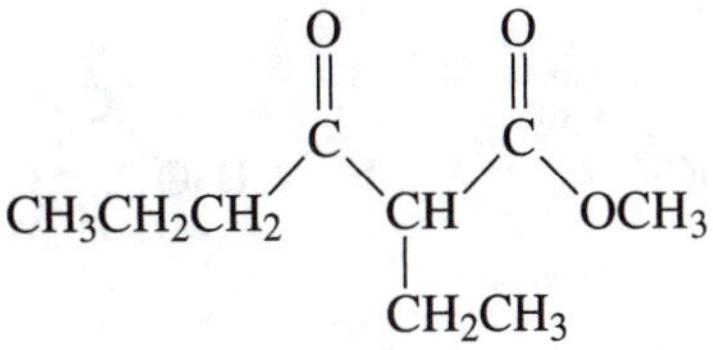

4. **a.** 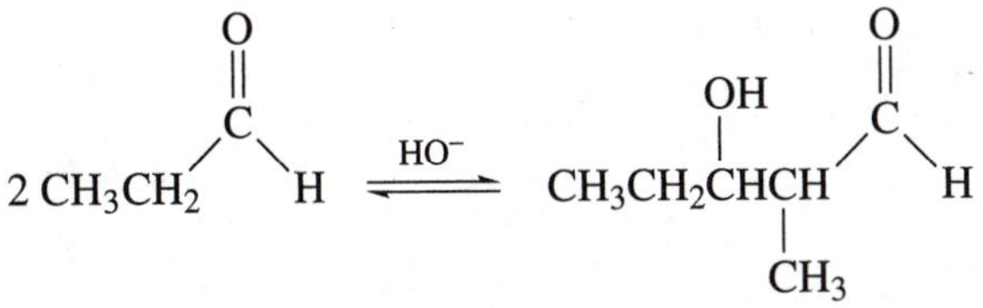

b. 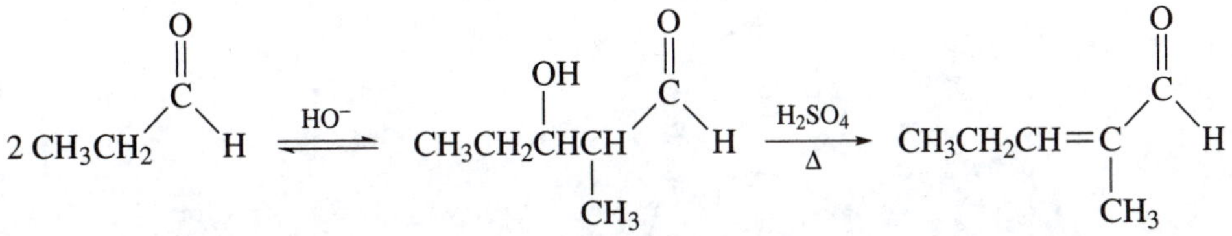

c. 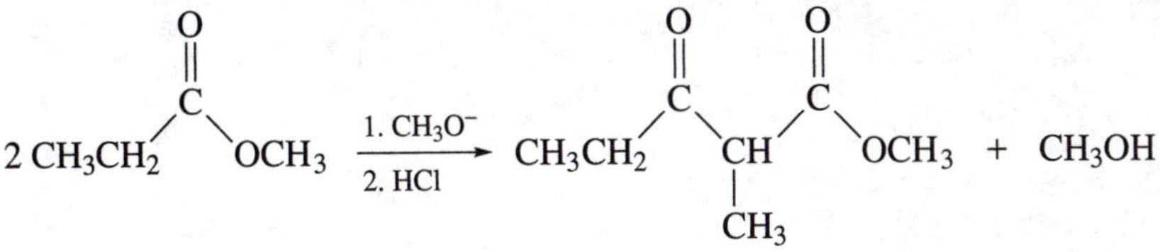

5.

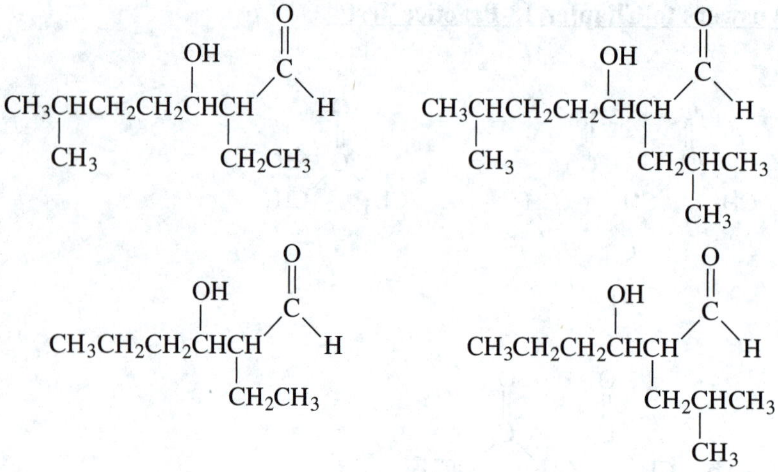

6. **a.**

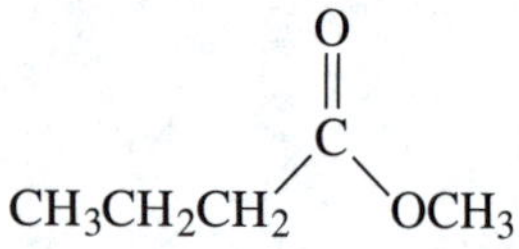

b.

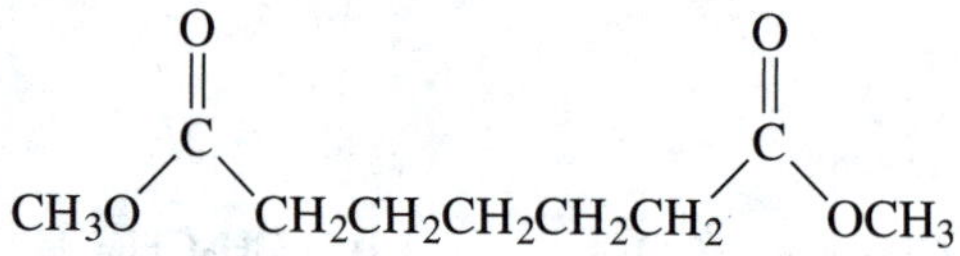

Important Terms

alkane	a hydrocarbon that contains only single bonds.
free radical (radical)	an atom or a molecule with an unpaired electron.
halogenation reaction	the reaction of an alkane (or other organic compounds) with a halogen.
heterolytic bond cleavage (heterolysis)	breaking a bond with the result that both bonding electrons stay with one of the previously bonded atoms.
homolytic bond cleavage (homolysis)	breaking a bond with the result that each of the atoms that formed the bond gets one of the bonding electrons.
initiation step	the step in which radicals are created and/or the step in which the radical needed for the first propagating step is created.
peroxide	a compound with an O—O bond.
primary alkyl radical	an alkyl radical with the unpaired electron on a primary carbon.
propagation step	in the first of a pair of propagation steps, a radical reacts to produce another radical that reacts in the second propagation step to produce the radical that was the reactant in the first propagation step.
radical (often called a free radical)	an atom or a molecule with an unpaired electron.
radical chain reaction	a reaction in which radicals are formed and react in repeating propagating steps.
radical inhibitor	a compound that traps radicals.
radical initiator	a compound that creates radicals.
radical substitution reaction	a substitution reaction that has a radical intermediate.
saturated hydrocarbon	a hydrocarbon that contains only single bonds (it is saturated with hydrogen).
secondary alkyl radical	an alkyl radical with the unpaired electron on a secondary carbon.
termination step	two radicals combine to produce a molecule in which all the electrons are paired.
tertiary alkyl radical	an alkyl radical with the unpaired electron on a tertiary carbon.

Solutions to Problems

1.

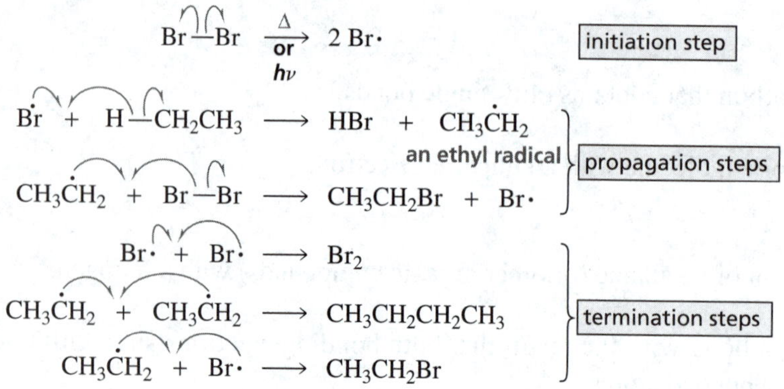

$$Br \overset{\frown\frown}{-} Br \xrightarrow[\text{or}]{\Delta} 2\ Br\cdot \qquad \boxed{\text{initiation step}}$$

$$Br\cdot\ +\ H\overset{\frown}{-}CH_2CH_3 \longrightarrow HBr\ +\ CH_3\dot{C}H_2$$
an ethyl radical

$$CH_3\dot{C}H_2\ +\ Br\overset{\frown}{-}Br \longrightarrow CH_3CH_2Br\ +\ Br\cdot$$
$\left.\rule{0pt}{24pt}\right\}$ $\boxed{\text{propagation steps}}$

$$Br\cdot\ +\ Br\cdot \longrightarrow Br_2$$

$$CH_3\dot{C}H_2\ +\ CH_3\dot{C}H_2 \longrightarrow CH_3CH_2CH_2CH_3$$
$\left.\rule{0pt}{34pt}\right\}$ $\boxed{\text{termination steps}}$

$$CH_3\dot{C}H_2\ +\ Br\cdot \longrightarrow CH_3CH_2Br$$

2.

$$Cl_2 \xrightarrow{h\nu} 2\ Cl\cdot \qquad \}\ \text{initiation}$$

$$Cl\cdot\ +\ Cl\cdot \longrightarrow Cl_2$$

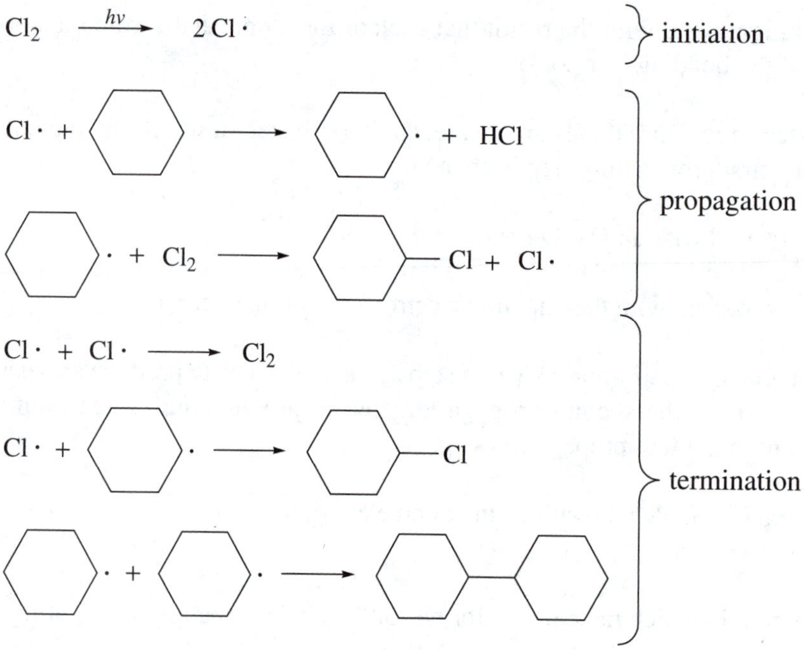

3.   Six secondary hydrogens

4.

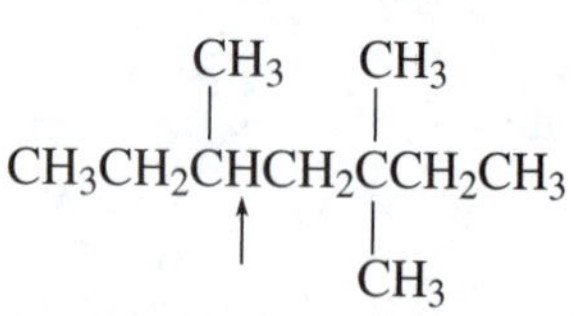

$$\underset{\displaystyle\uparrow}{CH_3CH_2\underset{\displaystyle\overset{|}{CH_3}}{CH}CH_2\underset{\displaystyle\overset{|}{CH_3}}{\overset{\overset{\textstyle CH_3}{|}}{C}}CH_2CH_3}$$

5.   **a.** 3   **c.** 3   **e.** 5
  b. 1   **d.** 5   **f.** 4

6.   Solved in the text.

7.   Chlorination, because the halogen is substituting for a primary hydrogen, and bromination would be more likely at the tertiary position.

8. Solved in the text.

9. Five products would be formed, because 2-chloro-3-methylbutane has an asymmetric center, so *R* and *S* stereoisomers will be formed.

0. To start peroxide formation, the chain-initiating radical removes a hydrogen atom from an α-carbon of the ether.

    a. **D** is most apt to form a peroxide, because removal of a hydrogen from an α-carbon forms a secondary radical.

    b. **B** is least apt to form a peroxide, because it does not have any hydrogens bonded to its α-carbons.

1. Four atoms (three carbons and one oxygen) share the unpaired electrons.

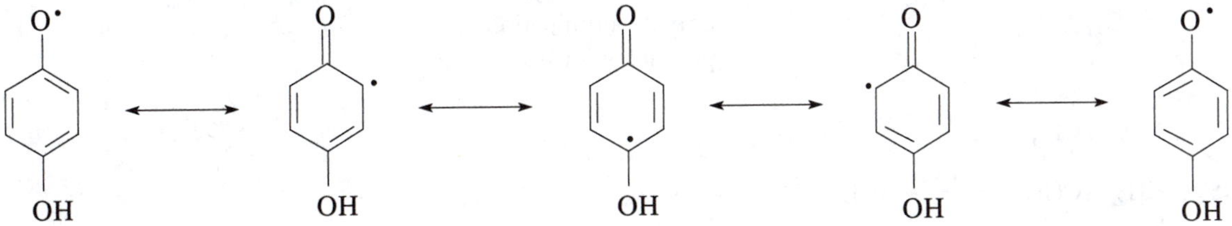

2.   **a.**

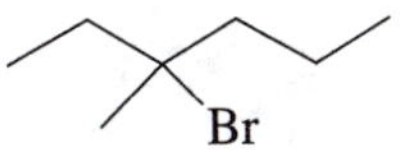

major product

    b.

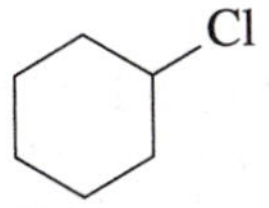

    c.

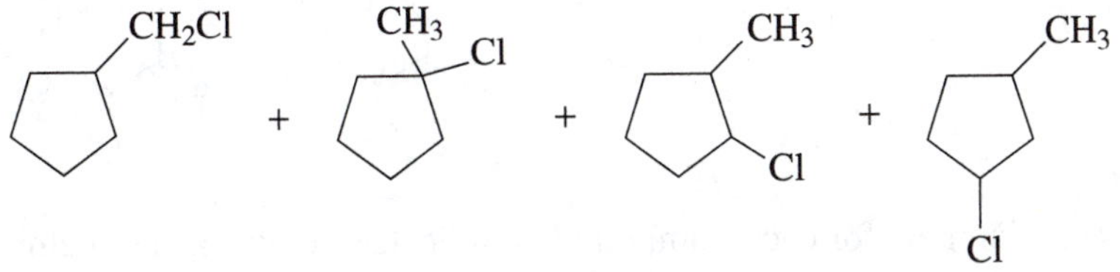

13.   **a.**

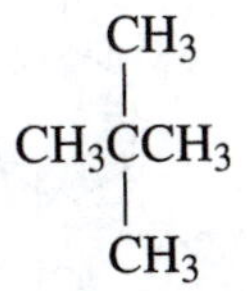

dimethylpropane

    b. 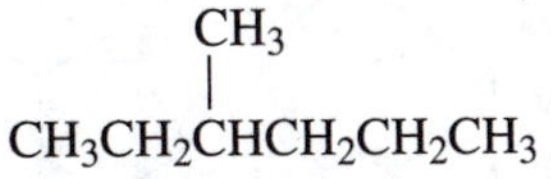 $CH_3CH_2CHCH_2CH_2CH_3$ (with CH_3 branch)

3-methylhexane

14. In each case, the major bromination product will be the one where bromine substitutes for a tertiary hydrogen.

    a. 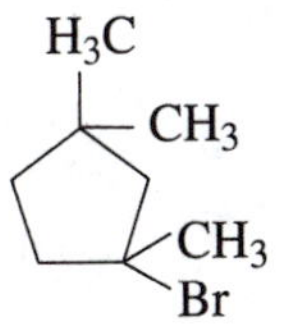   **b.** 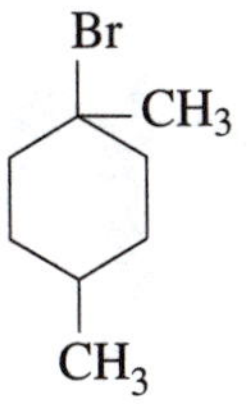   **c.**

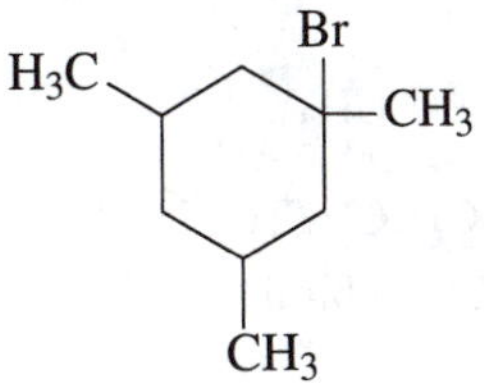

15.   **a.**   Bromination, because the halogen is substituting for a tertiary hydrogen.

b.   Because the molecule has only one kind of hydrogen, both chlorination and bromination will form only one monohalogenated product.

16.   **a.**

$$CH_3\overset{\underset{\displaystyle CH_3}{|}}{CH}CH_3 \xrightarrow[hv]{Br_2} CH_3\overset{\underset{\displaystyle Br}{|}}{\overset{\overset{\displaystyle CH_3}{|}}{C}}CH_3$$

b.

$$CH_3\overset{\underset{\displaystyle CH_3}{|}}{CH}CH_3 \xrightarrow[hv]{Br_2} CH_3\overset{\underset{\displaystyle Br}{|}}{\overset{\overset{\displaystyle CH_3}{|}}{C}}CH_3 \xrightarrow{HO^-} CH_2{=}\overset{\overset{\displaystyle CH_3}{|}}{C}CH_3$$

Because the alkyl halide is tertiary, only the elimination product is formed.

c.

$$CH_2{=}\overset{\overset{\displaystyle CH_3}{|}}{C}CH_3 \xrightarrow{HI} CH_3\overset{\underset{\displaystyle I}{|}}{\overset{\overset{\displaystyle CH_3}{|}}{C}}CH_3$$

answer to part **b**

17.   **a.**   Five monochlorination products are possible.

b.   The number of possible stereoisomers for each compound is indicated below each structure.

1 — 1   These do not have an asymmetric center.

4 — 4   These have two asymmetric centers.

2   This does not have an asymmetric center but has cis–trans isomers.

A total of 12 stereoisomers can be obtained.

18.   **a.**

$$CH_3\overset{\overset{\displaystyle CH_3}{|}}{CH}CH_3 \xrightarrow[hv]{Cl_2} \underset{\text{achiral}}{CH_3\overset{\overset{\displaystyle CH_3}{|}}{CH}CH_2Cl} + \underset{\text{achiral}}{CH_3\overset{\underset{\displaystyle Cl}{|}}{\overset{\overset{\displaystyle CH_3}{|}}{C}}CH_3}$$

b. 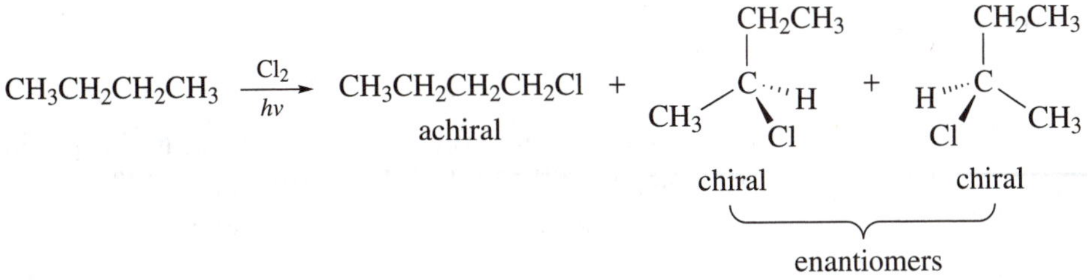 $CH_3CH_2CH_2CH_3 \xrightarrow[hv]{Cl_2} CH_3CH_2CH_2CH_2Cl$ +

achiral

chiral   chiral

enantiomers

9. Antioxidants are radical inhibitors—that is, they react with radicals and, thereby, prevent radical chain reactions. There are several OH groups in a catechin that, upon losing a hydrogen atom as a result of reacting with a reactive radical, form a radical that is stabilized by electron delocalization. This highly stabilized radical is sufficiently unreactive that it cannot damage cells by reacting with them. An example of one of the stabilized radicals is shown below.

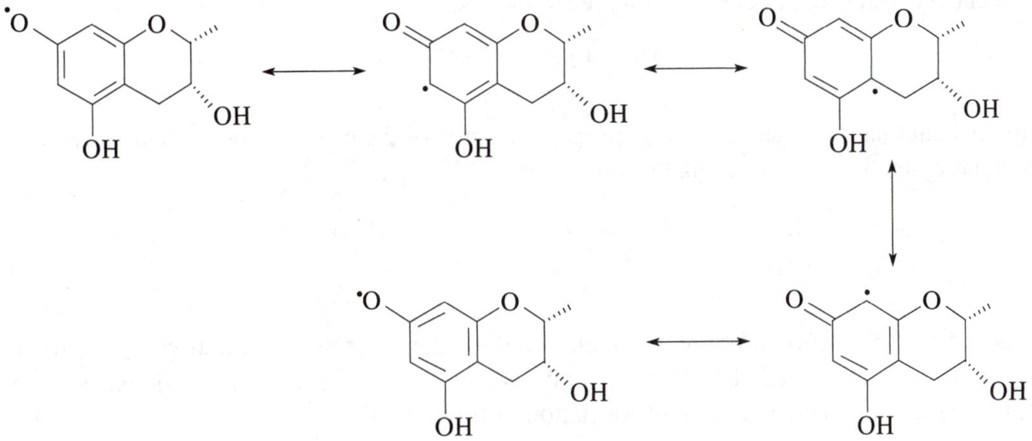

20.

$$CH_3CHCH_2CH_3$$
$$|$$
$$CH_3$$

2-methylbutane

2-Methylbutane forms two primary alkyl halides, one secondary alkyl halide and one tertiary alkyl halide. First, we need to calculate how much alkyl halide is formed from each primary hydrogen available, from each secondary hydrogen available, and from each tertiary hydrogen available.

a primary alkyl halide

$ClCH_2CHCH_2CH_3$
$|$
CH_3

Substitution of any one of six hydrogens leads to this product.
   percentage of this product that is formed = 36%
   percentage formed per hydrogen available = 36/6 = 6%

$CH_3CHCH_2CH_2Cl$
$|$
CH_3

Substitution of any one of three hydrogens leads to this product.
   percentage of this product that is formed = 18%
   percentage formed per hydrogen available = 18/3 = 6%

a secondary alkyl halide

$$CH_3CHCHCH_3$$
with Cl on the CH and CH₃ below

Substitution of any one of two hydrogens leads to this product.
percentage of this product that is formed = 28%
percentage formed per hydrogen available = 28/2 = 14%

a tertiary alkyl halide

$$CH_3CCH_2CH_3$$
with Cl on the C and CH₃ below

Substitution of the one tertiary hydrogen leads to this product.
percentage of this product that is formed = 18%
percentage formed per hydrogen available = 18/1 = 18%

From the above calculations, we see that at 300 °C, the relative rates of removal of a hydrogen atom from a tertiary, secondary, and primary carbocation are as follows:

$$18 : 14 : 6 = 3.0 : 2.3 : 1$$

21. The methyl radical that is created in the first propagation step of the bromination of methane reacts with Br in the second propagation step, forming bromomethane.

$$\cdot Br + CH_4 \longrightarrow \cdot CH_3 + HBr$$
$$\cdot CH_3 + Br_2 \longrightarrow CH_3Br + \cdot Br$$

If HBr is added to the reaction mixture, the methyl radical that is created in the first propagation step can react with Br_2 or with the added HBr. Because only reaction with Br_2 forms bromomethane (reaction with HBr re-forms methane), the overall rate of formation of bromomethane is decreased.

$$\cdot Br + CH_4 \longrightarrow \cdot CH_3 + HBr$$

$$\cdot CH_3 + Br_2 \longrightarrow CH_3Br + \cdot Br$$

$$\cdot CH_3 + HBr \longrightarrow CH_4 + \cdot Br$$

22.

Chapter 14 Practice Test

. How many monochlorinated products would be obtained from the monochlorination of the following alkanes? (Ignore stereoisomers.)

a.

d.

b.

e.

c.

. What is the first propagation step in the monochlorination of ethane?

. When (*S*)-2-bromopentane is brominated, 2,3-dibromopentanes are formed. Which of the following compounds are **not** formed?

. Draw the products that will be obtained from the monochlorination of the following alkane. (Ignore stereoisomers.)

Answers to Chapter 14 Practice Test

1.   **a.** 4   **b.** 4   **c.** 5   **d.** 4   **e.** 3

2. $CH_3CH_3 + \cdot Cl \longrightarrow CH_3\dot{C}H_2 + HCl$

3.

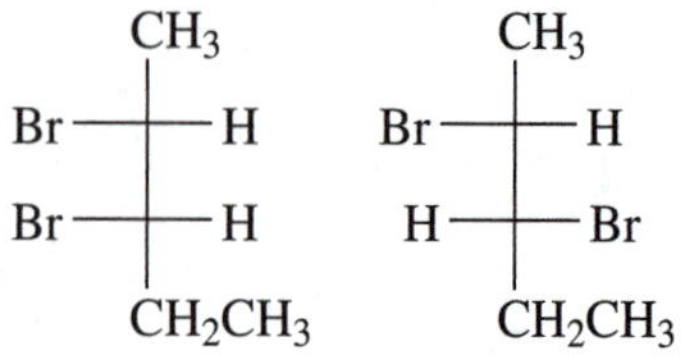

4.

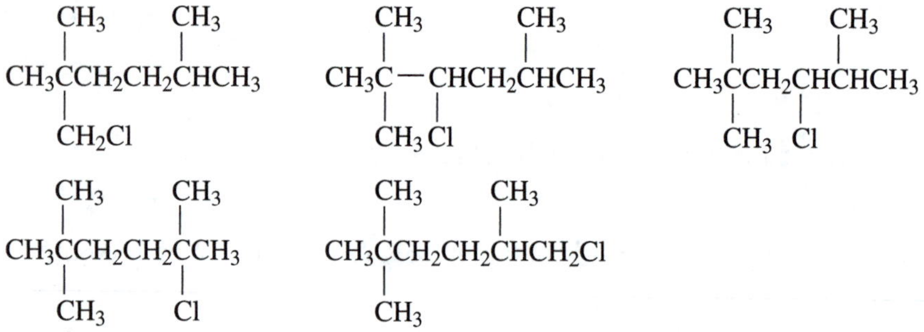

Important Terms

addition polymer (chain-growth polymer)	a polymer that is made by adding monomers to the growing end of a chain.
alternating copolymer	a copolymer in which two monomers alternate.
anionic polymerization	a chain-growth polymerization where the initiator is a nucleophile; the propagation site, therefore, is an anion.
aramide	an aromatic polyamide.
atactic polymer	a polymer in which the substituents are randomly oriented on the extended carbon chain.
biodegradable polymer	a polymer that can be degraded by microorganisms.
biopolymer	a polymer that is synthesized in nature.
block copolymer	a copolymer in which there are blocks of each kind of monomer within the polymer chain.
cationic polymerization	a chain-growth polymerization where the initiator is an electrophile; the propagation site, therefore, is a cation.
chain-growth polymer	see addition polymer.
chain transfer	a growing polymer chain reacts with a molecule XY in a manner that allows X · to terminate the chain, leaving behind Y · to initiate a new chain.
condensation polymer (step-growth polymer)	a polymer that is made by combining two molecules while removing a small molecule (usually water or an alcohol).
conducting polymer	a polymer that can conduct electricity down its backbone.
copolymer	a polymer formed using two or more different monomers.
cross-linking	connecting polymer chains by bond formation.
epoxy resin	a resin that is formed by mixing a low-molecular-weight prepolymer with a compound that forms a cross-linked polymer.
graft copolymer	a copolymer that contains branches of a polymer of one monomer grafted onto the backbone of a polymer made from another monomer.

347

head-to-tail addition	the head of one molecule is added to the tail of another molecule.
homopolymer	a polymer that contains only one kind of monomer.
isotactic polymer	a polymer in which all the substituents are on the same side of the fully extended carbon chain.
living polymer	a nonterminated chain-growth polymer that remains active. Therefore, the polymerization reaction can continue upon addition of more monomer.
materials science	the science of creating new materials that will have practical applications.
monomer	a repeating unit in a polymer.
oriented polymer	a polymer obtained by stretching out polymer chains and putting them back together in a parallel fashion.
plasticizer	an organic molecule that dissolves in a polymer and allows the polymer chains to slide by each other.
polyamide	a polymer in which monomers are connected by amide groups.
polycarbonate	a step-growth polymer in which the monomers are connected by carbonate groups.
polyester	a polymer in which monomers are connected by ester groups.
polymer	a large molecule made by linking monomers together.
polymer chemistry	the field of chemistry that deals with synthetic polymers; part of the larger discipline known as materials science.
polymerization	the process of linking up monomers to form a polymer.
polyurethane	a polymer in which monomers are connected by urethane groups.
propagating site	the reactive end of a chain-growth polymer.
radical polymerization	a chain-growth polymerization where the initiator is a radical; the propagation site, therefore, is a radical.
random copolymer	a copolymer with a random distribution of monomers.
ring-opening polymerization	a chain-growth polymerization that involves opening the ring of the monomer.
step-growth polymer (condensation polymer)	a polymer that is made by combining two molecules while removing a small molecule (usually water or an alcohol).

yndiotactic polymer   a polymer in which the substituents regularly alternate on both sides of the fully extended carbon chain.

ynthetic polymer   a polymer that is not synthesized in nature.

urethane (carbamate)   a compound with a carbonyl group that is both an amide and an ester.

$$\underset{RO}{}\overset{\displaystyle\overset{O}{\|}}{\underset{}{C}}\underset{NHR}{}$$

inyl polymer   a polymer in which the monomer is ethylene or a substituted ethylene.

ulcanization   increasing the flexibility of rubber by heating it with sulfur.

iegler–Natta catalyst   an aluminum–titanium initiator that controls the stereochemistry of a polymer.

Solutions to Problems

1. **a.** $CH_2\!\!=\!\!CHCl$

   b. $CH_2\!\!=\!\!CCH_3$ with substituent $\overset{|}{C}\!\!=\!\!O$ bearing $\overset{|}{OCH_3}$

   c. $CF_2\!\!=\!\!CF_2$

2. $-CH_2CH-CHCH_2CH_2CH-CHCH_2CHCH_2CH_2CHCH_2CH-$ (each CH bearing a phenyl ring)

3.

   $HO\!-\!OH \longrightarrow 2\ HO\cdot$

   $HO\cdot\ +\ CH_2\!\!=\!\!\overset{|}{\underset{Cl}{CH}} \longrightarrow HO\!-\!CH_2\overset{|}{\underset{Cl}{CH}}\cdot$

   $HO\!-\!CH_2\overset{|}{\underset{Cl}{CH}}\cdot\ +\ CH_2\!\!=\!\!\overset{|}{\underset{Cl}{CH}} \longrightarrow HO\!-\!CH_2\overset{|}{\underset{Cl}{CH}}CH_2\overset{|}{\underset{Cl}{CH}}\cdot$

   $HO\!-\!CH_2\overset{|}{\underset{Cl}{CH}}CH_2\overset{|}{\underset{Cl}{CH}}\cdot\ +\ CH_2\!\!=\!\!\overset{|}{\underset{Cl}{CH}} \longrightarrow HO\!-\!CH_2\overset{|}{\underset{Cl}{CH}}CH_2\overset{|}{\underset{Cl}{CH}}CH_2\overset{|}{\underset{Cl}{CH}}\cdot$

4. Because branching increases the flexibility of the polymer, beach balls are made from more highly branched polyethylene.

5. Decreasing ability to undergo cationic polymerization is in the same order as decreasing stability of the carbocation intermediate. (Electron donation increases the stability of the carbocation.)

   a. $CH_2\!\!=\!\!CH$ (benzene ring with OCH_3) $>$ $CH_2\!\!=\!\!CH$ (benzene ring with CH_3) $>$ $CH_2\!\!=\!\!CH$ (benzene ring with NO_2)

   donates electrons by resonance ... **withdraws electrons by resonance**

   b. $CH_2\!\!=\!\!CH\!-\!OCH_3\ >\ CH_2\!\!=\!\!CHCH_3\ >\ CH_2\!\!=\!\!CH\!-\!\overset{\overset{\textstyle O}{\|}}{C}CH_3$

   donates electrons by resonance ... **withdraws electrons by resonance**

Decreasing ability to undergo anionic polymerization is in the same order as decreasing stability of the carbanion intermediate. (Electron withdrawal increases the stability of the carbanion.)

a.

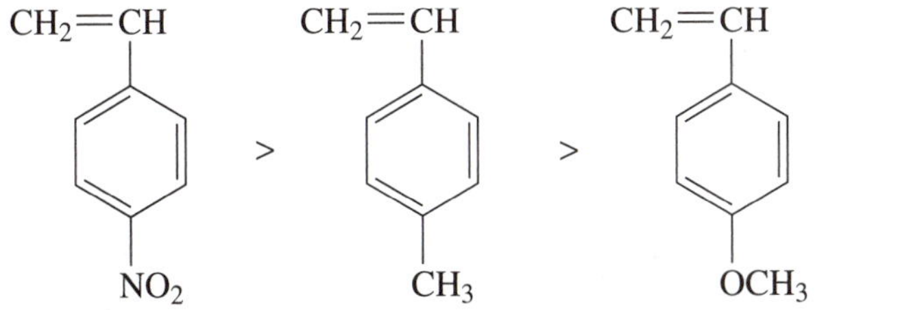

withdraws electrons
by resonance

donates electrons
by resonance

b. $CH_2=CHC\equiv N$ > $CH_2=CHCl$ > $CH_2=CHCH_3$

withdraws electrons
by resonance

withdraws
electrons inductively

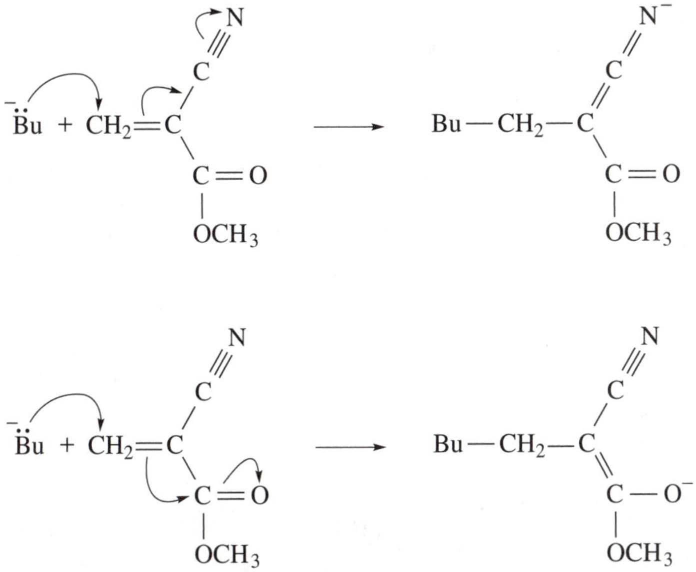

Methyl methacrylate does not undergo cationic polymerization because the carbocation propagating site would be very unstable, since the ester group is strongly electron-withdrawing.

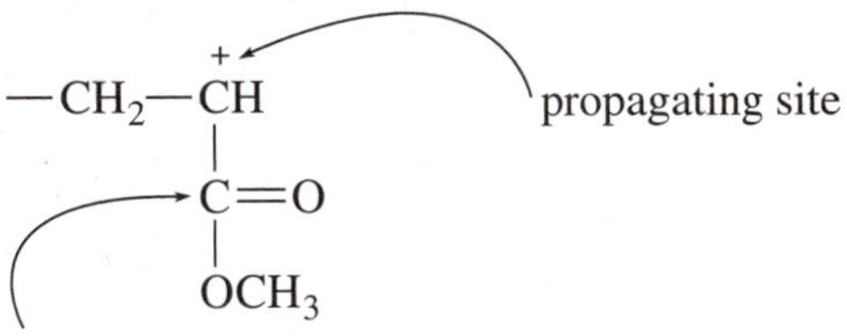

a strongly electron-withdrawing group

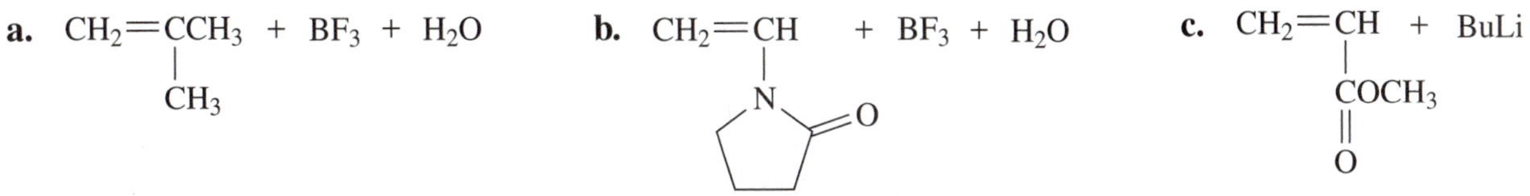

a. $CH_2=CCH_3 + BF_3 + H_2O$
        $\quad\quad\ \ |$
        $\quad\quad\ \ CH_3$

b. 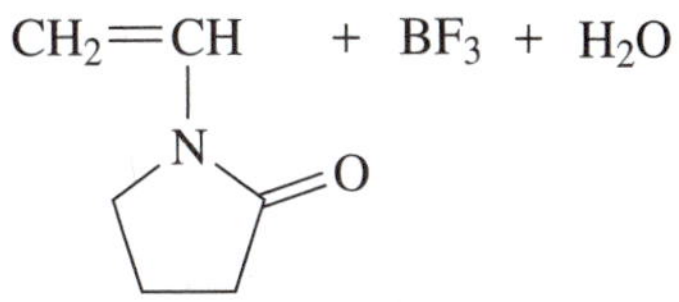 $CH_2=CH\ \ + BF_3 + H_2O$

c. $CH_2=CH\ \ + BuLi$
        $\quad\quad\ \ |$
        $\quad\quad\ \ COCH_3$
        $\quad\quad\ \ ||$
        $\quad\quad\ \ O$

10. In anionic polymerization, nucleophilic attack occurs at the less-substituted carbon because it is less sterically hindered; in cationic polymerization, nucleophilic attack occurs at the more-substituted carbon because the ring opens to give the more stable partial carbocation.

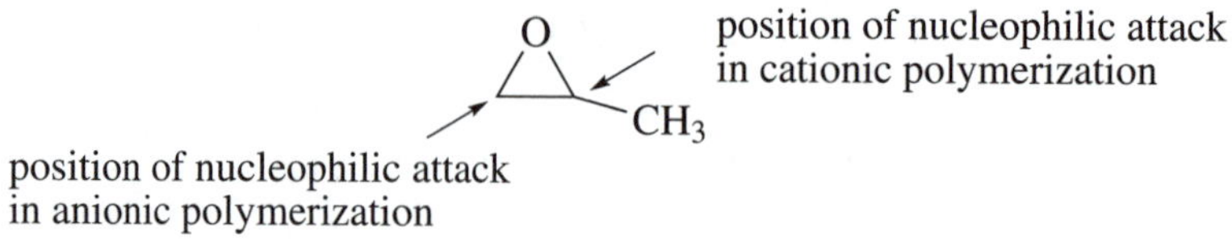

position of nucleophilic attack in cationic polymerization

position of nucleophilic attack in anionic polymerization

11. **a.**

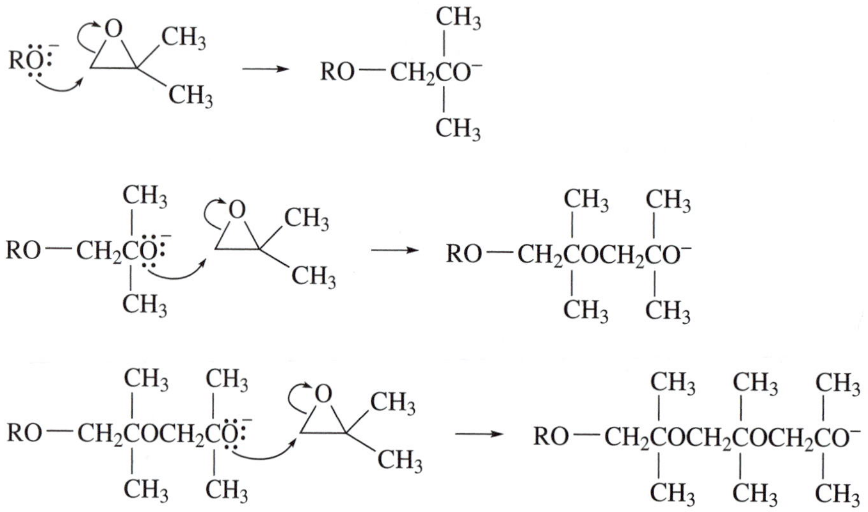

b.

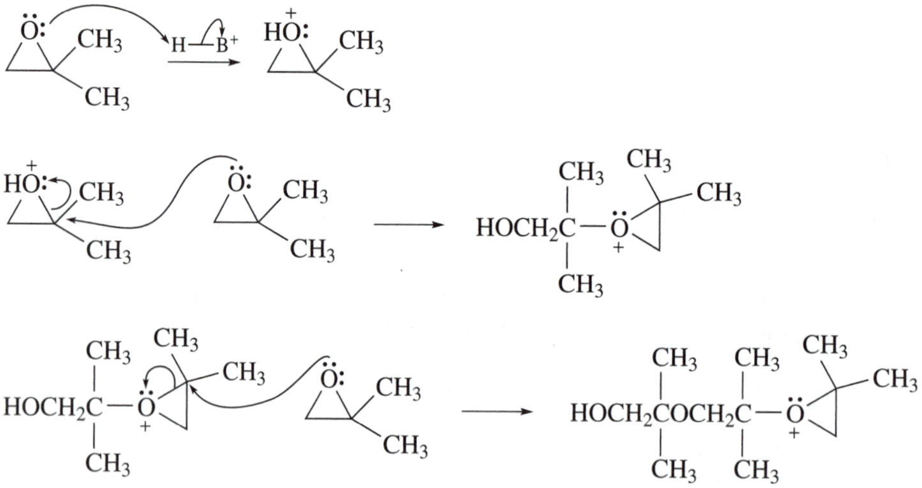

12.

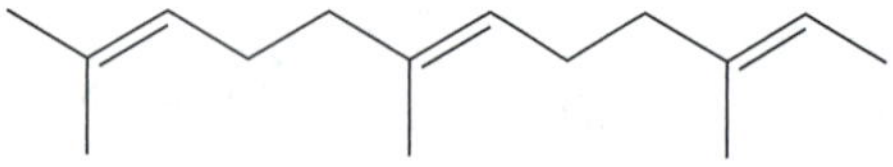

3.   a.

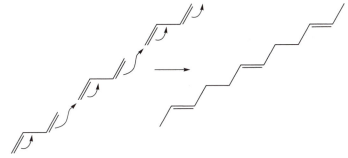

b.   CH$_2$=CH   CH$_2$=CH   CH$_2$=CH

            | | |

           CH=CH$_2$ CH=CH$_2$ CH=CH$_2$

—CH$_2$—CH—CH$_2$—CH—CH$_2$—CH—

            | | |

         CH=CH$_2$ CH=CH$_2$ CH=CH$_2$

4.   —CH$_2$—CH—CH$_2$—CH—CH$_2$—CH—CH$_2$—CH—

                 | | | |

            (phenyl)   C   (phenyl)   C

                        ‖   ‖

                        N   N

5.

                         O   O

                         ‖   ‖

a.   —NHCH$_2$CH$_2$CH$_2$CNHCH$_2$CH$_2$CH$_2$C—

             O   O   O   O

             ‖   ‖   ‖   ‖

b.   —NH(CH$_2$)$_4$NHCCH$_2$CH$_2$CNH(CH$_2$)$_4$NHCCH$_2$CH$_2$C—

6.

    O   O   O   O

    ‖   ‖   ‖   ‖

—C(CH$_2$)$_4$C$\left[\right.$NH(CH$_2$)$_6$NH—C(CH$_2$)$_4$C$\left.\right]_n$NH(CH$_2$)$_6$NH—

                ↓ H$_2$SO$_4$

         O   O

         ‖   ‖

    n HOC(CH$_2$)$_4$COH   + n H$_3\overset{+}{\text{N}}$(CH$_2$)$_6\overset{+}{\text{N}}$H$_3$

17. They hydrolyze to give salts of dicarboxylic acids and diols.

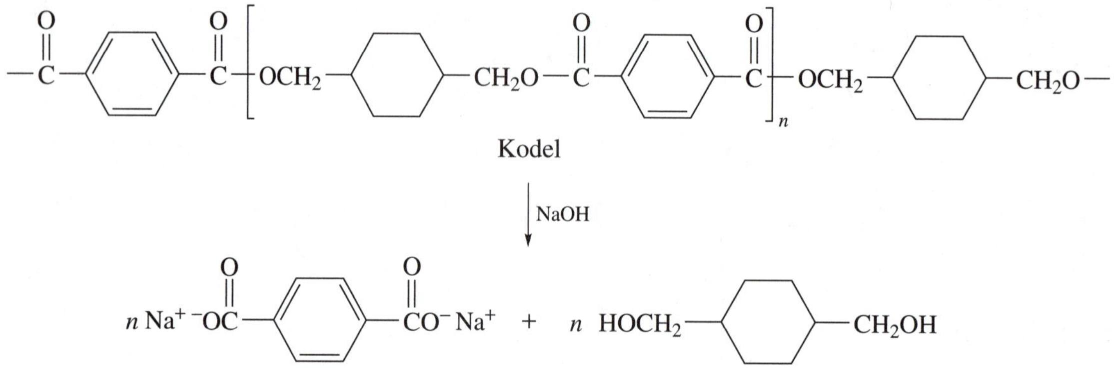

Kodel

$\downarrow$ NaOH

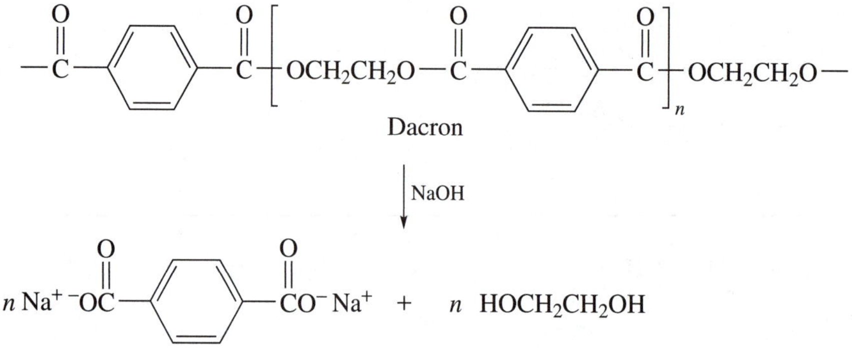

n Na$^+$ $^-$OC—⬡—CO$^-$ Na$^+$ + n HOCH$_2$—⬡—CH$_2$OH

Dacron

$\downarrow$ NaOH

n Na$^+$ $^-$OC—⬡—CO$^-$ Na$^+$ + n HOCH$_2$CH$_2$OH

18. **a.**

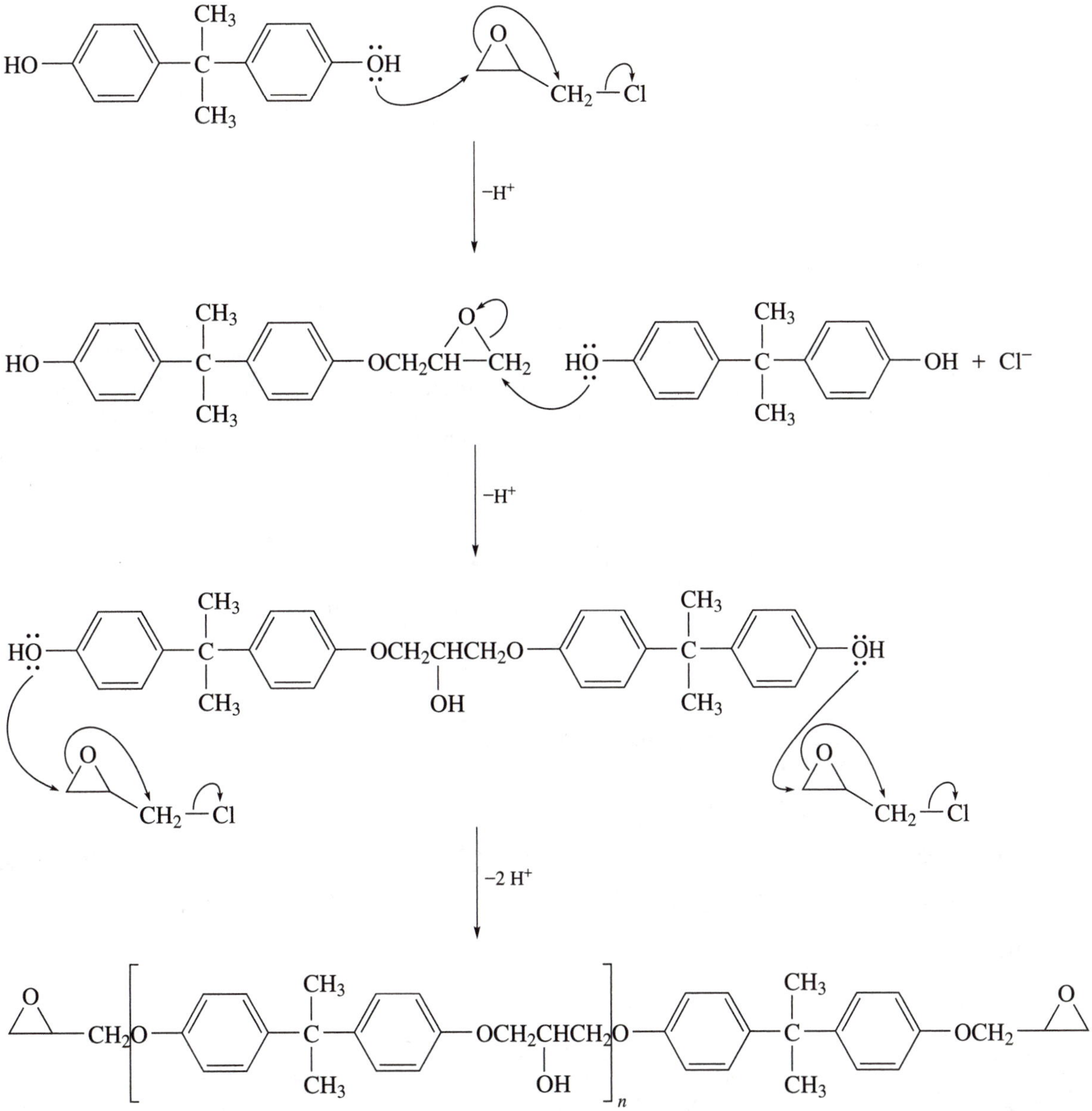

$-H^+$

$-H^+$

$-2\,H^+$

$+\ Cl^-$

b.

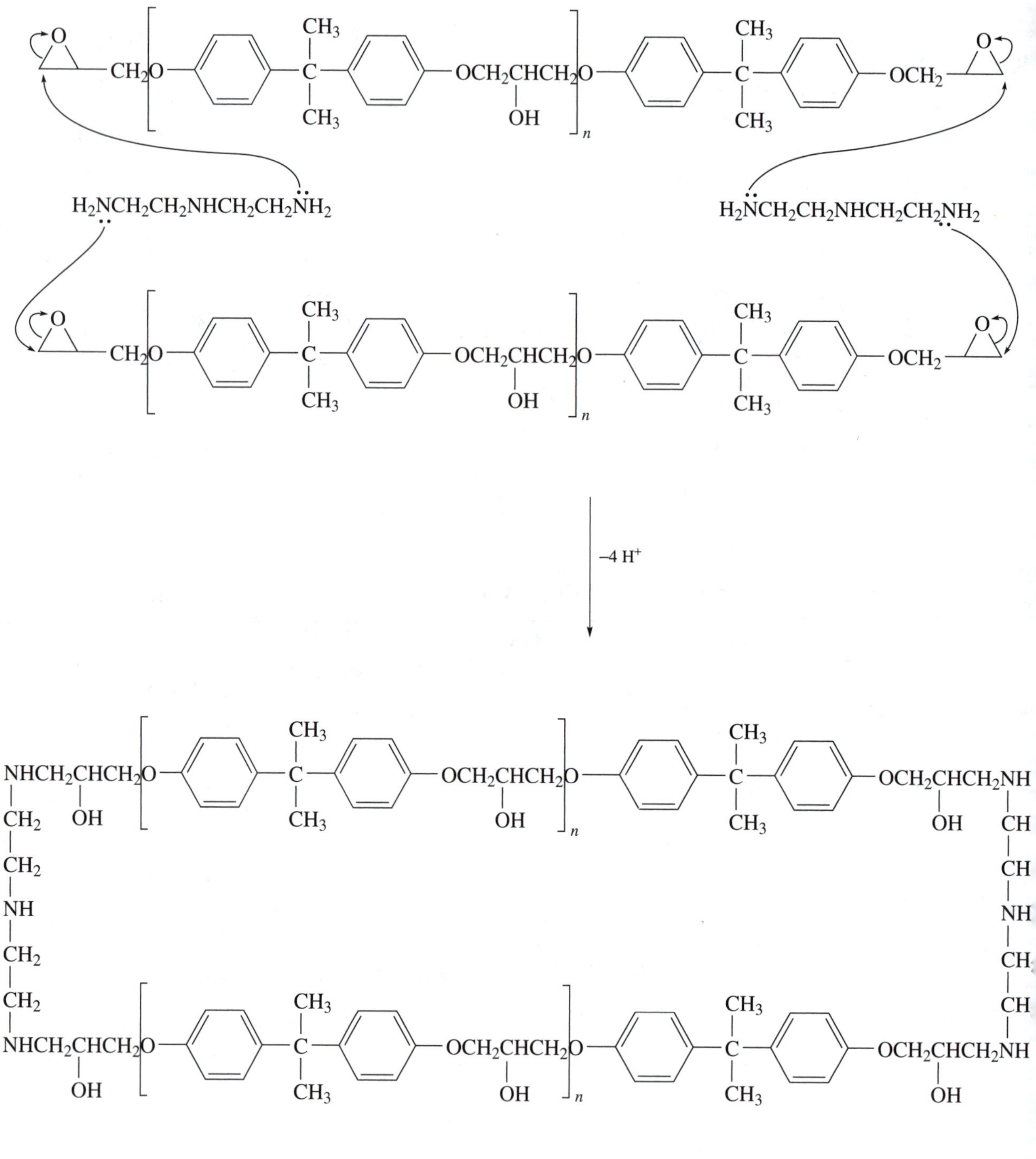

$H_2NCH_2CH_2NHCH_2CH_2\ddot{N}H_2$

$H_2\ddot{N}CH_2CH_2NHCH_2CH_2NH_2$

$-4\ H^+$

19. Formation of a protonated imine between formaldehyde and one amino group, followed by reaction with a second amino group, accounts for formation of the linkage that holds the monomers together.

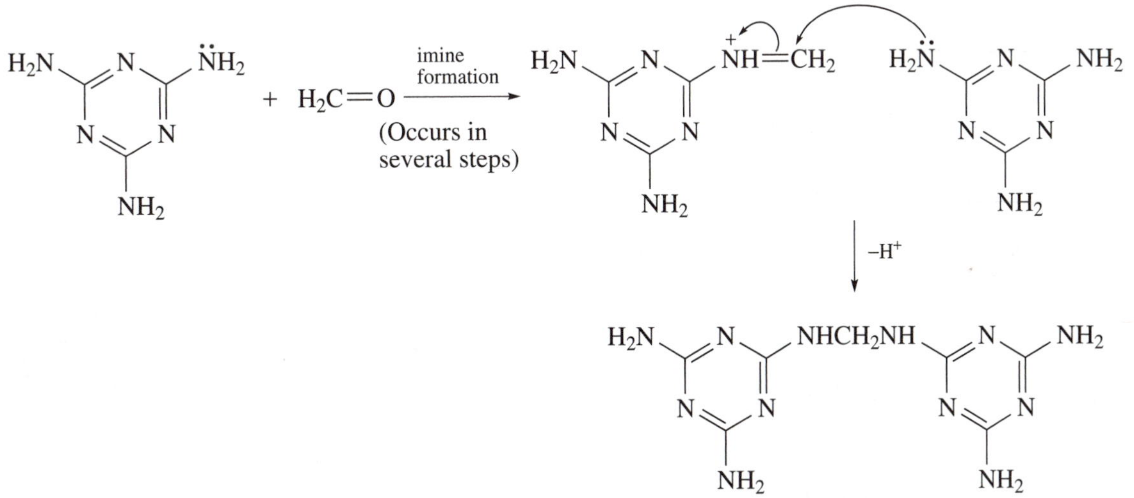

a dimer of Melmac

20. The polymer is stiffer because glycerol cross-links the polymer chains.

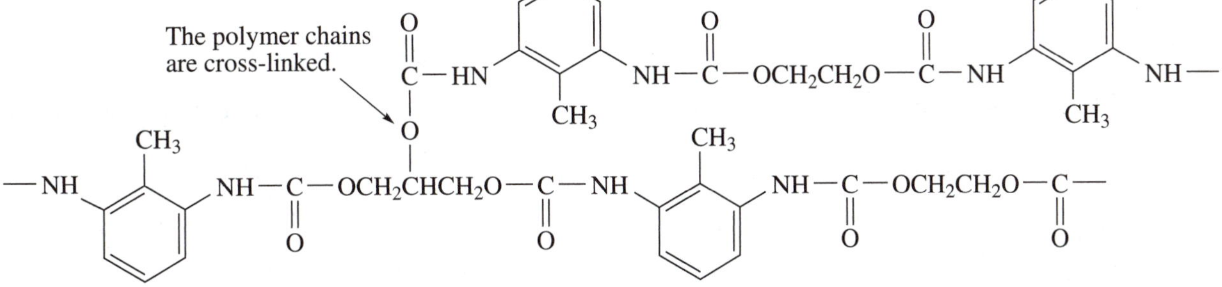

21.   a.

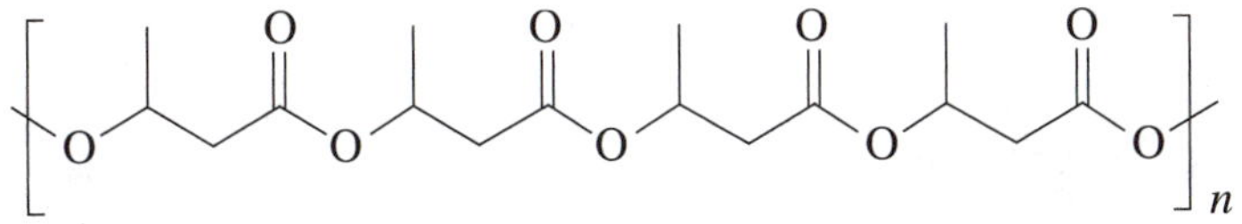

b.

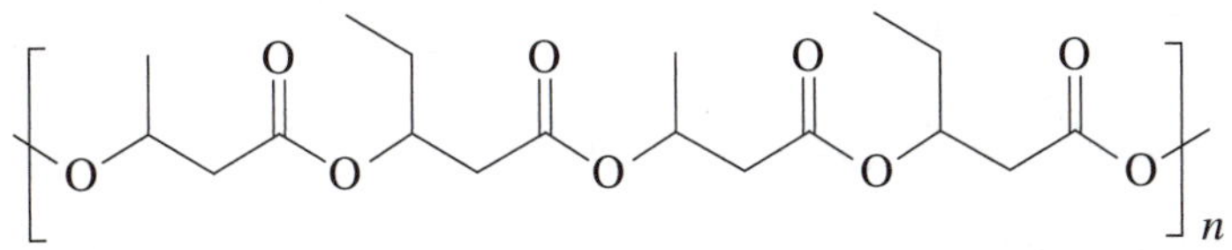

22.   a. —CH₂CHCH₂CHCH₂CH—   chain-growth polymer

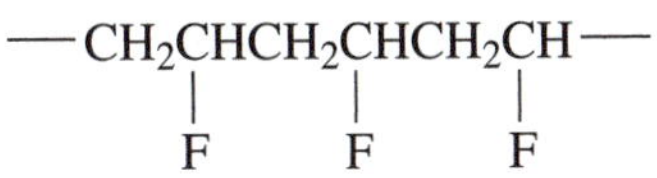

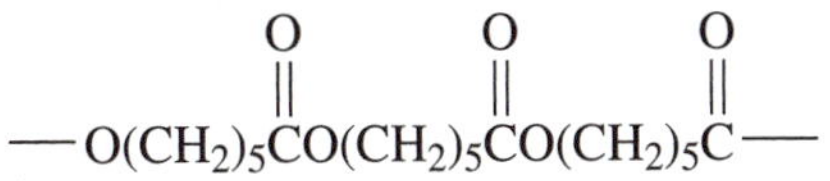

b. —O(CH₂)₅CO(CH₂)₅CO(CH₂)₅C—   step-growth polymer

c. —CH$_2$CHCH$_2$CHCH$_2$CH—   chain-growth polymer
         | | |
        CO$_2$H CO$_2$H CO$_2$H

d. —NH(CH$_2$)$_5$NHC(CH$_2$)$_5$CNH(CH$_2$)$_5$NHC(CH$_2$)$_5$C—   step-growth polymer

(each C bearing =O above)

e. —OCNH⟨benzene ring, CH$_3$ substituent⟩NHCOCH$_2$CH$_2$OCNH⟨benzene ring, CH$_3$ substituent⟩NHCO—   step-growth polymer

23.

a. —CH$_2$CH$_2$OCH$_2$CH$_2$N⟨piperazine ring⟩N—

b. —⟨benzene ring⟩—OCH$_2$CH$_2$CH$_2$O—⟨benzene ring⟩—N=CHCH=N—

24.

a. CH$_2$=CHCH$_2$CH$_3$

b. ⟨epoxide with O⟩CH$_3$

c. ClSO$_2$—⟨benzene ring⟩—SO$_2$Cl + H$_2$N(CH$_2$)$_6$NH$_2$

d. HOC—⟨benzene ring⟩—COH + HOCH$_2$CH$_2$OH (each C bearing =O)

e. CH$_2$=CCH=CH$_2$ with CH$_3$ on the second carbon

f. HO(CH$_2$)$_5$COH   (C bearing =O)

a, b, and **e** are chain-growth polymers.

c, d, and **f** are step-growth polymers.

25.

a. CH$_3$OCH$_2$CHOCH$_2$CHOCH$_2$CHOCH$_2$CHO—
                    | | | |
                   CH$_3$ CH$_3$ CH$_3$ CH$_3$

b. —CH$_2$CH—CH$_2$CH—CH$_2$CH—CH$_2$CH—
             | | | |
            OCH$_3$ OCH$_3$ OCH$_3$ OCH$_3$

c. —CH$_2$CH—CH$_2$CH—CH$_2$CH—CH$_2$CH—
             | | | |
            COCH$_3$ COCH$_3$ COCH$_3$ COCH$_3$
             ‖   ‖   ‖   ‖
             O   O   O   O

26. The polymer in the flask that contained a high-molecular-weight polymer and little material of intermediate molecular weight was formed by a chain-growth mechanism, whereas the polymer in the flask that contained mainly material of intermediate molecular weight was formed by a step-growth mechanism.

In a chain-growth mechanism, monomers are added to the growing end of a chain. This means that at any one time, there will be polymeric chains and monomers.

Step-growth polymerization is not a chain reaction; any two monomers can react. Therefore, high-molecular-weight material will not be formed until the end of the reaction when pieces of intermediate molecular weight can combine.

27.   **a.** Because it is a polyamide (and not a polyester), it is a nylon.

b. 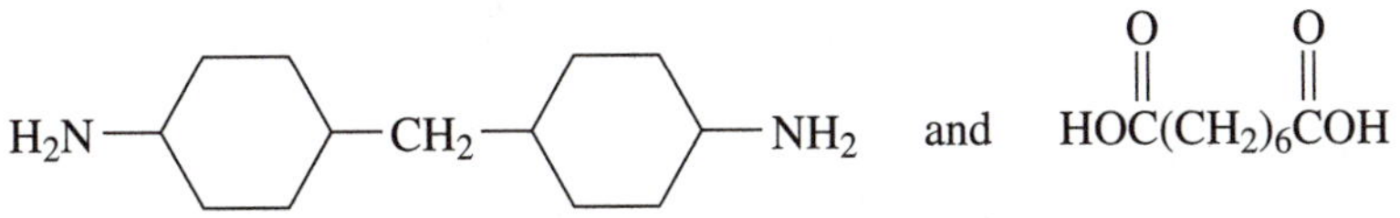 H_2N—⬡—CH_2—⬡—NH_2 and $HOC(CH_2)_6COH$ (with two C=O groups shown as O above the carbons)

28. Because 1,4-divinylbenzene has substituents on both ends of the benzene ring that can engage in polymerization, the polymer chains can become cross-linked, which increases the rigidity of the polymer.

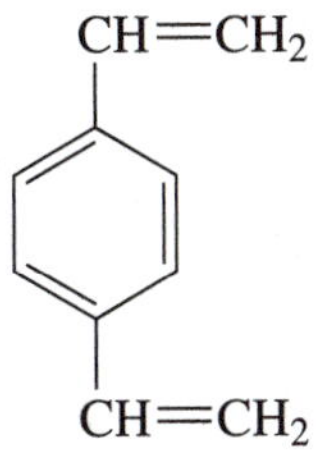

1,4-divinylbenzene

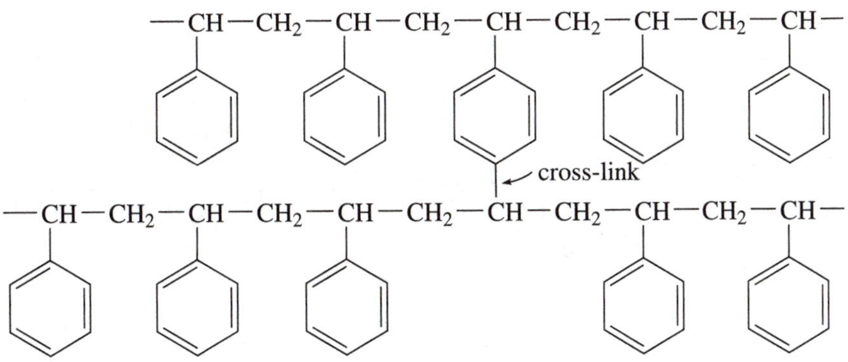

cross-link

29.   The presence of three OH groups in glycerol allows for extensive cross-linking during polymerization Glyptal gets its strength from this cross-linking.

terephthalic acid   +   glycerol

0. 3,3-Dimethyloxacyclobutane undergoes cationic polymerization by the following mechanism:

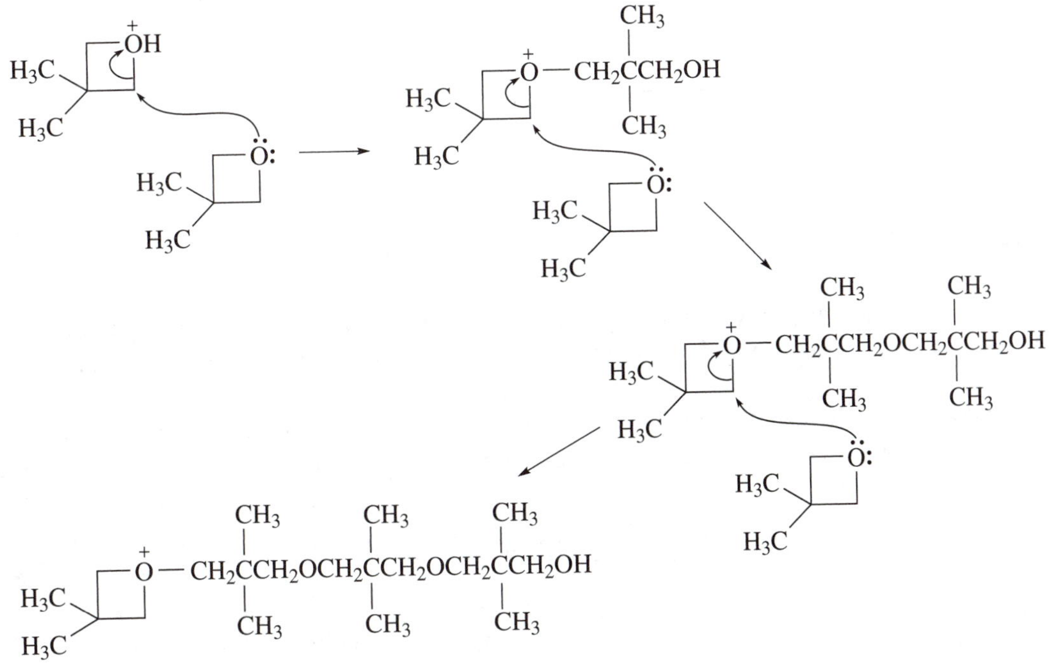

31. Both compounds can form esters via intramolecular or intermolecular reactions. The product of the intramolecular reaction is a lactone; the intermolecular reaction leads to a polymer.

5-Hydroxypentanoic acid reacts intramolecularly to form a six-membered ring lactone, whereas 6-hydroxyhexanoic acid reacts intramolecularly to form a seven-membered ring lactone.

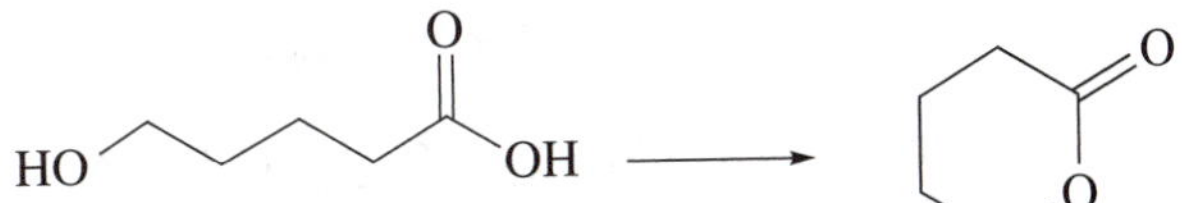

5-hydroxypentanoic acid

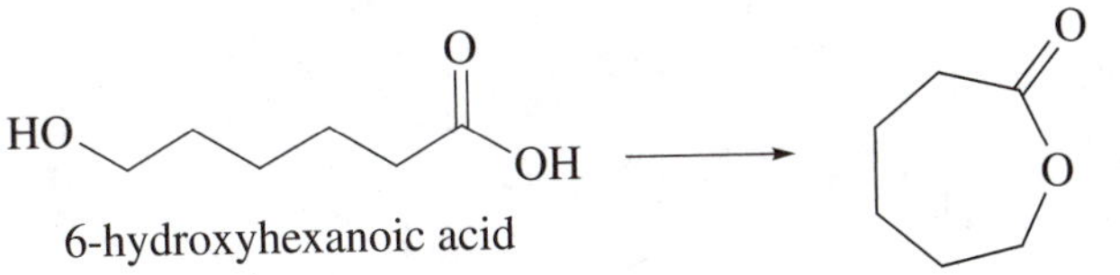

6-hydroxyhexanoic acid

The compound that forms more polymer will be the one that forms less lactone, because the two reactions compete with one another.

The six-membered-ring lactone is more stable and therefore has a more stable transition state for its formation, compared to a seven-membered ring lactone. Because it is easier for 5-hydroxypentanoic acid to form the six-membered-ring lactone than for 6-hydroxyhexanoic acid to form the seven-membered ring lactone, **6-hydroxyhexanoic acid** will form more polymer.

32.

$CH_2\!=\!CH\!-\!CH\!=\!O$ $-CH_2-CH=CH-O-$

    acrolein   repeating unit

$CH_2\!=\!CH$ $-CH_2-CH-$
        | |
       CH   CH
       ‖   ‖
       O   O

    acrolein   repeating unit

33. The plasticizer that keeps vinyl soft and pliable can vaporize over time, causing the polymer to becom brittle. For this reason, high-boiling materials are preferred as plasticizers over low-boiling materials.

34. Hydrolysis converts the ester substituents into alcohol substituents, which can react with ethylene oxid to graft a polymer of ethylene oxide onto the backbone of the alternating copolymer of styrene and viny acetate.

5. The desired polymer is an alternating copolymer of ethylene and 1,2-dibromoethylene.

$$CH_2{=}CH_2 \qquad \underset{\displaystyle \overset{|}{Br} \quad \overset{|}{Br}}{CH{=}CH}$$

6. a.

b. Delrin is a chain-growth polymer, because it is made by adding monomers to the end of a growing chain.

Chapter 15 Practice Test

1.   Draw a short segment of the polymer obtained from each of the following monomers. In each case, te whether it is a chain-growth polymer or a step-growth polymer.

   a. CH_2=CHCOOH

   b. 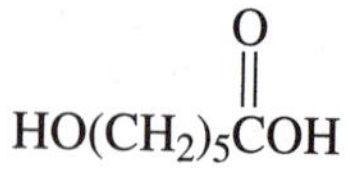 $HO(CH_2)_5\overset{\overset{\displaystyle O}{\|}}{C}OH$

   c. CH_2=CCl_2

   d. 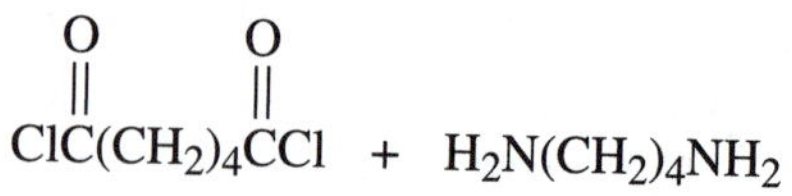 $Cl\overset{\overset{\displaystyle O}{\|}}{C}(CH_2)_4\overset{\overset{\displaystyle O}{\|}}{C}Cl$ + $H_2N(CH_2)_4NH_2$

2.   Draw the structure of the monomer or monomers used to synthesize the following polymers. In each case tell whether the polymer is a chain-growth polymer or a step-growth polymer.

   a. 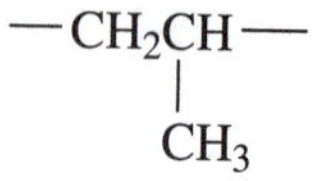 $-CH_2CH-$
               $\quad\quad\quad\quad|$
               $\quad\quad\quad\ CH_3$

   b. 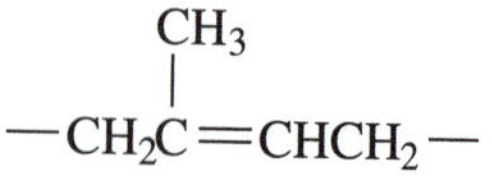 $\quad\quad\ \ CH_3$
            $\quad\quad\quad|$
            $-CH_2C$=$CHCH_2-$

   c. $-CH_2-O-$

   d. 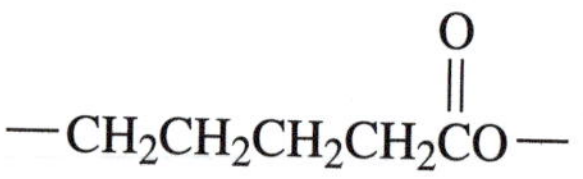 $\quad\quad\quad\quad\quad\quad\quad\quad\ O$
            $\quad\quad\quad\quad\quad\quad\quad\quad\ \|$
            $-CH_2CH_2CH_2CH_2CO-$

3.   Draw short segments of the polymers obtained from the following compounds under the given reaction conditions.

   a. 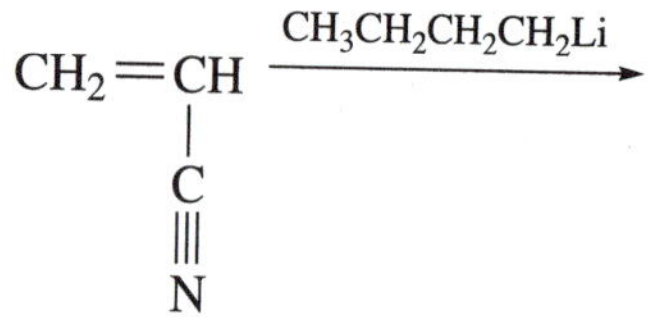 CH_2=CH $\xrightarrow{CH_3CH_2CH_2CH_2Li}$
               $\quad\quad\quad|$
               $\quad\quad\quad C$ 
               $\quad\quad\quad\|\|\|$
               $\quad\quad\quad N$

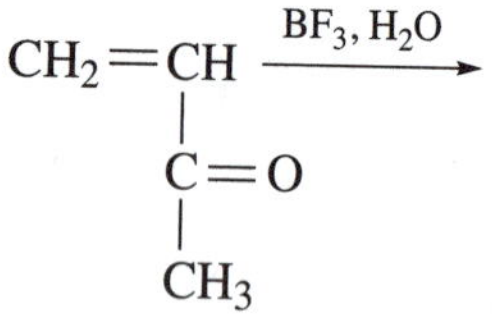

   b. CH_2=CH $\xrightarrow{BF_3,\ H_2O}$
               $\quad\quad\quad|$
               $\quad\quad\quad C$=O
               $\quad\quad\quad|$
               $\quad\quad\ CH_3$

   c. $H_2C\overset{O}{\overset{\diagup\ \diagdown}{-}}CHCH_2CH_3 \xrightarrow{CH_3O^-}$

4.   Explain why CH_2=CCl_2 does not form an isotactic, syndiotactic, or atactic polymer.

Answers to Chapter 15 Practice Test

a. $-CH_2-CH-CH_2-CH-CH_2-CH-CH_2-CH-$ chain-growth

           COOH   COOH   COOH   COOH

b. $-O(CH_2)_5\overset{\text{O}}{\overset{\|}{C}}O(CH_2)_5\overset{\text{O}}{\overset{\|}{C}}O(CH_2)_5\overset{\text{O}}{\overset{\|}{C}}O-$ step-growth

c. $-CH_2-\overset{\text{Cl}}{\underset{\text{Cl}}{C}}-CH_2-\overset{\text{Cl}}{\underset{\text{Cl}}{C}}-CH_2-\overset{\text{Cl}}{\underset{\text{Cl}}{C}}-CH_2-CH-$ chain-growth

d. $-\overset{\text{O}}{\overset{\|}{C}}(CH_2)_4\overset{\text{O}}{\overset{\|}{C}}NH(CH_2)_4NH\overset{\text{O}}{\overset{\|}{C}}(CH_2)_4\overset{\text{O}}{\overset{\|}{C}}NH(CN_2)_4NH-$ step-growth

2.

a. $CH_2{=}CH$ chain-growth

        |

       CH_3

b. $CH_2{=}\overset{\text{CH}_3}{\overset{|}{C}}-CH{=}CH_2$

    chain-growth

c. $CH_2{=}O$ chain-growth

d. $HOCH_2CH_2CH_2CH_2\overset{\text{O}}{\overset{\|}{C}}OH$

    step-growth

3.

a. $CH_3CH_2CH_2CH_2-CH_2-CH-CH_2-CH-CH_2-CH-$

                           $\overset{C}{\underset{N}{\|\|\|}}$ $\overset{C}{\underset{N}{\|\|\|}}$ $\overset{C}{\underset{N}{\|\|\|}}$

b. $CH_3-CH-CH_2-CH-CH_2-CH-$

         | | |

       $C{=}O$ $C{=}O$ $C{=}O$

         | | |

       CH_3 CH_3 CH_3

c. $CH_3O-CH_2CH-O-CH_2CH-O-CH_2CH-O-$

                 | | |

               CH_2CH_3 CH_2CH_3 CH_2CH_3

4. Because there are two chlorine substituents, the chlorine substituents cannot all be on the same side of the carbon chain, they cannot alternate on both sides of the chain, and they cannot be randomly oriented.

$$-CH_2-\overset{\text{Cl}}{\underset{\text{Cl}}{C}}-$$

Important Terms

aldose	a polyhydroxy aldehyde.
anomeric carbon	the carbon in a cyclic sugar that is the carbonyl carbon in the straight-chain form.
anomers	two cyclic sugars that differ in configuration only at the carbon that is the carbonyl carbon in the straight-chain form.
bioorganic compound	an organic compound that is found in a biological system.
carbohydrate	a sugar, a saccharide. Naturally occurring carbohydrates have the D-configuration.
complex carbohydrate	a carbohydrate that contains two or more sugar molecules linked together; it can be hydrolyzed to simple sugars.
deoxy sugar	a sugar in which one of the OH groups has been replaced by a hydrogen.
disaccharide	a compound containing two sugar molecules linked together.
enediol rearrangement	a base-catalyzed reaction that interconverts monosaccharides.
epimerization	changing the configuration of a carbon by removing a proton and then reprotonating it.
epimers	monosaccharides that differ in configuration at only one carbon.
furanose	a five-membered ring sugar.
furanoside	a five-membered ring glycoside.
glycoprotein	a protein that is covalently bonded to an oligosaccharide.
glycoside	the acetal of a sugar.
N-glycoside	a glycoside with a nitrogen instead of an oxygen at the glycosidic linkage.
glycosidic bond	the bond between the anomeric carbon of one sugar and an alcohol residue of a second sugar in a glycoside.
α-1,4′-glycosidic linkage	a glycosidic linkage between the C-1 of one sugar and the C-4 of a second sugar with the oxygen atom at C-1 in the axial position.
α-1,6′-glycosidic linkage	a glycosidic linkage between the C-1 of one sugar and the C-6 of a second sugar with the oxygen atom at C-1 in the axial position.

-1,4′-glycosidic linkage	a glycosidic linkage between the C-1 of one sugar and the C-4 of a second sugar with the oxygen atom at C-1 in the equatorial position.
Haworth projection	a way to show the structure of a sugar in which the five- and six-membered rings are represented as being flat.
eptose	a monosaccharide with seven carbons.
exose	a monosaccharide with six carbons.
etose	a polyhydroxy ketone.
molecular recognition	the recognition of one molecule by another as a result of specific interactions.
monosaccharide	a single sugar molecule.
mutarotation	a slow change in optical rotation to an equilibrium value.
oligosaccharide	3 to 10 sugar molecules linked by glycosidic bonds.
oxocarbenium ion	an ion in which the positive charge is shared by a carbon and an oxygen.
pentose	a monosaccharide with five carbons.
polysaccharide	a compound containing 10 or more sugar molecules linked together.
pyranose	a six-membered ring sugar.
pyranoside	a six-membered ring glycoside.
simple carbohydrate	a single sugar molecule.
tetrose	a monosaccharide with four carbons.
triose	a monosaccharide with three carbons.

Solutions to Problems

1.   D-Ribose is an aldopentose.
      D-Sedoheptulose is a ketoheptose.
      D-Mannose is an aldohexose.

2.   Notice that an L-sugar is the mirror image of a D-sugar.

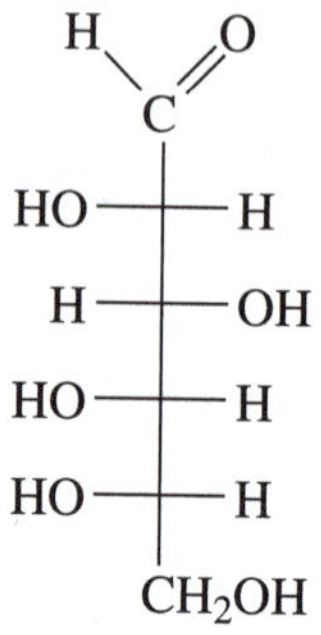

L-glucose 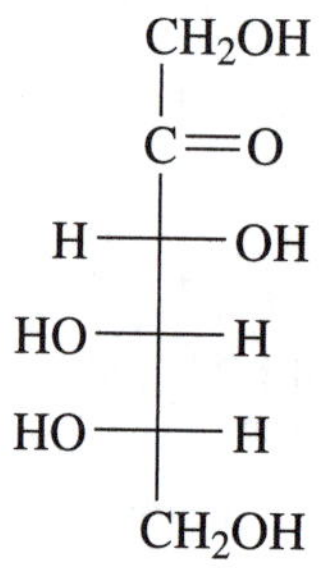   L-fructose

3.   **a.** Enantiomers, because they are mirror images.

      b. Diastereomers, because the configuration of one asymmetric center is the same in both and the configuration of the other asymmetric center is the opposite.

4.   **a.** D-ribose   **b.** L-talose   **c.** L-allose   **d.** L-ribose

5.   D-psicose

6.   **a.** A ketoheptose has four asymmetric centers ($2^4 = 16$ stereoisomers).

      b. An aldoheptose has five asymmetric centers ($2^5 = 32$ stereoisomers).

      c. A ketotriose does not have an asymmetric center; therefore, it has no stereoisomers.

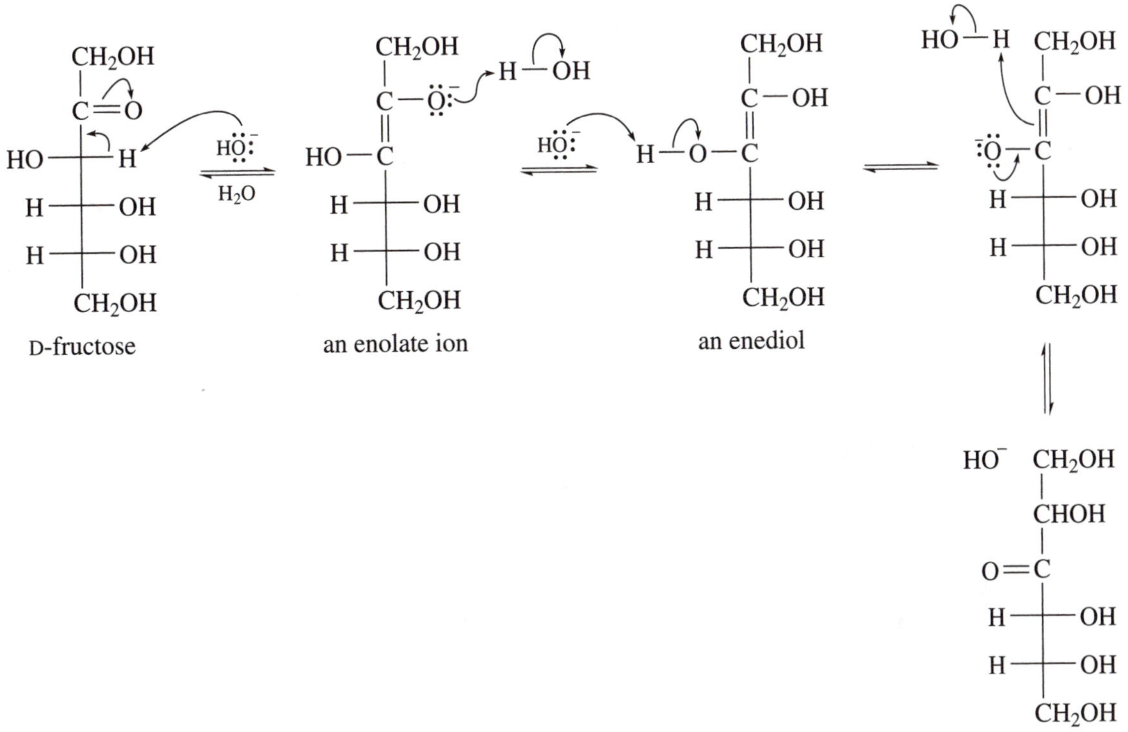

D-fructose   an enolate ion   an enediol

Removal of an α-hydrogen creates an enol that can enolize back to the ketone (using the OH at C-2) or can enolize to an aldehyde (using the OH at C-1). The aldehyde has a new asymmetric center (indicated by an *); one of the epimers is D-glucose, and the other is D-mannose.

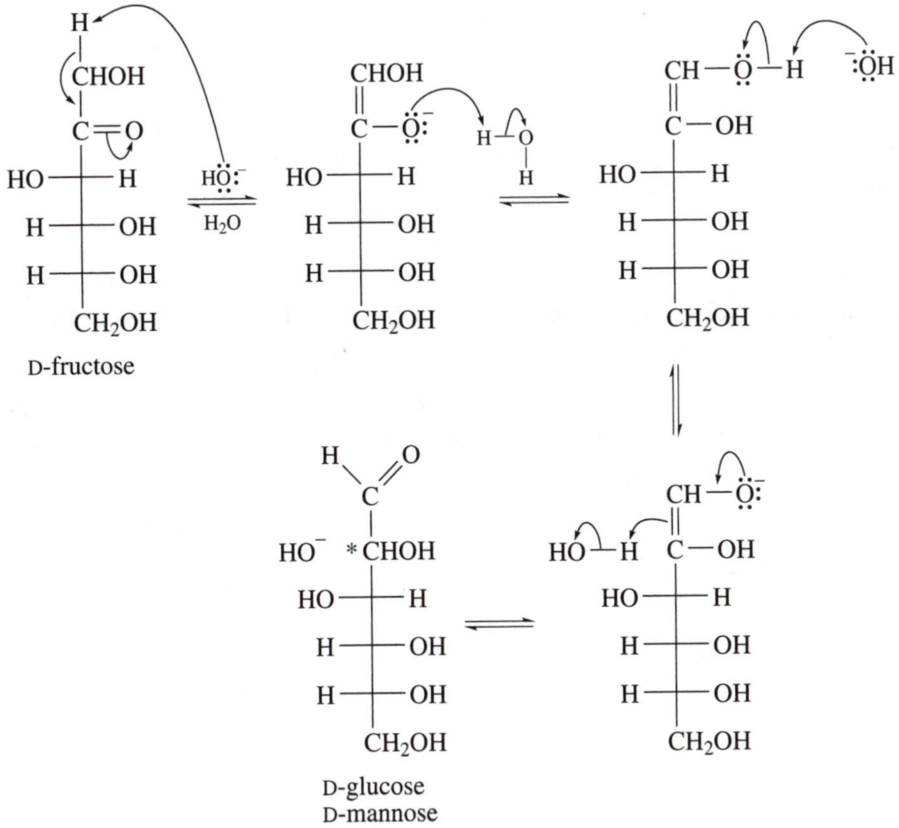

D-fructose

D-glucose
D-mannose

9. The hemiacetals in **a** and **b** have one asymmetric center; therefore, each has two stereoisomers. Th
hemiacetals in **c** and **b** have two asymmetric centers; therefore, each has four stereoisomers.

    a. Solved in the text.

    b.

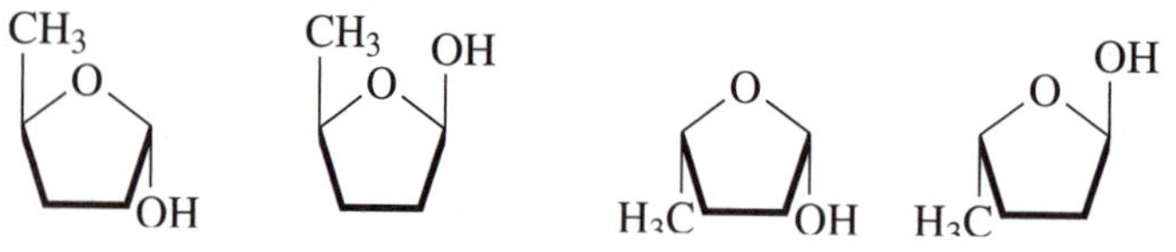

    c.

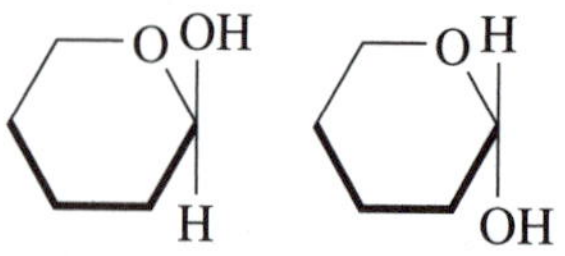

    d.

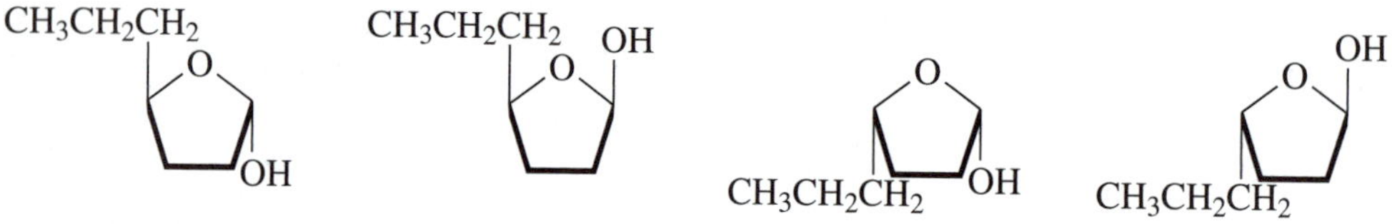

10.   **a.** 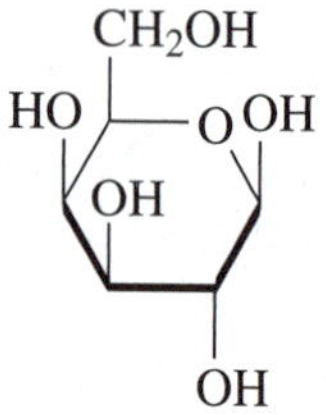   **b.** 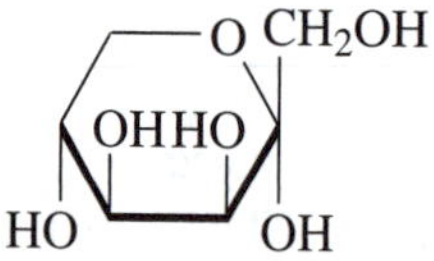   **c.**

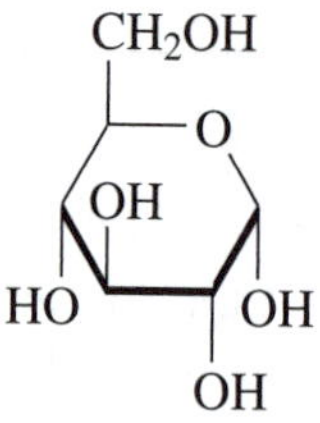

    D-glucose

11. In the chair conformation, an α-anomer has the anomeric OH group in the axial position and a β-anome
has the anomeric OH group in the equatorial position. Glucose has all of its OH groups in equatorial posi
tions. Now this question can be answered easily.

    a. Solved in the text.

    b. D-Idose differs in configuration from D-glucose at C-2, C-3, and C-4. Therefore, the OH groups at C-2
        C-3, and C-4 in β-D-idose are in axial positions.

    c. D-Allose is a C-3 epimer of D-glucose. Therefore, the OH group at C-3 is in the axial position and
        because it is the α-anomer, the OH group at C-1 (the anomeric carbon) is also in the axial position.

12.

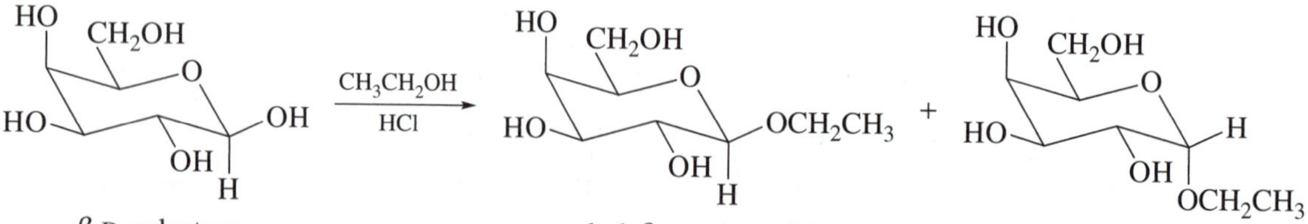

    β-D-galactose   ethyl β-D-galactoside   ethyl α-D-galactoside

$\xrightarrow[\text{HCl}]{\text{CH}_3\text{CH}_2\text{OH}}$

3. **a.** Solved in the text.

b. methyl α-D-galactoside (or methyl α-D-galactopyranoside)

4. If more than a trace amount of acid is used, the amine that acts as a nucleophile when it forms the *N*-glycoside becomes protonated, and a protonated amine is not a nucleophile.

5.

specific rotation of glucose $+$ specific rotation of fructose $= -22.0$

$+52.7 +$ specific rotation of fructose $= -22.0$

specific rotation of fructose $= -22.0 + (-52.7)$

specific rotation of fructose $= -74.7$

6. **a.** Amylose has α-1,4'-glycosidic linkages, whereas cellulose has β-1,4'-glycosidic linkages.

b. Amylose has α-1,4'-glycosidic linkages, whereas amylopectin has both α-1,4'-glycosidic linkages and α-1,6'-glycosidic linkages.

c. Glycogen and amylopectin have the same kind of linkages, but glycogen has a higher frequency of α-1,6'-glycosidic linkages.

d. Cellulose has a hydroxy group at C-2, whereas chitin has an *N*-acetyl amino group at that position.

7. **a.** People with type O blood can receive blood only from other people with type O blood, because types A, B, and AB blood have sugar components that type O blood does not have.

b. People with type AB blood can give blood only to other people with type AB blood, because type AB blood has sugar components that types A, B, and O blood do not have.

8.
C-2 epimer $=$ D-mannose   C-4 epimer $=$ D-galactose
C-3 epimer $=$ D-allose   C-5 epimer $=$ L-idose

9. **a.** D-ribose and L-ribose, D-arabinose and L-arabinose, D-xylose and L-xylose, D-lyxose and L-lyxose

b. D-ribose and D-arabinose, D-xylose and D-lyxose, L-ribose and L-arabinose, L-xylose and L-lyxose

20.   They are all optically active except for D-allose and D-galactose, which are not optically active because they each have a plane of symmetry.

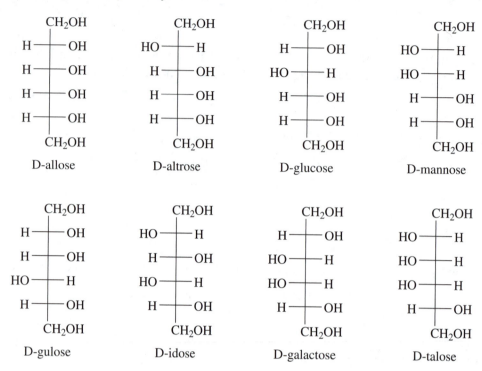

	CH₂OH	CH₂OH	CH₂OH	CH₂OH

D-allose   D-altrose   D-glucose   D-mannose

D-gulose   D-idose   D-galactose   D-talose

21.   The aldohexoses are D-talose and D-galactose; the ketohexoses are D-tagatose and D-sorbose.

D-tagatose

D-talose   D-galactose   D-tagatose   D-sorbose

2. The specific rotation would be –112.2, because α-L-glucose is the enantiomer of α-D-glucose.

3.

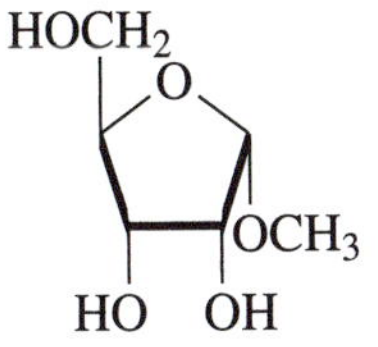

methyl α-D-ribofuranose

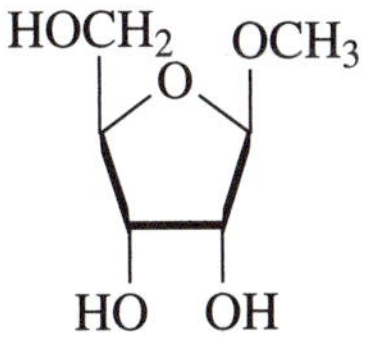

methyl β-D-ribofuranose

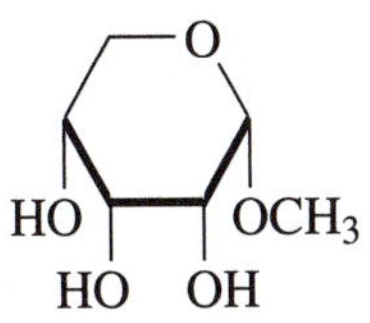

methyl α-D-ribopyranose

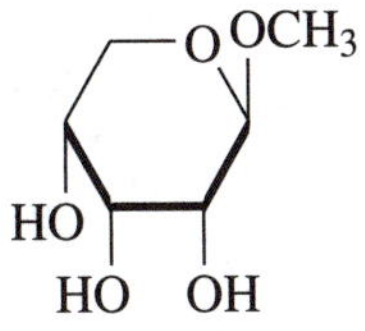

methyl β-D-ribopyranose

4. Maltose and lactose undergo mutarotation, because each has a hemiacetal linkage (at the anomeric carbon on the right-hand side of the molecule). The hemiacetal can open to an aldehyde and reclose to give either the α- or the β- configuration at that anomeric carbon.
Sucrose does not undergo mutarotation, because it does not have a hemiacetal linkage (its two anomeric carbons are bonded together).

5. She can take a sample of one of the sugars and reduce it with sodium borohydride. If the product is optically active, the sugar was D-lyxose; if the product is not optically active, the sugar was D-xylose.

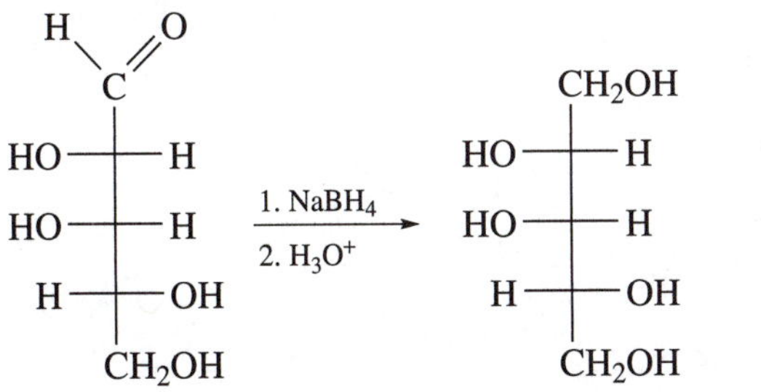

D-lyxose →(1. NaBH₄ 2. H₃O⁺)→ optically active does not have a plane of symmetry

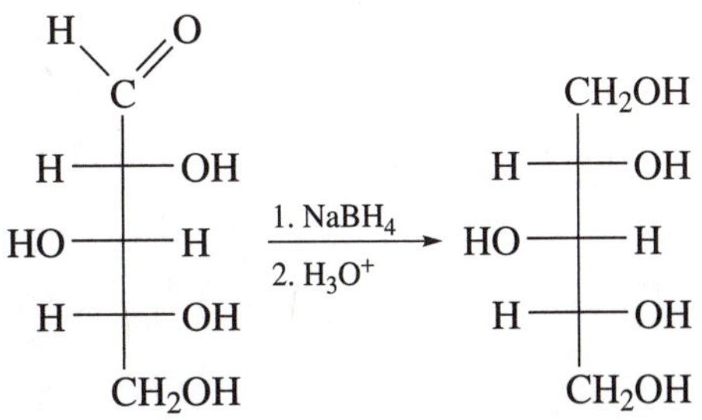

D-xylose →(1. NaBH₄ 2. H₃O⁺)→ optically inactive has a plane of symmetry

6. a. methyl β-D-sorboside
  b. ethyl β-D-guloside
  c. methyl α-D-idoside

27.   **a.**   Recall that what is down in a Haworth projection is on the right in a Fischer projection.

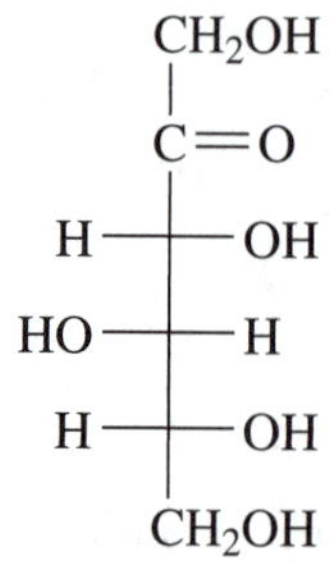

$$
\begin{array}{c}
CH_2OH \\
| \\
C{=}O \\
H{-}\!\!-{-}OH \\
HO{-}\!\!-{-}H \\
H{-}\!\!-{-}OH \\
CH_2OH
\end{array}
$$

b.   The monosaccharide has the opposite configuration that glucose has at C-3 and C-4.

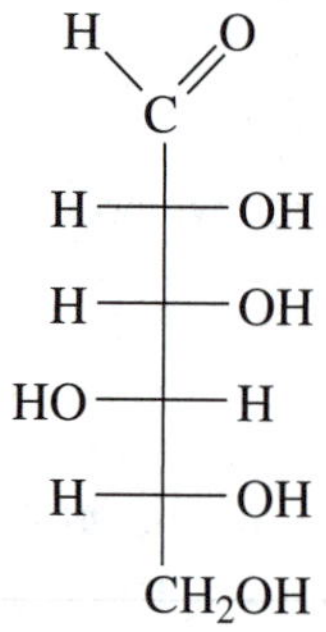

$$
\begin{array}{c}
H\diagdown \ \diagup O \\
C \\
H{-}\!\!-{-}OH \\
H{-}\!\!-{-}OH \\
HO{-}\!\!-{-}H \\
H{-}\!\!-{-}OH \\
CH_2OH
\end{array}
$$

c.   The monosaccharide has the opposite configuration that glucose has at C-2, C-3, and C-4.

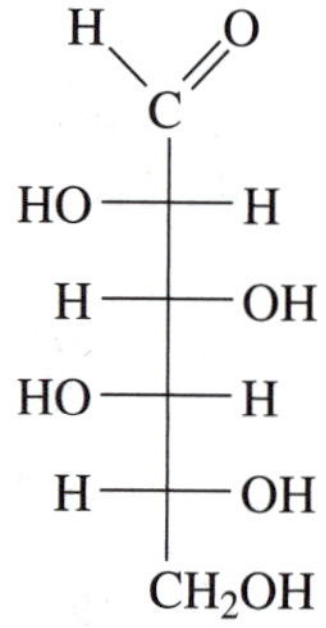

$$
\begin{array}{c}
H\diagdown \ \diagup O \\
C \\
HO{-}\!\!-{-}H \\
H{-}\!\!-{-}OH \\
HO{-}\!\!-{-}H \\
H{-}\!\!-{-}OH \\
CH_2OH
\end{array}
$$

28.

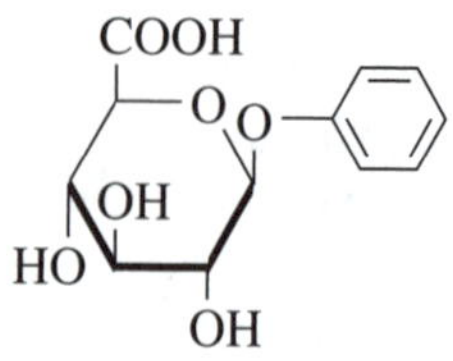

the β-D-glucuronide 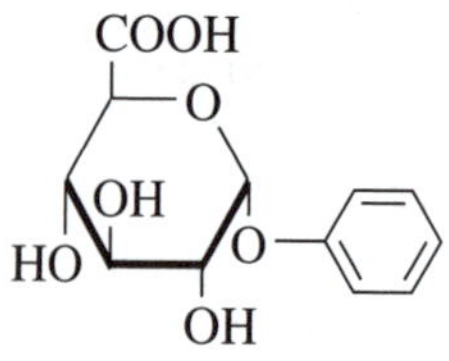   the α-D-glucuronide

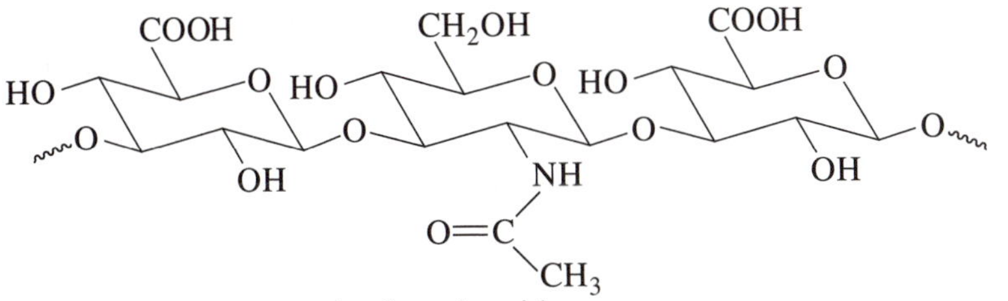

9.

hyaluronic acid

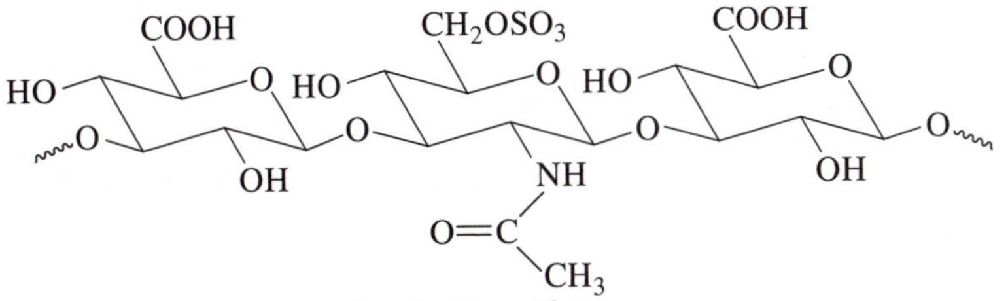

9.

chondroitin sulfate

1.   A monosaccharide with a molecular weight of 150 must have five carbons (five C's = 60, five O's = 80, and 10 H's = 10, for a total of 150). All aldopentoses are optically active. Therefore, the compound must be a ketopentose. The following is the only ketopentose that would not be optically active.

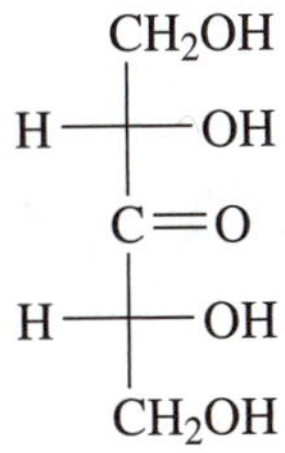

32.

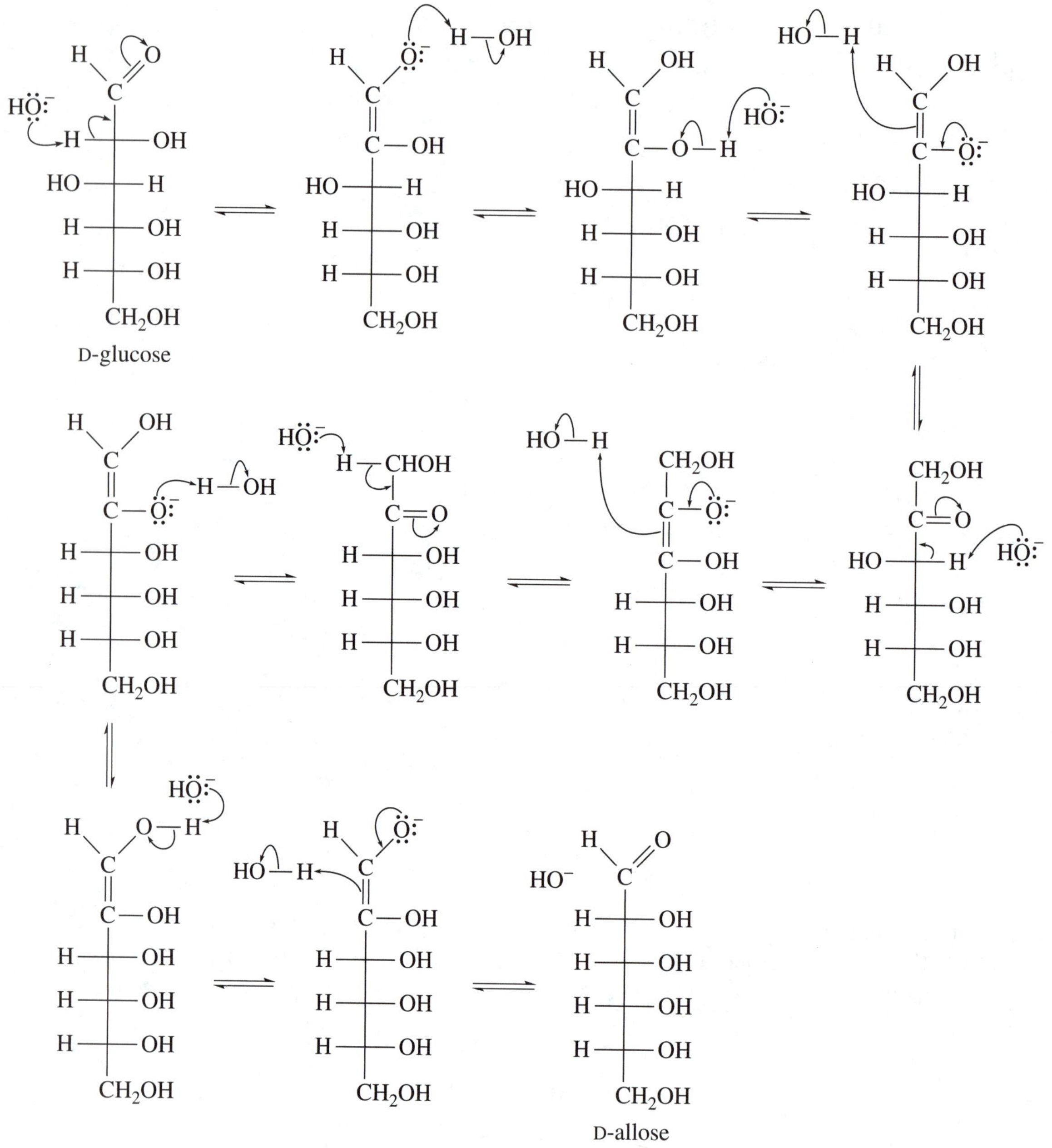

D-glucose

D-allose

3.

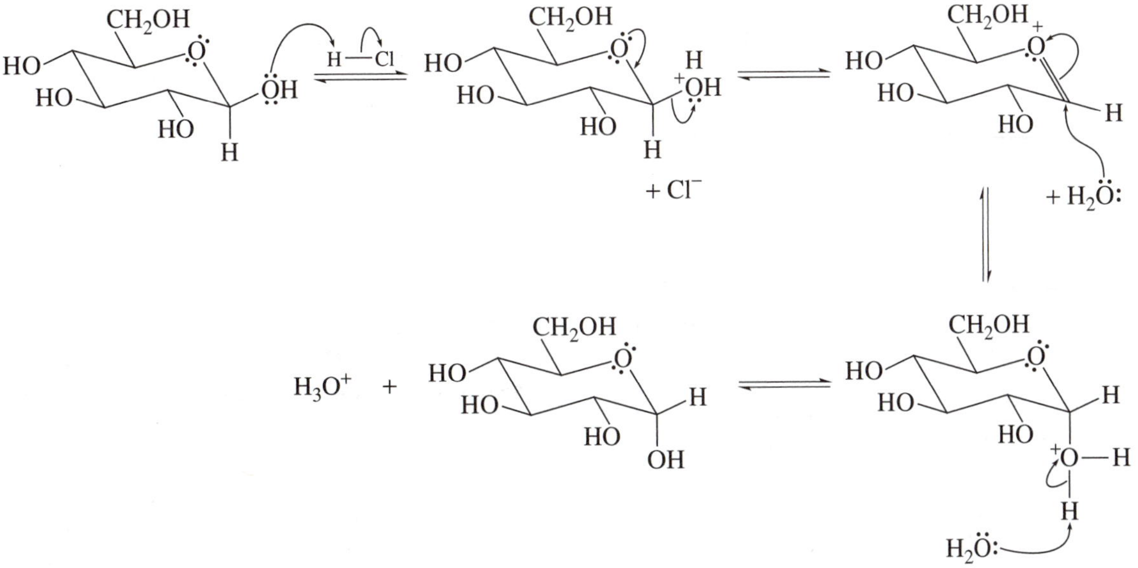

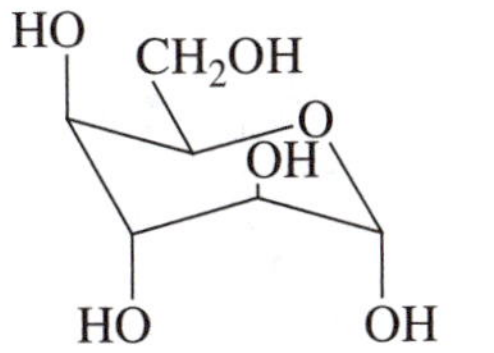

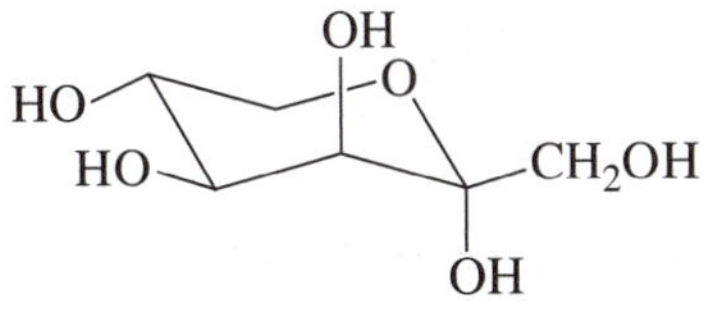

4.

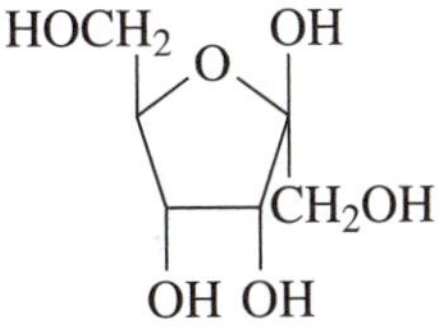

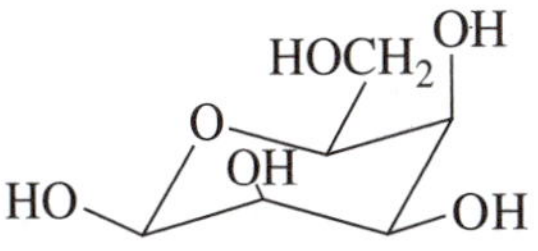

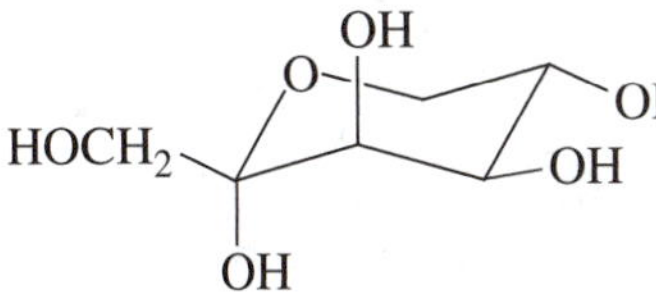

a.

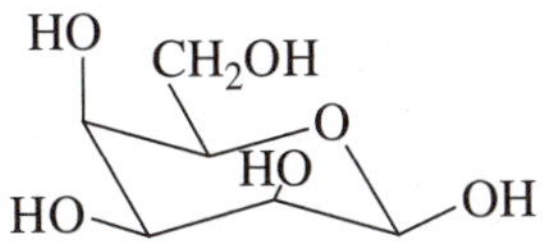

c.

e.

b.

d.

f.

35.

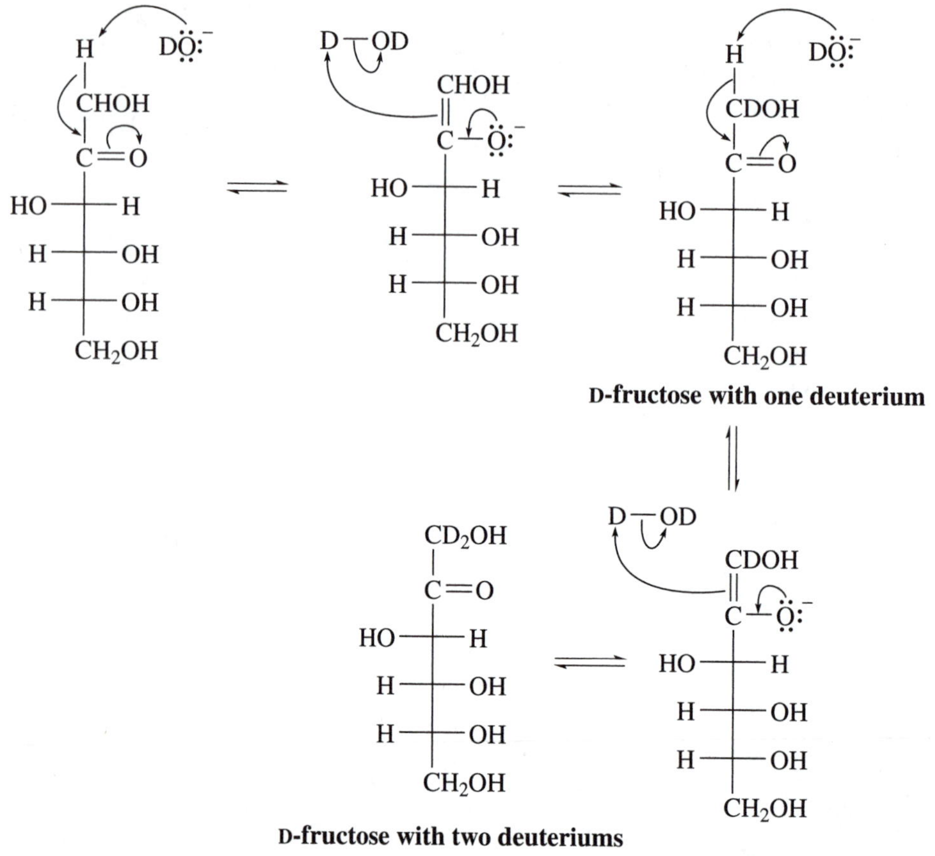

D-fructose with one deuterium

D-fructose with two deuteriums

36.

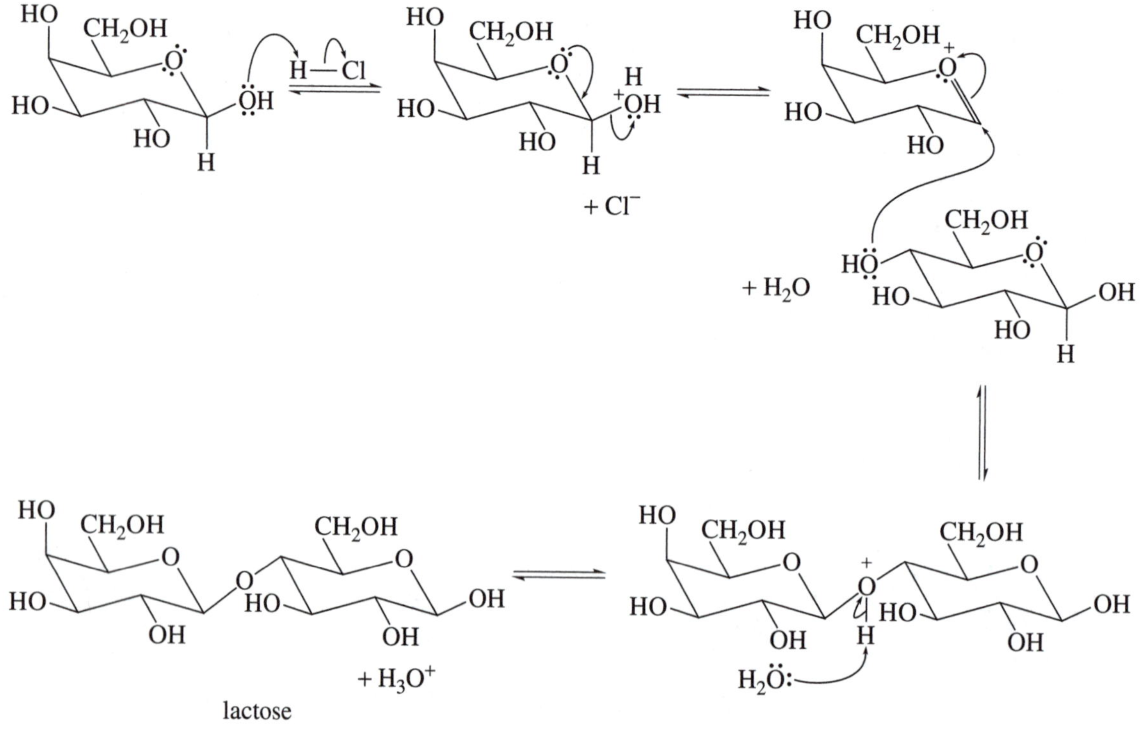

lactose

7. A proton is more easily lost from the C-3 OH group, because the electrons that are left behind when the proton is removed are delocalized onto an oxygen. When a proton is removed from the C-2 OH group, the electrons that are left behind are delocalized onto a carbon. Because oxygen is more electronegative than carbon, a negatively charged oxygen is more stable than a negatively charged carbon. Recall that the more stable the base, the stronger its conjugate acid.

8.

The β-anomer will also be formed when H_2O adds to the top of the plane of the oxocarbenium ion.

39.   Let A = the fraction of D-glucose in the α-form and B = the fraction of D-glucose in the β-form.

$$A + B = 1$$
$$B = 1 - A$$

specific rotation of A = 112.2
specific rotation of B = 18.7
specific rotation of the equilibrium mixture = 52.7
specific rotation of the mixture = specific rotation of A $\times$ fraction of D-glucose in the α-form +
                                specific rotation of B $\times$ fraction of D-glucose in the β-form

$$52.7 = 112.2\,A + (1 - A)\,18.7$$
$$52.7 = 112.2\,A + 18.7 - 18.7\,A$$
$$34.0 = 93.5\,A$$
$$A = 0.36$$
$$B = 0.64$$

This calculation shows that 36% is in the α-form and 64% is in the β-form.

40.   D-Altrose will most likely exist as a furanose, because
(1) the furanose is particularly stable since all the large substituents are trans to each other, and
(2) the pyranose has two of its OH groups in the unstable axial position.

D-altrofuranose   D-altropyranose

41.   In the case of D-idose, the chair conformer with both the OH substituent at C-1 and the CH$_2$OH substituent in axial positions (which is necessary for the formation of the anhydro form) has the OH substituents at C-2, C-3, and C-4 in equatorial positions. Thus, this is a relatively stable conformer, because three of the five large substituents are in the more stable equatorial position.

In the case of D-glucose, the chair conformer with both the OH substituent at C-1 and the CH$_2$OH substituent in axial positions has the OH substituents at C-2, C-3, and C-4 in axial positions. This is a relatively unstable conformer, because all the large substituents are in less stable axial positions.

anhydro form of D-idose   anhydro form of D-glucose

Therefore, a large percentage of D-idose but only a small percentage of D-glucose exists in the anhydro form at 100 °C.

Chapter 16 Practice Test

Draw the product(s) of the following reaction:

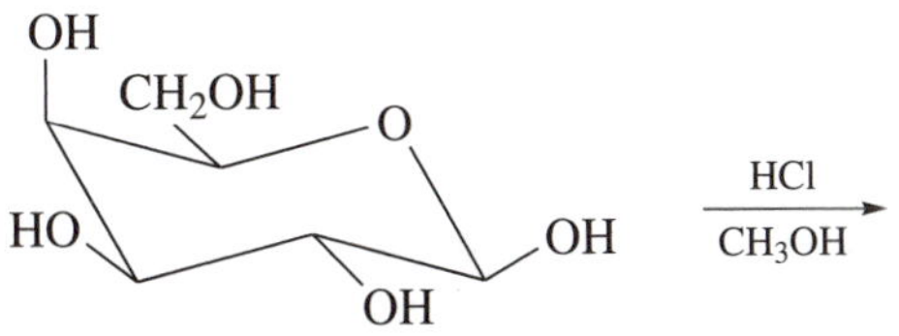

$$\xrightarrow[\text{CH}_3\text{OH}]{\text{HCl}}$$

Indicate whether each of the following statements is true or false:

a. Glycogen contains α-1,4'- and β-1,6'-glycosidic linkages.   T   F

b. D-Mannose is a C-1 epimer of D-glucose.   T   F

c. D-Glucose and L-glucose are anomers.   T   F

d. D-Erythrose and D-threose are diastereomers.   T   F

If the aldehyde group were reduced to an alcohol, which of the compounds would be optically active?

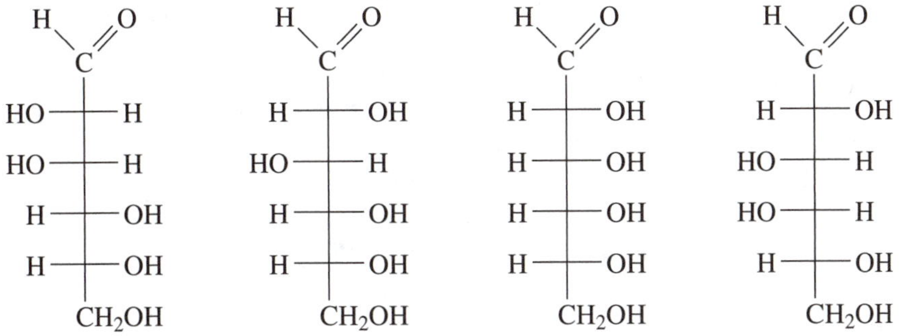

When crystals of D-fructose are dissolved in a basic aqueous solution, two aldohexoses are obtained. Identify the aldohexoses.

Draw three tetroses that form the same enol.

D-Talose and _____ are reduced to the same compound.

What is the main structural difference between amylose and cellulose?

8.   What aldohexose is the C-3 epimer of D-glucose?

9.   Draw the most stable chair conformer of β-D-allose.

Answers to Chapter 16 Practice Test

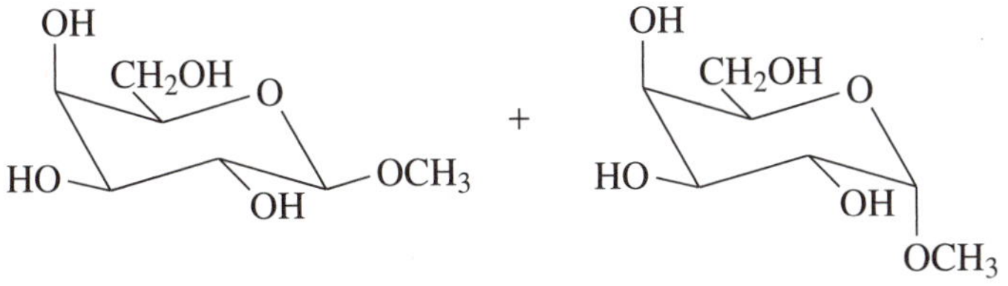

+

a. Glycogen contains α-1,4$'$ and β-1,6$'$-glycosidic linkages.   F
b. D-Mannose is a C-1 epimer of D-glucose.   F
c. D-Glucose and L-glucose are anomers.   F
d. D-Erythrose and D-threose are diastereomers.   T

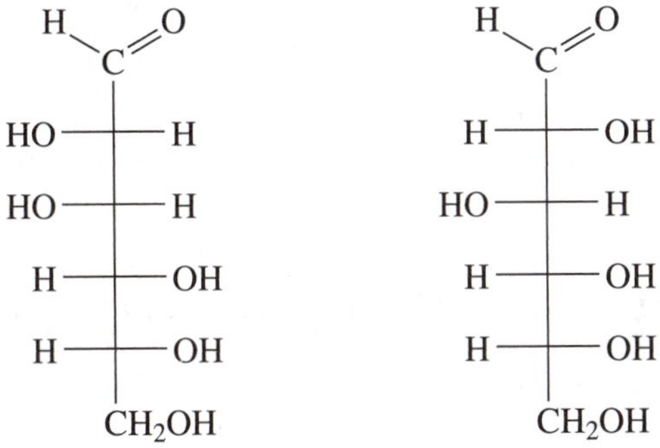

D-mannose and D-glucose

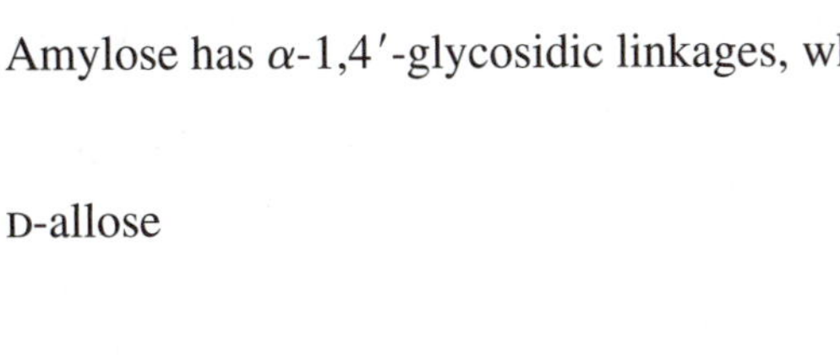

6. D-altrose

Amylose has α-1,4$'$-glycosidic linkages, whereas cellulose has β-1,4$'$-glycosidic linkages.

D-allose

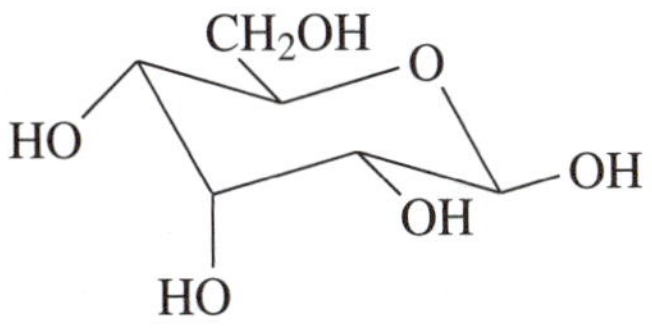

Important Terms

amino acid	an α-amino carboxylic acid. Naturally occurring amino acids have the L-configuration.
D-amino acid	the configuration of an amino acid drawn in a Fischer projection with the carboxyl group on top, the hydrogen on the left, and the amino group on the right.
L-amino acid	the configuration of an amino acid drawn in a Fischer projection with the carboxyl group on top, the hydrogen on the right, and the amino group on the left.
amino acid analyzer	an instrument that automates the ion-exchange separation of amino acids.
antiparallel β-pleated sheet	a type of secondary structure in which the adjacent hydrogen-bonded peptide chains in a β-pleated sheet run in opposite directions.
cation-exchange resin	a resin that binds cations.
coil conformation (loop conformation)	the part of a protein that is highly ordered but not in an α-helix or a β-pleated sheet.
C-terminal amino acid	the terminal amino acid of a peptide (or protein) that has a free carboxyl group.
denaturation	the destruction of the highly organized secondary and tertiary structures of a protein.
dipeptide	two amino acids linked together by an amide bond.
disulfide	a compound with an S—S bond.
disulfide bridge	a disulfide (S—S) bond formed by two cysteine residues in a peptide or protein.
Edman's reagent	phenyl isothiocyanate; the reagent used to determine the N-terminal amino acid of a polypeptide.
electrophoresis	a technique that separates amino acids on the basis of their pI values.
endopeptidase	an enzyme that hydrolyzes a peptide bond that is not at the end of a peptide chain.
essential amino acid	an amino acid that humans must obtain from their diet because they either cannot synthesize it at all or cannot synthesize it in adequate amounts.
exopeptidase	an enzyme that hydrolyzes a peptide bond at the end of a peptide chain.
fibrous protein	a water-insoluble protein that has its polypeptide chains arranged in bundles.

globular protein	a water-soluble protein that tends to have a roughly spherical shape.
α-helix	the backbone of a polypeptide coiled in a right-handed spiral with hydrogen bonding occurring within the helix.
hydrophobic interactions	interactions between nonpolar groups. These interactions increase stability by decreasing the amount of structured water (increasing entropy).
interchain disulfide bridge	a disulfide bridge between two cysteine residues in different peptide chains.
intrachain disulfide bridge	a disulfide bridge between two cysteine residues in the same peptide chain.
ion-exchange chromatography	a technique that uses a column packed with an insoluble resin to separate compounds on the basis of their charge and polarity.
isoelectric point (pI)	the pH at which there is no net charge on an amino acid.
loop conformation (coil conformation)	see coil conformation.
N-terminal amino acid	the terminal amino acid of a peptide (or protein) that has a free amino group.
oligomer	a protein with more than one peptide chain.
oligopeptide	three to ten amino acids linked by amide bonds.
paper chromatography	a technique that separates amino acids based on polarity.
parallel β-pleated sheet	a type of secondary structure in which the adjacent hydrogen-bonded peptide chains in a β-pleated sheet run in the same direction.
partial hydrolysis	a technique that hydrolyzes only some of the peptide bonds in a polypeptide.
peptidase	an enzyme that catalyzes the hydrolysis of a peptide bond.
peptide	a polymer of amino acids linked together by amide bonds.
peptide bond	the amide bond that links the amino acids in a peptide or protein.
β-pleated sheet	a type of secondary structure in which the backbone of a polypeptide extends in a zigzag structure with hydrogen bonding between neighboring chains.
polypeptide	many amino acids linked by amide bonds.
primary structure	the sequence of amino acids and the location of the disulfide bridges in a protein.
protein	a naturally occurring polymer of 40 to 4000 amino acids linked together by amide bonds.

quaternary structure a description of the way in which the individual polypeptide chains of an oligomeric protein are arranged with respect to one another.

secondary structure a description of the conformation of the backbone of a protein.

side chain the substituent attached to the α-carbon of an amino acid.

structural protein a protein that gives strength to a biological structure.

subunit an individual chain of an oligomeric protein.

tertiary structure a description of the three-dimensional arrangement of all the atoms in a protein.

thin-layer chromatography a technique that separates compounds on the basis of their polarity.

tripeptide three amino acids linked by amide bonds.

zwitterion a compound with a negative charge and a positive charge on nonadjacent atoms.

olutions to Problems

a. When the imidazole ring is protonated, the double-bonded nitrogen is the one that accepts the proton. The lone-pair electrons on the single-bonded nitrogen are delocalized and therefore are not available to be protonated.

In contrast, the lone-pair electrons on the double-bonded nitrogen are not delocalized, because the constraints of the ring will not allow two adjacent double bonds.

Furthermore, if the lone pair on the single-bonded nitrogen were protonated, the compound would lose its aromaticity.

In contrast, when the double-bonded nitrogen is protonated, the protonated compound is still aromatic.

b. The lone-pair electrons on the double-bonded nitrogen are protonated, because the lone-pair electrons on the other nitrogens are delocalized and therefore cannot be protonated. In addition, protonation of the double-bonded nitrogen leads to a highly resonance-stabilized conjugate acid.

2.   Threonine and isoleucine are the only two amino acids that have more than one asymmetric center. Each h
two asymmetric centers.

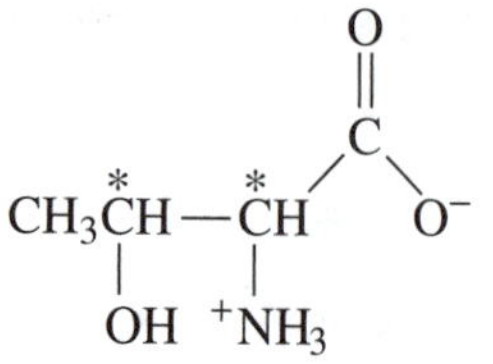

threonine 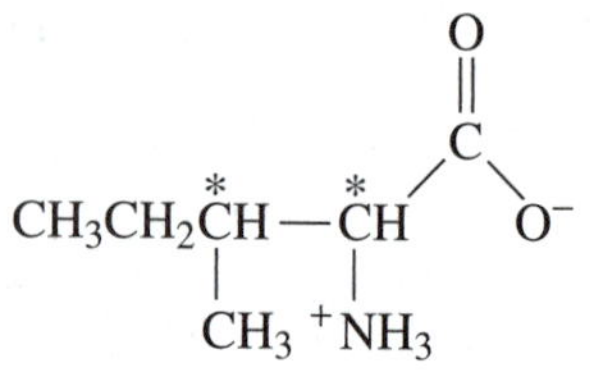   isoleucine

3.   An amino acid is insoluble in diethyl ether (a relatively nonpolar solvent) because an amino acid exists as
highly polar zwitterion at neutral pH. In contrast, carboxylic acids and amines are less polar, because the
either are neutral or have a single charge depending on the extent of dissociation in diethyl ether.

4.   The electron-withdrawing $^+NH_3$ substituent on the α-carbon increases the acidity of the carboxyl group.

5.   **a.** Solved in the text.   **c.** 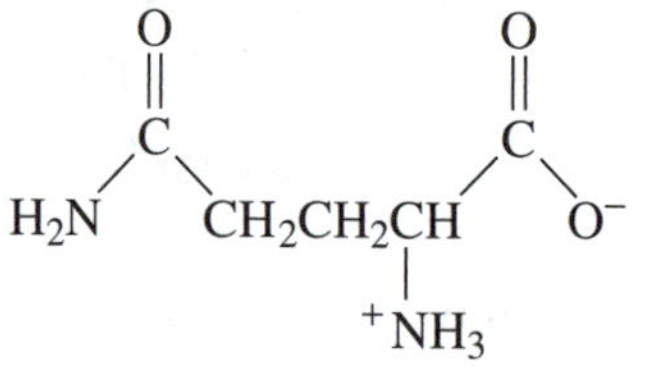   **e.**

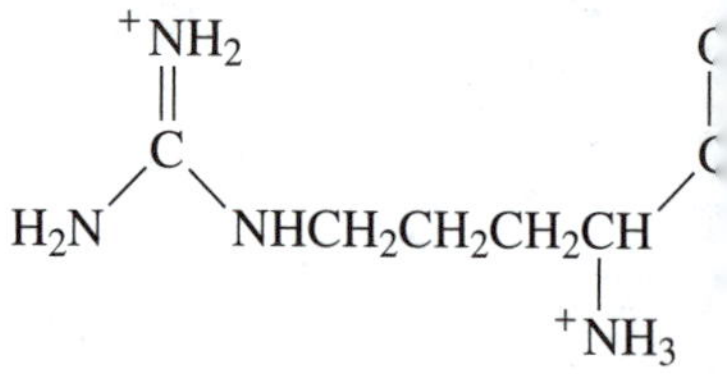

b. 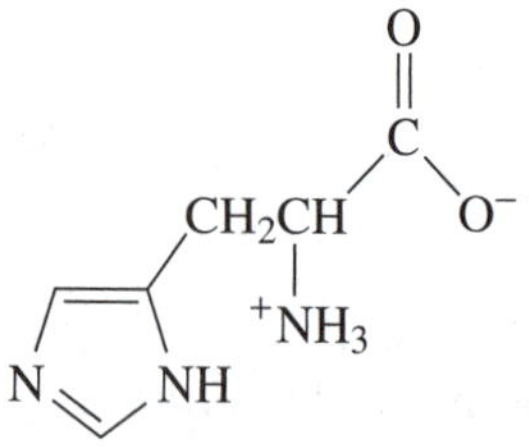   **d.** 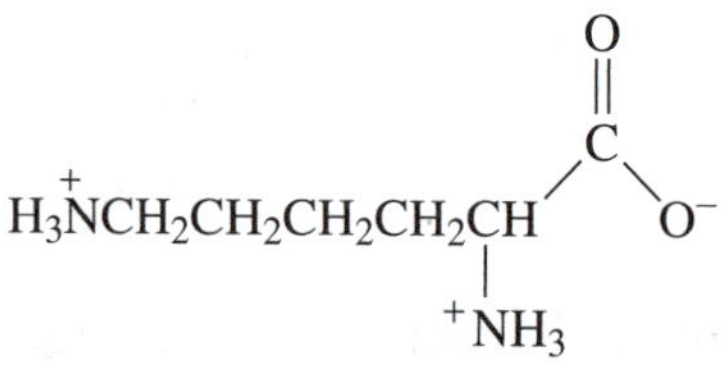   **f.**

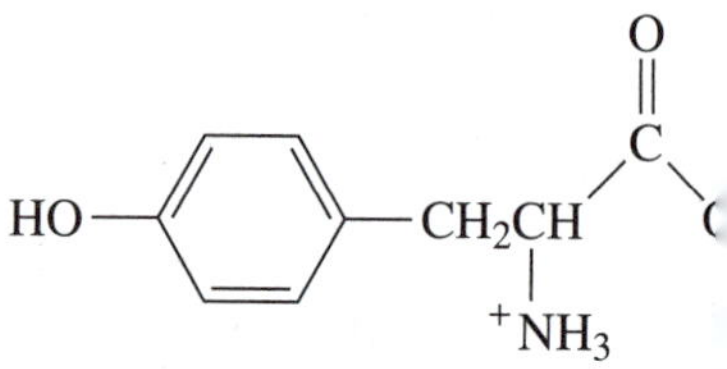

6.   **a.** 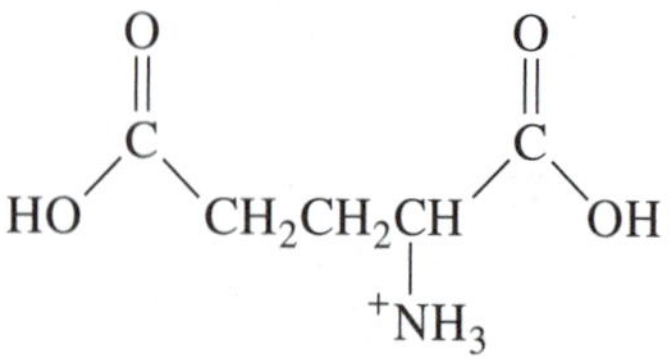   **c.**

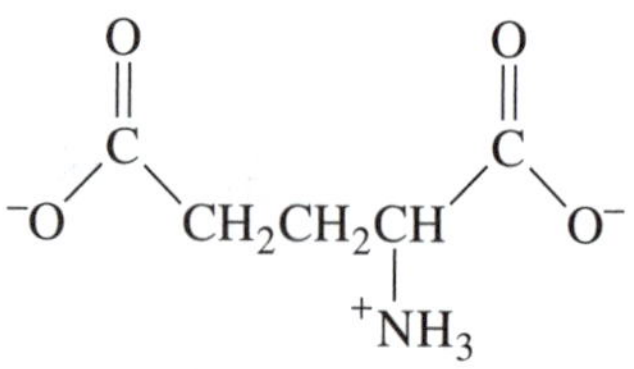

b. 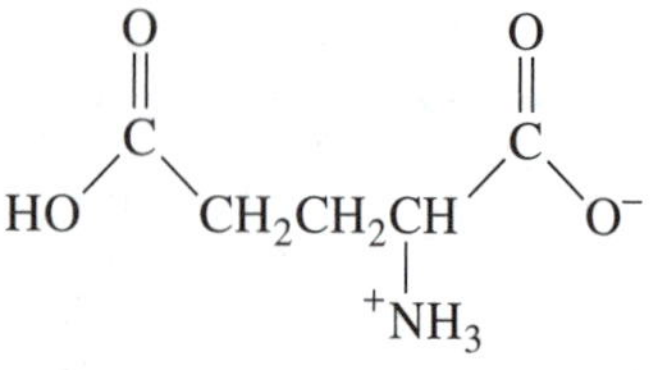   **d.**

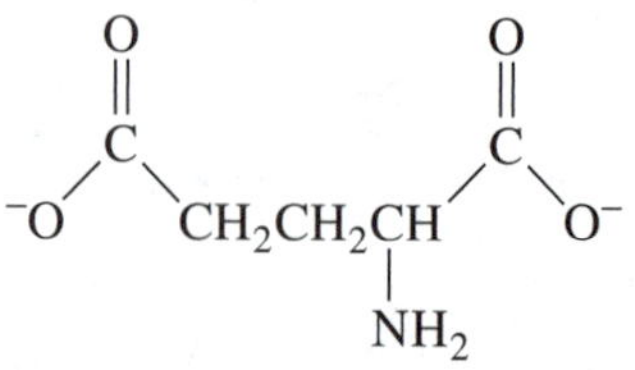

.

a. asparagine pI $= \dfrac{2.02 + 8.84}{2} = \dfrac{10.86}{2} = 5.43$

b. arginine pI $= \dfrac{9.04 + 12.48}{2} = \dfrac{21.52}{2} = 10.76$

c. serine pI $= \dfrac{2.21 + 9.15}{2} = \dfrac{11.36}{2} = 5.68$

d. aspartate pI $= \dfrac{2.09 + 3.86}{2} = \dfrac{5.95}{2} = 2.98$

.

a. aspartate (pI $= 2.98$)

b. arginine (pI $= 10.76$)

.

CH_3CH — with CH_3 branch — $C(=O)H$ the R group of the aldehyde is the same as the R group of the amino acid

10. Leucine and isoleucine both have C_4H_9 side chains and, therefore, have the same polarity. Consequently, the spots for both amino acids appear at the same place on the chromatographic plate. Therefore, the chromatographic plate has one fewer spot than the number of amino acids.

11. Cation-exchange chromatography releases amino acids in order of their pI values. The amino acid with the lowest pI is released first because, at a given pH, it will be the amino acid with the highest concentration of negative charge, and negatively charged molecules are not bound by the negatively charged resin. The relatively nonpolar resin will release polar amino acids before nonpolar amino acids.

a. Asp (pI $= 2.98$) is more negative at pH $= 4$ than is Ser (pI $= 5.68$).
b. Ser is more polar than Ala.
c. Val is more polar than Leu. (Val has fewer carbons than Leu.)
d. Tyr is more polar than Phe.

12. Because the amino acid analyzer contains a cation-exchange resin (it binds cations), the less positively charged the amino acid, the less tightly it is bound to the column. Using buffer solutions of increasingly higher pH to elute the column causes the amino acids bound to the column to become increasingly less positively charged, so they can be released from the column.

13.   **a.**   The following reactions show that pyruvic acid forms alanine, oxaloacetic acid forms aspartate, and
α-ketoglutarate forms glutamate.

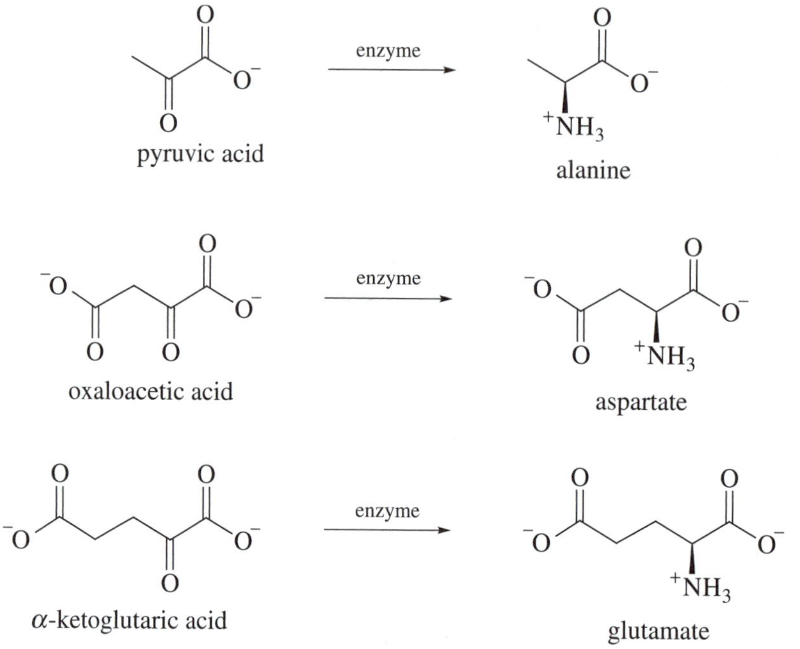

pyruvic acid → alanine

oxaloacetic acid → aspartate

α-ketoglutaric acid → glutamate

If reductive amination is carried out in the cell, only the L-isomer of each amino acid will be formed.

b.   Imine formation is best carried at a pH about 1.5 units lower than the pK_a of the protonated
ammonium—that is, at about pH = 8. Therefore, the carboxyl groups will be in their basic forms. If
reductive amination is carried out in the laboratory, both the D- and L-isomers (a racemic mixture) of
each amino acid will be formed.

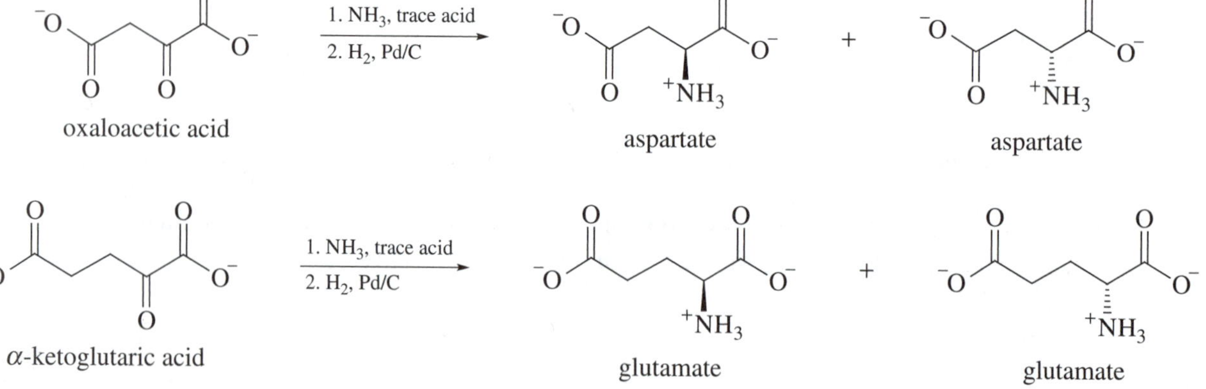

pyruvic acid →   1. NH₃, trace acid   2. H₂, Pd/C   →   alanine   +   alanine

oxaloacetic acid →   1. NH₃, trace acid   2. H₂, Pd/C   →   aspartate   +   aspartate

α-ketoglutaric acid →   1. NH₃, trace acid   2. H₂, Pd/C   →   glutamate   +   glutamate

4.   Notice that the R group attached to the Br is the same as the R group attached to the α-carbon of the amino acid.

$$R—Br \quad \text{corresponds to} \quad R—\underset{\overset{|}{^+NH_3}}{C}HCOO^-$$

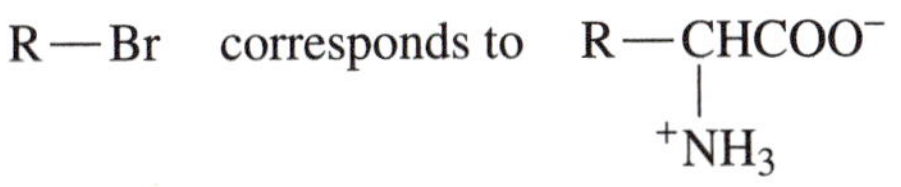

a. leucine   **b.** methionine

5.   Notice that the R group attached to the carbonyl group of the aldehyde is the same as the R group attached to the α-carbon of the amino acid.

$$R—\overset{\overset{O}{\|}}{C}H \quad \text{corresponds to} \quad R—\underset{\overset{|}{^+NH_3}}{C}HCOO^-$$

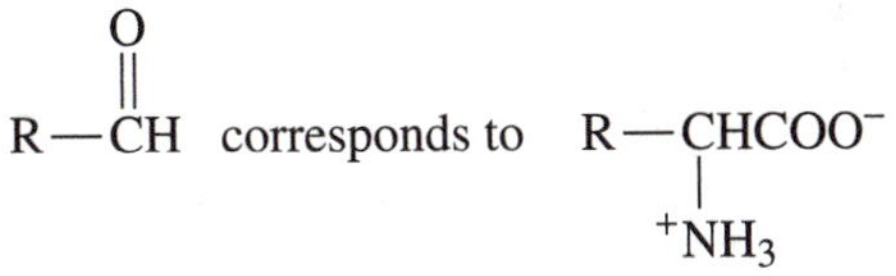

a. alanine   **b.** isoleucine   **c.** leucine

16.   Convert the amino acids into esters using $SOCl_2$ followed by ethanol. Then treat the esters with pig liver esterase. Because the enzyme hydrolyzes only esters of L-amino acids, the products will be the L-amino acid, ethanol, and the ester of the D-amino acid. These compounds can be readily separated. After they are separated, the D-amino acid can be obtained by acid-catalyzed hydrolysis of the ester of the D-amino acid. This separation technique is called a kinetic resolution, because the enantiomers are separated (resolved) as a result of reacting at different rates in the enzyme-catalyzed reaction.

17.

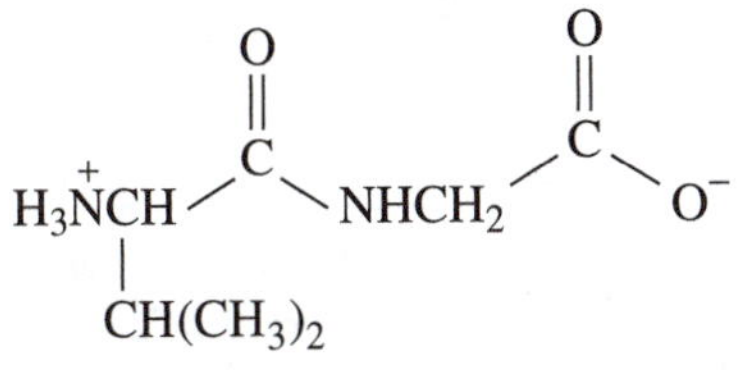

Val-Gly 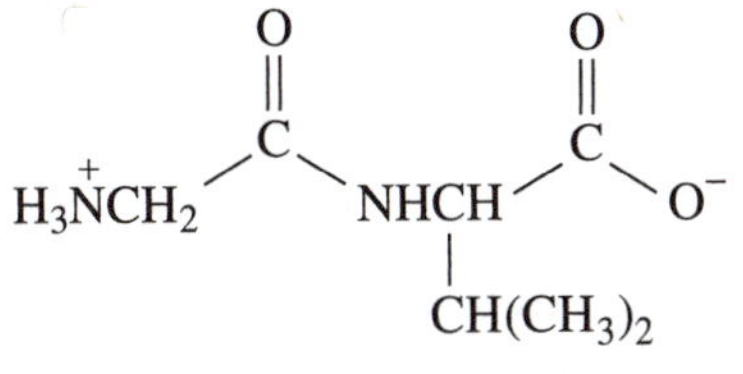   Gly-Val

18.

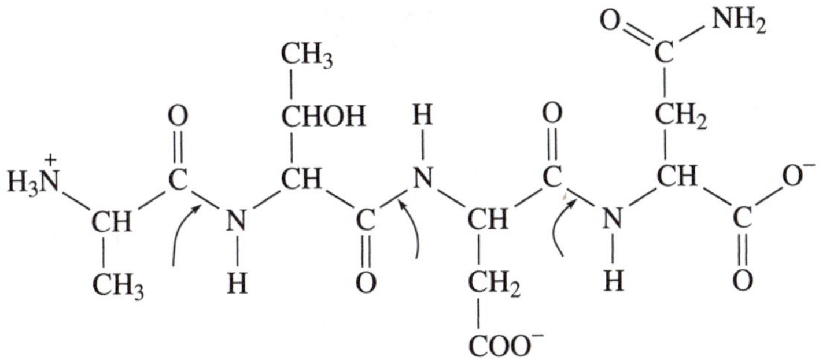

the peptide bonds are indicated by arrows

19. The bonds on either side of the α-carbon can freely rotate. In other words, the bond between the α-carbo[n] and the carbonyl carbon and the bond between the α-carbon and the nitrogen (the bonds indicated by a[r]rows) can freely rotate. The bond between the C and N (the peptide bond) cannot rotate, because it ha[s] partial double-bond character.

20.   **a.**   glutamate, cysteine, and glycine

   b.   In forming the amide bond between glutamate and cysteine, the amino group of cysteine reacts wit[h] the γ-carboxyl group of glutamate rather than with its α-carboxyl group.

21.   Because insulin has two peptide chains, treatment with Edman's reagent would release two PTH-amin[o] acids in approximately equal amounts.

22.   Knowing that the N-terminal amino acid is Gly, look for a peptide fragment that contains Gly.

   "Fragment 6" tells you that the second amino acid is Arg.

   "Fragment 5" tells you that the next two are Ala-Trp or Trp-Ala.

   "Fragment 4" tells you that Glu is next to Ala, so the third and fourth amino acids must be Trp-Ala and th[e] fifth is Glu.

   "Fragment 7" tells you that the sixth amino acid is Leu.

   "Fragment 8" tells you that the next two are Met-Pro or Pro-Met.

   "Fragment 3" tells you that Pro is next to Val, so the seventh and eighth amino acids must be Met-Pro an[d] the ninth is Val.

   "Fragment 2" tells you that the last amino acid is Asp.

   Gly-Arg-Trp-Ala-Glu-Leu-Met-Pro-Val-Asp

3. Cysteine can react with cyanogen bromide, but the sulfur would not be positively charged, so it would be a poor leaving group. In addition, the lactone will not be formed, because it would have a strained four-membered ring. Without lactone formation, the imine would not be formed, so cleavage cannot occur.

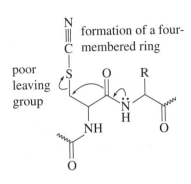

24. **a.** His-Lys   Leu-Val-Glu-Pro-Arg   Ala-Gly-Ala

     b. Leu-Gly-Ser-Met-Phe-Pro-Tyr   Gly-Val

25. Solved in the text.

26. The data from treatment with Edman's reagent and carboxypeptidase A identify the first and last amino acids.

    Leu __ __ __ __ __ __ Ser

The data from cleavage with cyanogen bromide identify the position of Met and identify the other amino acids in the pentapeptide and tripeptide but not their order.

                                           ⌐ cleavage with cyanogen bromide

               Arg, Lys, Tyr   Arg, Phe

    Leu __ __ __ Met | __ __ Ser

The data from treatment with trypsin put the remaining amino acids in the correct positions.

    Leu   Tyr   Lys | Arg | Met   Phe   Arg | Ser

27. It would fold so that its nonpolar residues are on the outside of the protein in contact with the nonpolar membrane and its polar residues are on the inside of the protein.

28.   A protein folds to maximize the number of polar groups on the surface of the protein and the number of nonpolar groups on the inside of the protein.

    a.   A cigar-shaped protein has the greatest surface-area-to-volume ratio, so it has the highest percentage of polar amino acids.

    b.   A subunit of a hexamer would have the smallest percentage of polar amino acids, because part of the surface of the subunit can be on the inside of the hexamer and therefore have nonpolar amino acids on its surface.

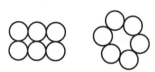

a scheme of two possible quaternary structures of a hexamer

29.   **a.** $H_3\overset{+}{N}CH_2CH_2CH_2CH_2-CH\underset{\overset{|}{\overset{+}{N}H_3}}{\overset{\displaystyle O \atop \displaystyle \parallel \atop \displaystyle C}{\diagup}}O^-$ **c.** $HO-\!\!\left\langle\!\!\bigcirc\!\!\right\rangle\!\!-CH_2-CH\underset{\overset{|}{\overset{+}{N}H_3}}{\overset{\displaystyle O \atop \displaystyle \parallel \atop \displaystyle C}{\diagup}}O^-$

    b. $H_2N\overset{\overset{\displaystyle +NH_2}{\displaystyle \parallel}}{C}NHCH_2CH_2CH_2-CH\underset{\overset{|}{\overset{+}{N}H_3}}{\overset{\displaystyle O \atop \displaystyle \parallel \atop \displaystyle C}{\diagup}}O^-$

30.   The pK_a of the OH group can be ignored because it is ~15, so there will be no ionization until the pH is ~ 13. Thus, it does not have to be considered in calculating the pI.

$$HOCH_2CH\underset{\overset{|}{\overset{+}{N}H_3}}{\overset{\displaystyle O \atop \displaystyle \parallel \atop \displaystyle C}{\diagup}}OH \qquad \frac{2.21 + 9.15}{2} = \frac{11.36}{2} = 5.68$$

$pK_a = 2.21$ $pK_a = 9.15$

31.   **a.**   pH $= 9.60$ **b.**   pH $= 5.97$ (its pI)   **c.**   pH $= 2.34$

32.   **a.**   Val-Arg-Gly-Met-Arg-Ala   Ser

    b.   Ser-Phe-Lys-Met   Pro-Ser-Ala-Asp

    c.   Arg   Ser-Pro-Lys   Lys   Ser-Glu-Gly

33.   As an amino acid moves from a solution with a pH equal to its pI to a more basic solution, the amino acid becomes more and more negatively charged. Because asparagine has a lower pI than leucine, in a solution of pH $= 7.4$, asparagine has moved farther from its pI than has leucine. Asparagine, therefore, will have a higher percentage of negative charge at pH $= 7.4$.

4. **a.** (structure: HO–C(=O)–CH₂CH(⁺NH₃)–C(=O)–OH)

c. (structure: ⁻O–C(=O)–CH₂CH(⁺NH₃)–C(=O)–O⁻)

b. (structure: HO–C(=O)–CH₂CH(⁺NH₃)–C(=O)–O⁻)

d. (structure: ⁻O–C(=O)–CH₂CH(NH₂)–C(=O)–O⁻)

35. Alanine will exist predominantly as a zwitterion in an aqueous solution with pH > 2.34 and pH < 9.69.

36. The student is correct. At the pI, the total of the positive charges on the tripeptide's amino groups must be one to balance the one negative charge of the carboxylate group. When the pH of the solution is equal to the pK_a of a lysine residue, the three lysine groups each have one-half a positive charge for a total of one and one-half positive charges. Thus, the solution must be more basic than this in order to have just one positive charge.

37. **a.** The carboxyl group of the aspartic acid side chain is a stronger acid than the carboxyl group of the glutamic acid side chain because the carboxyl group of the aspartic acid side chain is closer to the electron-withdrawing protonated amino group.

b. The protonated lysine side chain is a stronger acid than the protonated arginine side chain. The protonated arginine side chain has less of a tendency to lose a proton because its positive charge is delocalized over three nitrogens.

38. Since the mixture of amino acids is in a solution of pH = 5, **His** will have an overall positive charge and **Glu** will have an overall negative charge. **His,** therefore, will migrate to the cathode, and **Glu** will migrate to the anode.

Ser is more polar than **Thr** (both have OH groups, but **Thr** has an additional carbon).

Thr is more polar than **Met** (**Met** has SCH₂CH₂ instead of CHOH). **Met** is more polar than **Leu** (**Leu** has four carbons, whereas **Met** has three carbons and a sulfur).

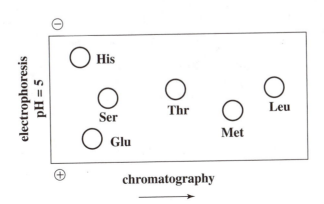

39.   We know that "Fragment 3" (Leu-Pro-Phe) is at the C-terminal end of the polypeptide. Thus, the only question we need to answer is, which of the two fragments obtained by cleavage with cyanogen bromide, which begin with Gly and end with Met, is nearest the C-terminal end of the polypeptide? The answer is obtained from "Fragment 4" obtained from trypsin cleavage. The Met nearest to the C-terminal end must be preceded by Arg or Lys. Therefore, the polypeptide has the following sequence:

Gly-Leu-Tyr-Phe-Lys-Ser-Met-Gly-Leu-Tyr-Lys-Val-Ile-Arg-Met-Leu-Pro-Phe

40.   Each compound has two groups that can act as a buffer, one amino group and one carboxyl group. Thus the compound in higher concentration (0.2 M glycine) will be a more effective buffer.

41.   Groups that are not fully charged at the given pH are shown in the form that predominates at that pH. For example, tyrosine has a pK_a = 9.11, so at pH = 7 it is shown in its acidic (neutral) form.

a.

b.

c.

d.

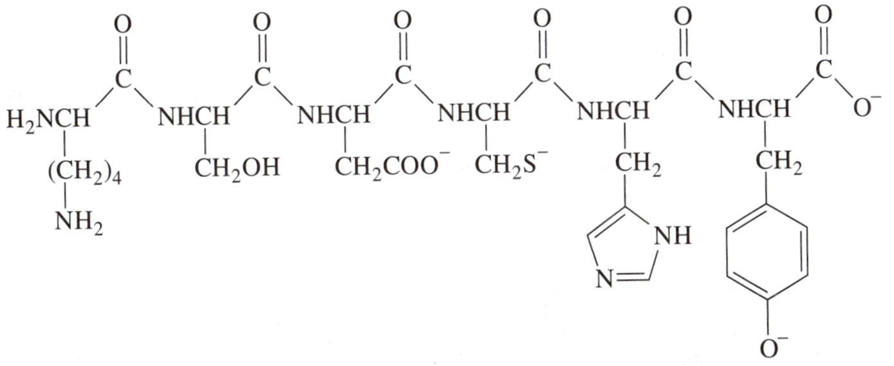

42. **a.** Trypsin cleaves at Arg and Lys. There are two possible primary structures:

Val-Gly-Asp-Lys-Leu-Glu-Pro-Ala-Arg-Ala-Leu-Gly-Asp

or

Leu-Glu-Pro-Ala-Arg-Val-Gly-Asp-Lys-Ala-Leu-Gly-Asp

The two possible primary structures can be distinguished by Edman's reagent. Edman's reagent would release Val in one case and Leu in the other.

b. Trypsin cleaves at Arg and Lys. There are two possible primary structures:

Ala-Glu-Pro-Arg-Ala-Met-Gly-Lys-Val-Leu-Gly-Glu

or

Ala-Met-Gly-Lys-Ala-Glu-Pro-Arg-Val-Leu-Gly-Glu

The two possible primary structures can be distinguished by treatment with cyanogen bromide. Cyanogen bromide would cleave one of the possible polypeptides into two hexamers and the other into a dimer and a decamer.

43.

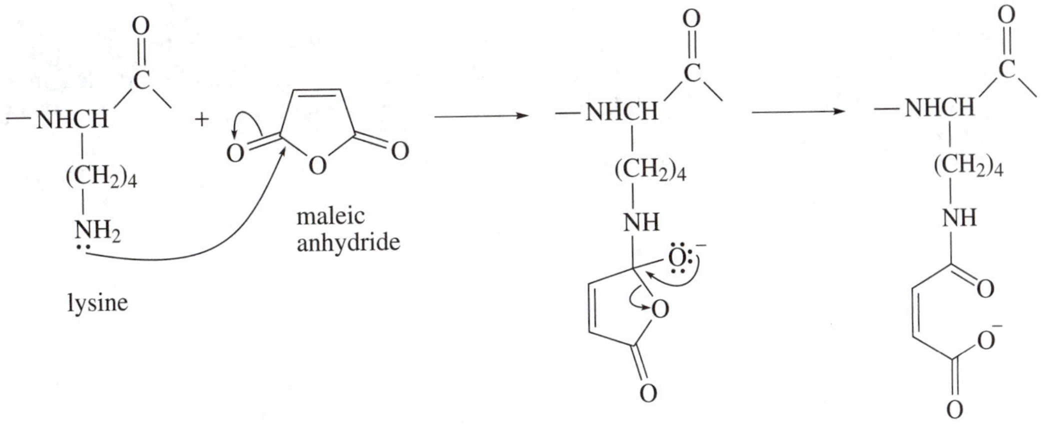

44.   **a.**   When the polypeptide is treated with maleic anhydride, lysine reacts with maleic anhydride (see Problem 43), but the amino group of arginine is not sufficiently nucleophilic to react with maleic anhydride. Therefore, trypsin will cleave only at arginine residues because the enzyme will no longer recognize lysine residues.

b.   Four fragments will be obtained from the polypetide. Remember that trypsin will not cleave the Arg–Pro bond.

c.   The N-terminal end of each fragment will be positively charged because of the $^+NH_3$ group. The C-terminal end will be negatively charged because of the COO^- group. Arginine residues will be positively charged. Aspartate and glutamate residues will be negatively charged. Lysine residues will be negatively charged because they are attached to the maleic acid group.

Elution order: **A > D > C > B**

A   Gly-Ala-Asp-Ala-Leu-Pro-Gly-Ile-Leu-Val-Arg   overall charge = 0
   +   –   +–

B   Asp-Val-Gly-Lys-Val-Glu-Val-Phe-Glu-Ala-Gly-Arg   overall charge = –3
 +–   –   –   –   +–

C   Ala-Glu-Phe-Lys-Glu-Pro-Arg   overall charge = –2
  +   –   –   –   + –

D   Leu-Val-Met-Lys-Val-Glu-Gly-Arg-Pro-Val-Ala-Ala-Gly-Leu-Trp   overall charge = –1
  +   –   –   +   –

45.   First, mark off where the chains would have been cleaved by chymotrypsin (C-side of Phe, Trp, and Tyr).

Val-Met-Tyr │-Ala-Cys-Ser-Phe ┤-Ala-Glu-Ser

Ser-Cys-Phe│ -Lys-Cys-Trp │-Lys-Tyr ┤Cys-Phe ┤Arg-Cys-Ser

Then, from the fragments given, you can determine where the disulfide bridges are in the original intact peptide. "Fragment 2" has two Phe, two Cys, and one Ser. Therefore, the first and fourth fragments of the second row must be connected by a disulfide bond. "Fragment 5" provides the evidence for the disulfide bond between the two chains.

Val-Met-Tyr-Ala-Cys-Ser-Phe-Ala-Glu-Ser
                │
                S
                │
                S
                │
Ser-Cys-Phe-Lys-Cys-Trp-Lys-Tyr-Cys-Phe-Arg-Cys-Ser
    └————————S—S————————┘

46. a.

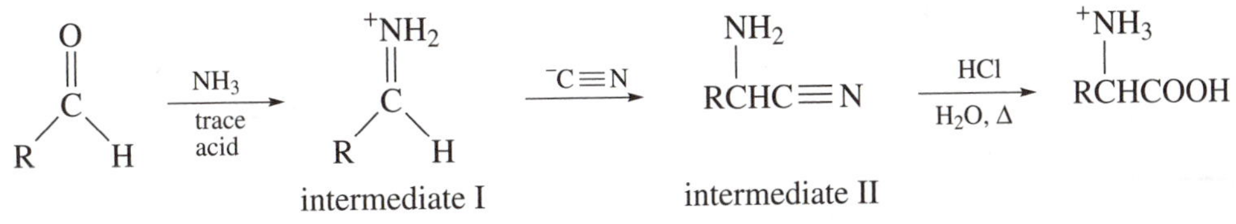
NH3
trace
acid
intermediate I
C≡N
intermediate II
HCl
H2O, Δ
RCHCOOH

b.

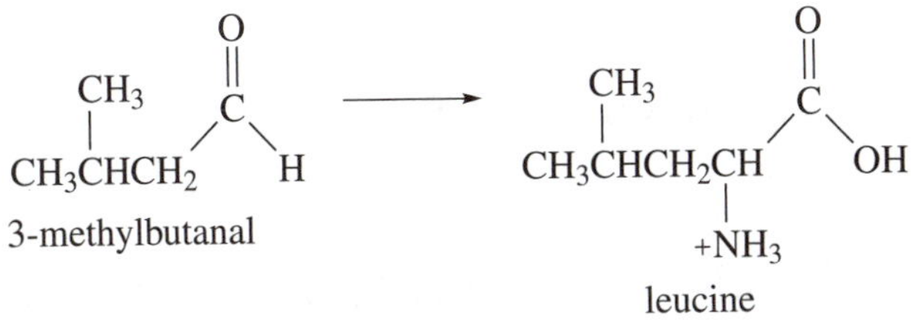
3-methylbutanal
leucine

c.

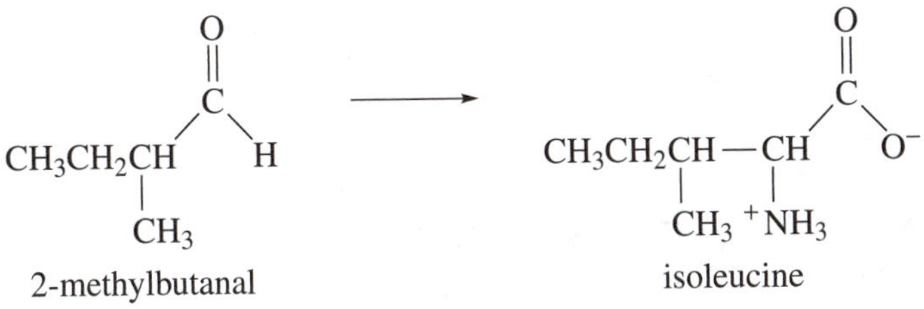
2-methylbutanal
isoleucine

47. Ser-Glu-Leu-Trp-Lys-Ser-Val-Glu-His-Gly-Ala-Met

From the experiment with carboxypeptidase A, we know the C-terminal amino acid is Met.

"Fragment 12" tells us the amino acid adjacent to Met is Ala.

"Fragment 5" tells us the next amino acid is Gly.

"Fragment 2" tells us the next amino acid is His.

"Fragment 7" tells us the next amino acid is Glu.

"Fragment 10" tells us the next amino acid is Val.

"Fragment 3" tells us the next amino acid is Ser.

"Fragment 9" tells us the next amino acid is Lys.

"Fragment 1" tells us the next amino acid is Trp.

"Fragment 8" tells us the next amino acid is Leu.

"Fragment 11" tells us the next amino acid is Glu.

"Fragment 4" tells us the next (first) amino acid is Ser.

48.   The pK_a of the carboxylic acid group of glycylglycine is higher than the pK_a of the carboxylic acid grou of the glycine, because the positively charged ammonium group of the glycine is more strongly electron withdrawing than the amide group of glycylglycine. This causes glycine to be a stronger acid and therefor have a lower pK_a.

$$H_3\overset{+}{N}CH_2\text{—C(=O)—OH} \qquad\qquad H_3\overset{+}{N}CH_2\text{—C(=O)—NHCH}_2\text{—C(=O)—OH}$$

lower pK_a

The pK_a of the ammonium group of the glycylglycine is lower than the pK_a of the ammonium group c glycine, because the amide group of the glycylglycine is more strongly electron-withdrawing than the can boxylate group of glycine.

$$H_3\overset{+}{N}CH_2\text{—C(=O)—O}^- \qquad\qquad H_3\overset{+}{N}CH_2\text{—C(=O)—NHCH}_2\text{—C(=O)—O}^-$$

lower pK_a

49.   You would (correctly) expect serine and cysteine to have lower pK_a values than alanine, since a hydroxy methyl and a thiomethyl group are more electron-withdrawing than a methyl group. Because oxygen is more electronegative than sulfur, you would expect serine to have a lower pK_a than cysteine. The fac that cysteine has a lower pK_a than serine can be explained by stabilization of serine's carboxyl proton by hydrogen bonding to the β-OH group of serine, which causes it to have less of a tendency to be removed by a base.

$$H_3\overset{+}{N}\text{—HC—C(=O)—O}\cdots\text{H}$$
$$\text{H}_2\text{C—O—H}$$
$$\text{H}$$

0. **a.**

(CH₃)₂CH / C=O / H — an aldehyde —NH₃ trace acid→ (CH₃)₂CH / C=NH / H — an imine —⁻C≡N / HCl→ (CH₃)₂CHCH—C≡N / ⁺NH₃ —HCl, H₂O / Δ→ (CH₃)₂CHCH—C(=O)OH / ⁺NH₃ valine

b. (CH₃)₂CHC(=O)—C(=O)OH —NH₃ excess / trace acid→ (CH₃)₂CHC(=NH)—C(=O)O⁻ —H₂ / Pd/C→ (CH₃)₂CHCH(NH₂)—C(=O)O⁻ valine

c. C₂H₅OC(=O)—CH(Br)—COC₂H₅ α-bromomalonic ester + potassium phthalimide (with N⁻K⁺) → phthalimide-N—C(H)(COC₂H₅)(COC₂H₅)

—CH₃CH₂O⁻→ phthalimide-N—C⁻(COC₂H₅)(COC₂H₅) CH₃CH₂OH

—CH₃CHBr / CH₃→ phthalimide-N—C(COC₂H₅)(COC₂H₅)(CH₃CH / CH₃) Br⁻

—HCl, H₂O / Δ→ phthalic acid (COOH, COOH) + CO₂ + H₃N⁺—CH(CH(CH₃)₂)—C(=O)OH valine + 2 CH₃CH₂OH

51. Tyrosine and cysteine each have two groups that are neutral in their acidic forms and negatively charged in their basic forms. Unlike those of other amino acids that have similarly ionizing groups, the pK_a value of one of the two similarly ionizing groups in tyrosine and in cysteine is close to the pK_a value of the group that ionizes differently. Therefore, the group that ionizes differently cannot be ignored in calculating the pI as it can in other amino acids that have three pK_a values.

52.   In each case, the two adjacent amino acids each have a negative charge. Two adjacent side chains will b
close to one another if they are in a helix, so they will repel each other if they have the same charge.

53.   A proline residue cannot fit into a helix because the bond between the proline nitrogen and the α-carbo
cannot rotate since it is in a ring. Not being able to rotate about this bond makes proline unable to fit into
helix.

54.   **a.**   Acid-catalyzed hydrolysis indicates that the peptide contains 12 amino acids.

— — — — — — — — — — — —

   b.   Treatment with Edman's reagent indicates that Val is the N-terminal amino acid.

<u>Val</u> — — — — — — — — — — —

   c.   Treatment with carboxypeptidase A indicates that Ala is the C-terminal amino acid.

<u>Val</u> — — — — — — — — — — <u>Ala</u>

   d.   Treatment with cyanogen bromide indicates that Met is the fifth amino acid, with Arg, Gly, and Ser i
an unknown order in positions 2, 3, and 4.

<u>Val</u> — — — <u>Met</u> — — — — — — <u>Ala</u>
    Arg, Gly, Ser

   e.   Treatment with trypsin indicates that Arg is the third amino acid, and Ser is second, Gly is fourth, Ty
is sixth, and Lys is seventh. Since Lys is in the terminal fragment, cleavage did not occur at Lys, so Pro
must be at lysine's cleavage site, but we don't know whether Lys-Pro comes before or after Phe and Ser

<u>Val</u> <u>Ser</u> <u>Arg</u> <u>Gly</u> <u>Met</u> <u>Tyr</u> <u>Lys</u> — — — — <u>Ala</u>
                    Lys-Pro, Phe, Ser

   f.   Treatment with chymotrypsin indicates that Phe is the tenth amino acid and Ser is the eleventh.

<u>Val</u> <u>Ser</u> <u>Arg</u> <u>Gly</u> <u>Met</u> <u>Tyr</u> <u>Lys</u> <u>Lys</u> <u>Pro</u> <u>Phe</u> <u>Ser</u> <u>Ala</u>

55.   **a.**   **1.**   Tyr-Gly-Gly-Phe-Met-Thr-Ser-Gly-Lys
Ser-Gln-Thr-Pro-Leu-Val-Thr-Leu-Phe-Lys-
Asn-Ala-Ile-Ile-Lys, Asn-Ala-Tyr-Lys, Lys, and Gly-Glu

   2.   Tyr-Gly-Gly-Phe-Met
Thr-Ser-Gly-Lys-Ser-Gln-Thr-Pro-Leu-Val-Thr-Leu-Phe-Lys-Asn-Ala-Ile-Ile-Lys-Asn-Ala-
Tyr-Lys-Lys-Gly-Glu

   3.   Tyr
Gly-Gly-Phe
Met-Thr-Ser-Glu-Lys-Ser-Gln-Thr-Pro-Leu-Val-Thr-Leu-Phe
Lys-Asn-Ala-Ile-Ile-Lys-Asn-Ala-Tyr
Lys-Lys-Gly-Glu

   b.   N-terminal end: Tyr-Gly-Gly-Phe-Met
C-terminal end: Tyr-Lys-Lys-Gly-Glu or Tyr-Lys-Lys-Glu-Gly

6. Because the native enzyme has four disulfide bridges, we know that the denatured enzyme has eight cysteine residues. The first cysteine has a one-in-seven chance of forming a disulfide bridge with the correct cysteine. The first cysteine of the next pair has a one-in-five chance, and the first cysteine of the third pair has a one-in-three chance.

$$\frac{1}{7} \times \frac{1}{5} \times \frac{1}{3} = 0.0095$$

If disulfide bridge formation were entirely random, the recovered enzyme should have 0.95% of its original activity. The fact that the enzyme the chemist recovered had 80% of its original activity supports his hypothesis that disulfide bridges form after the minimum energy conformation of the protein has been achieved. In other words, disulfide bridge formation is not random, but is determined by the tertiary structure of the protein.

7. The spot marked with an **X** is the peptide that is different in the normal and mutant polypeptides. The spot is closer to the cathode and farther to the right, indicating that the substituted amino acid in the mutant has a greater pI and is less polar.

The fingerprints are those of hemoglobin (normal) and sickle-cell hemoglobin (mutant). In sickle-cell hemoglobin, a glutamate in the normal polypeptide is substituted with a valine. This agrees with our observation that the substituted amino acid is less negative and more nonpolar.

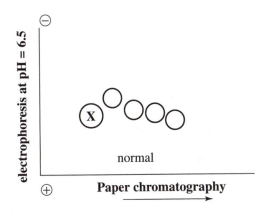

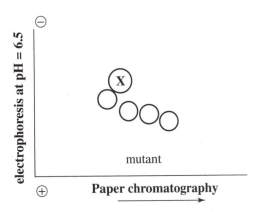

Chapter 17 Practice Test

1. Draw the structure of the following amino acids at pH = 7:

a. glutamic acid **b.** lysine **c.** isoleucine **d.** arginine **e.** asparagine

2. Draw the form of histidine that predominates at each of the following:

a. pH = 1 **b.** pH = 4 **c.** pH = 8 **d.** pH = 11

3. Answer the following:

a. Alanine has a pI = 6.02 and serine has a pI = 5.68. Which would have the higher concentration o positive charge at pH = 5.50?

b. Which amino acid is the only one that does not have an asymmetric center?

c. Which are the two most nonpolar amino acids?

d. Which amino acid has the lowest pI?

4. Why does the carboxyl group of alanine have a lower pK_a than the carboxyl group of propanoic acid?

$pK_a = 2.2$

alanine

$pK_a = 4.7$

propanoic acid

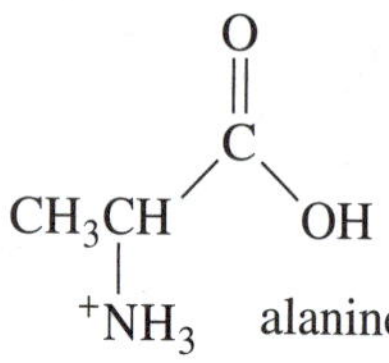

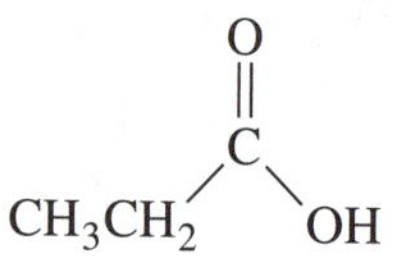

5. Indicate whether each of the following statements is true or false:

a. A cigar-shaped protein has a greater percentage of polar residues than a spherical protein. T F

b. Naturally occurring amino acids have the L-configuration. T F

c. There is free rotation about a peptide bond. T F

6. What compound is obtained from mild oxidation of cysteine?

7. Define each of the following:

a. the primary structure of a protein

b. the tertiary structure of a protein

c. the quaternary structure of a protein

Identify the spots.

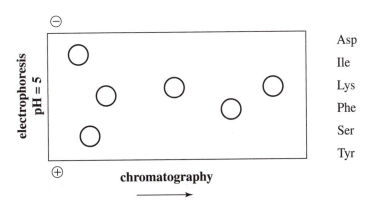

Asp
Ile
Lys
Phe
Ser
Tyr

Calculate the pI of each of the following amino acids:

a. phenylalanine (pK_as $= 2.16, 9.18$)   b. arginine (pK_as $= 2.17, 9.04, 12.48$)

0. From the following information, determine the primary sequence of the decapeptide:

a. Acid hydrolysis gives Ala, 2 Arg, Gly, His, Ile, Lys, Met, Phe, Ser.

b. Reaction with Edman's reagent liberated Ala.

c. Reaction with carboxypeptidase A liberated Ile.

d. Reaction with cyanogen bromide (cleaves on the C-side of Met)

   1. Gly, 2 Arg, Ala, Met, Ser

   2. Lys, Phe, Ile, His

e. Reaction with trypsin (cleaves on the C-side of Arg and Lys)

   1. Arg, Gly

   2. Ile

   3. Phe, Lys, Met, His

   4. Arg, Ser, Ala

f. Reaction with thermolysin (cleaves on the N-side of Leu, Ile, Phe, Trp, Tyr)

   1. Lys, Phe

   2. 2 Arg, Ser, His, Gly, Ala, Met

   3. Ile

11. Describe how acetaldehyde can be converted to alanine.

12. Draw the mechanism for the conversion of a thiol to a disulfide in a basic solution of Br_2.

406   Chapter 17

Answers to Chapter 17 Practice Test

1.   **a.** $^-O-\overset{\overset{\displaystyle O}{\|}}{C}-CH_2CH_2\underset{\underset{\displaystyle +NH_3}{|}}{CH}-\overset{\overset{\displaystyle O}{\|}}{C}-O^-$

   d. $H_2N-\overset{\overset{\displaystyle +NH_2}{\|}}{C}-NHCH_2CH_2CH_2\underset{\underset{\displaystyle +NH_3}{|}}{CH}-\overset{\overset{\displaystyle O}{\|}}{C}-O^-$

   b. $\overset{+}{H_3}NCH_2CH_2CH_2CH_2\underset{\underset{\displaystyle +NH_3}{|}}{CH}-\overset{\overset{\displaystyle O}{\|}}{C}-O^-$

   e. $H_2N-\overset{\overset{\displaystyle O}{\|}}{C}-CH_2\underset{\underset{\displaystyle +NH_3}{|}}{CH}-\overset{\overset{\displaystyle O}{\|}}{C}-O^-$

   c. $CH_3CH_2\underset{\underset{\displaystyle +NH_3}{|}}{\overset{\overset{\displaystyle CH_3}{|}}{CH}CH}-\overset{\overset{\displaystyle O}{\|}}{C}-O^-$

2.   **a.** imidazolium $CH_2\underset{\underset{\displaystyle +NH_3}{|}}{CH}-\overset{\overset{\displaystyle O}{\|}}{C}-OH$

   c. imidazole $CH_2\underset{\underset{\displaystyle +NH_3}{|}}{CH}-\overset{\overset{\displaystyle O}{\|}}{C}-O^-$

   b. imidazolium $CH_2\underset{\underset{\displaystyle +NH_3}{|}}{CH}-\overset{\overset{\displaystyle O}{\|}}{C}-O^-$

   d. imidazole $CH_2\underset{\underset{\displaystyle NH_2}{|}}{CH}-\overset{\overset{\displaystyle O}{\|}}{C}-O^-$

3.   **a.** Alanine, because it is farther away from its pI.   **c.** leucine and isoleucine

   b. glycine   **d.** aspartic acid

4.   The electron-withdrawing protonated amino group causes the carboxyl group of alanine to have a lower pK_a.

5.   **a.** A cigar-shaped protein has a greater percentage of polar residues than a spherical protein.   T

   b. Naturally occurring amino acids have the L-configuration.   T

   c. There is free rotation about a peptide bond.   F

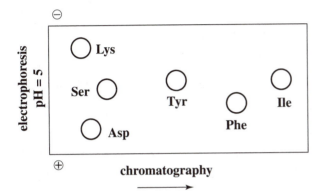

cystine

a. The sequence of the amino acids and the location of the disulfide bonds in the protein.

b. The three-dimensional arrangement of all the atoms in the protein.

c. A description of the way the subunits of an oligomer are arranged in space.

(electrophoresis pH = 5, ⊖)

Lys

Ser ○   ○ Tyr

Asp   ○ Phe   Ile

⊕ chromatography →

9.

a. $\dfrac{2.16 + 9.18}{2} = \dfrac{11.34}{2} = 5.67$

b. $\dfrac{9.04 + 12.48}{2} = \dfrac{21.52}{2} = 10.76$

10. <u>Ala</u> <u>Ser</u> <u>Arg</u> <u>Gly</u> <u>Arg</u> <u>Met</u> <u>His</u> <u>Phe</u> <u>Lys</u> <u>Ile</u>

11.

$$\underset{H}{\overset{CH_3}{\diagdown}}C=O \xrightarrow[\text{NH}_3]{\substack{\text{trace} \\ \text{acid}}} \underset{H}{\overset{CH_3}{\diagdown}}C=NH \xrightarrow[\text{HCl}]{^-C\equiv N} CH_3-\underset{^+NH_3}{\underset{|}{CH}}-C\equiv N \xrightarrow[\Delta]{\text{HCl, H}_2\text{O}} CH_3-\underset{^+NH_3}{\underset{|}{CH}}\overset{O}{\overset{\|}{C}}{\diagdown}OH$$

12.

$$R-SH \underset{\text{H}_2\text{O}}{\overset{\text{HO}^-}{\rightleftharpoons}} R-\ddot{\underset{..}{S}}\!:^- \xrightarrow{Br-Br} R-S-Br \xrightarrow{R-\ddot{S}:^-} R-S-S-R + Br^-$$

Br⁻

How Enzymes Catalyze Reactions • The Organic Chemistry of the Vitamins

Important Terms

acid catalyst	a catalyst that increases the rate of a reaction by donating a proton.
active site	a pocket or cleft in an enzyme where the substrate is bound.
acyl-enzyme intermediate	an amino acid residue of an enzyme that has been acylated while catalyzing reaction.
amino acid side chain	the substituent on the α-carbon of an amino acid.
base catalyst	a catalyst that increases the rate of a reaction by removing a proton.
biotin	the coenzyme required by enzymes that catalyze carboxylation of a carbon adjacent to an ester or a keto group.
catalyst	a substance that increases the rate of a reaction without itself being consumed in the overall reaction.
coenzyme	a cofactor that is an organic molecule.
coenzyme A	a thiol used by biological organisms to form thioesters.
coenzyme B_{12}	the coenzyme required by enzymes that catalyze certain rearrangement reactions.
competitive inhibitor	a compound that inhibits an enzyme by competing with the substrate for binding at the active site.
dehydrogenase	an enzyme that carries out an oxidation reaction by removing hydrogen from the substrate.
electrophilic catalyst	an electrophile that facilitates a reaction.
electrostatic catalysis	the stabilization of a charge by an opposite charge.
enzyme	a protein that is a catalyst.
flavin adenine dinucleotide (FAD)	a coenzyme required in certain oxidation reactions. It is reduced to $FADH_2$, which is a coenzyme required in certain reduction reactions.
heterocyclic compound	a cyclic compound in which one or more of the ring atoms is an atom other than carbon.
induced fit model	a model that describes the specificity of an enzyme for its substrate: the shape of the active site does not become completely complementary to the shape of the substrate until after the enzyme has bound the substrate.

408

tramolecular catalysis	catalysis in which the catalyst that facilitates the reaction is part of the molecule undergoing reaction.
poate	a coenzyme required in certain oxidation reactions.
ck-and-key model	a model that describes the specificity of an enzyme for its substrate: the substrate fits the enzyme like a key fits into a lock.
echanism-based inhibitor (uicide inhibitor)	an inhibitor that inactivates an enzyme by undergoing part of the normal catalytic mechanism.
olecular recognition	the recognition of one molecule by another as a result of specific interactions—for example, the specificity of an enzyme for its substrate.
icotinamide adenine inucleotide (NAD$^+$)	a coenzyme required in certain oxidation reactions. It is reduced to NADH, which is a coenzyme required in certain reduction reactions.
icotinamide adenine inucleotide phosphate NADP$^+$)	a coenzyme that is reduced to NADPH, which is a coenzyme required in certain reduction reactions.
ucleophilic catalysis	catalysis that occurs as a result of a nucleophile forming a covalent bond with one of the reactants.
ucleophilic catalyst	a catalyst that increases the rate of a reaction by acting as a nucleophile.
ucleotide	a heterocycle attached in the β-position to the anomeric carbon of a phosphorylated ribose.
yridoxal phosphate	the coenzyme required by enzymes that catalyze certain transformations of amino acids.
elative rate	the relative rate is obtained by dividing the actual rate constant by the rate constant of the slowest reaction in the group being compared.
ubstrate	the reactant of an enzyme-catalyzed reaction.
uicide inhibitor (mechanism-based nhibitor)	a compound that inactivates an enzyme by undergoing part of its normal catalytic mechanism.
etrahydrofolate (THF)	the coenzyme required by enzymes that catalyze a reaction that donates a group containing a single carbon to its substrate.
hiamine pyrophosphate TPP)	the coenzyme required by enzymes that catalyze a reaction that transfers an acyl group from one species to another.
ransamination	a reaction in which an amino group is transferred from one compound to another.

transimination

the reaction of a primary amine with an imine to form a new imine and a prima
amine derived from the original imine.

vitamin

a substance needed in small amounts for normal body function that the boc
cannot synthesize or cannot synthesize in adequate amounts.

vitamin KH$_2$

the coenzyme required by the enzyme that catalyzes the carboxylation of glutama
side chains.

olutions to Problems

Because the reacting groups in the trans isomer are pointed in opposite directions, they cannot react in an intramolecular reaction. Because they can react only via an intermolecular pathway, they will have approximately the same rate of reaction as they would have if the reacting groups were in separate molecules. Consequently, the relative rate would be expected to be close to one.

NAM would contain ^{18}O, because it is the ring that would undergo nucleophilic attack by $H_2{}^{18}O$.

Solved in the text.

4.   Because arginine extends farther into the binding pocket, it must be the one that forms direct hydrogen bonds. Lysine, which is shorter, needs the mediation of a water molecule in order to engage in bond formation with aspartate.

5.   **2, 3,** and **4** are bases, so they can help remove a proton.

6.   In the absence of an enzyme, D-fructose forms both D-glucose and D-mannose as a result of an enediol rearrangement. Both C-2 epimers are formed, because a new asymmetric center is formed at C-2 and it can have either the *R* or the *S* configuration. Enzyme-catalyzed reactions are typically highly stereoselective—the enzyme catalyzes the formation of a single stereoisomer. Thus, D-fructose forms only D-glucose in the presence of the enzyme that catalyzes the enediol rearrangement.

7.   Because **2** is a primary amine, it can form an imine. Notice that **1** cannot form an imine because the lone pair on the NH_2 group is delocalized onto the oxygen, so this NH_2 group is not a nucleophile. The N in **3** is not nucleophilic because its lone pair is delocalized (the lone pair is part of indole's π cloud).

8.

The positively charged nitrogen of the protonated imine can accept the electrons that are left behind when the C3—C4 bond breaks.

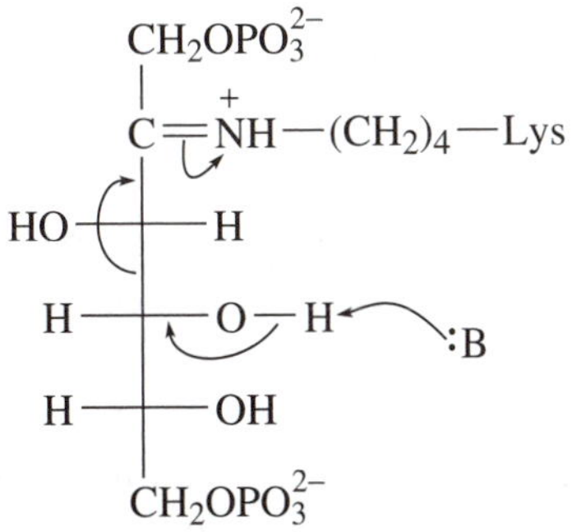

$$CH_2OPO_3^{2-}$$

C=NH—(CH₂)₄—Lys

HO——H

H——O—H   :B

H——OH

$$CH_2OPO_3^{2-}$$

In the absence of the protonated imine, the electrons would be delocalized onto a neutral oxygen (see Problem 8). The neutral oxygen is not as electron-withdrawing as the positively charged nitrogen of the protonated imine. In other words, imine formation makes it easier to break the C3—C4 bond.

0. The alcohol is oxidized to a ketone.

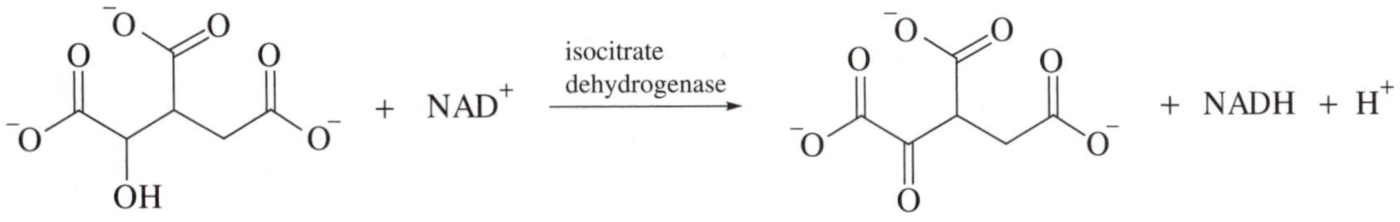

$$+ \text{NAD}^+ \xrightarrow{\text{isocitrate dehydrogenase}} + \text{NADH} + \text{H}^+$$

1. The ketone is reduced to an alcohol.

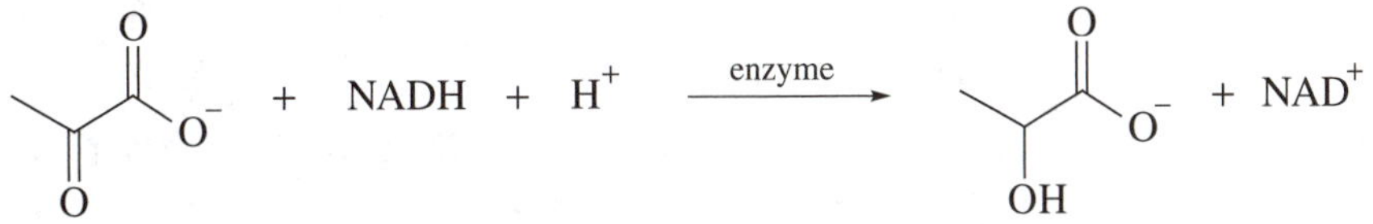

$$+ \text{NADH} + \text{H}^+ \xrightarrow{\text{enzyme}} + \text{NAD}^+$$

2. **a.** FAD has seven conjugated double bonds (indicated by *).

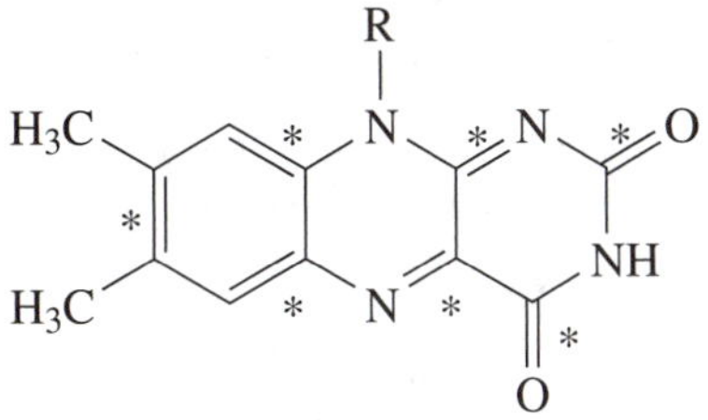

R

H₃C, N, N, O

H₃C, N, NH

O

b. FADH$_2$ has three conjugated double bonds (on the left of the molecule). It also has two conjugate double bonds (on the right of the molecule) that are isolated from the other three.

13.   The nitrogen that is the stronger base is the one more apt to be protonated. N-1 is a weaker base than N-5 because the π electrons that belong to N-1 can be delocalized onto an oxygen, so protonation on nitrogen will not occur when a nucleophile adds to C-10a. The π electrons that belong to N-5 cannot be delocalized onto an electronegative atom, so protonation on nitrogen will occur when a nucleophile adds to C-4a. Nucleophilic addition, therefore, occurs at the position (4a) that results in the stronger base (N-5) being protonated.

14.   Solved in the text.

15.   When a proton is removed from the methyl group at C-8, the electrons that are left behind can be delocalized onto the oxygen at the 2-position or onto the oxygen at the 4-position.

When a proton is removed from the methyl group at C-7, the electrons that are left behind can be delocalized only onto carbons that, being less electronegative than oxygen, are less able to accommodate the electrons.

6. Notice that the only difference in the mechanisms for pyruvate decarboxylase and acetolactate synthase is the species the two-carbon fragment is transferred to: a proton in the case of pyruvate carboxylase and pyruvate in the case of acetolactate synthase.

acetolactate

17.   Notice that the only difference in this reaction and that in Problem 16 is that the species to which the two-carbon fragment is transferred has an ethyl group in place of the methyl group.

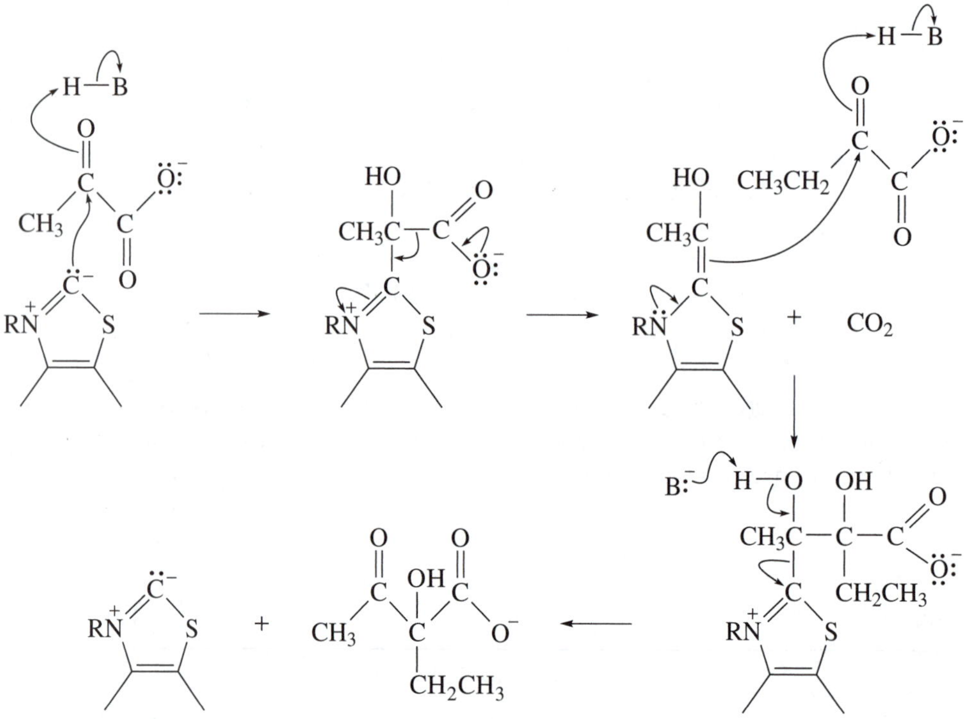

18.

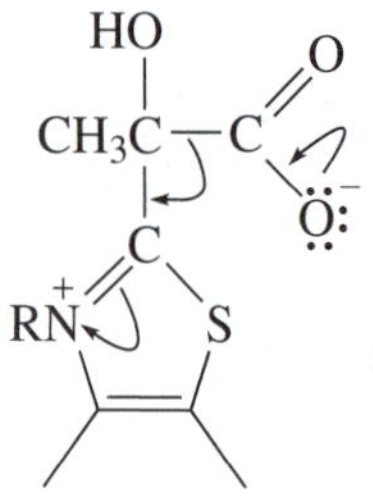

pyruvate-TPP
intermediate

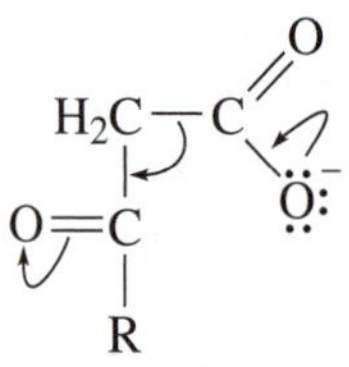

a β-keto acid

19.   **a.** 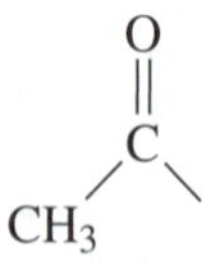   **b.**

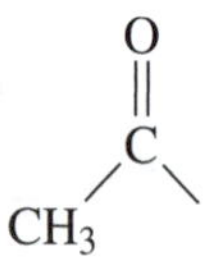

20.   Solved in the text.

21.   Nine, because one mole of arachidic acid is made from one mole of acetyl-CoA and nine moles of malonyl-CoA.

22.   Solved in the text.

3. The α-keto group that accepts the amino group from pyridoxamine is converted into an amino group.

a.

pyruvate

alanine

b.

oxaloacetate

aspartate

4. The compound on the right is more easily decarboxylated, because the electrons left behind when CO_2 is eliminated are delocalized onto the positively charged nitrogen of the pyridine ring. The electrons left behind if the other compound is decarboxylated cannot be delocalized.

5. The first step (after transimination) in all amino acid transformations catalyzed by PLP is removal of a substituent from the α-carbon of the amino acid. The electrons left behind when the substituent is removed are delocalized onto the positively charged nitrogen of the pyridine ring. If the ring nitrogen is not protonated, it will be less attractive to the electrons.

6. The hydrogen of the OH substituent forms a hydrogen bond with the nitrogen of the imine linkage (see Problem 25). This puts a partial positive charge on the nitrogen, which makes it easier for the amino acid to add to the imine carbon in the transimination reaction that attaches the amino acid to the coenzyme. It also makes it easier to remove a substituent from the α-carbon of the amino acid. If the OH substituent is replaced by an OCH_3 substituent, there is no longer a proton available to form the hydrogen bond.

27.

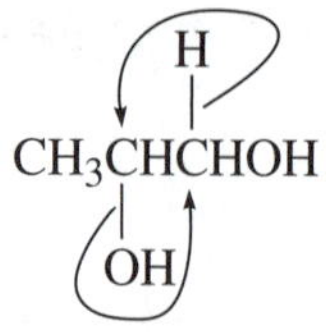

CH₃CHCHOH

28. It is called tetrahydrofolate (THF) because it is formed by adding four hydrogens to folate (folic acid).

29.

$$\underset{^+NH_3}{HSCH_2CH_2\overset{\overset{O}{\|}}{C}HCO^-} + N^5\text{-methyl-THF} \longrightarrow \underset{\text{methionine } ^+NH_3}{CH_3SCH_2CH_2\overset{\overset{O}{\|}}{C}HCO^-} + THF$$

30. They differ only in the circled part of the molecule.

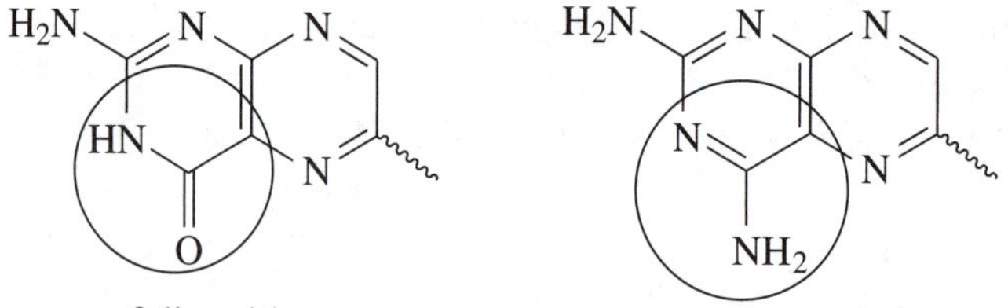

folic acid   aminopterin

31. The methyl group in thymidine comes from the methylene group of N^5,N^{10}-methylene-THF, followed by the addition of a hydride ion from the coenzyme.

32. The following compound will lose HBr more rapidly, because the negatively charged oxygen is in position to act as an intramolecular base catalyst.

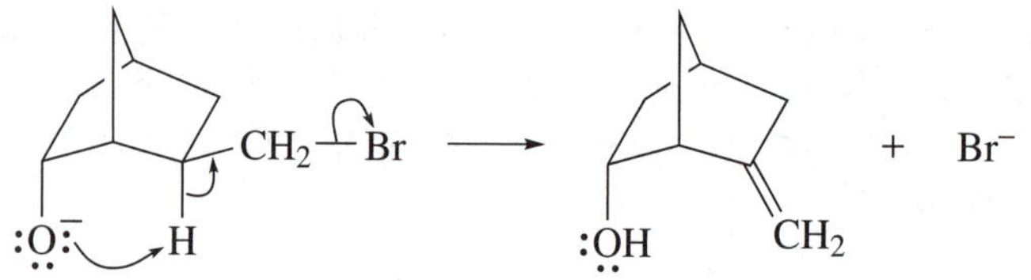

33. The following compound will form an anhydride more rapidly, because it forms a five-membered ring anhydride, which is less strained than the seven-membered ring anhydride formed by the other compound. The greater stability of the five-membered ring product causes the transition state leading to its formation to be more stable than the transition state leading to formation of the seven-membered ring product.

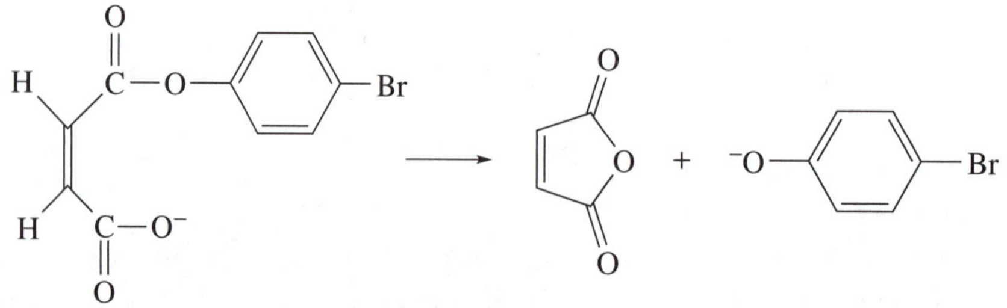

4.   **a.**   niacin (vitamin B$_3$)

   b.   riboflavin (vitamin B$_2$)

   c.   pyridoxine (vitamin B$_6$)

   d.   folic acid (folate)

5.   **a.**   thiamine pyrophosphate (TPP)

   b.   FAD oxidizes dihydrolipoate back to lipoate.

   c.   NAD$^+$ oxidizes FADH$_2$ back to FAD.

   d.

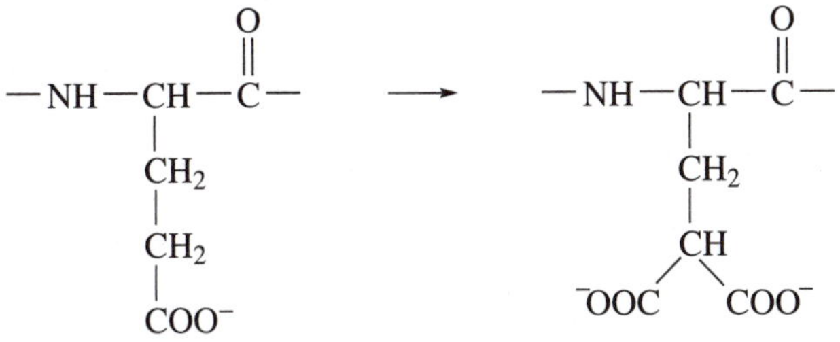

$$-NH-CH-\underset{\underset{CH_2}{|}}{C}- \quad\longrightarrow\quad -NH-CH-\underset{\underset{CH}{|}}{C}-$$

   e.   biotin and vitamin KH$_2$

   f.   Biotin carboxylates a carbon adjacent to a carbonyl group (that is, an α-carbon).
            Vitamin KH$_2$ carboxylates the γ-carbon of a glutamate.

6.   **a.**   acetyl-CoA carboxylase; biotin

   b.   dihydrolipoyl dehydrogenase; FAD

   c.   methylmalonyl-CoA mutase; coenzyme B$_{12}$

   d.   lactate dehydrogenase; NADH (This enzyme is named according to the reaction it catalyzes in the reverse direction.)

   e.   aspartate transaminase; pyridoxal phosphate

   f.   propionyl-CoA carboxylase; biotin

7.   The side chains of D-Arg and D-Lys are not positioned to bind correctly at the active site. However, they would be able to bind at a mirror image of the active site.

8. $RCH{=}CHCSR + FADH_2 \longrightarrow RCH_2CH_2CSR + FAD$

9.   **a.**   pyridoxal phosphate   **b.**   biotin

40.

41.   In order to break the C3—C4 bond, the carbonyl group has to be at the 2-position, so it can accept the electrons from the C3—C4 bond; the carbonyl group at the 1-position in glucose cannot accept those electrons. Therefore, glucose must isomerize to a ketose, so the carbonyl group will be at the 2-position.

2.

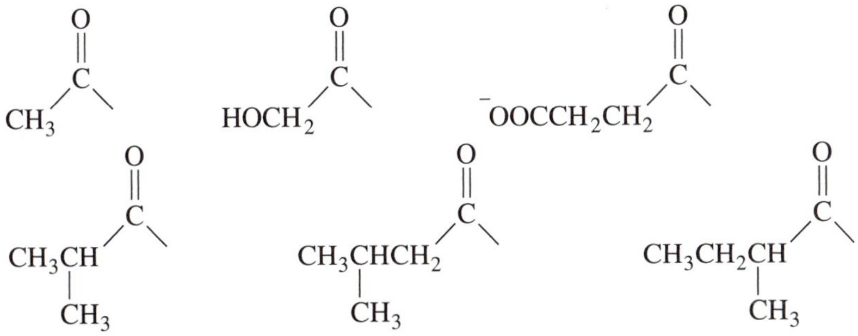

3.

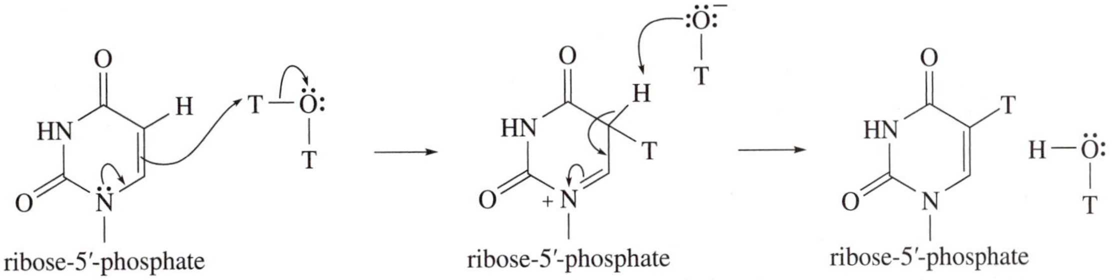

ribose-5′-phosphate   ribose-5′-phosphate   ribose-5′-phosphate

4. **a.** thiamine pyrophosphate, lipoate, coenzyme A, FAD, NAD⁺, indicated by numbers 1–5 in the mechanism below

b.

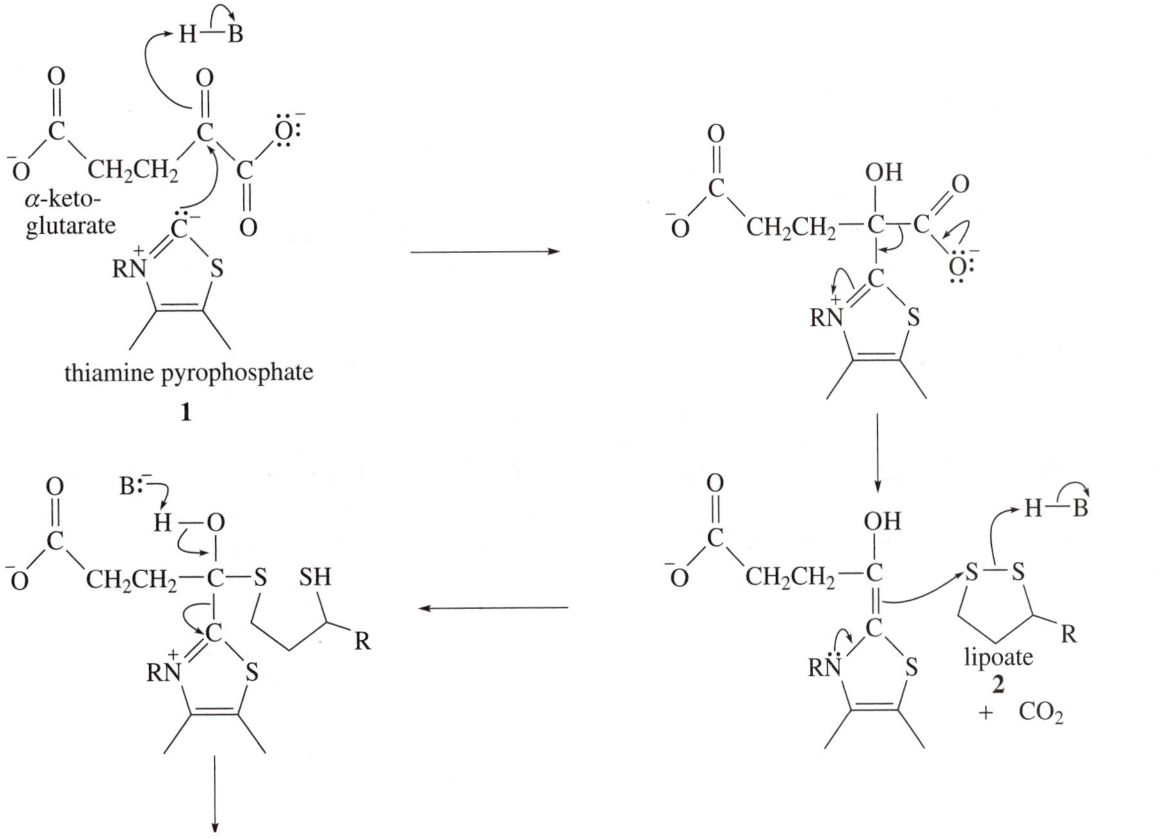

α-keto-glutarate

thiamine pyrophosphate
1

lipoate
2
+ CO_2

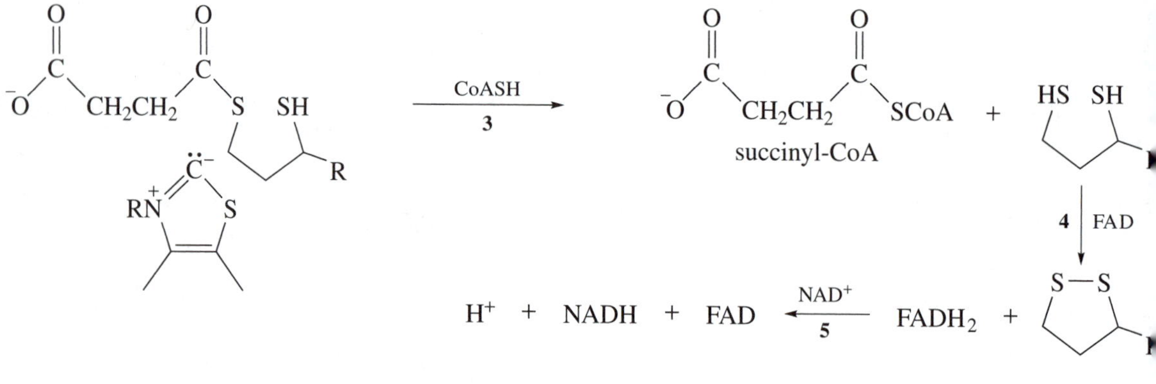

45. The product of transamination of an amino acid has a carbonyl group in place of the amino group.

a.

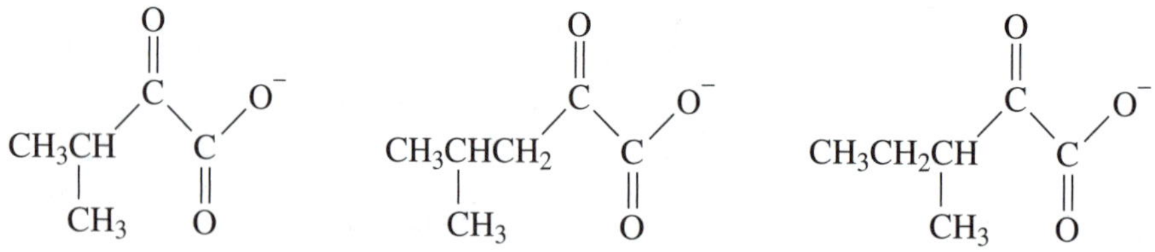

derived from Val   derived from Leu   derived from Ile

b.

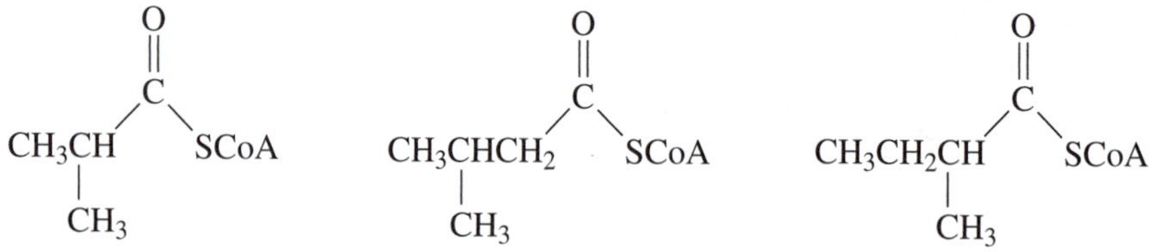

c. The reaction catalyzed by the enzyme is identical to the reaction catalyzed by the pyruvate dehydrogenase complex. Therefore, they both require the same coenxymes: thiamine pyrophosphate, lipoate, coenzyme A, FAD, NAD$^+$.

d. The disease can be treated by a diet low in branched-chain amino acids.

46. Cysteine residues react with iodoacetic acid because a thiol is a good nucleophile and iodine is a good leaving group. If a cysteine residue is at the active site of an enzyme, adding a substituent to the sulfur in this way could interfere with the enzyme's being able to bind the substrate, or it could interfere with positioning of a group that catalyzes the reaction. Adding a substituent to cysteine might also cause a conformational change in the enzyme that could destroy its activity.

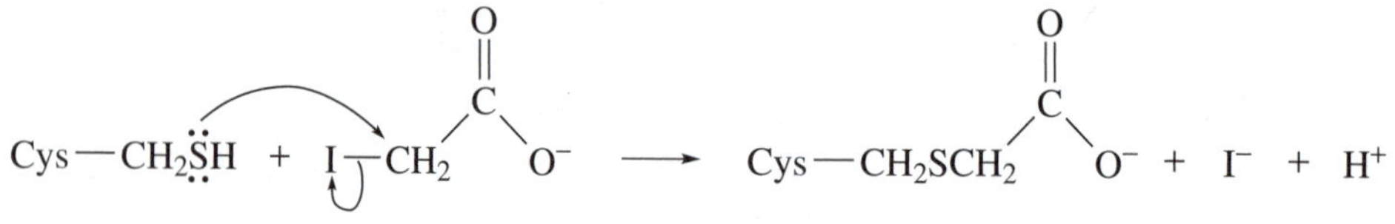

Chapter 18 Practice Test

. Indicate whether each of the following statements is true or false:

a. A catalyst increases the equilibrium constant of a reaction.   T   F

b. An acid catalyst donates a proton to the substrate, and a base catalyst removes a proton from the substrate.   T   F

c. The reactant of an enzyme-catalyzed reaction is called a substrate.   T   F

. In lysozyme, glutamate 35 is a catalyst that is active in its acid form. Explain its catalytic function.

. Using arrows, show the first step in the mechanism for chymotrypsin.

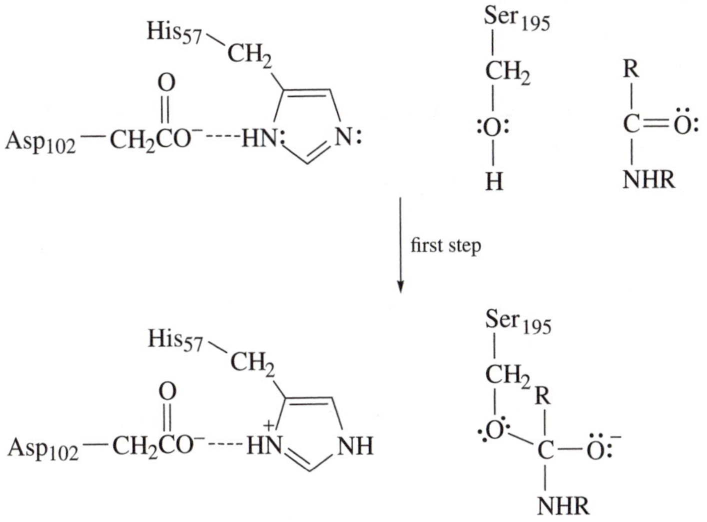

a. What kind of catalyst is histidine in this step?

b. What kind of catalyst is serine in this step?

c. How does aspartate catalyze the reaction?

4. What two coenzymes put carboxyl groups on their substrates?

5. Show the mechanism for NADPH reducing its substrate.

6. Draw the structure of the compound obtained when the following amino acid undergoes transamination:

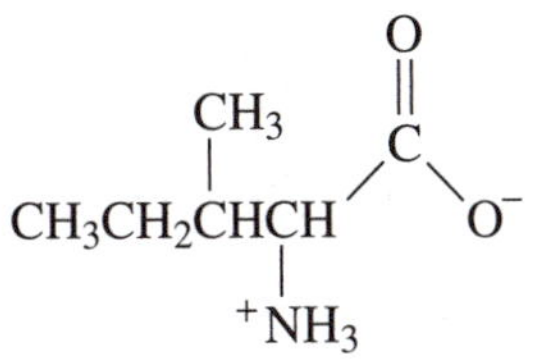

7. What is the first step in the reaction of the substrate with coenzyme B_{12} in an enzyme-catalyzed reaction that requires coenzyme B_{12}?

8.   What coenzyme is required for each of the following enzyme-catalyzed reactions?

$$CH_3CH_2-\overset{\overset{O}{\|}}{C}-SCoA \xrightarrow{enzyme} CH_3\underset{\underset{COO^-}{|}}{CH}-\overset{\overset{O}{\|}}{C}-SCoA \xrightarrow{enzyme} CH_2\underset{\underset{COO^-}{|}}{CH_2}-\overset{\overset{O}{\|}}{C}-SCoA$$

9.   Show the enzyme-catalyzed reaction that requires vitamin KH_2 as a coenzyme.

10.   Draw the product of the enzyme-catalyzed reaction that requires biotin and whose substrate is acetyl-CoA

11.   **a.** Other than the substrate, enzyme, and coenzyme, what three additional reagents are needed by a reaction that requires biotin as a coenzyme?
      b. What is the function of each of these reagents?

12.   What is the function of FAD in the pyruvate dehydrogenase system?

13.   Draw the structures of the two products obtained from the following transamination reaction:

$$HO-\langle\text{ring}\rangle-CH_2\underset{\underset{+NH_3}{|}}{CH}-\overset{\overset{O}{\|}}{C}-O^- + {}^-O-\overset{\overset{O}{\|}}{C}-CH_2CH_2-\overset{\overset{O}{\|}}{\underset{\underset{O}{\|}}{C}}-\overset{\overset{O}{\|}}{C}-O^- \xrightarrow{transamination}$$

14.   Indicate whether each of the following statements is true or false:
      a. Vitamin B_1 is the only water-insoluble vitamin that has a coenzyme function.   T   F
      b. $FADH_2$ is a reducing agent.   T   F
      c. Thiamine pyrophospate is vitamin B_6.   T   F
      d. Vitamin K is a water-soluble vitamin.   T   F
      e. Lipoic acid is covalently bound to its enzyme by an amide linkage.   T   F

Answers to Chapter 18 Practice Test

a. A catalyst increases the equilibrium constant of a reaction.   F

b. An acid catalyst donates a proton to the substrate, and a base catalyst removes a proton from the substrate.   T

c. The reactant of an enzyme-catalyzed reaction is called a substrate.   T

It protonates the leaving group to make it a better leaving group.

first step

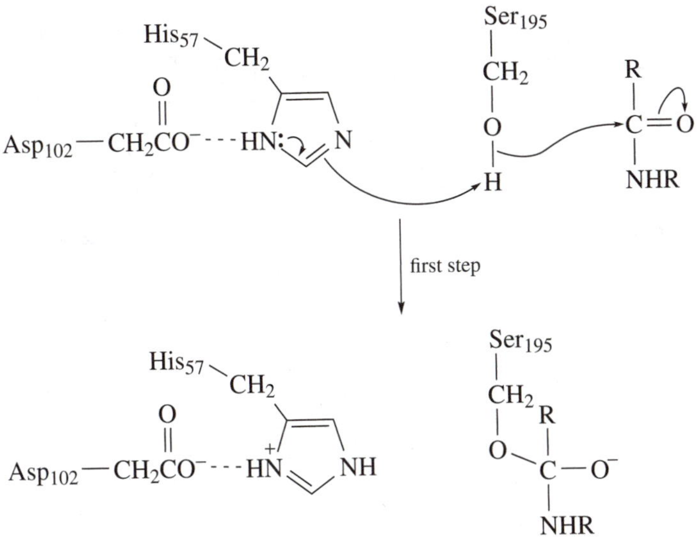

a. base catalyst

b. nucleophilic catalyst

c. It stabilizes the positive charge on histidine.

4. biotin and vitamin KH_2

5.

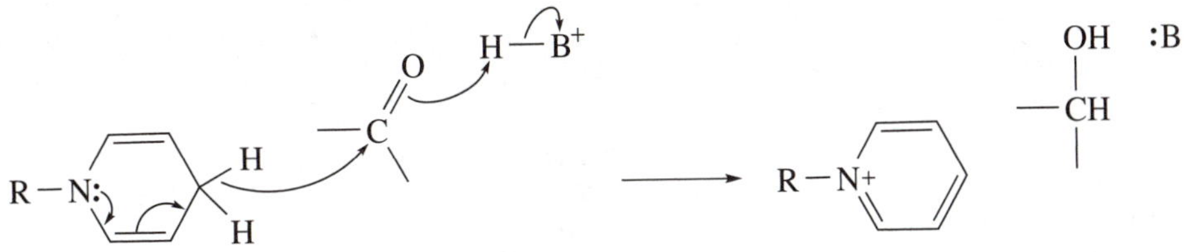

6.

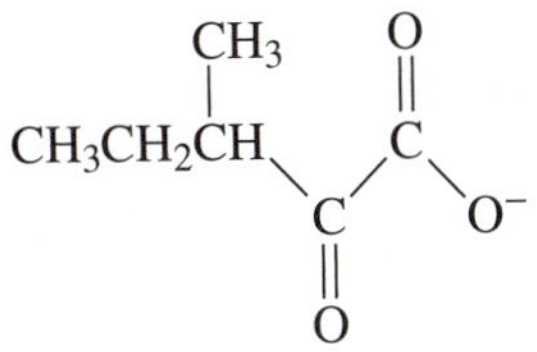

7.   The first step is removing the hydrogen atom that is going to change places with a group on an adjace[nt] carbon.

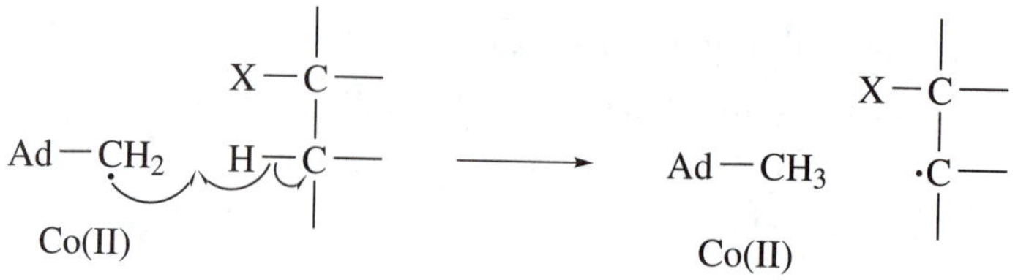

8.   first coenzyme = biotin

second coenzyme = coenzyme B$_{12}$

9.

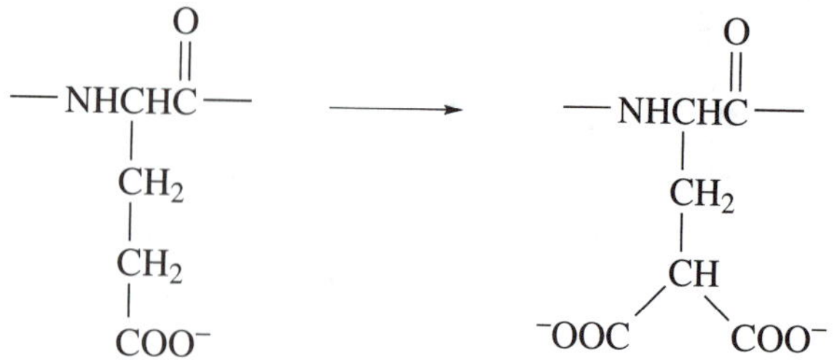

10.

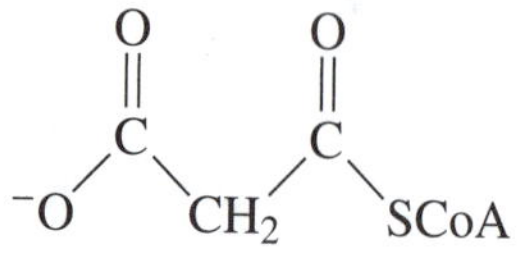

11.   **a.**   ATP, Mg^{2+}, HCO$_3^-$

b.   ATP activates bicarbonate (HCO$_3^-$) by putting a good leaving group (phosphate) on it. Mg^{2+} complexes with ATP in order to reduce the negative charge on ATP so that it can react with a nucleophile. HCO$_3^-$ is the source of the COO$^-$ group that is put on the reactant.

12.   FAD oxidizes dihydrolipoate to lipoate.

13.

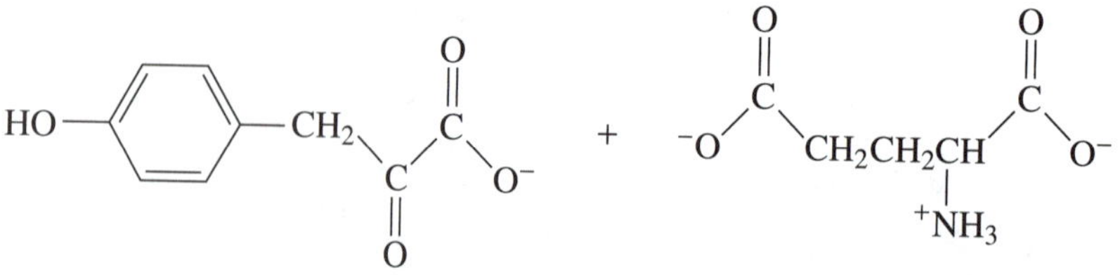

14.   **a.**   Vitamin B$_1$ is the only water-insoluble vitamin that has a coenzyme function.   F

b.   FADH$_2$ is a reducing agent.   T

c.   Thiamine pyrophosphate is vitamin B$_6$.   F

d.   Vitamin K is a water-soluble vitamin.   F

e.   Lipoic acid is covalently bound to its enzyme by an amide linkage.   T

CHAPTER 19
The Organic Chemistry of the Metabolic Pathways

Important Terms

acyl adenylate

$$\underset{R}{}\!\!\!\overset{\displaystyle O}{\underset{\displaystyle \|}{C}}\!\!-\!O\!-\!\underset{\displaystyle O^-}{\overset{\displaystyle O}{\underset{\displaystyle \|}{P}}}\!\!-\!O\!-\!\text{adenosine}$$

acyl phosphate

$$\underset{R}{}\!\!\!\overset{\displaystyle O}{\underset{\displaystyle \|}{C}}\!\!-\!O\!-\!\underset{\displaystyle O^-}{\overset{\displaystyle O}{\underset{\displaystyle \|}{P}}}\!\!-\!O^-$$

acyl pyrophosphate

$$\underset{R}{}\!\!\!\overset{\displaystyle O}{\underset{\displaystyle \|}{C}}\!\!-\!O\!-\!\underset{\displaystyle O^-}{\overset{\displaystyle O}{\underset{\displaystyle \|}{P}}}\!\!-\!O\!-\!\underset{\displaystyle O^-}{\overset{\displaystyle O}{\underset{\displaystyle \|}{P}}}\!\!-\!O^-$$

allosteric activator/ inhibitor   a compound that activates/inhibits an enzyme by binding to a site on the enzyme other than the active site.

anabolism   the reactions living organisms carry out that result in the synthesis of complex bio-molecules from simple precursor molecules.

catabolism   the reactions living organisms carry out to provide energy and simple precursor molecules for synthesis.

citric acid cycle   a series of reactions that convert the acetyl group of acetyl-CoA into two molecules of CO_2 and a molecule of CoASH.

feedback inhibitor   a compound that inhibits a step at the beginning of the pathway for its biosynthesis.

glycolysis   the series of reactions that converts glucose into two molecules of pyruvate.

gluconeogenesis   the synthesis of glucose from pyruvate.

high-energy bond   a bond that releases a great deal of energy when it is broken.

metabolism   reactions living organisms carry out in order to obtain the energy they need and to synthesize the compounds they require.

β-oxidation   a repeating series of four reactions that convert a fatty acyl-CoA molecule into molecules of acetyl-CoA.

oxidative phosphorylation   the fourth stage of catabolism in which NADH and $FADH_2$ are oxidized back t
NAD^+ and FAD: for each NADH that is oxidized, 2.5 ATPs are formed; for eac
$FADH_2$ that is oxidized, 1.5 ATPs are formed.

phosphoanhydride bond   the bond that holds two phosphoric acid molecules together.

phosphoryl transfer reaction   the transfer of a phosphate group from one compound to another.

regulatory enzyme   an enzyme that catalyzes an irreversible reaction near the beginning of a pathway
thereby allowing independent control over degradation and synthesis.

Solutions to Problems

1. Because palmitic acid has 16 carbons and the acyl group of acetyl-CoA has 2 carbons, eight molecules of acetyl-CoA are formed from one molecule of palmitic acid.

2. Seven: one mole of NADH will be obtained from each of the following rounds of β-oxidation.

3. The resonance contributor on the right shows that the β-carbon of the α,β-unsaturated carbonyl compound has a partial positive charge. The nucleophilic OH group, therefore, is attracted to the β-carbon.

4.

5. a. conversion of glucose to glucose-6-phosphate (the first step)
   conversion of fructose-6-phosphate to fructose-1,6-bisphosphate (the third step)

   b. conversion of 1,3-bisphosphoglycerate to 3-phosphoglycerate (the seventh step)
   conversion of phosphoenolpyruvate to pyruvate (the tenth step)

6. The reaction that follows the oxidation of glyceraldehyde-3-phosphate to 1,3-bisphosphoglycerate (the conversion of 1,3-bisphosphoglycerate to 3-phosphoglycerate) is highly exergonic. Therefore, as 1,3-bisphosphoglycerate is converted to 3-phosphoglycerate, glyceraldehyde-3-phosphate will be converted to 1,3-bisphosphoglycerate to replenish it.

7. Two: each molecule of D-glucose is converted to two molecules of glyceraldehyde-3-phosphate, and each molecule of glyceraldehyde-3-phosphate requires one molecule of NAD^+ for it to be converted to one molecule of pyruvate.

8.   acetaldehyde reductase

9.   a ketone

10.

11.

alanine   pyruvate

12.   Protonated histidine ($pK_a = 6.0$) is not strong enough an acid to fully protonate the OH group to make it a good leaving group (H_2O) that would be able to leave in the first step of the elimination reaction, which is required for an E1 reaction. Therefore, it is an E2 reaction with protonated histidine acting as a general-acid catalyst to protonate the OH group as it departs.

13.   a secondary alcohol

14.   citrate and isocitrate (and the alkene intermediate generated during the conversion of citrate to isocitrate)

15.

16.   **a.**   The conversion of one molecule of glycerol to dihydroxyacetone phosphate consumes one molecule of ATP. The conversion of dihydroxyacetone phosphate to pyruvate produces two molecules of ATP. Therefore, one molecule of ATP is obtained from the conversion of one molecule of glycerol to pyruvate.

   b.   One NADH is formed from the conversion of one molecule of glycerol to dihydroxyacetone phosphate, and one NADH is formed from the conversion of dihydroxyacetone phosphate to pyruvate. Each NADH forms 2.5 ATP in the fourth stage of catabolism. So, when the fourth stage of catabolism is included, six molecules ($2.5 + 2.5 + 1 = 6$) of ATP are obtained from the conversion of one molecule of glycerol to pyruvate.

17.   **a.**   glycerol kinase   **b.**   phosphatidic acid phosphatase

8.   The first step is an S$_N$2 reaction; the second step is a nucleophilic addition elimination reaction.

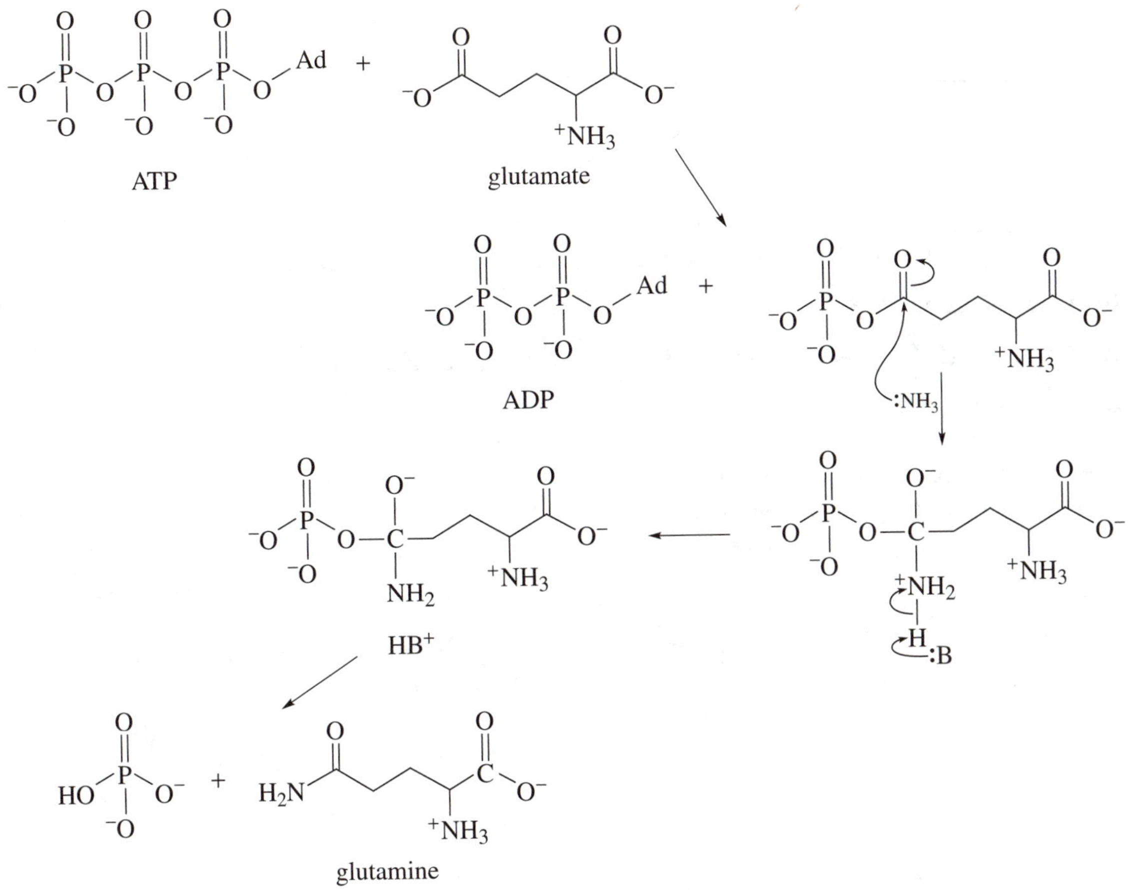

ATP   glutamate

ADP

HB$^+$

glutamine

19.   **a.** catabolic   **b.** catabolic

20.

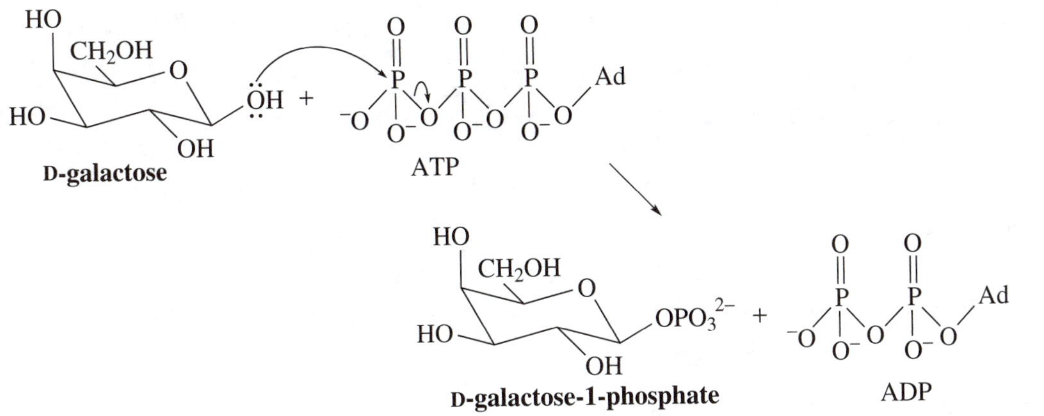

D-galactose   ATP

D-galactose-1-phosphate   ADP

21. The hydrogen on the α-carbon.

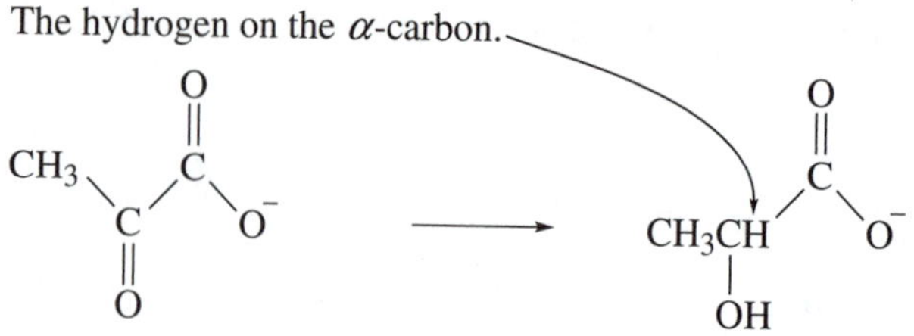

CH₃–C(=O)–C(=O)–O⁻ ⟶ CH₃CH(OH)–C(=O)–O⁻

pyruvate   lactate

22.   **a.**   reactions 1 and 3 (ADP is phosphorylated to ATP in reactions 7 and 10)

   b.   reactions 2, 5, and 8

   c.   reaction 6 (NAD+ is reduced when it oxidizes glyceraldehyde-3-phosphate)

   d.   reaction 9

23.   the conversion of citrate to isocitrate
the conversion of fumarate to (S)-malate

24.

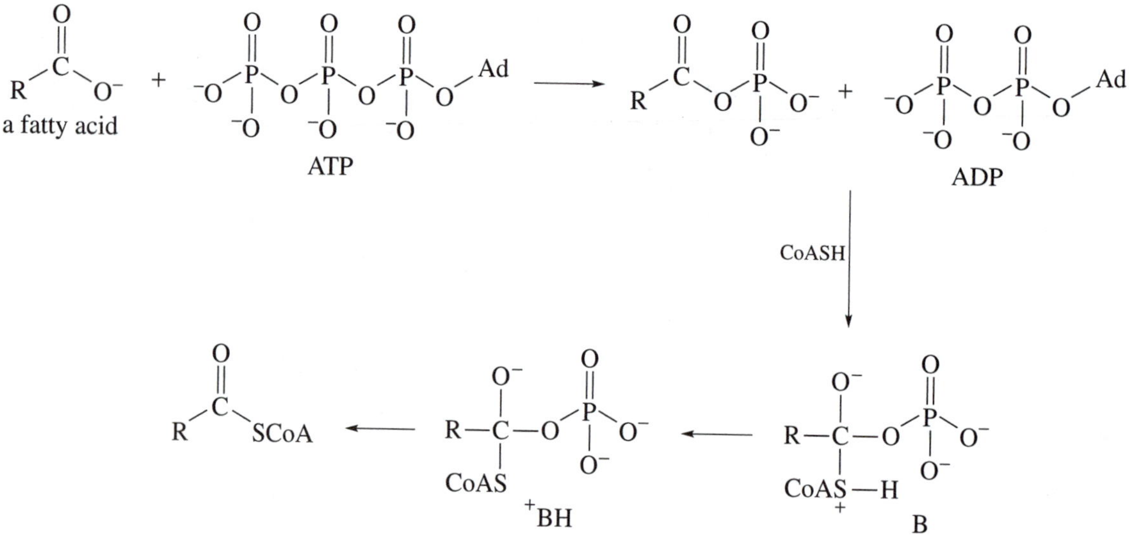

R–C(=O)–O⁻ + ATP ⟶ R–C(=O)–O–P + ADP

a fatty acid

CoASH

R–C(=O)–SCoA ⟵ R–C(O⁻)(CoAS)–O–P ⟵ R–C(O⁻)(CoAS–H)–O–P

          ⁺BH   B

25.

⁻O–C(=O)–CH₂–CH₂–C(=O)–C(=O)–O⁻ ──reduction──▶ ⁻O–C(=O)–CH₂–CH₂–CH(OH)–C(=O)–O⁻

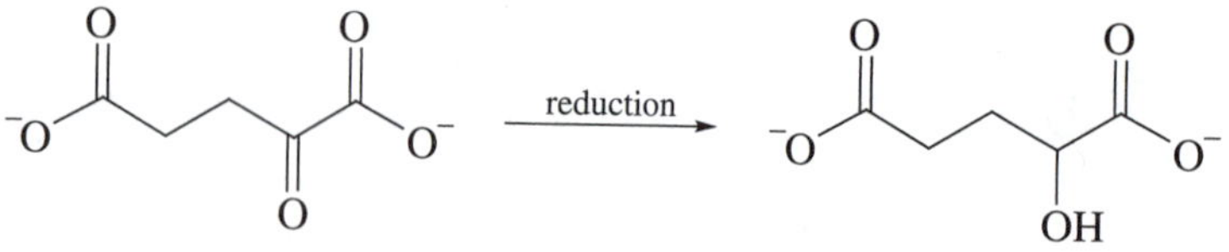

α-ketoglutarate

26.   The label will be on the phosphate group that is attached to the enzyme (phosphoglycerate mutase) that catalyzes the isomerization of 3-phosphoglycerate to 2-phosphoglycerate.

27.   Succinyl-CoA synthetase: this enzyme catalyzes a reaction of succinyl-CoA; the reverse reaction would be the synthesis of succinyl-CoA.

8.   If you examine the mechanism for the isomerism of glucose-6-phosphate to fructose-6-phosphate, you can see that C-1 in D-glucose is also C-1 in D-fructose.

Now, if you examine the mechanism for the aldolase-catalyzed cleavage of fructose-1,6-bisphosphate to form glyceraldehyde-3-phosphate, you can see which carbons in D-glucose correspond to the carbons in dihydroxyacetone phosphate and D-glyceraldehyde-3-phosphate.

Then, you will see how the carbons in dihydroxyacetone phosphate correspond to the carbons in D-glyceraldehyde-3-phosphate.

Finally, we see how the carbons in D-glyceraldehyde-3-phosphate correspond to the carbons in pyruvate. Therefore, C-3 and C-4 of glucose each become a carboxyl group in pyruvate.

29.   Pyruvate loses its carboxyl group when it is converted to ethanol. Because the carboxyl group is C-3 or C-4 of glucose, half of the ethanol molecules contain C-1 and C-2 of glucose and the other half contain C-5 and C-6 of glucose.

30.   At the beginning of a fast, blood glucose levels would be normal.
After a 24-hour fast, blood glucose levels would be very low, because both dietary glucose and glycogen have been depleted and glucose cannot be synthesized as a result of the deficiency of fructose-1,6-bisphosphatase.

31.   The conversion of pyruvate to lactate is a reversible reaction. Lactate can be converted back to pyruvate by oxidation.

The conversion of pyruvate to acetaldehyde is not a reversible reaction because it is a decarboxylation. Th$\blacksquare$ CO_2 cannot be put back onto acetaldehyde.

pyruvate   acetaldehyde

32. The β-oxidation of a molecule of a 16-carbon fatty acyl-CoA will form eight molecules of acetyl-CoA.

33. Each molecule of acetyl-CoA forms 2 molecules of CO_2. Therefore, the 8 molecules of acetyl-CoA obtained from a molecule of a 16-carbon fatty acyl-CoA will form 16 molecules of CO_2.

34. No ATP is formed from β-oxidation.

35. Each molecule of acetyl-CoA that is cleaved from the 16-carbon fatty acyl-CoA forms one molecule o$\blacksquare$ $FADH_2$ and one molecule of NADH. Since a 16-carbon fatty acyl-CoA undergoes seven cleavages, seve$\blacksquare$ molecules of $FADH_2$ and seven molecules of NADH are formed from the 16-carbon fatty acyl-CoA.

36. Because each NADH forms 2.5 molecules of ATP and each $FADH_2$ forms 1.5 molecules of ATP in oxidative phosphorylation, the 7 molecules of NADH form 17.5 molecules of ATP and the 7 molecules of $FADH_2$ form 10.5 molecules of ATP. Therefore, 28 molecules of ATP are formed.

37. We have seen that each molecule of acetyl-CoA that enters the citric acid cycle forms 10 molecules of ATP. A molecule of a 16-carbon fatty acid will form 8 molecules of acetyl-CoA. These will form 80 molecules of ATP. When these are added to the number of ATP molecules formed from the NADH and $FADH_2$ generated in β-oxidation, we see that 108 molecules of ATP are formed from complete metabolism of a 16-carbon saturated fatty acyl-CoA.

38. Each molecule of glucose, while being converted to two molecules of pyruvate, forms two molecules of ATP and two molecules of NADH.

The two molecules of pyruvate form two molecules of NADH while being converted to two molecules of acetyl-CoA.

Each molecule of acetyl-CoA that enters the citric acid cycle forms three molecules of NADH, one molecule of $FADH_2$, and one molecule of ATP. Thus, the two molecules of acetyl-CoA obtained from glucose form six molecules of NADH, two molecules of $FADH_2$, and two molecules of ATP.

Therefore, each molecule of glucose forms 4 molecules of ATP, 10 molecules of NADH ($2 + 2 + 6$), and 2 molecules of $FADH_2$.

Since each NADH forms 2.5 molecules of ATP and each $FADH_2$ forms 1.5 molecules of ATP, 1 molecule of glucose forms $4 + (10 \times 2.5) + (2 \times 1.5)$ molecules of ATP. That is, each molecule of glucose forms 32 molecules of ATP.

9. Pyruvate can be converted to alanine (transamination), oxaloacetate (carboxylation), lactate (reduction), and acetyl-CoA (by the pyruvate dehydrogenase complex).

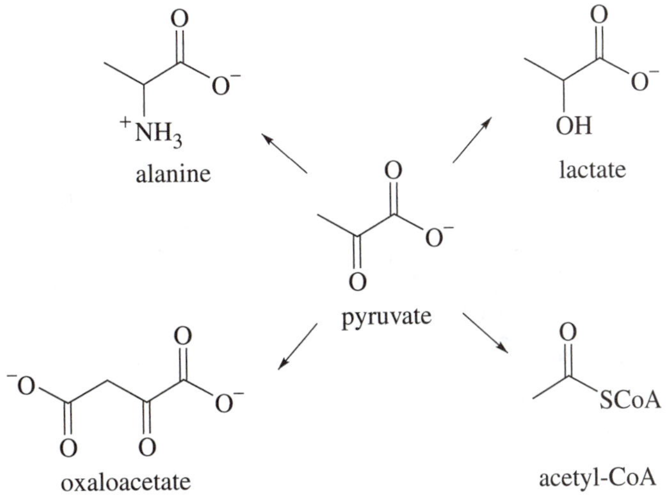

alanine

lactate

pyruvate

oxaloacetate

acetyl-CoA

0. The conversion of propionyl-CoA to methylmalonyl-CoA requires biotin (vitamin H).
The conversion of methylmalonyl-CoA to succinyl-CoA requires coenzyme B_{12} (vitamin B_{12}).

1. In Problem 28 we saw how the carbons in D-glyceraldehyde-3-phosphate correspond to the carbons in pyruvate.

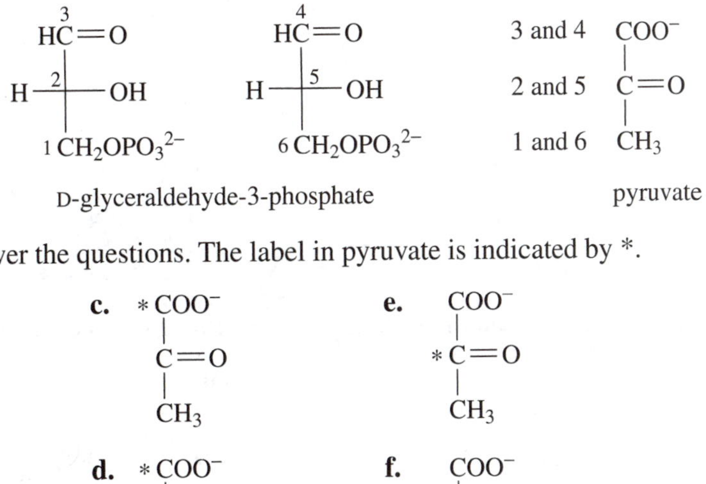

$^3HC=O$	$^4HC=O$	3 and 4 COO^-
$H-^2\!\!-OH$	$H-^5\!\!-OH$	2 and 5 $C=O$
$_1CH_2OPO_3^{2-}$	$_6CH_2OPO_3^{2-}$	1 and 6 CH_3
D-glyceraldehyde-3-phosphate		pyruvate

Now we can answer the questions. The label in pyruvate is indicated by *.

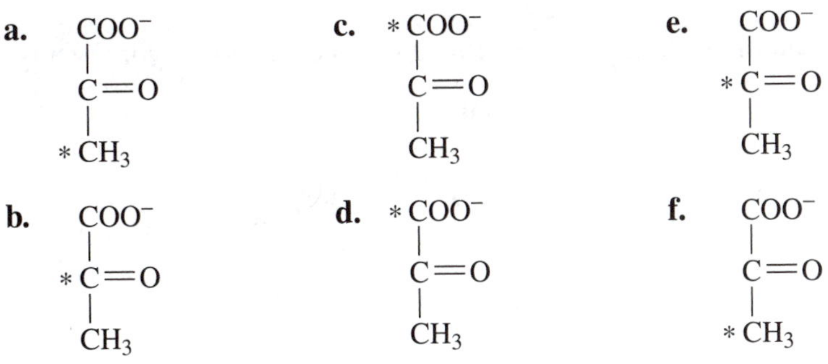

a.
$$COO^-$$
$$C=O$$
$$*CH_3$$

c.
$$*COO^-$$
$$C=O$$
$$CH_3$$

e.
$$COO^-$$
$$*C=O$$
$$CH_3$$

b.
$$COO^-$$
$$*C=O$$
$$CH_3$$

d.
$$*COO^-$$
$$C=O$$
$$CH_3$$

f.
$$COO^-$$
$$C=O$$
$$*CH_3$$

42.

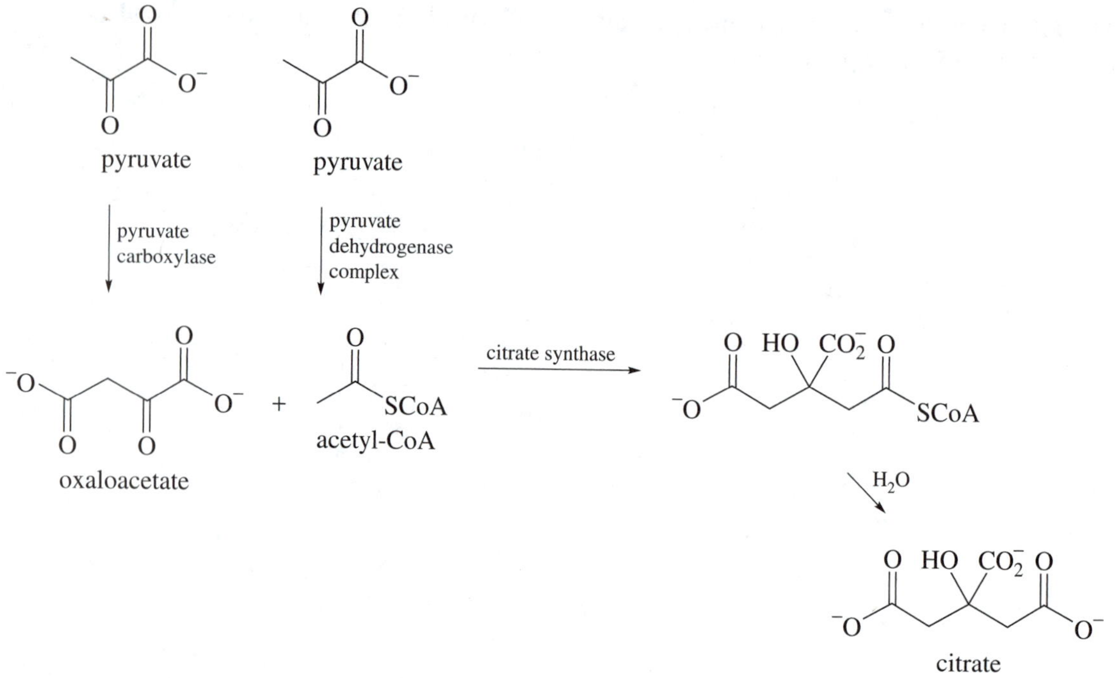

pyruvate   pyruvate

pyruvate   pyruvate
carboxylase   dehydrogenase
                  complex

citrate synthase

oxaloacetate   acetyl-CoA

H_2O

citrate

43.   A Claisen condensation between two molecules of acetyl-CoA forms acetoacetyl-CoA that, when hydrolyzed, forms acetoacetate.

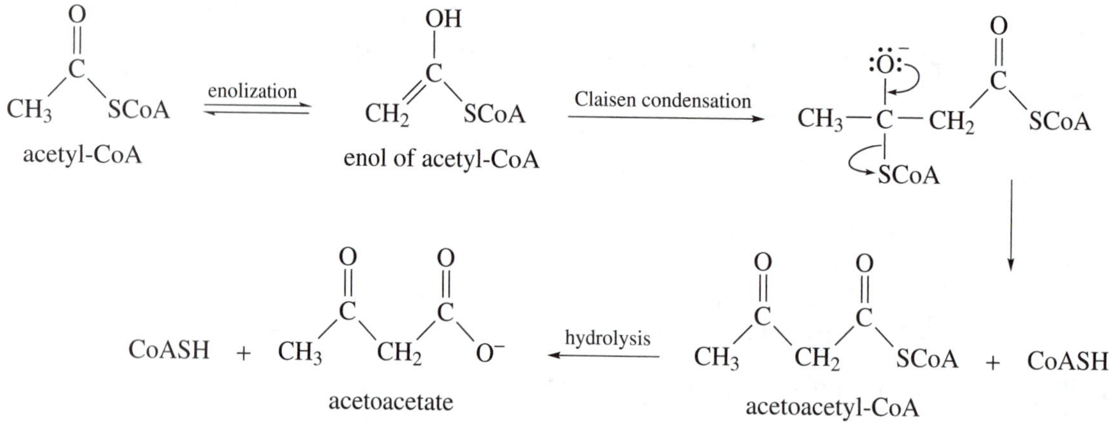

acetyl-CoA   enol of acetyl-CoA

enolization

Claisen condensation

CoASH + CH₃   CH₂   O⁻

hydrolysis

acetoacetate

acetoacetyl-CoA

Acetoacetate can undergo decarboxylation to form acetone, or it can be reduced to 3-hydroxybutyrate.

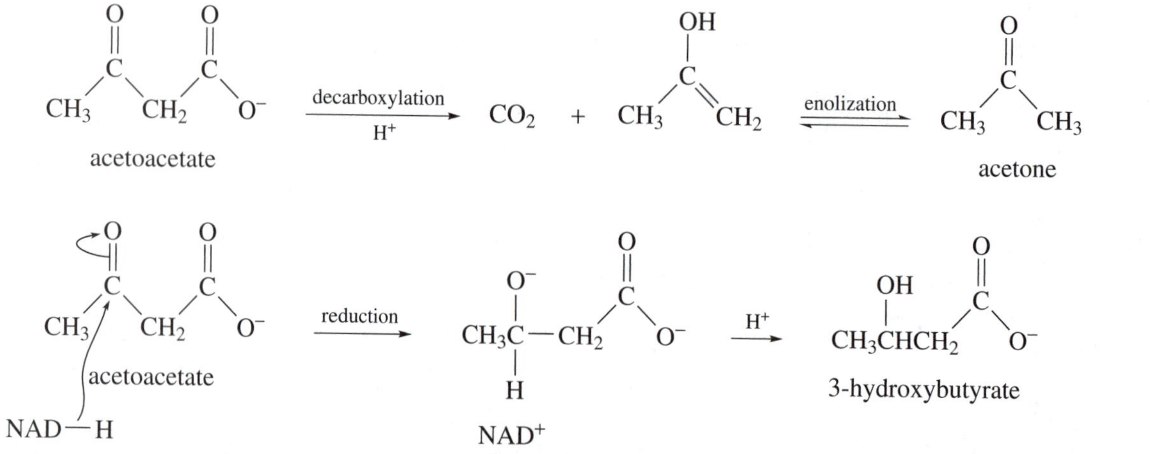

acetoacetate

decarboxylation
H^+

CO_2 + CH₃   CH₂

enolization

CH₃   CH₃

acetone

acetoacetate

reduction

H^+

3-hydroxybutyrate

NAD—H

NAD⁺

4. From the mechanisms for the conversion of fructose-1,6-bisphosphate to glyceraldehyde-3-phosphate and dihydroxyacetone phosphate, and the conversion of dihydroxyacetone phosphate to glyceraldehyde-3-phosphate, you can see that the label (*) was at C-1 in glyceraldehyde-3-phosphate.

5. **a.** UDP-galactose and UDP-glucose are C-4 epimers. NAD^+ oxidizes the C-4 OH group of UDP-galactose to a ketone. When NADH reduces the ketone back to an OH, it attacks the sp^2 carbon from above the plane, forming the C-4 epimer of the starting material.

b. The enzyme is called an epimerase because it converts a compound into an epimer (in this case, a C-4 epimer).

46. Because the compound that would react in the second step with the activated carboxylic acid group is excluded from the incubation mixture, the reaction between the carboxylate ion and ATP will come to equilibrium.

If radioactively labeled pyrophosphate is put into the incubation mixture, ATP will become radioactive if the mechanism involves attack on the α-phosphorus because pyrophosphate is a reactant in the reverse reaction that forms ATP.

ATP will not become radioactive if the mechanism involves attack on the β-phosphorus because pyrophosphate is not a reactant in the reverse reaction that forms ATP. (In other words, because pyrophosphate is not a product of the reaction, it cannot become incorporated into ATP in the reverse reaction.)

attack on the α-phosphorus

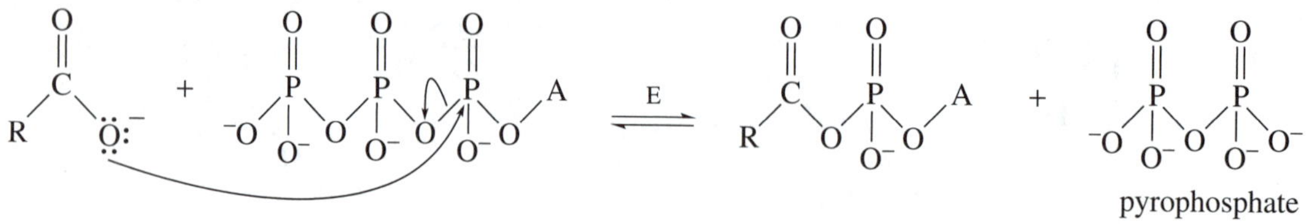

pyrophosphate

attack on the β-phosphorus

AMP

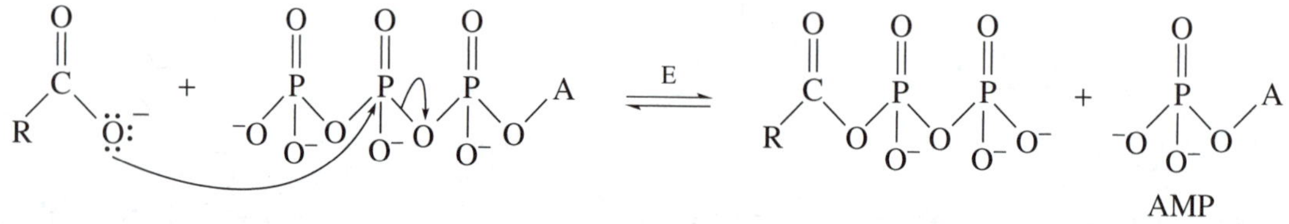

47.   If radioactive AMP is added to the reaction mixture, the results will be opposite. If the mechanism involves attack on the α-phosphorus, ATP will not become radioactive, because AMP is not a reactant in the reverse reaction that forms ATP. If the mechanism involves attack on the β-phosphorus, ATP will become radioactive, because AMP is a reactant in the reverse reaction that forms ATP.

Chapter 19 Practice Test

Draw a structure for each of the following:

a. the intermediate formed when a nucleophile (RO^-) attacks the γ-phosphorus of ATP

b. an acyl pyrophosphate

c. pyrophosphate

Fill in the six blanks in the following scheme:

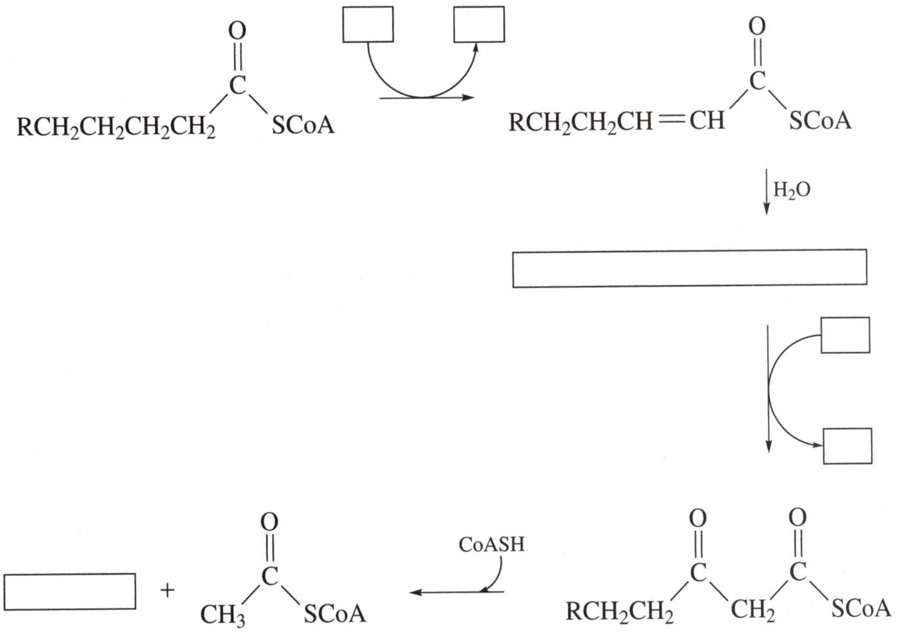

3. Which of the following are not citric acid cycle intermediates: fumarate, acetate, citrate?

4. Which provide energy to the cell: anabolic reactions or catabolic reactions?

5. What compounds are formed when proteins undergo the first stage of catabolism?

6. What compound is formed when fatty acids undergo the second stage of catabolism?

7. Indicate whether each of the following statements is true or false:

		T	F
a.	Each molecule of $FADH_2$ forms 2.5 molecules of ATP in the fourth stage of catabolism.	T	F
b.	$FADH_2$ is oxidized to FAD.	T	F
c.	NAD^+ is oxidized to NADH.	T	F
d.	Acetyl-CoA is a citric acid cycle intermediate.	T	F

Answers to Chapter 19 Practice Test

1. **a.**

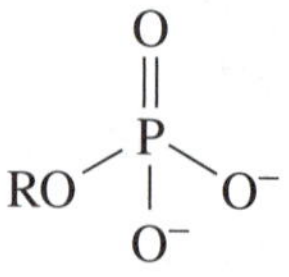

$$RO-\overset{\overset{\displaystyle O}{\|}}{\underset{\underset{\displaystyle O^-}{|}}{P}}-O^-$$

b.

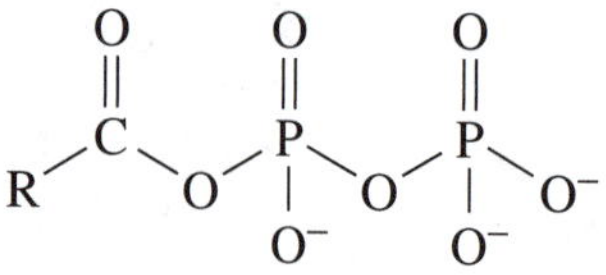

$$R-\overset{\overset{\displaystyle O}{\|}}{C}-O-\overset{\overset{\displaystyle O}{\|}}{\underset{\underset{\displaystyle O^-}{|}}{P}}-O-\overset{\overset{\displaystyle O}{\|}}{\underset{\underset{\displaystyle O^-}{|}}{P}}-O^-$$

c.

$$^-O-\overset{\overset{\displaystyle O}{\|}}{\underset{\underset{\displaystyle O^-}{|}}{P}}-O-\overset{\overset{\displaystyle O}{\|}}{\underset{\underset{\displaystyle O^-}{|}}{P}}-O^-$$

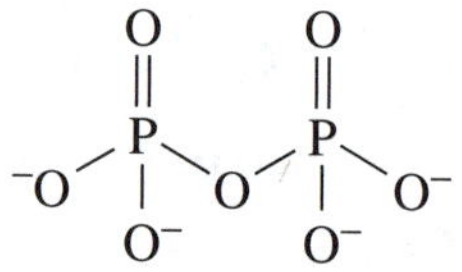

2.

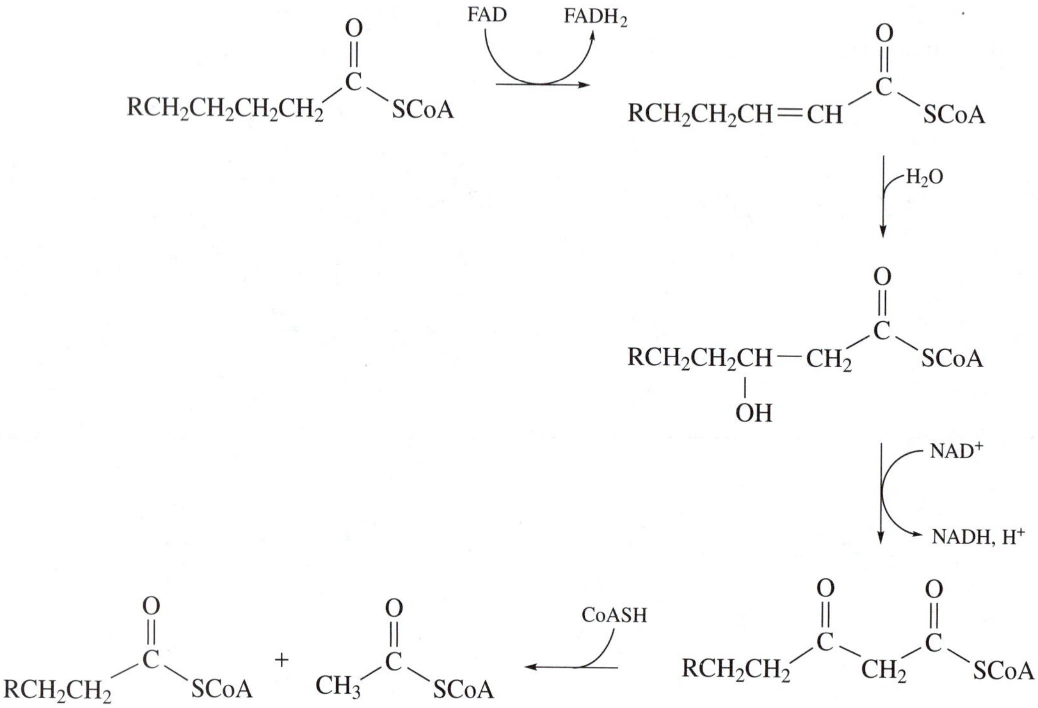

3.   acetate

4.   catabolic reactions

5.   amino acids

6.   acetyl-CoA

7.   **a.**   Each molecule of $FADH_2$ forms 2.5 molecules of ATP in the fourth stage of catabolism.   F
         b. $FADH_2$ is oxidized to FAD.   T
         c. NAD^+ is oxidized to NADH.   F
         d.   Acetyl-CoA is a citric acid cycle intermediate.   F

CHAPTER 20
The Organic Chemistry of Lipids

Important Terms

cholesterol	a steroid that is the precursor of all other steroids.
dimethylallyl pyrophosphate	a compound needed for the biosynthesis of terpenes and biosynthesized from isopentenyl pyrophosphate.
diterpene	a terpene that contains 20 carbons.
fat	a triester of glycerol that exists as a solid at room temperature.
fatty acid	a long-chain carboxylic acid.
isopentenyl pyrophosphate	the starting material for the biosynthesis of terpenes.
lipid	a water-insoluble compound found in a living system.
lipid bilayer	two layers of phosphoacylglycerols arranged so that their polar heads are on the outside and their nonpolar fatty acid chains are on the inside.
membrane	the material that surrounds the cell in order to isolate its contents.
micelle	a spherical aggregation of molecules, each with a long hydrophobic tail and a polar head, arranged so that the polar head points to the outside of the sphere.
mixed triglyceride	a triglyceride in which the fatty acid components are different.
monoterpene	a terpene that contains 10 carbons.
oil	a triester of glycerol that exists as a liquid at room temperature.
phosphoglyceride (phosphoacylglycerol)	formed when two OH groups of glycerol form esters with fatty acids and the terminal OH group is part of a phosphodiester.
phospholipid	a lipid that contains a phosphate group.
polyunsaturated fatty acid	a fatty acid with more than one double bond.
prostaglandins	lipids that regulate a variety of physiological responses
saponification	hydrolysis of a fat under basic conditions.
sesquiterpene	a terpene that contains 15 carbons.
simple triglyceride	a triglyceride in which the fatty acid components are the same.

441

soap   a sodium or potassium salt of a fatty acid.

sphingolipid   a lipid that contains sphingosine instead of glycerol

squalene   a triterpene that is a precursor of steroid molecules.

terpene   a lipid isolated from a plant that contains carbon atoms in multiples of five.

terpenoid   a terpene that contains oxygen.

tetraterpene   a terpene that contains 40 carbons.

triterpene   a terpene that contains 30 carbons.

wax   an ester formed from a long straight-chain carboxylic acid and a long straight chain alcohol.

olutions to Problems

a. Stearic acid has the higher melting point, because it has two more methylene groups (giving it a greater surface area) than palmitic acid.

b. Palmitic acid has the higher melting point, because it does not have any carbon–carbon double bonds, whereas palmitoleic acid has a cis double bond that prevents the molecules from packing closely together.

c. Oleic acid has the higher melting point, because it has one double bond, while linoleic acid has two double bonds, which give greater interference to close packing of the molecules.

Glyceryl tripalmitate has a higher melting point, because the carboxylic acid components are saturated and can, therefore, pack more closely together than the unsaturated carboxylic acid components of glyceryl tripalmitoleate.

To be optically inactive, the fat must have a plane of symmetry. In other words, the fatty acids at C-1 and C-3 must be identical. Therefore, stearic acid must be at C-1 and C-3.

$$
\begin{array}{l}
CH_2-O-\overset{\displaystyle O}{\overset{\|}{C}}-(CH_2)_{16}CH_3 \\
\;\;| \\
CH-O-\overset{\displaystyle O}{\overset{\|}{C}}-(CH_2)_{10}CH_3 \\
\;\;| \\
CH_2-O-\overset{\displaystyle O}{\overset{\|}{C}}-(CH_2)_{16}CH_3
\end{array}
$$

To be optically active, the fat must not have a plane of symmetry. Therefore, the two stearic acid groups must be attached to adjacent alcohol groups.

$$
\begin{array}{l}
CH_2-O-\overset{\displaystyle O}{\overset{\|}{C}}-(CH_2)_{16}CH_3 \\
\;\;| \\
CH-O-\overset{\displaystyle O}{\overset{\|}{C}}-(CH_2)_{16}CH_3 \\
\;\;| \\
CH_2-O-\overset{\displaystyle O}{\overset{\|}{C}}-(CH_2)_{10}CH_3
\end{array}
$$

Solved in the text.

Because the interior of a membrane is nonpolar and the surface of a membrane is polar, integral proteins will have a higher percentage of nonpolar amino acids.

The bacteria could synthesize phosphoacylglycerols with more saturated fatty acids because these triacyl-glycerols would pack more tightly in the lipid bilayer and, therefore, would have higher melting points and be less fluid.

Membranes must be kept in a semifluid state in order to allow transport across them. Cells closer to the hoof of an animal are going to be in a colder average environment than cells closer to the body. Therefore, the cells closer to the hoof have a higher degree of unsaturation to give them a lower melting point so that the membranes will not solidify at the colder temperature.

9.   **a.** The sphingomyelins can differ in the fatty acid component of the amide and have either choline ethanolamine attached to the phosphate group.

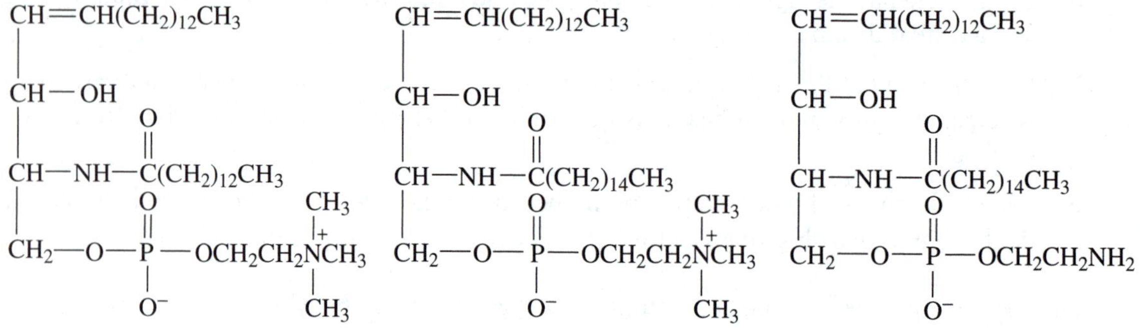

$$CH{=}CH(CH_2)_{12}CH_3$$
$$CH{-}OH$$
$$CH{-}NH{-}\overset{O}{\overset{\|}{C}}(CH_2)_{12}CH_3$$
$$CH_2{-}O{-}\overset{O}{\overset{\|}{P}}{-}OCH_2CH_2\overset{+}{N}CH_3$$

$$CH{=}CH(CH_2)_{12}CH_3$$
$$CH{-}OH$$
$$CH{-}NH{-}\overset{O}{\overset{\|}{C}}(CH_2)_{14}CH_3$$
$$CH_2{-}O{-}\overset{O}{\overset{\|}{P}}{-}OCH_2CH_2\overset{+}{N}CH_3$$

$$CH{=}CH(CH_2)_{12}CH_3$$
$$CH{-}OH$$
$$CH{-}NH{-}\overset{O}{\overset{\|}{C}}(CH_2)_{14}CH_3$$
$$CH_2{-}O{-}\overset{O}{\overset{\|}{P}}{-}OCH_2CH_2NH_2$$

b.

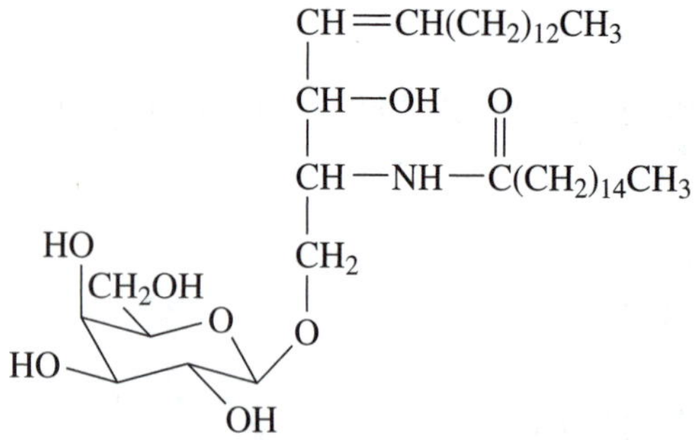

10. The fact that the tail-to-tail linkage occurs in the exact center of the molecule suggests that the two halve are synthesized (in a head-to-tail fashion) and then joined together in a tail-to-tail linkage.

tail-to-tail linkage

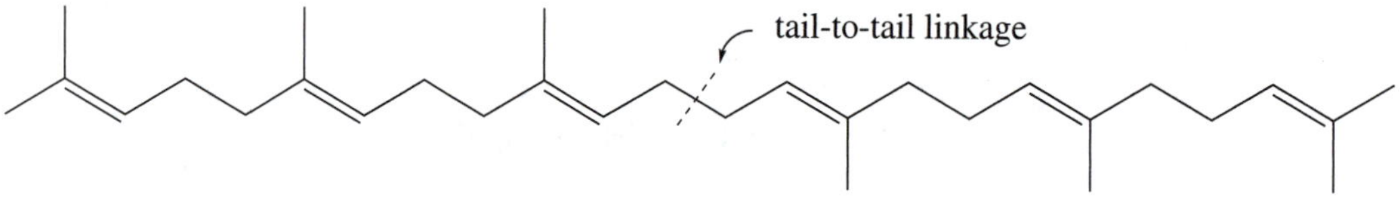

11. Squalene, lycopene, and β-carotene are all synthesized in the same way. In each case, two halves are syn thesized (in a head-to-tail fashion) and then joined together in a tail-to-tail linkage.

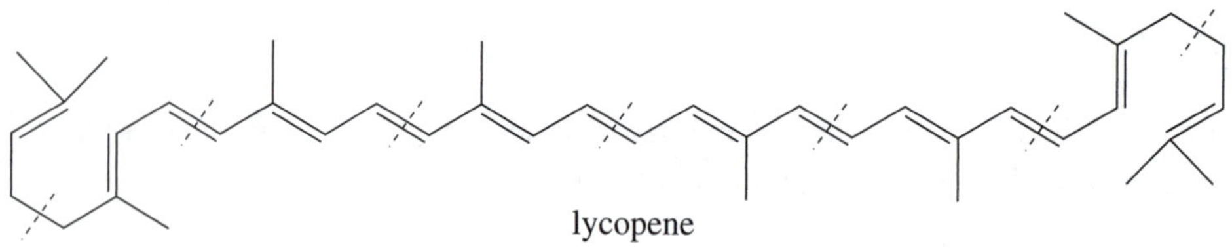

lycopene

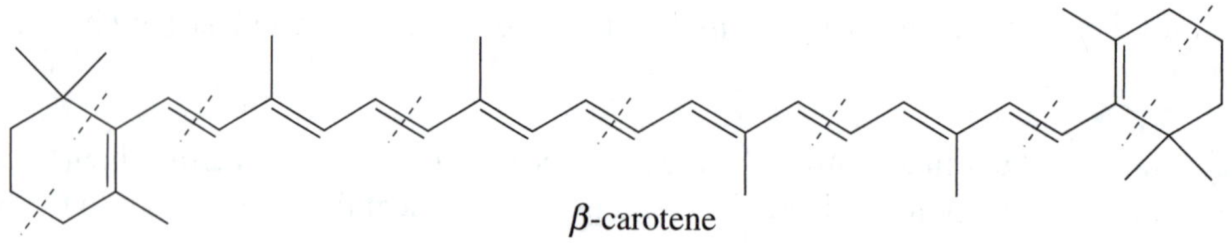

β-carotene

2.

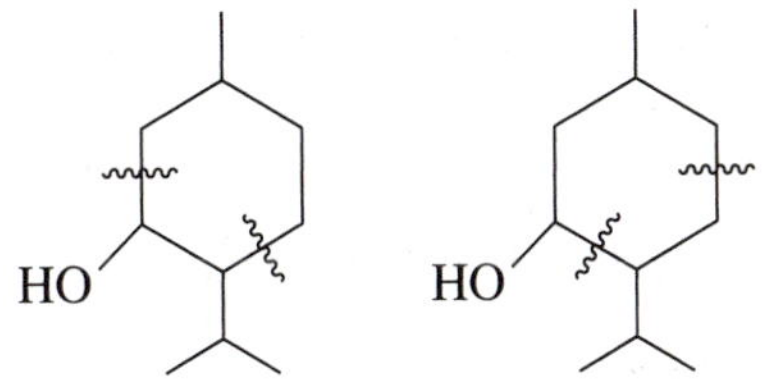

3.

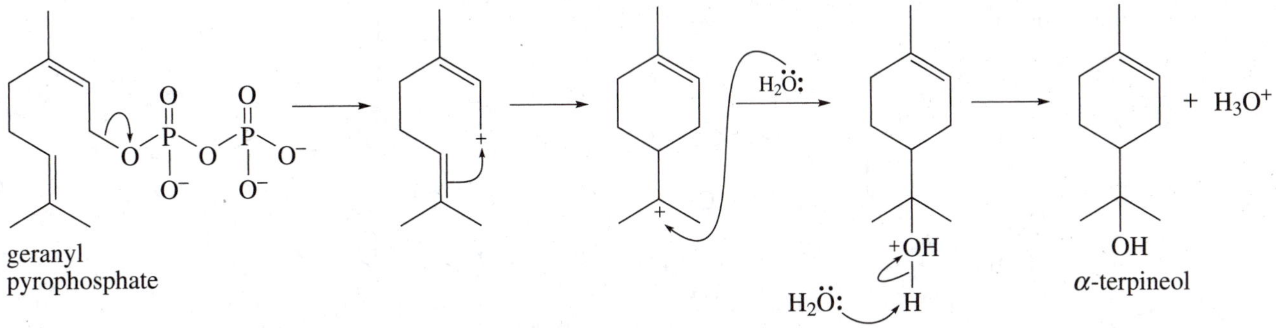

geranyl
pyrophosphate

α-terpineol

$+ \ H_3O^+$

4. It tells you that the reaction is an S_N1 reaction, because the fluoro-substituted carbocation is less stable than the nonfluoro-substituted carbocation (due to the strongly electron-withdrawing fluoro substituent), so it would form more slowly.

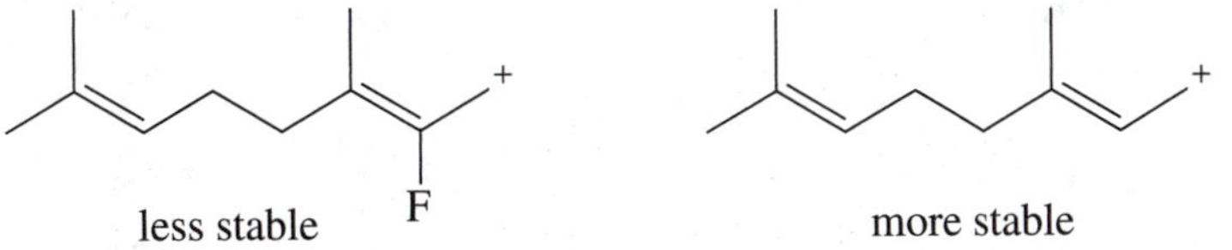

less stable   F   more stable

If the reaction had been an S_N2 reaction, the fluoro-substituted compound would have reacted more rapidly than the nonfluoro-substituted carbocation, because the electron-withdrawing fluoro substituent would make the compound more susceptible to nucleophilic attack.

15.   There are two 1,2-hydride shifts and two 1,2-methyl shifts. The last step is elimination of a proton.

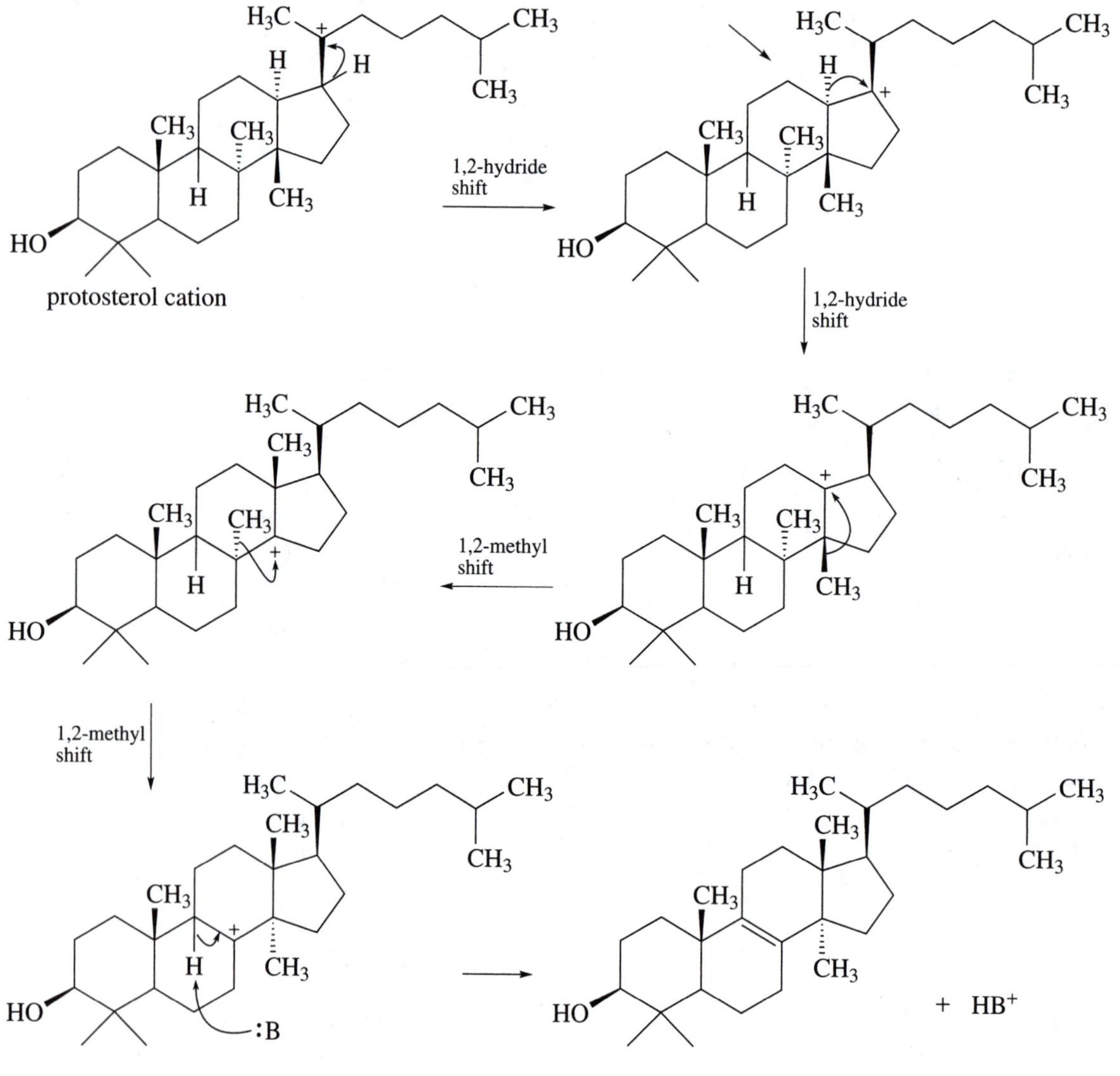

protosterol cation

1,2-hydride shift

1,2-hydride shift

1,2-methyl shift

1,2-methyl shift

+ HB$^+$

16.   Compared to testosterone, Dianabol has an OH group attached to the five-membered ring, and it has two double bonds in the ring on the far left.

17.   a.

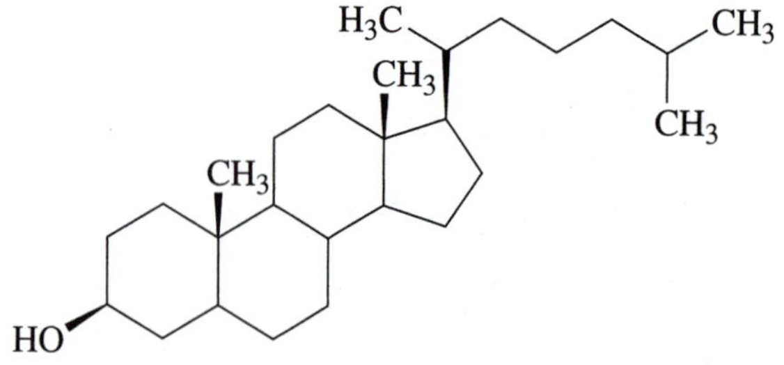

b.

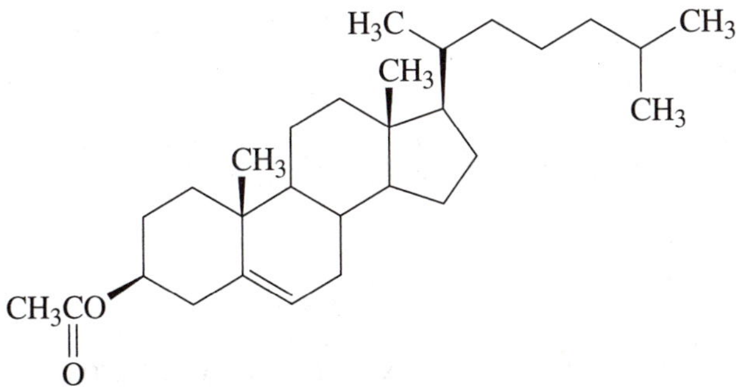

c.

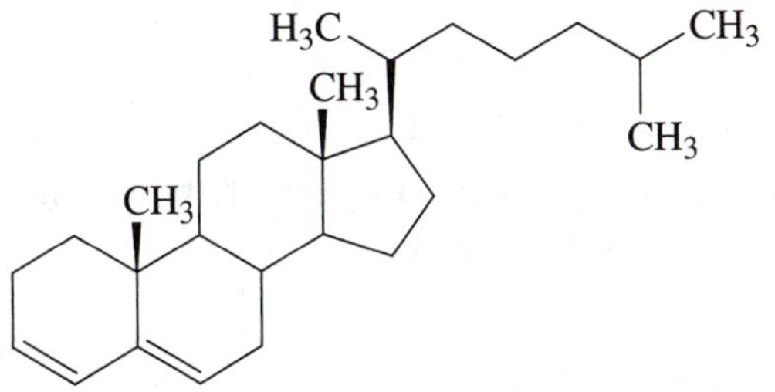

d.

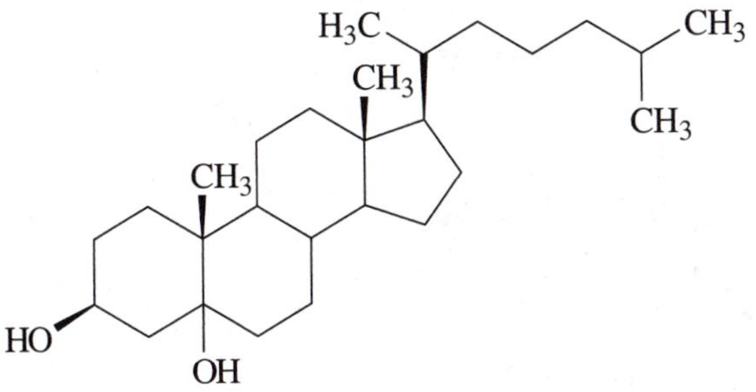

e.

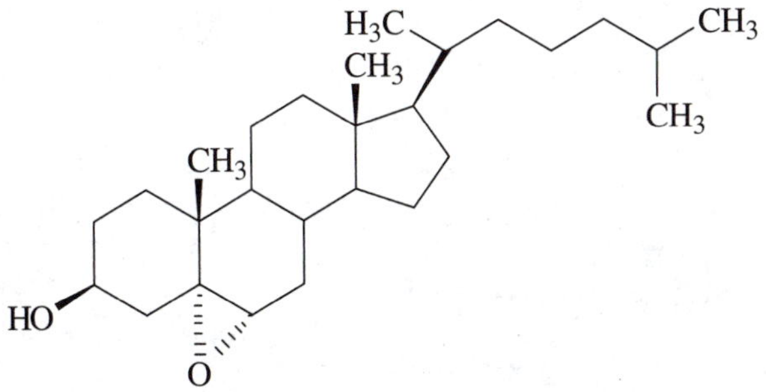

8. All triacylglycerols do not have the same number of asymmetric centers. If the carboxylic acid components at C-1 and C-3 of glycerol are not identical, the triacylglycerol has one asymmetric center (C-2). If the carboxylic acid components at C-1 and C-3 of glycerol are identical, the triacylglycerol has no asymmetric centers.

9.

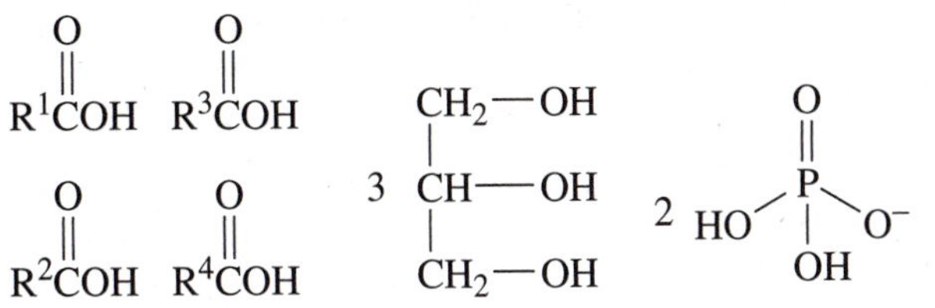

20.

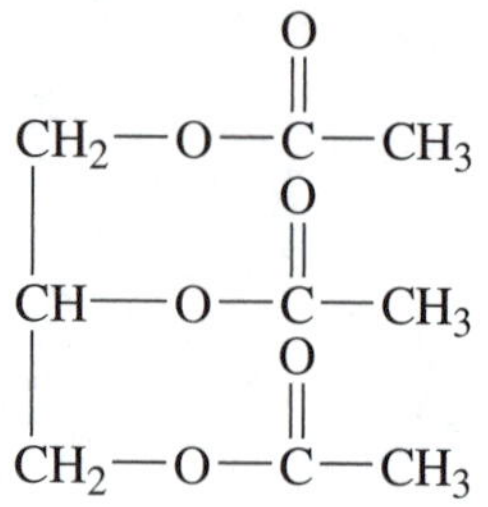

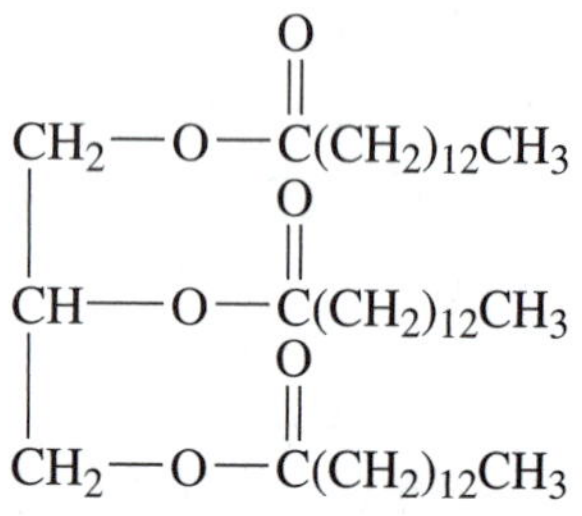

the trigylceride in nutmeg

The structure at left has a molecular formula $= C_9H_{14}O_6$ and a molecular weight $= 218$.

Subtracting 218 from the total molecular weight gives the molecular weight of the methylene (CH_2) groups in the triacylgcerol.

$$722 - 218 = 504$$

Dividing 504 by the molecular weight of a methylene group (14) will give the number of methylene groups.

$$\frac{504}{14} = 36$$

Since there are 36 methylene groups, each fatty acid in the triacylglycerol has 12 methylene groups.

21.

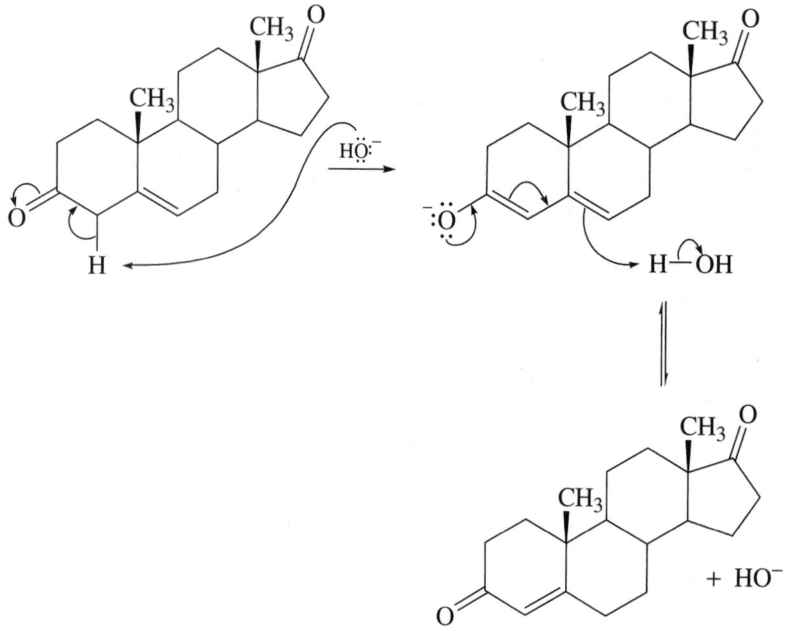

22.   **a.**   There are three triacylglycerols in which one of the fatty acid components is lauric acid and two ar myristic acid. Myristic acid can be at C-1 and C-3 of glycerol, in which case the triacylglycerol doe not have any asymmetric centers. If myristic acid is at C-1 and C-2 of glycerol, C-2 is an asymmetri center, and consequently, two enantiomers are possible for the compound.

b. There are six triacylglycerols in which one of the fatty acid components is lauric acid: one is myristic acid, and one is palmitic acid. The three possible arrangements are shown below (with the fatty acid components abbreviated as L, M, and P). Since each has an asymmetric center (indicated by an asterisk), each can exist as a pair of enantiomers for a total of six triacylglycerols.

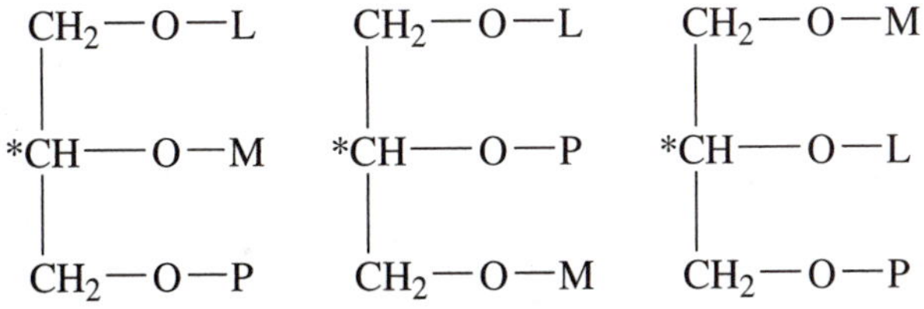

3.

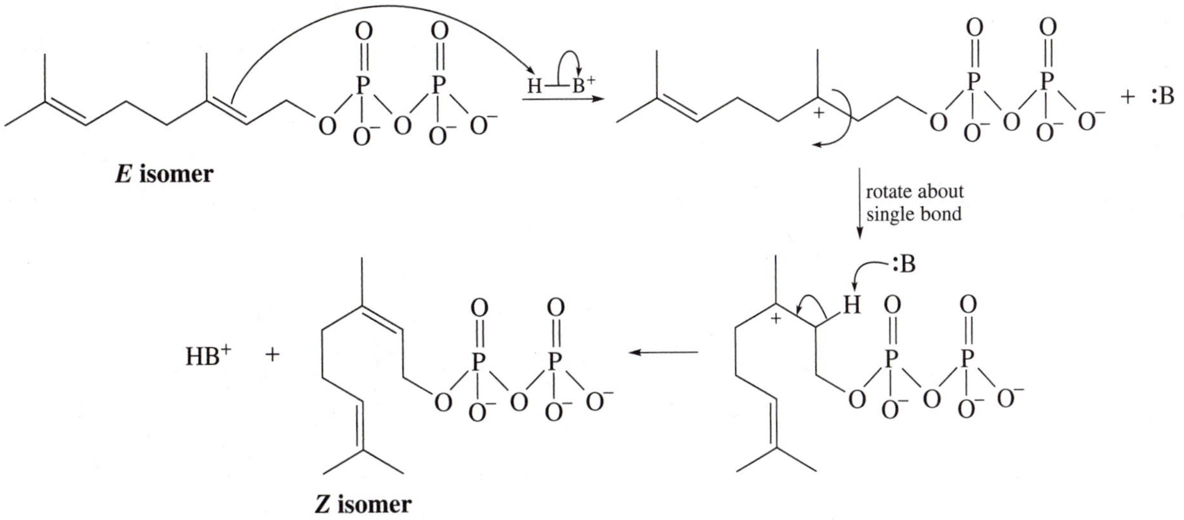

4.

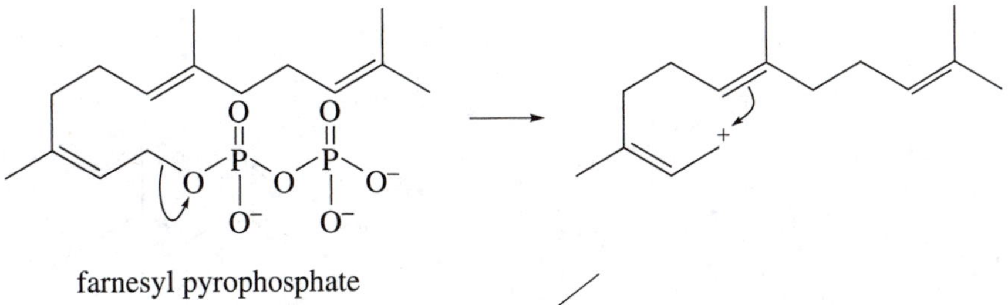

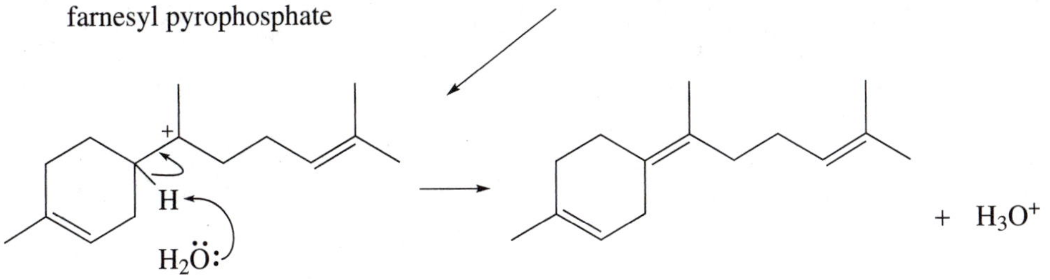

25.   The mechanism for the conversion of farnesyl pyrophosphate to eudesmol:

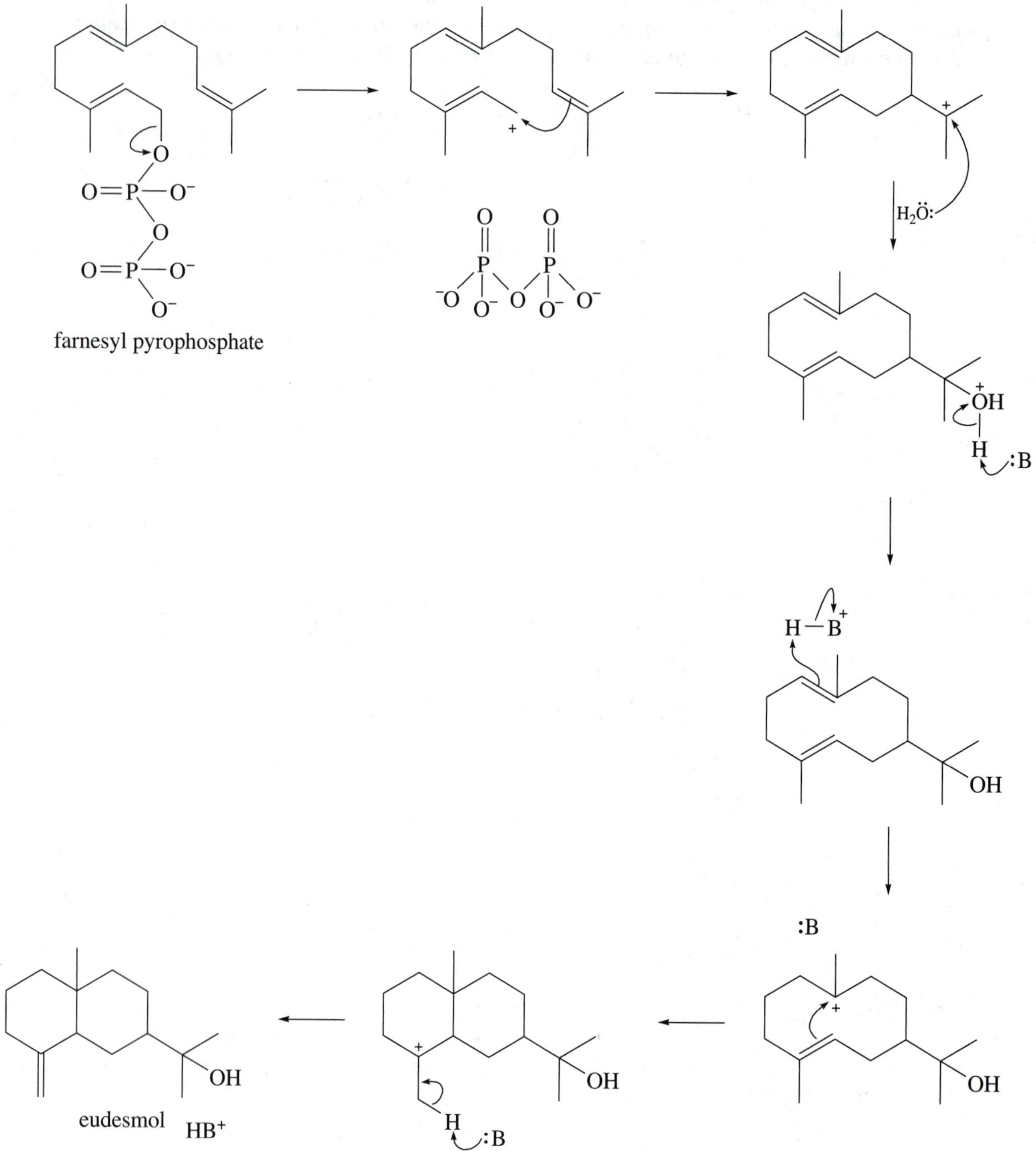

farnesyl pyrophosphate

eudesmol   HB⁺

Chapter 20 Practice Test

. Explain why the melting points of fats are higher than those of oils.

. Mark off the isoprene units in squalene.

squalene

3. How many isoprene units does a triterpene have?

4. Draw the structure of a phosphatidylethanolamine.

5. Indicate whether each of the following statements is true or false:

a. Lipids are insoluble in water.   T   F

b. Cholesterol is the precursor of all other steroids.   T   F

c. Saturated fatty acids have higher melting points than do unsaturated fatty acids.   T   F

d. Fats have a higher percentage of saturated fatty acids than do oils.   T   F

Answers to Chapter 20 Practice Test

1.   Because fats are composed primarily of saturated fatty acids, the fat molecules can pack closely togethe
which gives them higher melting points.

2.

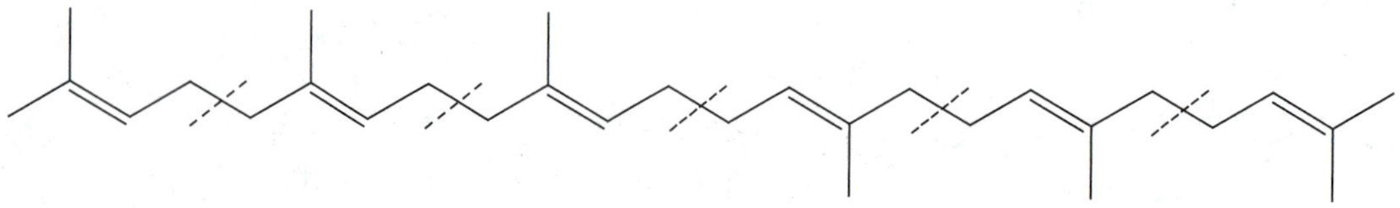

squalene

3.   six

4.

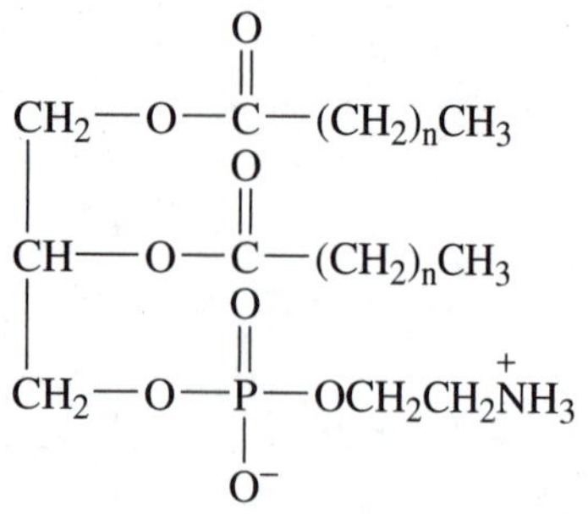

5.   **a.** Lipids are insoluble in water.   T

b. Cholesterol is the precursor of all other steroids.   T

c. Saturated fatty acids have higher melting points than do unsaturated fatty acids.   T

d. Fats have a higher percentage of saturated fatty acids than do oils.   T

Important Terms

Anticodon	the three bases at the bottom of the middle loop in a tRNA.
Base	a nitrogen-containing heterocyclic compound (a purine or a pyrimidine) found in DNA and RNA.
Codon	a sequence of three bases in mRNA that specifies the amino acid to be incorporated into a protein.
Deamination	a hydrolysis reaction that results in the removal of ammonia.
Deoxyribonucleic acid (DNA)	a polymer of deoxyribonucleotides.
Deoxyribonucleotide	a nucleotide where the sugar component is D-2-deoxyribose.
Dinucleotide	two nucleotides linked by a phosphodiester bond.
Double helix	the term used to describe the secondary structure of DNA.
Gene	a segment of DNA that codes for a inheritable trait.
Gene therapy	a technique that inserts a synthetic gene into the DNA of an organism defective in that gene.
Genetically modified organism (GMO)	an organism whose genetic material has been altered.
Genetic code	the amino acid specified by each three-base sequence of mRNA.
Genetic engineering	recombinant DNA technology, in which DNA molecules are attached to DNA in a host cell and allowed to replicate.
Human genome	the total DNA of a human cell.
Major groove	the wider and deeper of the two alternating grooves in DNA.
Minor groove	the narrower and more shallow of the two alternating grooves in DNA.
Nucleic acid	a chain of five-membered ring sugars linked by phosphodiester groups with each sugar bearing a heterocyclic amine at the anomeric carbon in the β-position. The two kinds of nucleic acids are DNA and RNA.
Nucleoside	a heterocyclic base (purine or pyrimidine) bonded to the anomeric carbon of a sugar (D-ribose or D-2-deoxyribose) in the β-position.

nucleotide   a nucleoside with one of its OH groups bonded to a phosphate group via a ester linkage.

oligonucleotide   three to ten nucleotides linked by phosphodiester bonds.

phosphodiester   a species in which two of the OH groups of phosphoric acid have been con verted to OR groups.

polynucleotide   many nucleotides linked by phosphodiester bonds.

primary structure   the sequence of bases in the nucleic acid.

pyrosequencing   an automated technique used to sequence DNA; it detects the identity of th base that adds to the DNA primer.

replication   the synthesis of identical copies of DNA.

restriction endonuclease   an enzyme that cleaves DNA at a specific base sequence.

restriction fragment   a fragment that is formed when DNA is cleaved by a restriction endonuclease

ribonucleic acid (RNA)   a polymer of ribonucleotides.

ribonucleotide   a nucleotide where the sugar component is D-ribose.

ribosomal RNA (rRNA)   the structural component of ribosomes, the particles on which protein synthe sis takes place.

semiconservative replication   the mode of replication that results in a daughter molecule of DNA having on of the original DNA strands plus a newly synthesized strand.

sense strand   the strand in DNA that is not read during transcription; it has the sam sequence of bases as the synthesized mRNA strand, except that the mRNA ha Us in place of the Ts in DNA.

stacking interactions   van der Waals interactions between the mutually induced dipoles of adjacen pairs of bases in DNA.

stop codon   a codon that says "stop protein synthesis here."

template strand   the strand in DNA that is read during transcription.

transcription   the synthesis of mRNA from a DNA blueprint.

transfer RNA (tRNA)   a single-stranded RNA molecule that carries an amino acid to be incorporated into a protein.

translation   the synthesis of a protein from an mRNA blueprint.

Solutions to Problems

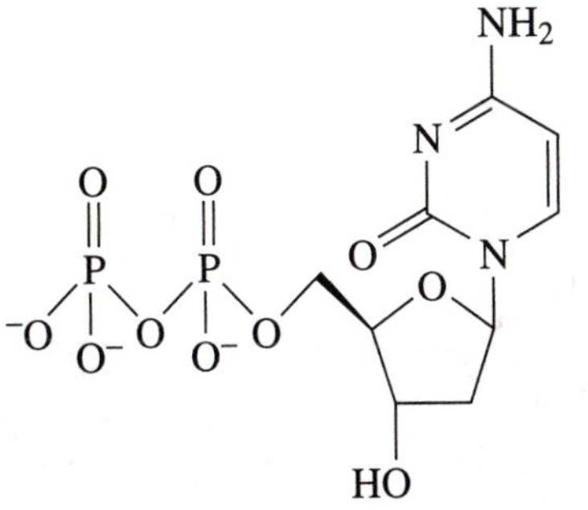

a.

dCDP

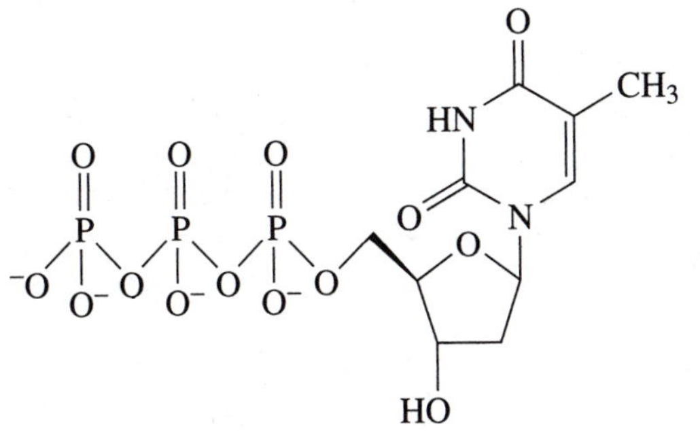

d.

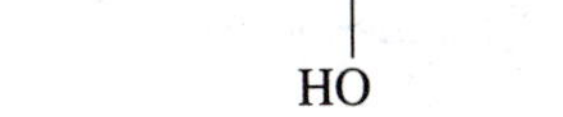

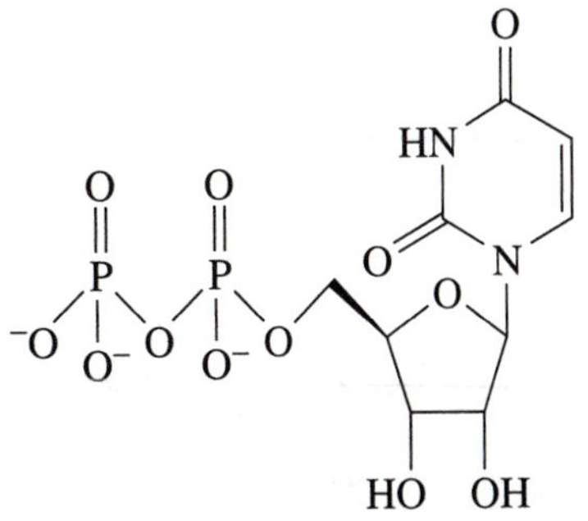

UDP

b.

dTTP

e.

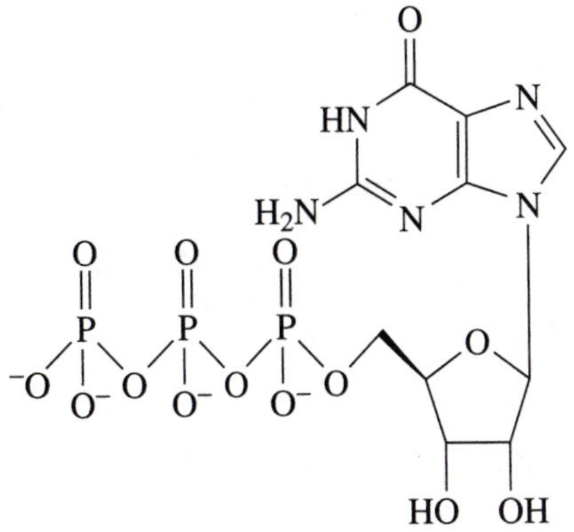

guanosine 5′-triphosphate
GTP

c.

dUMP

f.

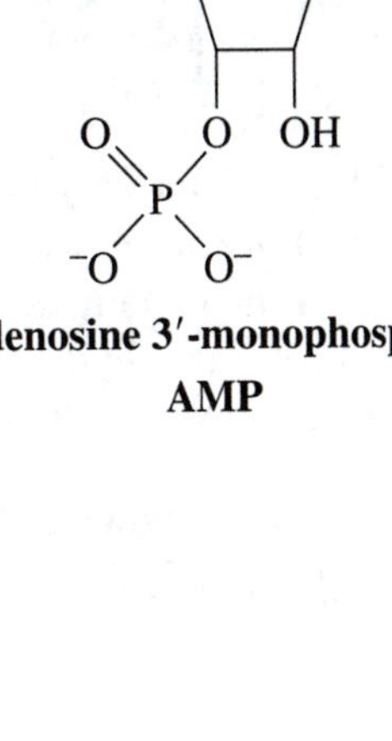

adenosine 3′-monophosphate
AMP

2.   **a.** 3′—C—C—T—G—T—T—A—G—A—C—G—5′
b. guanine

3.   Notice that when a nucleophile attacks the phosphorus of a diester, the π bond breaks. However, when nucleophile attacks the phosphorus of an anhydride, a σ bond breaks in preference to the π bond.

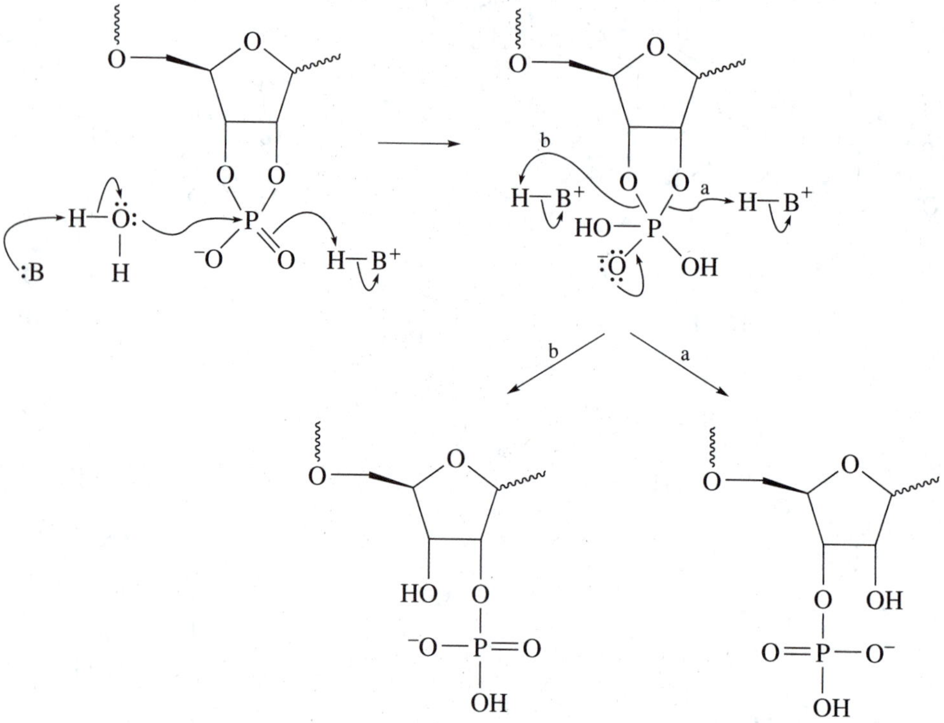

4.

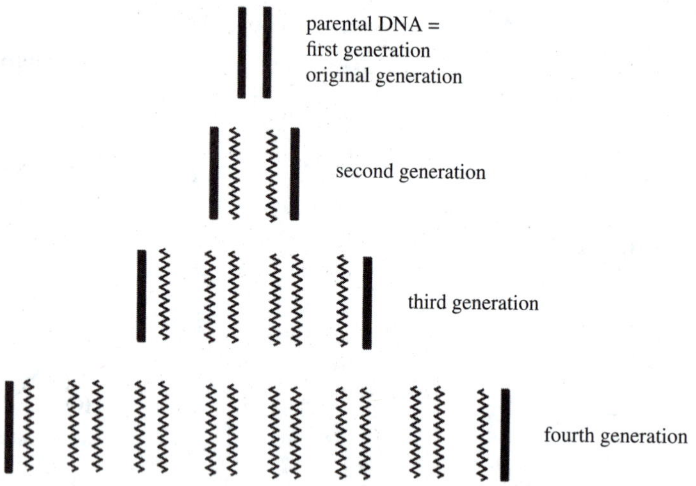

parental DNA =
first generation
original generation

second generation

third generation

fourth generation

5.   Thymine and uracil differ only in that thymine has a methyl substituent that uracil does not have (thymine is 5-methyluracil). Because thymine and uracil both have the same groups in the same positions that can participate in hydrogen bonding, they will both call for the incorporation of the same purine. Because thymine and uracil form one hydrogen bond with guanine and two with adenine, they will both incorporate adenine in order to maximize hydrogen bonding.

6.   Because methionine is known to be the first base incorporated into the heptapeptide, the mRNA sequence is read beginning at AUG, since that is the only codon that codes for methionine.

Met-Asp-Pro-Val-Ile-Lys-His

7.   Met-Asp-Pro-Leu-Leu-Asn

It does not cause protein synthesis to stop because the sequence UAA does not occur within a triplet. The reading frame causes the triplets to be AU**U** and **AA**A. In other words, the U is at the end of a triplet and the next triplet starts with AA.

A change in the third base of a codon would be least likely to cause a mutation because the third base is variable for many amino acids. For example, CUU, CUC, CUA, and CUG all code for leucine.

0. The sequence of bases in the template strand of DNA specifies the sequence of bases in mRNA, so the bases in the template strand and the bases in mRNA are complementary. Therefore, the sequence of bases in the sense strand of DNA is identical to the sequence of bases in mRNA, except wherever there is a U in mRNA, there is a T in the sense strand of DNA.

$$5'—G—C—A—T—G—G—A—C—C—C—C—G—T—T—$$
$$A—T—T—A—A—A—C—A—C—3'$$

1.

	Met	Asp	Pro	Val	Ile	Lys	His
codons	AUG	GAU	CCU	GUU	AUU	AAA	CAU
		GAC	CCC	GUC	AUC	AAG	CAC
			CCA	GUA	AUA		
			CCG	GUG			

anticodons

Note that the anticodons are stated in the $5' \rightarrow 3'$ direction. For example, the anticodon of AUG is stated as CAU (not UAC).

codons   5′ A U G 3′
anticodons   3′ U A C 5′

CAU	AUC	AGG	AAC	AAU	UUU	AUG
	GUC	GGG	GAC	GAU	CUU	GUG
		UGG	UAC	UAU		
		CGG	CAC			

12.

adenine   hypoxanthine

guanine   xanthine

13. Deamination involves hydrolyzing an imine linkage to a carbonyl group and ammonia, so cytosine can b deaminated to uracil.

Thymine does not have an amino substituent on the ring, which means that it cannot form an imine an therefore it cannot be deaminated.

thymine

14. **a** is the only sequence that has a chance of being recognized by a restriction endonuclease, because it is th only one that has the same sequence of bases in the $5' \rightarrow 3'$ direction that the complementary strand has i the $5' \rightarrow 3'$ direction.

$$^{5'}A—C—G—C—G—T^{3'}$$
$$^{3'}T—G—C—G—C—A^{5'}$$

15. a. b. c. d.

16. Lys-Val-Gly-Tyr-Pro-Gly-Met-Val-Val

17. 5′—GAC—CAC—CAT—TCC—GGG—GTA—GCC—AAC—TTT—3′

18. 5′—AAA—GTT—GGC—TAC—CCC—GGA—ATG—GTG—GTC—3′

9.   **a.** Ile   **b.** Asp   **c.** Val   **d.** Val

20.   The third base in each codon has some variability.

mRNA   5′-GG(UCA or G)UC(UCA or G)CG(UCA or G)GU(UCA or G)CA(U or C)GA(A or G)-3′

or AG(U or C)   AG(A or G)

DNA
template   3′-CC(AGT or C)AG(AGT or C)GC(AGT or C)CA(AGT or C)GT(A or G)CT(T or C)-5′

or TC(A or G)   TC(T or C)

sense   5′-GG(TCA or G)TC(TCA or G)CG(TCA or G)GT(TCA or G)CA(T or C)GA(A or G)-3′
or AG(T or C)   AG(A or G)

Notice that Ser and Arg are two of three amino acids that can be specified by six different codons.

21.

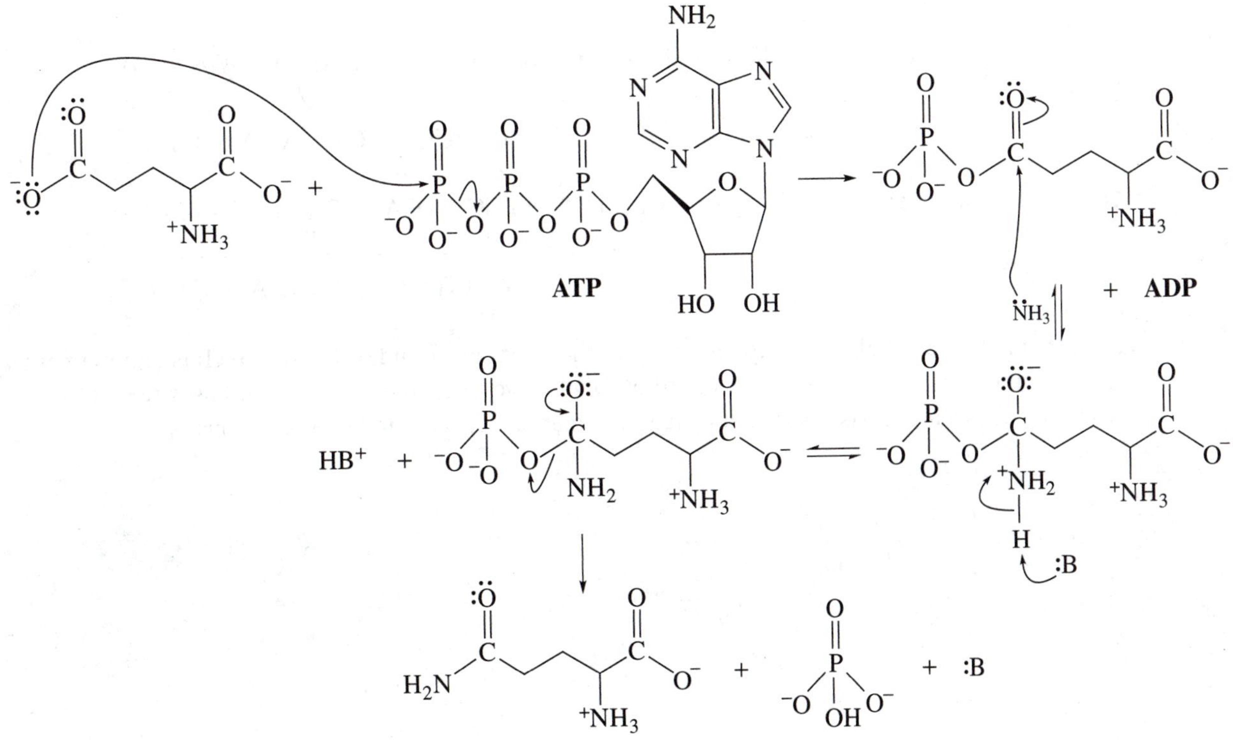

22.   A segment of DNA with 18 base pairs has 36 bases. If there are 7 cytosines, there are 7 guanines, accounting for 14 bases. Therefore, 22 (36 − 14 = 22) are adenines and thymines.

a.   11 thymines   **b.**   7 guanines

23.

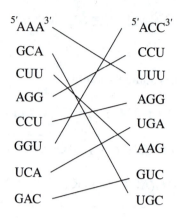

24.

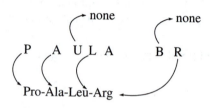

| **mRNA** | CC(UCA or G)GC(UCA or G)CU(UCA or G)CG(UCA or G) |

or UU(A or G)   AG(A or G)

| **DNA (sense strand)** | CC(TCA or G)GC(TCA or G)CT(TCA or G)CG(TCA or G) |

or TT(A or G)   AG(A or G)

Note that because mRNA is complementary to the template strand of DNA, which is complementary to the sense strand, mRNA and the sense strand of DNA have the same sequence of bases (except DNA has a T where RNA has a U). Also note that Leu and Arg are each specified by six codons.

5. A hydrogen bond acceptor is indicated by **A**; a hydrogen bond donor is indicated by **D**.
The **A** and **D** designations show that the maximum number of hydrogen bonds that can form is two
between thymine and adenine and three between cytosine and guanine. Notice that uracil and thymine have
the same **A** and **D** designations.

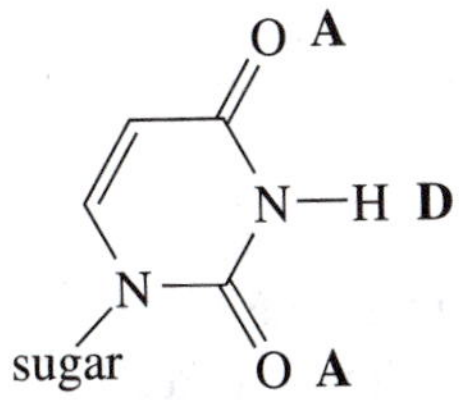

uracil 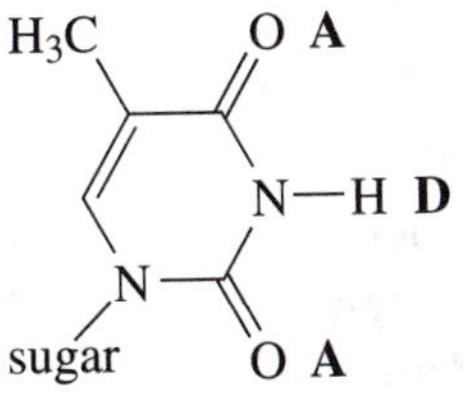   thymine 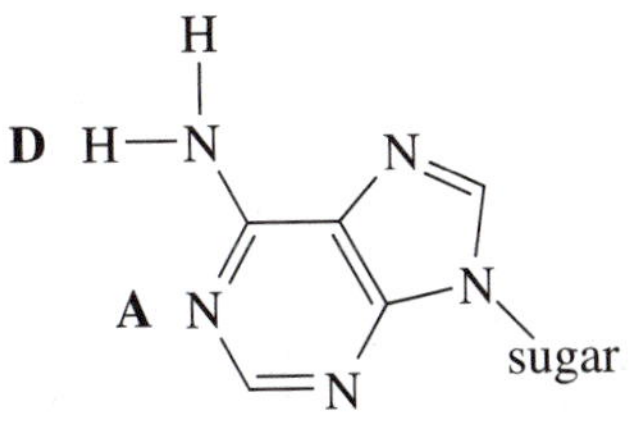   adenine

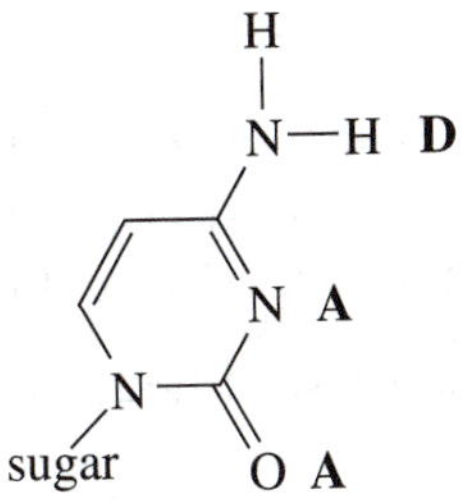

cytosine 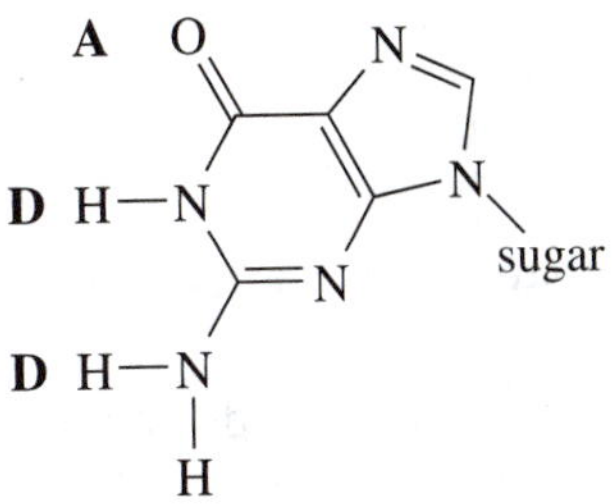   guanine

26. If the bases existed in the enol form, no hydrogen bonds could form between the bases unless one of the
bases were shifted vertically.

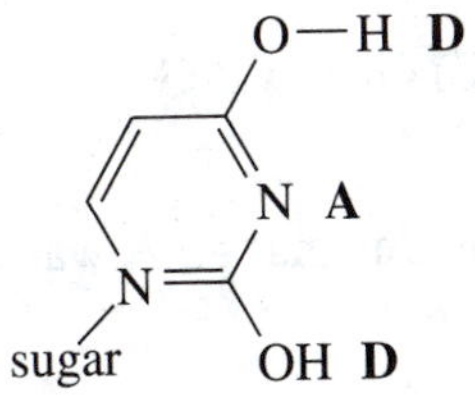

uracil 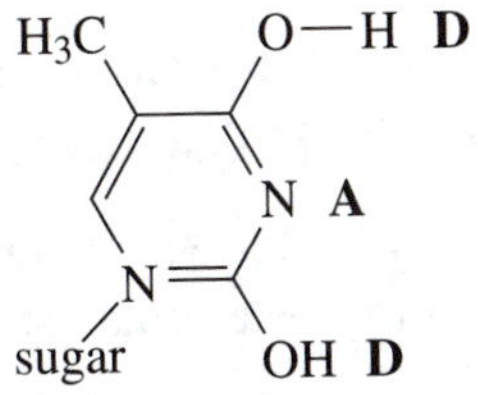   thymine 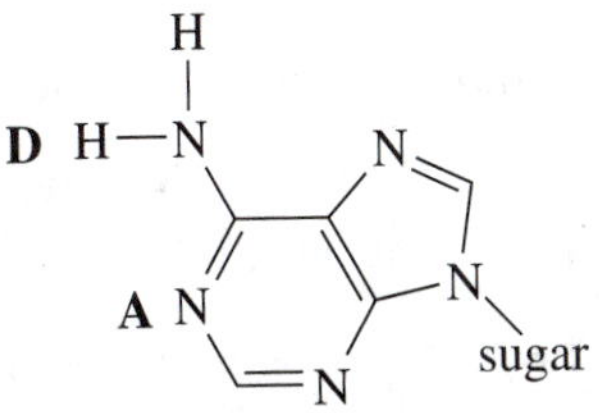   adenine

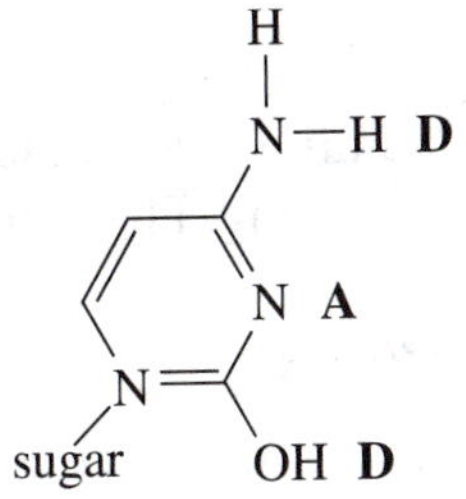

cytosine 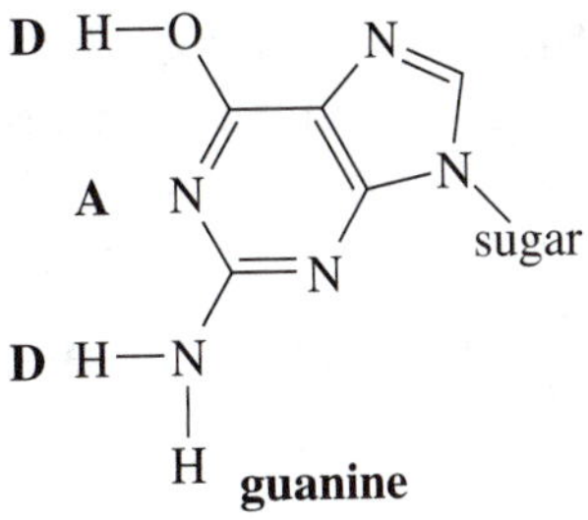   guanine

27.   **a.** CC and GG   **c.** CA and TG

CA and TG are formed in equal amounts, since A pairs with T and C pairs with G. (The dinucleotides a
written in the $5' \rightarrow 3'$ direction. For example, part **f** is not a correct answer because it has A pairing with
and T pairing with T.)

$$\begin{array}{cc} 5' & \text{C} \;\; \text{A} \;\; 3' \\ & | \;\;\; | \\ 3' & \text{G} \;\; \text{T} \;\; 5' \end{array}$$

28.   UUU = phenylalanine   UUG = leucine   GGU = glycine
GGG = glycine   UGU = cysteine   GUG = valine
                           GUU = valine   UGG = tryptophan

29.   The number of different possible codons using four nucleotides is $(4)^n$, where n is the number of letter
(nucleotides) in the code.

$$\begin{aligned} \text{for a two-letter code:} &\quad (4)^2 = 16 \\ \text{for a three-letter code:} &\quad (4)^3 = 64 \\ \text{for a four-letter code:} &\quad (4)^4 = 256 \end{aligned}$$

Because there are 20 amino acids that must be specified, a two-letter code would not provide enoug
codons.
A three-letter code provides enough codes for all the amino acids and also provides the necessary sto
codons.
A four-letter code provides many more codes than would be needed.

30.   AZT is incorporated into DNA when the $3'$-OH group of the last nucleotide incorporated into the growin
chain of DNA attacks the α-phosphorus of AZT-triphosphate instead of a normal nucleotide. When AZT
is thus incorporated into DNA, DNA synthesis stops because AZT does not have a $3'$-OH group that ca
react with another nucleotide triphosphate.

31.   The normal and mutant peptides would have the following base sequence in their mRNA:

normal: CA(AG) UA(UC) GG(UCAG) AC(UCAG) CG(UCAG) UA(UC) GU(UCAG)
mutant: CA(AG) UC(UCAG) GA(AG) CC(UCGA) GG(UCGA) AC(UCAG)

a.   The middle nucleotide (A) in the second triplet was deleted. This means that an A was deleted in the
sense strand of DNA or a T was deleted in the template strand of DNA.

b.   The mRNA for the mutant peptide has an unused $3'$-terminal two-letter code, U(UCAG). The las
amino acid in the octapeptide of the normal fragment is leucine, so its last triplet is UU(AG) or
CU(UCAG).

This means that the triplet for the last amino acid in the mutant is U(UCAG)(UC) and that the last
amino acid in the mutant is one of the following: Phe, Ser, Tyr, or Cys.

32.   If deamination does not occur, the mRNA sequence will be

AUG—UCG—CUA—AUC, which will code for the following tetrapeptide:

Met-Ser-Leu-Ile

Deamination of a cytosine results in a uracil.
If the cytosines are deaminated, the mRNA sequence will be

AUG—UUG—UUA—AUU, which will code for the following tetrapeptide:

Met-Leu-Leu-Ile

The only cytosine that will change the amino acid that is incorporated into the peptide is the first one. Therefore, this is the cytosine that could cause the most damage to an organism if it were deaminated.

33. 5-Bromouracil is incorporated into DNA in place of thymine because of their similar size. Thymine pairs with adenine via two hydrogen bonds. 5-Bromouracil exists primarily in the enol form. The enol cannot form any hydrogen bonds with adenine, but it can form two hydrogen bonds with guanine. Therefore, 5-bromouracil pairs with guanine. Because 5-bromouracil causes guanine to be incorporated instead of adenine into newly synthesized DNA strands, it causes mutations.

thymine   adenine   5-bromouracil   H   guanine

34. It requires energy to break the hydrogen bonds that hold the two chains together, so an enormous amount of energy would be required to unravel the chain completely. However, as the new nucleotides that are incorporated into the growing chain form hydrogen bonds with the parent chain, energy is released, and this energy can be used to unwind the next part of the double helix.

35. The ribosome, the particle on which protein synthesis occurs, has a binding site for the growing peptide chain and a binding site for the next amino acid to be incorporated into the chain.

peptide binding site   amino acid binding site

In protein synthesis, all peptide bonds are formed by the reaction of an amino acid with a peptide, except the first peptide bond, which has to be formed by the reaction of two amino acids. Therefore, for the synthesis of the first peptide bond, the first (N-terminal) amino acid has to have a peptide bond that will fit into the peptide binding site.

The formyl group of *N*-formylmethionine will provide the peptide group that will be recognized by the peptide binding site, and the second amino acid will be bound in the amino acid binding site.

N-formylmethionine

Chapter 21 Practice Test

1. Is the following compound dTMP, UMP, dUMP, or dUTP?

2. If one of the strands of DNA has the following sequence of bases running in the $5' \rightarrow 3'$ direction, what is the sequence of bases in the complementary strand?

$$5'—A—C—T—T—G—C—A—T—3'$$

3. What base is closest to the 5'-end in the complementary strand?

4. Indicate whether each of the following statements is true or false:

   a. Guanine and cytosine are purines.   T   F

   b. The 3'-OH group allows RNA to be easily cleaved.   T   F

   c. The number of As in DNA is equal to the number of Ts.   T   F

   d. rRNA carries the amino acid that will be incorporated into a protein.   T   F

   e. The template strand of DNA is the one transcribed to form RNA.   T   F

   f. The 5'-end of DNA has a free OH group.   T   F

   g. The synthesis of proteins from an RNA blueprint is called transcription.   T   F

   h. A nucleotide consists of a base and a sugar.   T   F

   i. RNA contains Ts, and DNA contains Us.   T   F

5. Which of the following base sequences would most likely be recognized by a restriction endonuclease?

   1. ACGCGT   3. ACGGCA   5. ACATCGT

   2. ACGGGT   4. ACACGT   6. CCAACC

What would be the sequence of bases in the mRNA obtained from the following segment of DNA?

sense strand
5′—G—C—A—T—G—G—A—C—C—C—C—G—T—3′
3′—C—G—T—A—C—C—T—G—G—G—G—C—A—5′
template strand

Which of the following pairs of dinucleotides will occur in equal amounts in DNA?

(Nucleotides are always written in the 5′ → 3′ direction.)

CA and GT   CG and AT
CG and GG   CA and TG

Answers to Chapter 21 Practice Test

1.   dUMP

2. $5'-A-T-G-C-A-A-G-T-3'$

3.   A

4.   a. Guanine and cytosine are purines.   F
      b. The 3′-OH group allows RNA to be easily cleaved.   F
      c. The number of As in DNA is equal to the number of Ts.   T
      d. rRNA carries the amino acid that will be incorporated into a protein.   F
      e. The template strand of DNA is the one transcribed to form RNA.   T
      f. The 5′-end of DNA has a free OH group.   F
      g. The synthesis of proteins from an RNA blueprint is called transcription.   F
      h. A nucleotide consists of a base and a sugar.   F
      i. RNA contains Ts, and DNA contains Us.   F

5.   only **1**

6. $5'-G-C-A-U-G-G-A-C-C-C-C-G-U-3'$

7.   CA and TG